AMPHIBIANS IN DECLINE

CANADIAN STUDIES OF A GLOBAL PROBLEM

Edited By

David M. Green

HERPETOLOGICAL CONSERVATION

NUMBER ONE

Printed in the United States of America

Society for the Study of Amphibians and Reptiles, Saint Louis, Missouri, U.S.A.

In association with:

Canadian Association of Herpetologists, Montréal, Québec, Canada
Canadian Declining Amphibian Populations Working Group (DAPCAN), Victoria, British Columbia, Canada

Cover photograph: Gray treefrog (*Hyla versicolor*), by Martin Ouellet.

Title page illustration: Oregon spotted frog (*Rana pretiosa*); photo by David M. Green; converted to line drawing using Corel® Photo-paint.

Publisher's Cataloging-in-Publication Data

Green, David M. [editor], 1953–

Amphibians in decline: Canadian studies of a global problem / edited by David M. Green; foreword by W. Ronald Heyer

Herpetological Conservation, Volume 1

p. cm.

Includes bibliographical references and index

ISBN 0-916984-40-0

1. Herpetology—conservation. 2. Amphibians—Canada. 3. Declining amphibian populations—Canada.

QL 641.H47 1997

597.6 97-67612

CIP

Contents

Contributors

Yves Bachand, St. Lawrence Valley Natural History Society, 21 111 Bord-du-Lac, Ste.-Anne-de-Bellevue, Québec H9X 1C0, Canada.

Michael Berrill, Department of Biology, Trent University, Peterborough, Ontario K9J 7B8, Canada.

Susan Bertram, Department of Biology, Trent University, Peterborough, Ontario K9J 7B8, Canada. *Present address*: Department of Zoology, Arizona State University, Tempe, Arizona 85287, USA

Christine A. Bishop, Canadian Wildlife Service, Box 5050, Burlington, Ontario L4R 4A6, Canada.

James P. Bogart, Department of Zoology, University of Guelph, Guelph, Ontario N1G 2W1, Canada.

Joël Bonin, Canadian Wildlife Service, Québec Region, P.O.Box 10100, Ste.-Foy, Québec G1V 4H5, Canada, and St. Lawrence Valley Natural History Society, 21 111 Bord-du-Lac, Ste.-Anne-de-Bellevue, Québec H9X 1C0, Canada. *Present Address*: Redpath Museum, McGill University, 859 Sherbrooke St., W., Montréal, Québec H3A 2K6, Canada.

Ronald J. Brooks, Department of Zoology, University of Guelph, Guelph, Ontario N1G 2W1, Canada.

Maria Helena Caetano, Departmento de Zoologia e Antropologia, Faculdade de Ciências, Bloco C-2, Campo Grande, 1700 Lisboa, Portugal

Douglas Clay, Fundy National Park, P.O. Box 40, Alma, New Brunswick E0A 1B0, Canada.

Wanda J. Cook, Department of Zoology, University of Guelph, Guelph, Ontario N1G 2W1, Canada.

Rhéaume Courtois, Ministère de l'environnement et de la faune, Direction de la faune et des habitats, 150 boul. René-Lévesque est, Québec, Québec G1R 4Y1, Canada.

Graham J. Crawshaw, Metropolitan Toronto Zoo, 361A Old Finch Avenue, Scarborough, Ontario M1B 5K7, Canada.

Claude Daigle, Ministère de l'environnement et de la faune, Direction de la faune et des habitats, 150 boul. René-Levesque, 5e étage, Québec, Québec G1R 4Y1, Canada.

Theodore M. Davis, Department of Biology, University of Victoria, Victoria, British Columbia V8W 2Y2, Canada.

Jean-Luc DesGranges, Canadian Wildlife Service, Québec Region, P.O.Box 10100, Ste.-Foy, Québec G1V 4H5, Canada.

Andrew Didiuk, Saskatchewan Amphibian Monitoring Project, P.O. Box 1574, Saskatoon, Saskatchewan S7K 3R3, Canada.

Linda A. Dupuis, Centre for Applied Conservation Biology, Faculty of Forestry, University of British Columbia, Vancouver, British Columbia V6T 1Z4, Canada.

Michael A. Fournier, Canadian Wildlife Service, P.O. Box 637, Yellowknife, Northwest Territories X1A 2N5, Canada.

François Gagné, St. Lawrence Centre, Environment Canada, 105 McGill St., suite 400, Montréal, Québec H2Y 2E7, Canada.

David A. Galbraith, Redpath Museum, McGill University, 859 Sherbrooke St. W., Montréal, Québec H3A 2K6, Canada. *Present address*: Royal Botanical Gardens, P.O. Box 399, Hamilton, Ontario L8N 3H8, Canada.

Mary E. Gartshore, R.R. #1, Walsingham, Ontario N0E 1X0, Canada.

David M. Green, Redpath Museum, McGill University, 859 Sherbrooke St. W., Montréal, Québec H3A 2K6, Canada.

Stephen J. Hecnar, Department of Biological Sciences, University of Windsor, Windsor, Ontario N9B 3P4, Canada.

W. Ronald Heyer, Biodiversity Programs, NHB Mail Stop 180, Smithsonian Institution, Washington, DC 20560, USA.

Janice D. James, Vertebrate Morphology Research Group, Department of Biological Sciences, and Kananaskis Research Stations, University of Calgary, 2500 University Dr. NW, Calgary, Alberta T2N 1N4, Canada.

Michel Lepage, Ministère de l'environnement et de la faune, Direction de la faune et des habitats, 150 boul. René-Levesque, 5e étage, Québec, Québec G1R 4Y1, Canada.

Raymond Leclair, Jr., Département de Chimie-Biologie, Université du Québec à Trois-Rivières, Québec G9A 5H7, Canada.

Leslie A. Lowcock, Redpath Museum, McGill University, 859 Sherbrooke St. W., Montréal, Québec H3A 2K6, Canada. Present Address: c/o Centre for Biodiversity and Conservation Biology, Royal Ontario Museum, 100 Queen's Park, Toronto, Ontario, M5S 2C6, Canada.

Duncan A. MacLeod, P.O. Box 3248, Station C, Ottawa, Ontario K1Y 4J5, Canada.

Sylvie Matte, St. Lawrence Valley Natural History Society, 21 111 Bord-du-Lac, Ste-Anne-de-Bellevue, Québec H9X 1C0, Canada.

John E. Maunder, Natural History Unit, Newfoundland Museum, P.O. Box 8700, St. Johns, Newfoundland A1B 4J6, Canada.

Donald F. McAlpine, Natural Sciences Division, New Brunswick Museum, 277 Douglas Ave., Saint John, New Brunswick E2K 1E5, Canada.

Lee Mennell, Box 105, Carcross, Yukon Y0B 1B0, Canada.

Robert W. Murphy, Centre for Biodiversity and Conservation Biology, Royal Ontario Museum, 100 Queen's Park, Toronto, Ontario M5S 2C6, Canada.

Stephen J. Nelson, Vertebrate Morphology Research Group, Department of Biological Sciences, and Kananaskis Research Stations, University of Calgary, 2500 University Dr. NW, Calgary, Alberta T2N 1N4, Canada.

Stan A. Orchard 1745 Bank Street, Victoria, British Columbia, V8R 4V7, Canada.

Martin Ouellet, Redpath Museum, McGill University, 859 Sherbrooke St. W., Montréal, Québec H3A 2K6, Canada.

Kristiina Ovaska, Renewable Resources Consulting Services Ltd., #214 Marine Technology Centre, 9865 West Saanich Rd., Sidney, British Columbia V8L 3S1, Canada.

Cynthia A. Paszkowski, Department of Zoology University of Alberta, Edmonton, Alberta T6G 2E9, Canada.

Bruce Pauli, Canadian Wildlife Service, Environment Canada, 100 Gamelin Blvd #9, Hull, Québec K1A 0H3, Canada.

Karen E. Pettit, Canadian Wildlife Service, Box 5050, Burlington, Ontario L4R 4A6, Canada. *Present Address*: Department of Animal Science, University of British Columbia, Vancouver, British Columbia, V6T 1Z4, Canada.

G. Lawrence Powell, Vertebrate Morphology Research Group, Department of Biological Sciences, and Kananaskis Research Stations, University of Calgary, 2500 University Dr. NW, Calgary, Alberta T2N 1N4, Canada.

Jean Rodrigue, Canadian Wildlife Service, Québec Region, P.O.Box 10100, Ste.-Foy, Québec G1V 4H5, Canada.

Anthony P. Russell, Vertebrate Morphology Research Group, Department of Biological Sciences, and Kananaskis Research Stations, University of Calgary, 2500 University Dr. NW, Calgary, Alberta T2N 1N4, Canada.

Leslie A. Rye, Department of Zoology, University of Guelph, Guelph, Ontario N1G 2W1, Canada.

Carolyn N.L. Seburn, Department of Zoology University of Alberta, Edmonton, Alberta T6G 2E9, Canada. *Present address*: Seburn Ecological Service, 920 Mussell Road, R.R. #1, Oxford Mills, Ontario K0G 1S0, Canada.

David C. Seburn, Department of Geography University of Alberta, Edmonton, Alberta T6G 2E9, Canada. *Present address*: Seburn Ecological Service, 920 Mussell Road, R.R. #1, Oxford Mills, Ontario K0G 1S0, Canada.

Timothy F. Sharbel, Redpath Museum, McGill University, 859 Sherbrooke St. W., Montréal, Québec H3A 2K6, Canada. *Present address*: Arbeitsgruppe Michiels, Max-Planck Institut für Verhaltensphysiologie, D-82319 Seewiesen (Post Starnberg), Germany.

Leonard J. Shirose, Department of Zoology, University of Guelph, Guelph, Ontario N1G 2W1, Canada.

Ruth Waldick, Department of Biology, Dalhousie University, Halifax, Nova Scotia B3H 4H7, Canada. *Present address*: Department of Biology, McMaster University, Hamilton, Ontario L8S 4K1, Canada.

Richard J. Wassersug, Department of Anatomy and Neurobiology, Dalhousie University, Halifax, Nova Scotia B3H 4H7, Canada.

Sheri M. Watson, Vertebrate Morphology Research Group, Department of Biological Sciences, and Kananaskis Research Stations, University of Calgary, 2500 University Dr. NW, Calgary, Alberta T2N 1N4, Canada.

Wayne F. Weller, Ontario Field Herpetologists, 66 High Street E., Suite 504, Mississauga, Ontario L5G 1K2, Canada.

Foreword

The Declining Amphibian Populations Task Force (DAPTF) had its origins at the First World Congress of Herpetology held in England in 1989. At that meeting, amphibian researchers traded observations about (then) recent and disturbing amphibian declines. Various meetings were held subsequently to ascertain whether there was enough concern to warrant some kind of coordinated response. Two conclusions were reached from these meetings: 1) although most of the evidence for amphibian declines was anecdotal, the number and the geographically dispersed nature of the informal reports indicated that the situation should be addressed and treated as a possible environmental emergency; and 2) an international working group should be established to determine the extent of the problem using scientifically defensible information. DAPTF is the resultant international working group organized within the Species Survival Commission of the World Conservation Union (IUCN). DAPTF immediately moved to set up a volunteer network of regional (or national) working groups. The groups were formed to determine whether amphibians were in decline within their regions and to report this information to a central DAPTF office. DAPTF could then assess whether there might be global causes for any observed amphibian declines. With regard to DAPTF's global mission, The Canadian Declining Amphibian Populations Task Force (DAPCAN) has served as one of these regional working groups.

The progress of DAPCAN has been of interest to DAPTF for several reasons. Amphibian populations at the edge of their geographical distributions are typically under greater environmental stress than more centrally located populations. If some global change was adversely effecting amphibian populations, the effects might be expected to be observed first in geographically marginal populations. All species of amphibians in Canada reach their northern geographical distribution limits somewhere within the country. Studies of Canadian amphibians provide reliable evidence of whether all recent amphibian declines are caused by global, rather than local, changes.

DAPCAN developed an organizational scheme which tied provincial and territorial working groups to a national co-ordinator with attendant annual national meetings. This scheme has been very successful in yielding timely results and is now used as a model for DAPTF regional working groups in other countries.

The original DAPTF charge to the regional working groups was to ascertain whether amphibian populations were in decline in each working group region, and if so, to identify the suspected causes. No suggestions for structure or function of the working groups were at first supplied. Thus, working groups were on their own to determine how amphibian declines could be documented and to determine what factors might be involved. In reality, the definition and documentation of amphibian declines is not at all straightforward. To its credit, DAPCAN struggled with these basic questions. This publication provides valuable case history information on how to approach studies of amphibian populations and, importantly, offers recommendations for the kinds of studies that are most useful in documenting amphibian declines. DAPCAN has been extremely successful and effective in its role as a DAPTF regional working group. This publication, the fulfillment of DAPCAN's mandate from DAPTF is an example of their success and sets a high standard for all subsequent DAPTF working group reports.

W. Ronald Heyer, Chair, DAPTF
Washington, D.C.
August, 1995

Preface

In 1982, I was a visiting Canadian postdoctoral fellow at the University of California, Berkeley, intent upon studying the genetic relationships of the five western North American species of frogs related to *Rana aurora* and *R. boylii*. This required catching a lot of frogs, which I enjoyed. It also required considerable travel all over the western United States and Canada so that I could find populations and catch those frogs. I enjoyed that, too. Sometimes catching the frogs was easy, especially as I went north from California, up the Pacific coast, and back home into British Columbia. But for some species and some populations, it was far harder than I anticipated. There were good 1950s and 1960s records and collections of *Rana boylii* and *Rana muscosa* from the mountains ringing Los Angeles. In 1982, where were they? If I could find frogs in Sonoma, Napa, or Contra Costa Counties north and east of San Francisco, I could catch them, but the records of even 20 yr previous gave no indication where frogs might still occur. *Rana muscosa* should have been abundant throughout the upper Sierra Nevada, but I had great difficulties finding viable populations where populations should have been. My colleagues knew no better than I did. In 1983, Dave Wake, director of the Museum of Vertebrate Zoology at Berkeley, and I took a trip to a lake in the Sierra Nevada where Dave had seen *Rana muscosa* for years and confidently predicted we would find them. No frogs. I could not know then that the *Rana pretiosa* I had caught two years before in the Little Campbell River in Surrey, British Columbia, would be the last ever captured there, nor could Dave Wake and I know on that sunny but frogless day in the Sierra Nevada in 1983 that other amphibian biologists around the world were registering the same kinds of observations.

The stories by now are famous, even clichéd. There is the evident disappearance of the Golden toad, *Bufo periglenes*, from Monteverde, Costa Rica, its only known locality. The Gastric-brooding frog, *Rheobatrachus silus*, appears to be gone from the only stream where it was known to exist, in Queensland, Australia. Frogs of any sort, especially *Rana aurora*, *R. boylii*, and *R. muscosa*, have virtually disappeared from southern California. Spotted frogs, *Rana pretiosa*, except in a few cases, no longer exist west of the Cascades Mountains of Oregon and Washington. Leopard frogs, *Rana pipiens*, once so common on the Canadian prairies as to support a commercial harvest, are now vanishingly rare. Rachel Carson's silent spring appears to be upon us!

Equally famous, among herpetologists, is the story of how observations such as these, shared over beer and chips at the First World Congress of Herpetology in 1989 in Canterbury, England, led to the world-wide initiative to investigate declines in amphibian populations. Arising from this and subsequent meetings, mainly in North America, the World Conservation Union (IUCN) established a Declining Amphibian Populations Task Force (DAPTF) under the aegis of its Species Survival Commission. By 1992, DAPTF, coordinated by Jim Vial, had set up its office in Corvallis, Oregon, and, later that year, held its first board meeting, chaired by Dave Wake, in St.

Mountain yellow-legged frog, Rana muscosa. *Photo by David M. Green.*

Helena, California. By 1992, working groups to investigate the problem of declining amphibians had been established across the United States, in Lower Central America, Western Europe, Taiwan, Sri Lanka, Australia, the United Kingdom, and Canada. Soon there would be working groups all over the globe. The first issue of FROGLOG, the DAPTF newsletter, appeared in March 1992 and had as its lead article "Canada Launches Major Initiative".

The Canadian Declining Amphibian Populations Working Group had its beginnings in October, 1991 when Christine Bishop, of the Canadian Wildlife Service, and Bob Johnson, of the Metro Toronto Zoo, invited amphibian biologists and conservation workers across Canada to a weekend meeting at the Canadian Center for Inland Waters in Burlington, Ontario. The purpose of the meeting was to discuss how best to monitor amphibian populations across the country, with a view to understanding distribution and trends in abundance in those populations. The meeting was not explicitly part of the new DAPTF initiative but it became clear that any Canadian plan to monitor amphibians could contribute to it in a substantive way. Consequently, those attending the Burlington meeting agreed to form a DAPTF working group to study declines in amphibian populations in Canada. The Canadian group was one of the first national or regional working groups of DAPTF to be organized. I took on the job of National Co-ordinator of the new working group. Don McAlpine and Stan Orchard became eastern and western regional co-ordinators, respectively, and co-ordinators for every province and territory were recruited.

The following year, 1992, I organized the second meeting of DAPCAN (as the Canadian Working Group came to be called) at the Redpath Museum, McGill University, in Montréal. There, the group agreed upon a mission statement and action plan (see sidebar) and ratified the framework of national, regional and provincial co-ordinators (Appendix II, this volume). The National Co-ordinator's role was to communicate between regional coordinators, subgroups, and the DAPTF office, establish and co-ordinate an *ad hoc* council to serve as advisory body and secretariat for DAPCAN, and direct and edit publications of the working group. The two Regional Co-ordinators (western and eastern) were detailed to communicate between provinces and the National Co-ordinator, compile reports on regional activities, and serve on the council. Provincial Co-ordinators were asked to oversee and organize survey efforts and establish inventories of projects in their provinces, report progress to their regional co-ordinators, and establish networks of volunteers for extensive surveys. In addition to these co-ordinators, DAPCAN established issue-based work groups to look into the historical data-base, disease, extensive monitoring, and intensive monitoring. To promote internal communication, DAPCAN agreed to publish semi-annual reports in the Canadian Association of Herpetologists' (CAH) Bulletin, compile a general mailing list with telephone and fax numbers, and organize annual meetings. DAPCAN determined to seek grants and government contracts for surveys, research, reports, and meetings and so established a funding subcommittee. A time-table was set for the completion of its tasks. An agreement to establish a protocol for continuation of the Working Group, or spin-off projects, was tabled.

Early in DAPCAN's existence, it elected not to concentrate upon the special problems of salvaging endangered species. DAPCAN specifically asked workers to identify amphibians in decline. To do this, the group promoted setting up a network of extensive surveys to establish locations and approximate numbers of amphibians. DAPCAN sought to develop and implement standardized report formats, and to organize compilation of data into summaries or atlases. We asked that historical records be examined for evidence of past distributions and numbers and to study the results of established intensive population studies for evidence of declines. With those data, we intended to compare present and past evidence to assess the possible nature and extent of declines.

DAPCAN further intended to examine causes of declines and encourage studies into normal population fluctuations. We wished to promote studies into possible effects of various anthropogenic factors, including pollutants, pesticides and toxins, pH, nutrient and sediment loading, drought, harvest, and survey methods. But, further, DAPCAN wanted to encourage research into the socio-economic factors of habitat degradation, which include industrial expansion, urban encroachment, agriculture, forestry, fisheries practices, and construction of highways, pipelines, dams, etc. The im-

Mission and goals of the Canadian Declining Amphibian Populations Working Group (DAPCAN), adopted October, 1992

Mission Statement

To determine the nature, extent and possible causes of declines in amphibians in Canada, and advocate means by which the declines can be avoided, halted and reversed.

Goals

Promote and coordinate research efforts aiming to document declines in amphibian populations

Promote the collection and analysis of evidence concerning possible causes of amphibian declines

Propose action to halt, avoid and reverse declines.

Promote establishment of long-term monitoring schemes on amphibian populations

Disseminate information on declines to the scientific community, government agencies, and the public

Communicate findings among individuals, conservation organizations, appropriate governmental agencies, and the IUCN/SSC DAP Task Force

pacts of epidemics and diseases, parasite load, introduced exotic organisms, and global climate change were also targeted as worthy of research. DAPCAN identified a need for long-term monitoring of populations and so sought to encourage and develop standardized protocols for long-term studies at selected sites. The group encouraged provincial groups to develop extensive long-term atlas and monitoring projects and promoted intensive population ecology projects by individuals.

Finally, DAPCAN agreed to compile written reports of the findings of the Working Group into an edited volume for publication, which is this book. DAPCAN's intent was to inform governments, land managers, conservation organizations, public educators, and the media of its results. But it also intended to distribute individual articles and abstracts, prepare press releases, prepare popular literature on Canadian amphibians, and, lastly, to communicate to the DAP Task Force.

DAPCAN set itself some lofty goals in Montréal. Not everything it attempted to promote has been achieved but the major facets of its initiative have produced results. DAPCAN deliberately set itself up as a co-ordinating body, not a research group *per se*, and this distinction was quickly recognized. Most of the provincial co-ordinators already had research projects either planned or under way and the initiative was seen as a collective effort with contributions gratefully accepted from all interested parties. The de-centralized organization ensured participation from every province and territory of Canada and reduced the workload on any single individual. The conferences and reports in the CAH Bulletin fostered discussion among participants. Research into Canadian amphibian populations was organized and communicated across the country.

All of the DAPCAN meetings have been the largest gatherings of Canadian herpetologists at any place, at any time. After the 1992 Montréal meeting, Stan Orchard organized DAPCAN III at the Royal British Columbia Museum in Victoria in 1993, and Bill Preston organized DAPCAN IV at the Manitoba Museum of Man and Nature in Winnipeg in 1994. At each meeting, participants discussed their findings in plenary sessions and took stock of how well the group was progressing towards its goal of determining the nature, extent, and cause of amphibian population declines in Canada. Each meeting featured contributed papers and open discussions, which ranged from tackling the basic question of whether or not declines exist to discussion of better land management practices to ensure continuity of populations and species. Minutes and reports of all the DAPCAN meetings have appeared in the Canadian Association of Herpetologists Bulletin and synopses have been faithfully reported in FROGLOG.

This book arises from the 4 DAPCAN conferences of 1991 to 1994 which gave the group its impetus and focus. It is not a symposium volume in any strict sense because much of the work entailed here was begun before DAPCAN got its start. Some of the chapters were not presented as papers at the DAPCAN conferences and many of the conference presentations have not found their way into this volume. A call for papers was issued in March, 1993, and 46 titles of prospective chapters were submitted. Not all of the offered titles materialized into manuscripts and, because all submitted chapters were peer-reviewed, not all submitted manuscripts were accepted for publication. Some chapters were specifically solicited by the editor but the book is far from being merely his creation. It is the product of the whole DAPCAN initiative and belongs equally to all who were, and still are, involved.

The various chapters in this book address many different facets of amphibian population biology and conservation, with a focus on Canada. Steve Hecnar begins the volume with an examination of amphibian pond communities on a landscape scale in southwestern Ontario. The demography of amphibian populations is, with few exceptions, poorly understood and the projects represented in this volume constitute a substantial corpus of badly needed information. These include the studies by Leonard Shirose and Ron Brooks on ranid frogs in Algonquin Park in Ontario, Raymond Leclair and Helena Caetano on *Notophthalmus viridescens* in the Mastigouche reserve of Québec, Larry Powell et al. on *Ambystoma macrodactylum* in the Alberta Rockies, my own work on the *Bufo fowleri* population at Long Point in Ontario, and the study by Sue Bertram and Mike Berrill on *Hyla versicolor* at Nogies Creek in Ontario. Carolyn Seburn et al. examine the crucial role of dispersal by focusing on *Rana pipiens* populations in Alberta, and Claude Daigle presents recent information on the status of *Pseudacris triseriata* in Québec, now considered rare in that province. Genetical aspects of amphibian population biology are covered in a chapter by Leslie Rye and Jim Bogart, who stress the need for genetic identifications of cryptic species, particularly those in the *Ambystoma jeffersonianum* complex, and a chapter by Tim Sharbel et al. who explore novel uses for cytogenetic methods, particularly flow cytometry, for assaying population structure.

There has been, until this day, extremely little discussion, or even exploration, of amphibian populations in remoter regions of northern and northeastern Canada. Grouped together in one chapter are new data on amphibians in Newfoundland and Labrador by John Maunder, the Northwest Territories by Mike Fournier, and the Yukon by Lee Mennell that form an original contribution to understanding a poorly known fauna. The historic status of species in other particular parts of Canada is further discussed in chapters by Andrew Diduik on Saskatchewan and Don McAlpine on New Brunswick who attempts a rare treatment of historical data to discern declines.

Several chapters in this volume address the methods and results of extensive monitoring of amphibians, putting into practice the hopes and recommendations expressed at the first DAPCAN meeting. Chapters by Michel Lepage et al., Joël Bonin et al., and Christine Bishop et al. test the survey methods that have been proposed for broad-scale monitoring of amphibian populations. They form a substantial contribution towards refining a methodology for this purpose. Sampling methods for terrestrial salamanders using artificial cover objects are examined by Ted Davis for western Canada and by Joël Bonin and Yves Bachand for eastern Canada and a method of surveying aquatic frogs is presented by Don McAlpine.

The critical impact of natural and anthropogenic stressors upon amphibian populations is the subject of several chapters in this volume. Forestry practices and the severe impact of forest harvesting upon amphibians are examined in the eastern and western parts of the country, respectively, in chapters by Linda Dupuis and Ruth Waldick. Probable effects of global warming and ultraviolet radiation on amphibian populations are examined by Kristiina Ovaska. Douglas Clay considers the effectiveness of *Ambystoma maculatum* as a bioindicator of acid precipitation. The effects of pesticides are studied in chapters by Michael Berrill et al. and Joël Bonin et al. while Graham Crawshaw discusses the prevalence of disease in amphibian populations. Richard Wassersug discusses the validity of attempting to survey amphibian larvae as indicators of population viability and David Galbraith discussions of the use and need for molecular methods in conservation biology. Finally, to conclude the volume, I offer a chapter discussing amphibian population declines in general, embodying the discussions and recommendations of the DAPCAN initiative and my own perspectives. The volume also contains, as an appendix by Wayne Weller and me, current information on the status of each of the 45 species of amphibians found in Canada.

Altogether, this volume represents a benchmark for investigations of amphibian population abundance in general, and for understanding amphibian populations in Canada in particular. It not only shows what a coordinated working group of individuals can accomplish to investigate declining populations, it provides fundamental information for attempts to reverse a disturbing trend. This volume is dedicated to all those whose concern for the well-being of amphibians and the health of this earth motivates them to expend their time and energy for the good of all.

This book could not have been produced without the dedication, patience, and contributions of all of those involved in the DAPCAN initiative. I thank all of them. I also thank Evelynne Barten, Jacqui Brinkman, Gillian Dell, Heather Gray, Reid McDougall, Linda Pactow, Anne Pelligrin, Jon Rabinowitz, Nic Schlecht, Enid Stiles, and Marla Stone for their help in proof-reading and managing manuscripts and the McGill University Translation Office, Martin Ouellet, and Hélène Joly for producing and reviewing the French abstracts. Finally, I thank the many reviewers of manuscripts for their time and diligence: Ronn Altig, Ray Ashton, Christine Bishop, Andrew Blaustein, James Bogart, Joël Bonin, Ronald Brooks, Bruce Bury, Janalee Caldwell, Francis Cook, Steve Corn, Rosemary Curley, Ted Davis, Richard DeGraaf, Andrew Didiuk, Kenneth Dodd, Sam Droege, Charles Drost, Linda Dupuis, Gary Fellers, David Galbraith, Douglas Gill, John R. Gold, Patrick Gregory, Russell Hall, Steven Hecnar, Paula Henry, Tom Herman, Ron Heyer, Robert Jaeger, Michael Klemens, Fred Kraus, Michael Lannoo, Clément Lanthier, Michel Lepage, John Maunder, Don McAlpine, Donald Newman, Stan Orchard, Martin Ouellet, James Platz, Larry Powell, Derek Roff, Anthony Russell, Leslie Rye, Carolyn Seburn, David Seburn, Diane Secoy, Brad Shaffer, Tim Sharbel, Leonard Shirose, Margaret Stewart, Brian Sullivan, James Taylor, Ruth Waldick, Richard Wassersug, Wayne Weller, Howard Whiteman, Henry Wilbur, Mary Wilson, Bruce Woodward, Richard Wyman, and Clifford Zeyl. All of them are friends of amphibians.

David M. Green
Montréal, Québec
August, 1995

©1997 by the Society for the Study of Amphibians and Reptiles
Amphibians in decline: Canadian studies of a global problem. David M. Green, editor.
Herpetological Conservation 1:1–15.

Chapter 1

AMPHIBIAN POND COMMUNITIES IN SOUTHWESTERN ONTARIO

STEPHEN J. HECNAR

Department of Biological Sciences, University of Windsor, Windsor, Ontario N9B 3P4, Canada.

ABSTRACT.—The status of amphibian communities at 174 ponds in southwestern Ontario was studied in 1992 and 1993. The study addressed whether amphibian declines have occurred historically, whether forest fragmentation affects amphibian communities, if regional dynamics are important in assessing amphibian status, if water acidity poses a threat, and whether predatory fish presence affects amphibian distribution. Patterns of species occurrence and richness differed among regions. Multiple regression determined that amount of woodlands surrounding ponds was the most important geographic variable affecting species richness. Estimates of turnover indicated the dynamic nature of all species, but differences among species and regions exist. In 2 of 3 regions net changes in species occurrence were negative. On a geographic scale, 36% of species decreased, 45% did not change, and 18% increased in occurrence. Ponds with predatory fish had low amphibian species richness. High pH and total alkalinity of pond water indicated that acidity would not affect amphibians. The present distribution and species richness patterns suggest that the amphibian fauna declined historically. Large-scale deforestation in the study area has contributed to the rarity of "woodland" species. The patterns observed indicated that a regional approach to assessing amphibian status is required.

RÉSUMÉ.—J'ai étudié des communautés d'amphibiens dans 174 étangs du sud-ouest de l'Ontario en 1992 et 1993 dans le but de déterminer si le déclin de leurs populations s'était toujours produit, si le morcellement de la forêt affectait leurs communautés, si la dynamique régionale jouait un rôle important dans l'évolution de leur statut, si l'acidité des eaux constituait une menace pour eux et si la présence de poissons prédateurs influençait leur répartition. La présence et l'abondance des espèces diffèrent selon les régions. L'analyse de régression multiple a permis de déterminer que la quantité des boisés autour des étangs était la variable géographique qui influençait le plus sur l'abondance des espèces. L'évaluation du renouvellement a mis en lumière le caractère dynamique de toutes les espèces quoiqu'il existe des différences entre les espèces et les régions. Dans 2 des 3 régions étudiées, les changements nets de fréquence des espèces ont été négatifs. Envisagées sous l'angle de l'échelle géographique, 36% des espèces ont diminué, 45% n'ont pas changé et 18% ont augmenté en fréquence. Les étangs où vivent des poissons prédateurs sont ceux où les espèces d'amphibiens sont les moins abondantes. Le pH élevé et l'alcalinité totale des eaux révèlent que l'acidité n'affecte pas les amphibiens. La répartition actuelle et l'abondance des espèces donnent à penser que la faune amphibienne a toujours connu un certain déclin. Le déboisement à grande échelle dans la zone étudiée a contribué à la rareté des espèces plutôt forestières. Compte tenu des observations effectuées, l'évaluation du statut des amphibiens exige une approche régionale.

Recent concern regarding an apparent global decline of amphibians has generated much interest (Barinaga 1990; Blaustein and Wake 1990; Phillips 1990; Wyman, 1990). Most hypotheses implicate either direct or indirect anthropogenic causes such as pollution (including acid rain and pesticides), habitat destruction (including urbanization and agriculture), global climate change (including the greenhouse effect and ozone depletion), and predation from introduced species. The evidence for declines is largely anecdotal and sometimes paradoxical in that not all species are declining, some species appear to be declining in apparently pristine areas, and some regions appear unaffected.

Investigating a problem as large as a world-wide decline of amphibians is made more difficult by our lack of detailed knowledge of amphibian biology, particularly population and community ecology. Much of our contemporary understanding of the mechanistic basis of community ecology comes from experiments using amphibians in artificial communities. Although these approaches are statistically rigorous, they have been criticized for lacking realism (Jaeger and Walls 1988). Studies in ecology have traditionally focused on interactions at local scales in an attempt to explain the structure of populations and communities. Many of the problems in ecology are problems of both spatial and temporal scale. Different patterns can emerge or be hidden depending on the scale of investigation (Callahan 1984; Wiens et al. 1986).

A growing interest in ecology concerns the relative roles of local versus regional processes and the role of history in structuring communities (Ricklefs 1987; Cornell and Lawton 1992; Ricklefs and Schluter 1993). If local amphibian assemblages form metapopulations (Gill 1978; Sjögren 1991) or metacommunities (Hanski and Gilpin 1991), understanding regional dynamics such as colonization and extinction processes is essential and necessary to investigate apparent declines. The few long-term studies that exist suggest that amphibian populations may undergo drastic fluctuations (Berven 1990; Pechmann et al. 1991; Blaustein et al. 1994) and that it may be difficult to distinguish declines from them (Pechmann et al. 1991). Although long-term studies are essential, the status of amphibian populations or communities cannot be determined by results obtained solely from local scales.

Habitat loss may be the most important factor that affects amphibians (Pechmann and Wilbur 1994; Blaustein et al. 1994). Habitat fragmentation results in less habitat available, increased isolation, and promotes external threats. In fragmented landscapes, woodlands are becoming recognized as important amphibian habitat (Strijbosch 1980; Loman 1988; Wederkinch 1988; Laan and Verboom 1990) because many species use woodland habitat for foraging or hibernation sites.

Amphibian breeding habitats occur along a gradient of pond duration (hydroperiod), where desiccation stresses predominate in ephemeral pools and high predation risk prevails in permanent ponds (Heyer et al. 1975; Wilbur 1980). Fish predation may play a key role in determining amphibian distribution or community composition (Heyer et al. 1975; Bradford et al. 1993; Brönmark and Edenhamn 1994).

Amphibians may be susceptible to toxic effects of water pollution because of their physiology and natural history (Vitt et al. 1990). Acidity has adverse effects on amphibians (Dunson et al. 1992), and acid rain has been implicated in amphibian decline (Harte and Hoffman 1989). Some regions may not be prone to acid deposition problems because of their geological characteristics.

Green frog, Rana clamitans. *Photo by Jacques Brisson.*

In 1992, I began to establish a network of ponds across southwestern Ontario with the goal of investigating and explaining patterns of species assembly in temperate zone amphibian communities on local and regional scales. The landscapes of southwestern Ontario have been highly altered by humans over the last century and comprise a study system that provides insight into long-term patterns of amphibian community change. Here, I report on the first 2 yr of a study of patterns of amphibian occurrence and diversity. Although short-term studies cannot determine if amphibians are declining currently, assessing contemporary patterns can help in determining if declines have occurred historically. Using an expanded multi-scaled approach in southwestern Ontario, I asked whether amphibian declines have occurred historically, whether forest fragmentation affects amphibian communities, if regional dynamics are important in assessing amphibian status, if water acidity poses a threat, and whether predatory fish presence affects amphibian distribution.

METHODS

Study Area

The study area is southwestern Ontario, defined as the land west of the Niagara escarpment and bordered by Lakes Huron, St. Clair, and Erie (Fig. 1). Southwestern Ontario consists of 42,962 km^2 of relatively flat to rolling terrain in the Great Lakes basin. Bedrock consists of sedimentary limestone, shale, and sandstone of Paleozoic origin, much of which is overlain by glacial lake deposits or glacial moraines. Climate ranges from moderate to cool temperate. The area has a diverse flora containing species of the Eastern Deciduous Forest and the deciduous-boreal ecotone of the Great Lakes-St. Lawrence Forest (Rowe 1972). The most common forest association in both districts is beech (*Fagus*)/maple (*Acer*).

The natural landscape has been highly altered by humans since the mid-1800s. Extensive forests and wetlands covered the area prior to European settlement, but by about 1850 maximal forest clearance for agriculture had occurred (Moss and Davis 1989). Forest cover in southwestern Ontario presently ranges from about 3% in the south to > 80% in the north. Original wetlands included hardwood swamps dominated by *Acer saccharinum* and *A. rubrum* and marshes dominated by *Typha* spp. (National Wetlands Working Group 1988). Extensive drainage of wetlands for agriculture began by the 1880s. Wetlands in pre-settlement times covered 68.9% of the southern portion of the area and 23.1% of the northern portion (Snell 1987). Today, wetlands cover 2.9 to 10.0% from south to north. Human population in the study area is now about 3,000,000 and most of the landscape remains under intensive agricultural use (Statistics Canada, 1991 census). Much of the present wetland habitat consists of artificial ponds constructed for agricultural use.

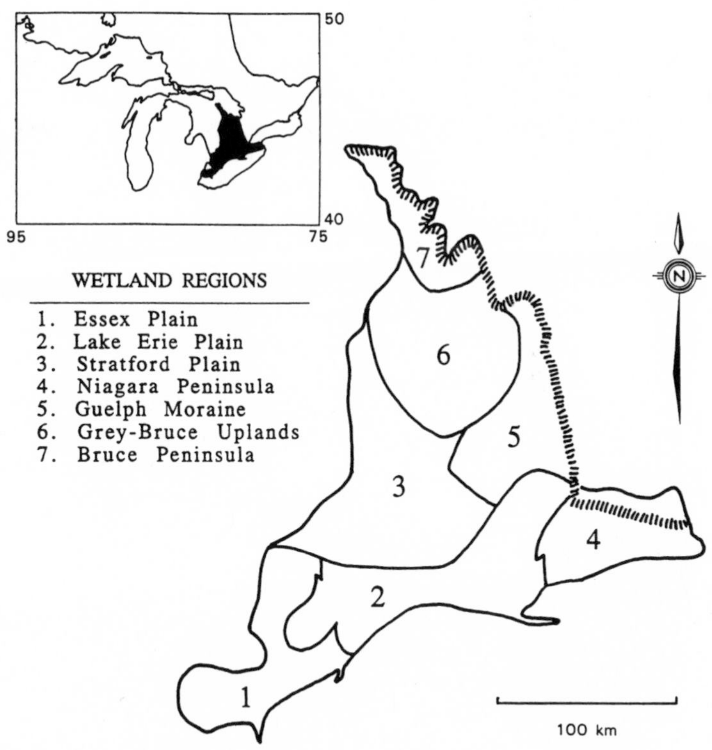

Figure 1 . *Study area, southwestern Ontario.*

For regional comparisons within the study area, I used wetland regions based on physiographic characteristics such as geology, drainage, and vegetation (Anonymous 1984). Using these study units represents a more natural partitioning of the landscape rather than using political boundaries or a grid system.

Study Ponds

In 1992 and 1993, I surveyed 174 ponds (119 surveyed in both years) in 5 of the 7 wetland regions, Essex Plain (n = 88), Lake Erie Plain (n = 2), Stratford Plain (n = 42), Grey-Bruce Uplands (n = 25), and the Bruce Peninsula (n = 17, surveyed in 1993 only). I did not include the Lake Erie Plain in regional comparisons because of the small sample size.

I used ponds as local study units, rather than other types of wetlands, because they are likely the most important wetland type presently in southwestern Ontario. Ponds are easily delimited spatially, they may act as functional islands, and they provide a study system capable of being analyzed on a geographic scale. I defined ponds as water bodies < 5 m deep. Of the ponds I studied, 87% were artificial, average depth was 1.91 ± 0.085 m, average area was 6594 ± 2674.9 m^2, and average age was 29.7 ± 1.52 yr. Amphibian communities could be compared among regions because ponds did not significantly differ in physical characteristics such as total area (ANOVA, $F_{3,160} = 1.469$, $P = 0.225$) or water depth (ANOVA, $F_{3,158} = 0.830$, $P = 0.479$).

I located ponds using topographic maps and by inquiring at Conservation Authorities, Provincial and National Parks. Some ponds were discovered by chance or by information provided by private landowners. After locating a pond, I added it to the survey list if permission for access was obtained. To avoid biased selection, all ponds accessed were included. Most of the study sites were semi-permanent to permanent ponds. I excluded small ephemeral pools, swamps, large marshes, lakes, rivers, creeks, and ditches.

Survey Methods and Species Pool

I conducted pond surveys from the commencement of amphibian activity in late March to late July in each year and proceeded generally from south to north to take advantage of the advancing spring weather. Because of the phenology of amphibian activity at breeding ponds, I made repeated surveys. Searches involved 3 to 7 people intensively searching the perimeter to approximately 10 m from the pond edge as well as wading, canoeing, and dip netting through the pond. If ponds were within or adjacent to wooded areas, I extended searches to check woody debris. I recorded a species as present if any age class was observed or if calls were heard on any visit during the year. Most ponds were visited on at least 3 occasions in spring and early summer of each year. During surveys I repeated circuits to combine day and night visits.

To test survey efficiency, I used 2 sites where the amphibian fauna was well known and examined cumulative species richness over a number of surveys. In both cases, species number did not increase from the 2nd to 7th surveys. I am confident that with the effort and methods used, the surveys allowed construction of accurate species lists, although transient individuals visiting ponds between surveys may have been missed.

To study species turnover, I estimated local extinction and colonization rate based on the presence or absence of particular species at local sites in each year. Connell and Sousa (1983) recommended a minimum of at least 1 complete turnover of individuals to assess population persistence. However, many amphibians are potentially long-lived (Duellman and Trueb 1986) and waiting years or decades before proclaiming extinction at an unoccupied site would underestimate true extinction rates. Additionally, if recolonization occurred during the time span, the local extinction would not be recognized.

Essentially the same pool of 12 pond-inhabiting amphibian species occurs in each wetland region of southwestern Ontario, permitting local and regional comparisons of communities. Two additional and regionally rare species have range boundaries which cross the study area. *Bufo fowleri* inhabits dune ponds at localized sites on Lake Erie in the Essex Plain, and *Rana septentrionalis* occurs in several stream systems in the northern half of the study area in the Grey-Bruce Uplands.

Geographic Variables

I measured maximum water depth (DEPTH, to nearest cm) and mapped ponds to determine area (AREA). Pond ages (AGE) were obtained in most cases by questioning landowners, and in some cases by noting the date a pond first appears on maps or aerial photographs. To determine most of the geographic features I used topographic maps and aerial photographs. I measured distances to woods (DISTW), highways (PROAD), roads (GROAD), other ponds (DISTP), potential corridors such as creeks, ditches, and rivers (CORR), and potential source areas such as large marshes, major rivers, and lakes (SOURCE) in the field (to 1 m) or on maps (to 25 m). Determining potential source areas was somewhat subjective but I tried to envision which sites would retain water under severe drought conditions. I determined percent woodland (WOODS) within a 2-km radius of each pond by using a circular dot grid on topographic maps. I chose the radius to be representative of the maximum annual overland dispersal for most species based on a literature survey. I calculated isolation (ISOL) as the mean distance to the nearest 5 other ponds or source areas. In the lab, I drew scale maps for each pond and calculated areas by planimetry. For human population density (URBAN) I used township census data (Statistics Canada 1993).

Water Acidity

I collected 4 water samples at each pond, 2 each in spring and summer, 1992. Samples were taken at about 15-cm depth in 1 L plastic sample bottles. To investigate the buffering capacity of the water to pH change I determined total alkalinity. Analyses were made using a US E.P.A. approved LaMotte™ Model A-25 portable test kit within 24 hr of collection. On each visit I took 4 pH readings with an Oakton™ Model WD-00624-20 pocket pH meter.

Fish Predation

Some fish species are well known predators on amphibians, whereas others are planktivorous or herbivorous and some are too small to prey on amphibians. Based on a literature survey, I classified ponds into 3 fish-type classes: no fish, fish, and predatory fish. Ponds classified as "fish" had only non-predatory fish present such as minnows, carp, and catfish. I considered "predatory fish" ponds to be those that had sunfish, bass, trout, pike, or perch.

Statistical Analyses

To meet the assumptions of parametric tests, I used transformations to normalize data and homogenize variances. All statistics followed Sokal and Rohlf (1981).

Results

Patterns of Occurrence and Species Richness

I encountered 12 amphibian species across the study area in 1992 and 1993 (Fig. 2). The frequency distribution was highly skewed; some species, such as *Rana clamitans* and *Rana pipiens*, were common, and others, such as *Ambystoma laterale* and *R. septentrionalis*, were rare. The frequency distributions did not differ between 1992 and 1993 ($G = 13.356$, 10 df, $P > 0.1$). The frequency distributions differed among regions (Fig. 3, Table 1). Patterns within regions did not differ between years in Essex ($G = 6.696$, 6 df, $P > 0.1$) or Stratford ($G = 2.749$, 6 df, $P > 0.5$), but did in Grey-Bruce ($G = 13.521$, 6 df, $P < 0.05$). The expected frequency distribution generated by the 1992 Grey-Bruce data was based on a small sample size, however.

Rana clamitans was the most common species in each region except the Bruce Peninsula where *R. pipiens* was most common. Differences in ranked occurrence for some species exist among regions (for example, for *R. pipiens* Bruce = 1, Grey-Bruce = 2, Stratford = 5, and Essex = 3). In Essex, species associated with woodlands (ie. *Pseudacris crucifer, Notophthalmus viridescens, Hyla versicolor, Rana sylvatica,* and *Ambystoma* spp.) have low occurrence. I found very low occurrence of *Rana catesbeiana* and *Ambystoma* spp. across the entire study area. Two Grey-Bruce sites in 1992 had single individuals of *R. septentrionalis,* which likely were transients.

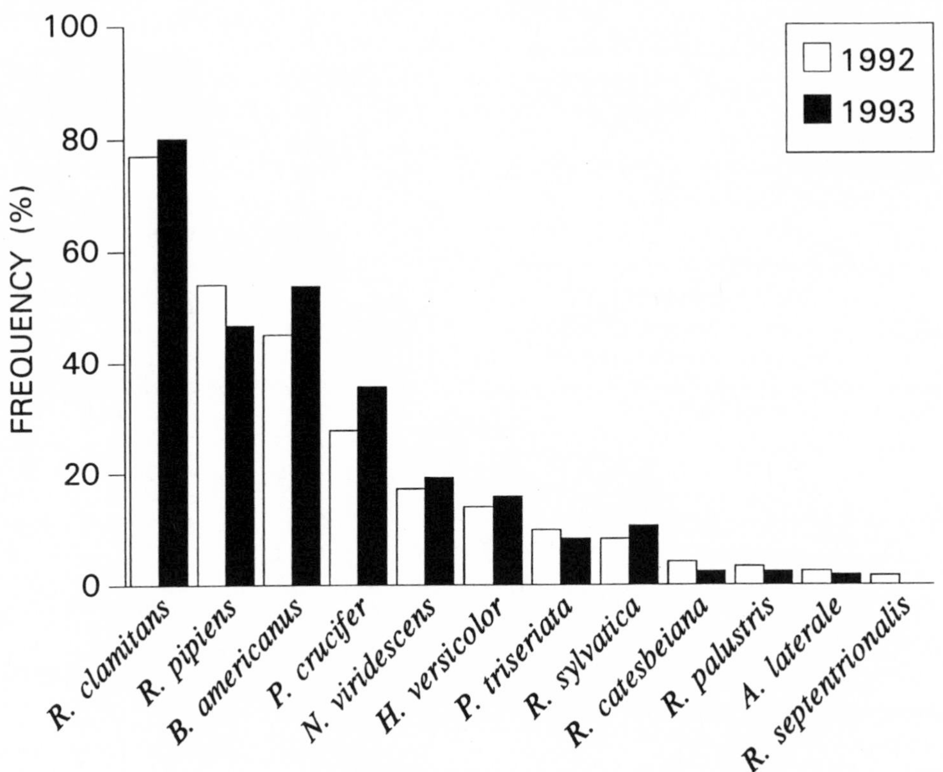

Figure 2. *Frequency of occurrence of amphibians observed at ponds in southwestern Ontario in 1992 and 1993.*

Table 1. *Comparison of occurrence distribution by region.*

	Bruce		Grey-Bruce		Stratford	
Region	G	df	G	df	G	df
Essex	9.97*	3	91.2***	8	136.1***	7
Stratford	13.5*	5	34.2***	7		
Grey-Bruce	22.4***	5				

*$P < 0.025$, ***$P < 0.001$

Mean local species richness (= α-diversity; Table 2) differed significantly among regions in 1992 ($F_{3,117} = 13.097$, $P < 0.001$) and 1993 ($F_{3,167} = 24.456$, $P < 0.001$). Multiple comparisons indicate that Essex has significantly lower α-diversity than each other region, but no differences occurred among Stratford, Grey-Bruce, and Bruce Peninsula (Table 3). α-diversity did not differ between 1992 and 1993 for the entire study area ($t = 0.078$, 117 df, $P = 0.938$) or for regions individually (Table 2).

Geographic Factors Affecting Species Diversity

Regression of local species richness over latitude indicates an increasing gradient from south to north ($F_{1,169} = 38.517$, $P < 0.001$, $r^2 = 0.186$). Of the 12 geographic variables, 5 were significantly correlated with local species richness in 1992 and 7 in 1993 (Table 4). In both years WOODS, PROAD, and URBAN, were positively related to species richness, and CORR, DISTW, and SOURCE were negatively related. Multiple regression of geographic variables resulted in a model which included WOODS, GROAD, and DISTW which accounted for 37% of the variance in 1992 (Table 5). The model for 1993 included WOODS, GROAD, PROAD, CORR, and AGE, and accounted for 28% of the variance (Table 5).

Species Turnover

In 1993, I made repeated surveys of 119 of the 122 ponds initially surveyed in 1992. The patterns of colonization of ponds and local extinctions suggested a highly dynamic system on all spatial scales.

Table 2. *Mean local amphibian species richness (α-diversity). No between year comparisons were significant.*

Region	1992 (n=122)	1993 (n=171)	df
Essex	2.00 ± 0.15	1.95 ± 0.13	75
Stratford	3.87 ± 0.27	3.93 ± 0.22	29
Grey-Bruce	3.08 ± 0.48	3.52 ± 0.32	11
Bruce		3.00 ± 0.31	

Table 3. *Multiple comparisons of local species richness among regions using Tukey's HSD test.*

Region	Bruce	Grey-Bruce	Stratford
Essex	0.281*	0.400***	0.509***
Stratford	–0.229	–0.110	
Grey-Bruce	–0.119		

* $P < 0.05$, *** $P < 0.001$

Table 4. *Correlation of local amphibian species richness (α-diversity) with geographic variables.*

Variable	1992 (n=121)	1993 (n=172)
AGE	0.116	0.014
AREA	0.170	0.195*
CORR	–0.212*	–0.238
DEPTH	–0.173	–0.081
DISTP	–0.100	–0.052
DISTW	–0.422***	–0.291**
GROAD	–0.122	–0.148
ISOL	–0.146	–0.071
PROAD	0.229*	0.274**
SOURCE	–0.071	–0.170
URBAN	0.278**	0.405***
WOODS	0.517***	0.470***

*$P < 0.05$, **$P < 0.01$, ***$P < 0.001$

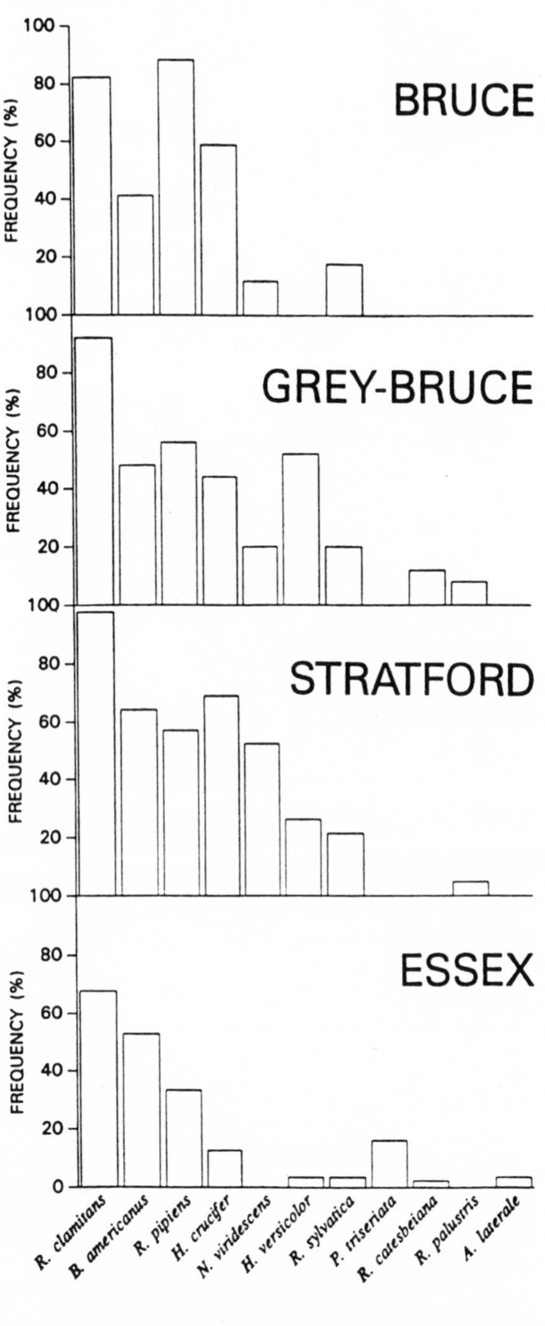

Figure 3. *Frequency of occurrence of amphibians by region in 1993.*

The mean local pond colonization and extinction rates did not differ among regions (Table 6). Comparing the mean colonization and extinction rates within regions suggests that Essex changed little but that extinction exceeded colonization in both Stratford and Grey-Bruce. The overall pattern for southwestern Ontario suggests that extinction rates are higher than colonization rates for 1992 to 1993.

Examining turnover rates for individual species suggests that all species are dynamic (Fig. 4), but differences among species and regions exist (Table 7). For example, *R. clamitans* had no turnover in Stratford, increased in occurrence in Essex, and decreased in Grey-Bruce. *Bufo americanus* had no turnover in Grey-Bruce, increased moderately in Stratford, and had a large increase in Essex. Occurrence of *R. pipiens* decreased in all regions but the magnitude of decrease varied from 5.3% to 27.8%.

Summarizing the net changes that occurred by species and region illustrates their complexity (Table 7). In Essex, 4 species did not change, 3 increased, and 2 decreased, suggesting that no overall annual decrease occurred. However, overall annual decreases are suggested in Stratford (1 with no change, 2 increased, 5 decreased) and Grey-Bruce (3 with no change, 1 increased, 4 decreased). Thus, 2 of 3 regions showed evidence of overall annual decrease. Across southwestern Ontario, 4 of 11 species (36%) showed an annual geographic decline, 5 (45%) had no change, and 2 (18%) increased. Some species, such as *Bufo americanus*, increased both regionally and geographically in both years. My field observations sug-

Table 5. *Multiple regression of amphibian species richness as a function of geographic variables in 1992 and 1993.*

Variable	Coefficient	SE	Standardized Coefficient	Tolerance	*t*
1992 ($F_{3,117} = 22.76$, $P < 0.001$, n = 121, $r^2 = 0.368$)					
Constant	1.93	0.148	0.000		13.1***
DISTW	–0.10	0.035	–0.265	0.609	–2.81**
GROAD	–0.11	0.039	–0.209	0.997	–2.84**
WOODS	0.01	0.003	0.360	0.608	3.82***
1993 ($F_{5,147} = 11.46$, $P < 0.001$, n = 153, $r^2 = 0.280$)					
Constant	1.57	0.161	0.000		9.74***
AGE	0.09	0.054	0.119	0.978	1.69
CORR	–0.05	0.022	–0.162	0.909	–2.20*
GROAD	–0.10	0.037	–0.188	0.914	–2.57*
PROAD	0.07	0.033	0.162	0.900	2.20*
WOODS	0.01	0.002	0.362	0.916	4.95***

$P < 0.05$, $^{**}P < 0.01$, $^{***}P < 0.001$

Table 6. *Comparison of rates of local colonization (ANOVA: $F_{2,116} = 2.067$, P = 0.131) and extinction (ANOVA: $F_{2,116} = 0.858$, P = 0.427) among regions.*

Region	n	Colonizations	Extinctions
Essex	77	0.25 ± 0.040	0.24 ± 0.037
Stratford	30	0.11 ± 0.031	0.20 ± 0.036
Grey-Bruce	12	0.05 ± 0.033	0.13 ± 0.048
Southwestern Ontario	119	0.19 ± 0.028	0.22 ± 0.026

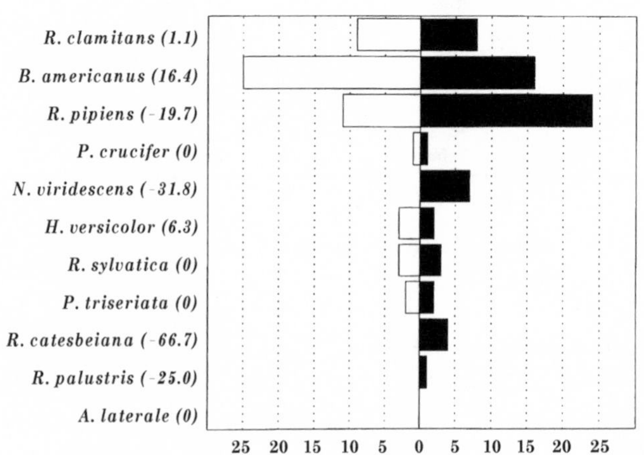

Figure 4. *Amphibian colonization and extinction in southwestern Ontario, 1992 to 1993. Clear bars are number of new ponds colonized; filled bars are number of local extinctions. Percent net change for each species is in parentheses.*

Table 7. *Species turnover expressed as net percent change. A dash (—) indicates that a species was not present in the region surveyed.*

Species	Southwestern Ontario	Region			Net Change
		Essex	Stratford	Grey-Bruce	
Rana clamitans	1.1	3.9	0	–9.1	0
Bufo americanus	16.4	24.2	5.9	0	+2
Rana pipiens	–19.7	–27.8	–5.3	–18.2	–3
Pseudacris crucifer	0	10.0	–5.3	0	0
Notophthalmus viridescens	–31.8	—	–30.0	–50.0	–2
Hyla versicolor	6.3	0	11.1	0	+1
Pseudacris triseriata	0	0	—	—	0
Rana sylvatica	0	0	–28.6	[a]	0
Rana catesbeiana	–66.7	–75.0	—	–50.0	–2
Rana palustris	–25.0	—	–33.3	—	–2
Ambystoma laterale	0	0	—	—	0
Net change	–1	+1	–3	–3	

[a]*R. sylvatica* present only in 1993; net change could not be calculated.

gested that reproduction and recruitment was high for toads in 1993. Species that declined both regionally and geographically include *R. catesbeiana*, *R. pipiens*, and *N. viridescens*.

Water Acidity

Basic statistics for southwestern Ontario and for individual regions indicated that pond water in the study area had high pH and high alkalinity (Table 8). Local species richness at ponds was not dependent on pH in either spring (regression: $F_{1,112} = 1.92$, $P = 0.169$, $r^2 = 0.017$) or summer ($F_{1,113} = 1.29$, $P = 0.258$, $r^2 = 0.011$). Similarly, species richness was not dependent on total alkalinity in either spring ($F_{1,112} = 1.179$, $P = 0.280$, $r^2 = 0.010$) or summer ($F_{1,113} = 1.678$, $P = 0.198$, $r^2 = 0.015$).

Table 8. *pH and total alkalinity (TAlk; mg/L) for ponds in southwestern Ontario and individual regions in 1992.*

Region	Variable	Spring		Summer	
		Mean	SE	Mean	SE
Essex	pH	8.2	0.08	8.3	0.07
	TAlk	180	11.1	203	17.2
Stratford	pH	7.9	0.06	8.2	0.09
	TAlk	203	12.4	204	13.1
Grey-Bruce	pH	7.9	0.20	8.3	0.11
	TAlk	222	21.7	219	18.8
Southwestern Ontario	pH	8.1	0.06	8.3	0.05
	TAlk	189	8.1	205	11.7

Fish Predation

The most commonly occurring predatory fish I observed were centrarchids (sunfish and basses). The 13 predatory fish species I observed were *Lepomis gibbosus* (pumpkinseed), *L. macrochirus* (bluegill), *L. cyanellus* (green sunfish), *Micropterus salmoides* (largemouth bass), *M. dolomieu* (smallmouth bass), *Pomoxis nigromaculatus* (black crappie), *Perca flavescens* (yellow perch), *Salmo gairdneri* (rainbow trout), *S. trutta* (brown trout), *Salvelinus fontinalis* (speckled trout), *Esox lucius* (northern pike), *Lepisosteus osseus* (longnose gar), and *Amia calva* (bowfin). Amphibian species richness was significantly lower at ponds having predatory fish present (2.9 ± 0.21) compared to ponds either lacking fish (3.9 ± 0.21) or having only non-predatory fish (4.0 ± 0.25; $F_{2,169} = 7.113$, $P = 0.001$). The proportion of ponds having predatory fish present differed among regions (Fig. 5). Predatory fish occurred most frequently in Essex, followed by Grey-Bruce and Stratford. For 5 of 10 species (*R. pipiens, P. crucifer, N. viridescens, H. versicolor*, and *Pseudacris triseriata)*, presence at a pond was dependent on the fish class (Table 9). Presence of *R. clamitans, B. americanus, R. sylvatica, R. catesbeiana*, and *Rana palustris* was independent of fish class. Regression of the G-scores (Table 4) as a function of adult body size indicates that large species are independent of fish class, whereas small species are dependent ($F_{1,8} = 15.647$, $P = 0.004$, $r^2 = 0.662$; Fig. 6).

Table 9. *Amphibian occurrence at ponds in relation to fish class present. Data are the number of ponds with the particular amphibian species present.*

	Pond fish class			
Species	No fish (n=73)	Fish (n=48)	Predatory Fish (n=51)	G (v=2)
Rana clamitans	61	41	45	0.52
Bufo americanus	44	29	35	1.06
Rana pipiens	48	36	24	8.61*
Pseudacris crucifer	36	19	7	18.2***
Notophthalmus viridescens	18	13	4	7.96*
Hyla versicolor	11	14	4	7.98*
Rana sylvatica	14	5	3	5.12
Pseudacris triseriata	14	1	1	14.9***
Rana catesbeiana	2	3	3	1.02
Rana palustris	3	0	2	2.93

*$P < 0.025$, ***$P < 0.001$

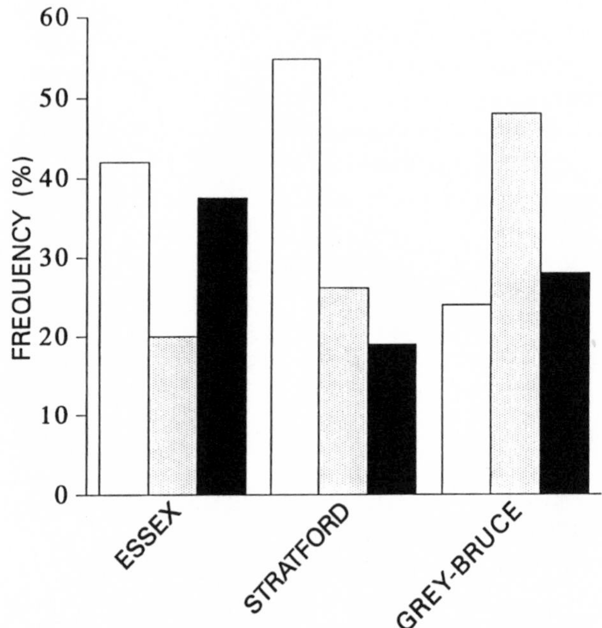

Figure 5. *Frequency of ponds by fish class. White bars indicate no fish present, stippled bars indicate non-predatory fish, and black bars indicate predatory fish.*

Discussion

The patterns of species occurrence, diversity, and turnover I observed in 1992 and 1993 indicated that amphibian pond communities in southwestern Ontario differ both spatially and temporally. A more than 40-fold difference in the percent occurrence exists between the most common and rarest species. Although pond communities in southwestern Ontario draw on the same species pool, different distribution patterns result among regions. Local species richness (α-diversity) also differs among regions, with a paradoxical trend of increasing richness from south to north. These patterns suggest that different processes or intensities of processes are acting or have acted on amphibians in each region. Other latitudinal trends across the study area include decreasing human population density and increasing forest cover and wetlands from south to north.

Amphibian species that use woodlands, such as *P. crucifer, R. sylvatica, H. versicolor, N. viridescens*, and ambystomatid salamanders, are conspicuously absent in local species lists and regionally rare in the Essex Plain. Distributional records and the occurrence of woodland species in relictual "islands," such as Rondeau Provincial Park, Walpole Island, and Point Pelee National Park, and scattered isolated populations indicate that these species were historically widespread. In pre-settlement times, hardwood swamps and marshes covered from 40 to > 60% of the land in the southern portion of the Essex Plain region, but presently only about 3% is forested and < 5% of wetlands remain (Snell 1987). This habitat loss and alteration has resulted in less complex and less diverse amphibian communities.

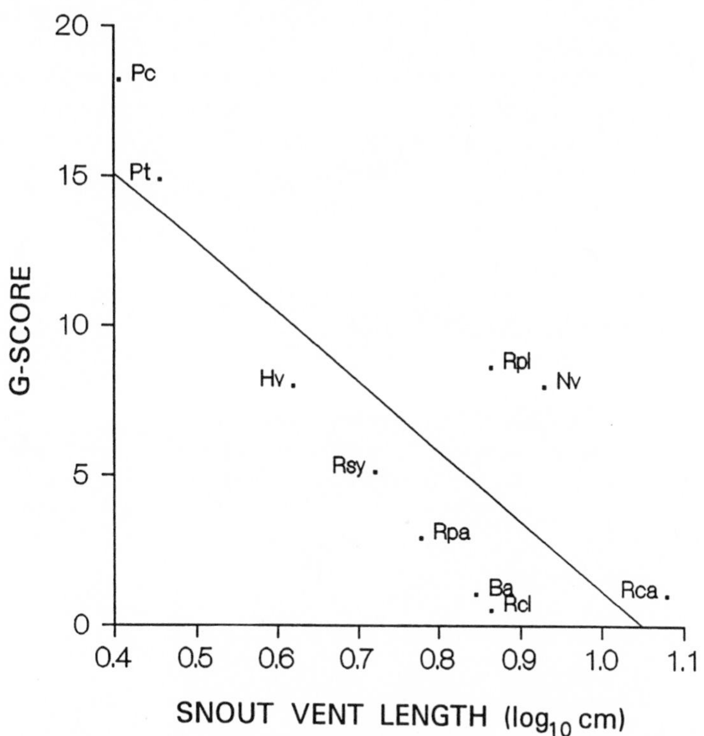

Figure 6. *Fish effect (G-score) versus adult body size (mid-point of range from Conant 1975).*

Rana clamitans is nearly ubiquitous in the ponds of southwestern Ontario. The reasons for its success may include its being a habitat generalist, or having some advantage in dispersal or reproduction. The proliferation of relatively deep ponds that were dug for agriculture may have allowed *R. clamitans* to increase their prevalence and abundance. Unfortunately, historical data for comparison do not exist. Anecdotal reports suggest that *R. pipiens* was the most abundant frog in the Essex Plain. During this study, *R. pipiens* declined in occurrence across all regions of southwestern Ontario. *Rana catesbeiana* was much rarer than I expected considering the nature of the ponds and the association of this species with farm ponds in eastern North America (Petranka et al. 1987). I found *R. catesbeiana* at only 5 of 174 ponds (2.9%) in 1992 and 1993. At Point Pelee National Park in the Essex region, *R. catesbeiana* was formerly abundant, but has not been detected since 1989. It remains common on Pelee Island, 30 km south of the Essex Plain. Other extant populations occur at Walpole Island (A. Bernier, pers. comm.), the Detroit River (R. Russell, pers. comm.), and Rondeau Provincial Park, Ontario (pers. obs.). My observations are consistent with the general pattern of long-term anecdotal observations in Ontario (L. McGillivray and M. Berrill, pers. comm.).

Perhaps the biggest surprise was the rarity of salamanders in southwestern Ontario. Although adults are rarely seen and larvae are hard to find in the field (L. Rye and J. Bogart, pers. comm.), I expected greater occurrence with the intensive search methods used. At sites in Rondeau Provincial Park and Pelee Island, I quickly determined salamander presence on most visits with my standard search protocol. Perhaps salamanders are not using the artificial ponds that now dominate the agricultural landscape, and they have become extremely rare. Another possibility is that salamanders either did not breed or had failed breeding attempts in both years. This had apparently occurred with a late freeze of ponds that destroyed eggs in Norfolk County in spring 1992 (M. Gartshore, pers. comm.).

Turnover was generally high for most amphibian species from 1992 to 1993, and net change indicated that extinctions predominated. Colonization and extinction rates differed among species and among regions. For each species some new ponds were colonized, local extinctions occurred, and presence at some ponds did not change. Net changes indicated overall decreases in Stratford and Grey-Bruce regions, which were the most diverse. Essex did not have an overall regional decline. In Essex the greatest impacts on amphibian diversity likely occurred historically, and the simplified communities today represent the worst case scenarios of extreme habitat alteration.

Pond water chemistry reflected the basic sedimentary geology of the area. The water was hard, alkaline, and well-buffered against changes in pH. The high pH and hardness also made the development of toxic species of metals unlikely. Although the effects of acidity on amphibians are well-documented (Freda et al. 1991; Dunson et al. 1992) and acid rain falls in southwestern Ontario, it cannot be implicated in amphibian declines south of the Canadian Shield because of the strong buffering capacity of the water. All pH values I observed were well above lethal levels reported for any amphibian species (Freda et al. 1991). Acid rain has also been ruled out as a potential cause in other geographic areas (Corn and Vertucci 1992; Bradford et al. 1994).

Fish predation is likely an important biotic factor on both local and regional scales. Ponds with predatory fish had low amphibian richness and certain amphibian species were less likely to be present. Low amphibian abundance, diversity, or absence of species has been reported in many studies (Efford and Mathias 1969; Heyer et al. 1975; Petranka 1983; Kats et al. 1988; Ireland 1989; Bradford et al. 1993). Heyer et al., (1975) considered fish to be the only predators capable of complete decimation of larval populations. Sexton and Phillips (1986) documented the invasion of fish from a river into 2 ponds following flooding. One pond received predatory fish and over time only 2 of 14 amphibian species remained. The other pond received non-predatory fish and no amphibian extirpations occurred. Ireland (1989) reported that *Ambystoma* spp. were more common in ponds without predatory fish and when co-occurring, fish were capable of preventing recruitment of metamorphs to the adult class.

In southwestern Ontario, *P. crucifer* and *P. triseriata* are the least likely species to coexist in ponds with predatory fish. Both species rarely encounter fish, have palatable larvae, and generally lack antipredatory behaviour (Kats et al. 1988). *Rana clamitans, B. americanus, R. catesbeiana*, and *R. palustris* are apparently not affected by the presence of predatory fish. All 4 species are common in farm ponds of eastern North America and the first 3 species have unpalatable larvae. In this study, species with larger adults coexisted with fish. This suggests that in addition to larval characteristics, small body size may be disadvantageous to temporary pond species that colonize fish ponds. While general patterns of amphibian presence with fish are apparent, few species show perfect fidelity to permanent or temporary ponds (Kats et al. 1988). Although the presence/absence patterns for temporary pond species are significant in southwestern Ontario, these species coexist with predatory fish at many localities, typically in microhabitats such as shallows and in emergent vegetation that serve as refugia.

Proportionately more ponds in Essex had predatory fish present, followed by Grey-Bruce and Stratford. This ranking of regions by predatory fish presence is the inverse of the regional mean amphibian species richness. Most ponds with predatory fish were isolated (no aquatic connections) and deliberately stocked with fish. Many landowners reported reduced amphibian populations and even extirpations following fish stocking. Thus a human-mediated biotic factor may be having consequences on a large spatial scale. The patterns of local distribution and regional occurrence in southwestern Ontario suggest that fish predation may alter local amphibian communities with cumulative regional effects.

Geographic factors such as the amount of woodlands, human population density, and distance to paved roads were positively correlated with amphibian species richness, whereas distance to woods, aquatic corridors, and source areas were negatively correlated. The best geographic models included amount of woodlands, proximity to woodlots, roads, and corridors, and pond age. These accounted for 28 to 37% of the variance in species richness. Beebee (1985) investigated amphibian diversity in 203 English ponds and ditches and found that pond level measurements, including water chemistry, were not as good predictors of diversity as variables related to the surrounding habitat.

Laan and Verboom (1990) investigated 77 ponds in an agricultural landscape in the Netherlands and found that distance to the woods and pond age were important factors affecting amphibian presence. Colonization of new ponds by rare species occurred only when distance from source ponds was < 1 km, suggesting the importance of dispersal. For new ponds (< 5 yr), age was the most important factor. For old ponds (> 7 yr), size, isolation, and depth were important. In 1993, I found a marginally significant area effect and an age effect in southwestern Ontario. I found that for ponds up to 10 yr old, richness increased significantly (unpublished data). Laan and Verboom (1990) studied amphibians in a highly fragmented agricultural landscape in the Netherlands and reported a mean species richness of 3.2, which is similar to the low mean regional species richness I observed (1.9 to 3.9).

In fragmented landscapes, woodlands are becoming recognized as important amphibian habitat (Strijbosch 1980; Loman 1988; Wederkinch 1988; Laan and Verboom 1990; this study). Studies also relate the importance of isolation to amphibian extinction (Gill 1978; Reh and Seitz 1990; Sjögren 1991; Bradford et al. 1993). The patterns of local and regional turnover (colonization and extinction) and the importance of geographic factors to amphibians suggest that studies from a metapopulation or metacommunity perspective would be highly enlightening and may uncover general patterns in amphibian ecology.

Clear patterns began to emerge in just 2 yr of study in southwestern Ontario. Local amphibian species richness was affected by the proximity and amount of woodlands and presence of fish predators, but not by acidity of the water. However, other factors potentially affecting amphibians such as agrochemicals have not been investigated thoroughly. It is also clear that amphibian populations and communities are highly dynamic and that strong regional differences exist.

The most important factor that has affected amphibian diversity in southwestern Ontario has occurred historically. The massive deforestation and wetland drainage of the 1800s destroyed much amphibian habitat. While the building of artificial ponds may have been beneficial (those not stocked with fish), the magnitude of habitat loss and change resulted in less diverse communities. Relictual islands, especially in the Essex Plain, remain at risk because of their isolation. For example, 3 amphibian species have become extinct at Point Pelee National Park since 1972. Some species have declined regionally and some on a geographic scale from 1992 to 1993. Current net turnover rates suggest that the regions with the highest amphibian diversity are having their diversity reduced by a predominance of extinctions. With the short-term mandate to report on the question of amphibian decline, it is difficult to reach conclusions with a high level of confidence, but given that wetland loss and forest fragmentation continue, amphibians will continue to decline in southwestern Ontario. Long-term data from local sites is required to determine if declines are presently occurring and to determine cause(s), but monitoring on large spatial scales is required to assess amphibian status. If a global decline is occurring, it is likely a cumulative effect of multiple regional causes. While long-term data are needed for confidence, delaying conservation initiatives may result in relaxation of diverse local communities to simple ones consisting of 1 or 2 amphibian species dependent on artificial ponds in an artificial landscape.

Acknowledgments

Thanks go to R.T. M'Closkey for supervising the project. D. Hecnar, T. Hecnar, D. Chalcraft, J. Cotter, D. Peterson, A. Plante, and R. Poulin assisted in the field and lab. J. Ciborowski, D. Haffner, and R. Russell provided technical advice. S. and V. Hecnar provided field accommodations and H. and B. Bober donated supplies. Many private landowners, Conservation Authorities, Ontario Provincial Parks, Parks Canada, and Agriculture Canada provided information and access to their properties. F. Schueler provided information on his work in the Bruce Peninsula. Two anonymous reviewers provided helpful comments. Funding was provided through a Natural Sciences and Engineering Research Council of Canada grant to R.T. M'Closkey, and a Parks Canada contract. The Canadian Association of Herpetologists, Friends of the Royal British Columbia Museum, and the University of Windsor have assisted with conference travel expenses.

Literature Cited

Anonymous. 1984. An evaluation system for wetlands of Ontario: south of the precambrian shield. 2nd ed. Ottawa: Ontario Ministry of Natural Resources, and Environment Canada.

Barinaga M. 1990. Where have all the froggies gone? Science 247:1033–1034.

Beebee TJC. 1985. Discriminant analysis of amphibian habitat determinants in south-east England. Amphibia-Reptilia 6:35–43.

Berven KA. 1990. Factors affecting population fluctuations in larval and adult stages of the wood frog (*Rana sylvatica*). Ecology 71:1599–1608.

Blaustein AR, Wake DB. 1990. Declining amphibian populations: a global phenomenon? Trends in Ecology and Evolution 5:203–204.

Blaustein AR, Wake DB, Sousa WP. 1994. Amphibian declines: judging stability, persistence, and susceptibility of populations to local and global extinctions. Conservation Biology 8:60–71.

Bradford DF, Tabatabai F, Graber DM. 1993. Isolation of remaining populations of the native frog, *Rana muscosa,* by introduced fishes in Sequoia and Kings Canyon National Parks, California. Conservation Biology 7:882–888.

Bradford DF, Gordon MS, Johnson DF, Andrews RD, Jennings WB. 1994. Acidic deposition as an unlikely cause for amphibian declines in the Sierra Nevada, California. Biological Conservation 69:155–161.

Brönmark C, Edenhamn P. 1994. Does the presence of fish affect the distribution of tree frogs (*Hyla arborea*)? Conservation Biology 8:841–845.

Callahan JT. 1984. Long-term ecological research. BioScience 34:363–367.

Conant R. 1975. A field guide to reptiles and amphibians of eastern and central North America. 2nd ed. Boston: Houghton Mifflin.

Connell JH, Sousa WP. 1983. On the evidence needed to judge ecological stability or persistence. American Naturalist 121:789–824.

Corn PS, Vertucci FA. 1992. Descriptive risk assessment of the effects of acidic deposition on Rocky Mountain amphibians. Journal of Herpetology 26:361–369.

Cornell HV, Lawton JH. 1992. Species interactions, local and regional processes, and the limits to the richness of ecological communities: a theoretical perspective. Journal of Animal Ecology 61:1–12.

Duellman WE, Trueb L. 1986. Biology of amphibians. New York: McGraw-Hill.

Dunson WA, Wyman RL, Corbett ES. 1992. A symposium on amphibian declines and habitat acidification. Journal of Herpetology 26:349–352.

Efford IE, Mathias JA. 1969. A comparison of two salamander populations in Marion Lake, British Columbia. Copeia 1969:723–736.

Freda J, Sadinski WJ, Dunson WA. 1991. Long term monitoring of amphibian populations with respect to the effects of acidic deposition. Water, Air, and Soil Pollution 55:445–462.

Gill DE. 1978. The metapopulation ecology of the red- spotted newt, *Notophthalmus viridescens* (Rafinesque). Ecological Monographs 48:145–166.

Hanski I, Gilpin M. 1991. Metapopulation dynamics: brief history and conceptual domain. Biological Journal of the Linnean Society 42:3–16.

Harte J, Hoffman E. 1989. Possible effects of acidic deposition on a Rocky Mountain population of the tiger salamander *Ambystoma tigrinum.* Conservation Biology 3:149–158.

Heyer WR, McDiarmid RW, Weigmann DL. 1975. Tadpoles, predation, and pond habitats in the tropics. Biotropica 7:100–111.

Ireland PH. 1989. Larval survivorship in two populations of *Ambystoma maculatum.* Journal of Herpetology 23:209–215.

Jaeger RG, Walls SC. 1988. On salamander guilds and ecological methodology. Herpetologica 44:111–119.

Kats LB, Petranka JW, Sih A. 1988. Antipredator defenses and the persistence of amphibian larvae with fishes. Ecology 69:1865–1870.

Laan R, Verboom B. 1990. Effects of pool size and isolation on amphibian communities. Biological Conservation 54:251–262.

Loman J. 1988. Breeding by *Rana temporaria:* the importance of pond size and isolation. Memoranda Society Fauna Flora Fennica 64:113–115.

Moss MR, Davis S. 1989. Evolving spatial interrelationships in the non-productive land-use components of rural Southern Ontario. University of Guelph Department of Geography, Occasional Papers in Geography 11:1–65.

National Wetlands Working Group. 1988. Wetlands of Canada. Ottawa: Sustainable Development Branch, Environment Canada; Montreal: Polyscience Publications. Ecological Land Classification Series Number 24.

Pechmann JHK, Scott DE, Semlitsch RD, Caldwell JP, Vitt LJ, Gibbons JW. 1991. Declining amphibian populations: the problem of separating human impacts from natural fluctuations. Science 253:892–895.

Pechmann JHK, Wilbur HM. 1994. Putting declining amphibian populations in perspective: natural fluctuations and human impacts. Herpetologica 50:65–84.

Petranka JW. 1983. Fish predation: a factor affecting the spatial distribution of a stream-breeding salamander. Copeia 1983:624–628.

Petranka JW, Katz LB, Sih A. 1987. Predator-prey interactions among fish and larval amphibians: use of chemical cues to detect predatory fish. Animal Behavior 35:420–425.

Phillips K. 1990. Where have all the frogs and toads gone? BioScience 40:422–424.

Reh W, Seitz A. 1990. The influence of land use on the genetic structure of populations of the common frog *Rana temporaria.* Biological Conservation 54:239–249.

Ricklefs RE. 1987. Community diversity: relative roles of local and regional processes. Science 235:167–171.

Ricklefs RE, Schluter D (editors). 1993. Species diversity in ecological communities: historical and geographical perspectives. Chicago: University of Chicago Press.

Rowe JS. 1972. Forest regions of Canada. Ottawa: Environment Canada. Canadian Forestry Service Publication Number 1300.

Sexton OJ, Phillips C. 1986. A qualitative study of fish-amphibian interactions in 3 Missouri ponds. Transactions of the Missouri Academy of Science 20:25–35.

Sjögren P. 1991. Extinction and isolation gradients in metapopulations: the case of the pool frog (*Rana lessonae).* Biological Journal of the Linnean Society 42:135–147.

Snell EA. 1987. Wetland distribution and conversion in southern Ontario. Ottawa: Environment Canada. Inland Waters and Lands Directorate, Working Paper Number 48.

Sokal RR, Rohlf FJ. 1981. Biometry. 2nd ed. New York: W.H. Freeman.

Statistics Canada. 1993. 1991 Census of Canada. Place name lists - Quebec and Ontario. Ottawa: Industry, Science, and Technology. Catalog Number 93-308.

Strijbosch H. 1980. Habitat selection by amphibians during their terrestrial phase. British Journal of Herpetology 6:93–98.

Vitt LJ, Caldwell JP, Wilbur HM, Smith DC. 1990. Amphibians as harbingers of decay. BioScience 40:418.

Wederkinch E. 1988. Population size, migration barriers, and other features of *Rana dalmatina* populations near Køge, Zealand, Denmark. Memoranda Society Fauna Flora Fennica 64:101–103.

Wiens JA, Addicott JF, Case TJ, Diamond J. 1986. Overview: the importance of spatial and temporal scale in ecological investigations. In: Diamond J, Case TJ, editors. Community ecology. New York: Harper and Row. p 145–153.

Wilbur HM. 1980. Complex life cycles. Annual Review of Ecology and Systematics 11:67–93.

Wyman RL. 1990. What's happening to the amphibians? Conservation Biology 4:350–352.

©1997 by the Society for the Study of Amphibians and Reptiles
Amphibians in decline: Canadian studies of a global problem. David M. Green, editor.
Herpetological Conservation 1:16–26.

Chapter 2

FLUCTUATIONS IN ABUNDANCE AND AGE STRUCTURE IN THREE SPECIES OF FROGS (ANURA: RANIDAE) IN ALGONQUIN PARK, CANADA, FROM 1985 TO 1993

LEONARD J. SHIROSE AND RONALD J. BROOKS

Department of Zoology, University of Guelph, Guelph, Ontario N1G 2W1, Canada

ABSTRACT.—Populations of bullfrogs (*Rana catesbeiana*), green frogs (*Rana clamitans*), and mink frogs (*Rana septentrionalis*) in a large "wilderness" zone in Algonquin Provincial Park, Ontario, Canada were monitored from May, 1985 through October, 1987, and again from May, 1991 through September, 1993. We tested whether populations of these species were declining in an area in which human disturbances and chemical contaminants were minimal and there was no commercial harvest or other significant disturbance by humans. Estimates of population size revealed short-term fluctuations in abundance in all three species, but no evidence of a long-term decline. The ratio of maximum population size/minimum population size was calculated for each species and ranged from 1.0 to 1.8 for individuals of age ≥ 1 yr post-transformation. In *R. catesbeiana*, fluctuations in the proportion of transforming individuals relative to older individuals and in the mean and maximum size of mature females suggest an unstable age structure caused by periods of increased mortality prior to 1985 and between 1987 and 1991. There is some evidence that these periods of increased mortality also affected *R. clamitans*, but not *R. septentrionalis*. We found periodic reassessment to be useful for monitoring trends and investigating variation in rate of growth, longevity, survivorship, sex ratio, size, and age at maturity. However, continuous long-term studies are required to determine whether variation in these parameters is cyclic and predictable enough to be useful in designing conservation strategies, and to test hypotheses about long-term processes in ecology and evolution.

RÉSUMÉ.—Des populations de ouaouarons (*Rana catesbeiana*), de grenouilles vertes (*Rana clamitans*) et de grenouilles du Nord (*Rana septentrionalis*) dans une grande zone «sauvage» du Parc provincial Algonquin en Ontario (Canada) ont été observées entre mai 1985 et octobre 1987 puis entre mai 1991 et septembre 1993. Nous avons cherché à déterminer si ces populations étaient en voie de disparition dans un secteur où les perturbations humaines et les contaminations chimiques étaient minimes, où il n'y avait aucune récolte commerciale ou autre perturbation humaine significative. Les évaluations de la taille de la population ont révélé des fluctuations à court terme au titre de l'abondance chez les trois espèces mais aucun déclin à long terme. Le rapport taille maximale de la population/taille minimale de la population a été calculé pour chaque espèce et se situe entre 1,0-1,8 pour les spécimens ≥ à un an après métamorphose. Chez *R. catesbeiana*, les fluctuations dans la proportion d'individus en cours de métamorphose par rapport aux individus plus vieux et la taille moyenne et maximale des femelles adultes révèlent une structure d'âge instable causée par des périodes de mortalité aiguë avant 1985 et entre 1987 et 1991. Il semble que ces périodes de mortalité aiguë aient également affecté *R. clamitans*, mais pas *R. septentrionalis*. Nous avons jugé la réévaluation périodique utile pour surveiller les tendances et étudier les variations du rythme de croissance, de la longévité, du taux de survie, du proportion mâle/femelle de la taille et de l'âge à la maturité. Toutefois, des études continue à long terme s'imposent pour déterminer si les variations de ces paramètres sont suffisamment cycliques et prévisibles pour pouvoir être intégrées à des stratégies de conservation et pour vérifier les hypothèses sur les procédés à long terme en écologie et en évolution.

Recently there has been concern that amphibian populations throughout the world have been declining (Barinaga 1990; Wake and Morowitz 1990; Pechmann et al. 1991; Wake 1991). Many of these declines have been dramatic and some may be accounted for by habitat loss, collecting (Barinaga 1990; Wyman 1990), or by normal climatic or population fluctuations. However, other declines seem to have occurred where there has been no obvious disturbance or climatic change (Wake and Morowitz 1990).

The decline of amphibian populations in many parts of the world is cause for concern for several reasons. In some ecosystems amphibians constitute a substantial fraction of the animal biomass (McCarrahar 1977; Rose and Armentrout 1974; Burton and Likens 1975). Because of the complex life cycles of many amphibians, a single species can act as both a primary consumer and as a secondary or tertiary consumer depending on the stage of the life cycle. The decline of amphibian populations may indicate a subtle deterioration of the ecosystem that supports them, especially in terms of water quality and the abundance and diversity of plants and invertebrates, and may precipitate a crisis in those species that depend on them. As well, many amphibian species are of economic importance to humans both directly as food, bait, and test subjects for researchers and educators, and indirectly as a factor controlling the abundance of invertebrate pests.

There appears to be no clear pattern in the decline of amphibian populations apart from the obvious effects of habitat destruction and modification (Brooks 1992a). The decline does not appear to affect all taxa in all locations and shows some signs of species-specificity with stable populations of some species co-occurring with declining populations of other species (Wake and Morowitz 1990). The task of understanding the decline in amphibian populations is not currently possible because of the dearth of long-term studies of the diversity and abundance of amphibians under unaltered conditions. This is particularly true in Ontario, for which there exist no long-term published data on amphibian numbers (Brooks 1992a). It is necessary to document the dynamics and range of variation in healthy populations before we can adequately assess the danger to populations under pressure from human activities or understand the pathology of populations in decline.

In this study, we present data on the abundance of 3 species of frogs, bullfrog (*Rana catesbeiana*), green frog (*R. clamitans*), and mink frog (*R. septentrionalis*), in central Ontario between 1985 and 1993. Monitoring sites should be chosen on the basis of location of key species, logistics, and, most important, strong evidence for long-term protection of the site as a relatively undisturbed research venue (Brooks 1992b). The Wildlife Research Area of Algonquin Provincial Park meets all the major criteria as a potential site for intensive long-term monitoring. Adverse anthropogenic effects on the study populations are minimal because access to the study site is restricted to authorized personnel, hunting and fishing are prohibited, and contaminant levels are low (Bishop et al. 1991). We selected *R. catesbeiana* as the focus for the study because of its high survivorship, recapture rate, and abundance relative to the other species and because populations of this species are reported to be declining in many areas (Baker 1942; Culley and Gravois 1971; Gibbs et al. 1971; Treanor and Nichola 1972; Treanor 1975; Vogt 1981; Bishop 1990; Berrill et al. 1992). The populations *R. clamitans* and *R. septentrionalis* were also monitored to investigate whether the close phylogenetic and ecological relationships among the 3 species and their syntopic distributions in the study area produce similarities in the pattern of abundance and age structure.

We tested the hypothesis of long-term decline in an undisturbed area by estimating abundance and changes in abundance of the 3 study species. We also examined adult body size and abundance of transforming individuals relative to older individuals to look for changes in age distribution or individual fitness.

Mink frog, Rana septentrionalis. *Photo by Claude Daigle*

Materials and Methods

Study Area

This study was conducted in the Wildlife Research Area (45°35'N, 78°30'W) of Algonquin Provincial Park, Ontario, Canada, from May, 1985 through October, 1987 and from May, 1991 through September, 1993. The study area included Lake Sasajewun (63 ha), an artificial impoundment dating from the early 1900s (Obbard, 1977, 1983), and the adjacent bogs and rivers (Fig. 1).

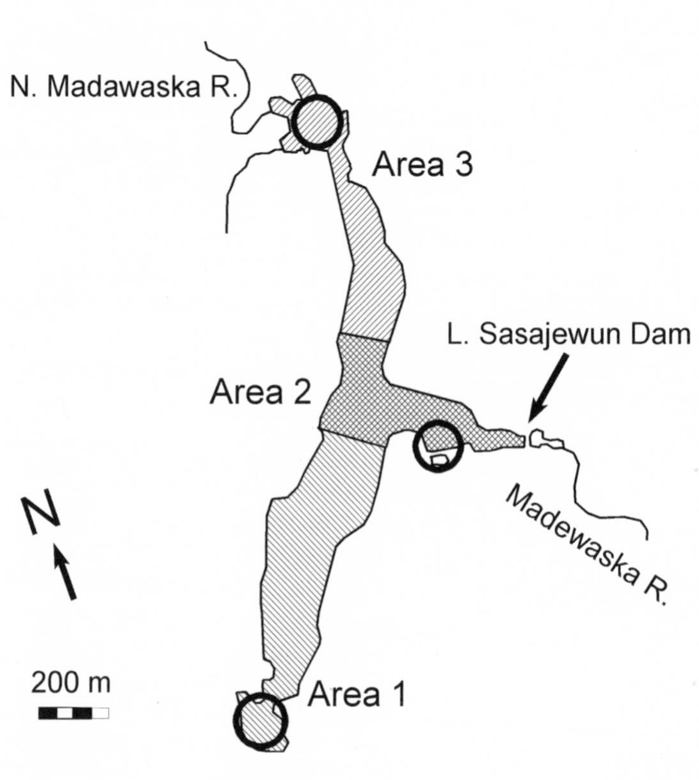

Figure 1. *Map of Lake Sasajewun in Algonquin Provincial Park, Ontario, indicating the locations of the 3 study areas. Circles indicate areas of population concentrations.*

Thick layers of soft mud and organic debris cover the lake bottom, and by mid-June the surface of much of the lake is covered by yellow water lily (*Nuphar variegatum*), white water lily (*Nymphaea odorata*), and water shield (*Brasenia schreberi*). In several bays the surface of the lake is covered by a vegetation mat consisting largely of sphagnum moss.

Lake Sasajewun was divided into 3 areas for sampling because logistic considerations precluded sampling the entire lake on any single day. Lake Sasajewun is T-shaped with an area of permanent vegetation cover at each terminus. The greatest concentrations of frogs were found in these terminal areas. The remainder of the lake was used primarily by chorusing males and gravid females. The extensive surface cover of floating annual vegetation from mid-June to the end of the summer allowed immature individuals and non-gravid females to use most of the lake. The terminal areas consistently maintained the highest concentrations of individuals throughout the summer. Each sampling area contained a single area of non-breeding habitat (an area of dense floating and emergent vegetation and high density of frogs) as well as a portion of the lake where breeding sites were located (Fig. 1).

Sampling

We sampled Lake Sasajewun populations weekly. No effort was made to standardize sampling effort because we found that catchability varied with climatic conditions. When conditions were favorable for collecting, we collected animals until 15 min of searching revealed no new uncaptured individuals. A flexible capture regime provides the large samples and high sampling efficiency required for mark-recapture estimates while minimizing searching for animals that are submerged and therefore unavailable for capture. During these periods most frogs were captured using dip nets during diurnal capture sessions. In 1987, 1991, 1992, and 1993, diurnal capture sessions were supplemented by nocturnal capture sessions in which bullfrogs (generally chorusing males, but also some females) were hand captured by night-lighting from a canoe. Nocturnal captures were made irregularly when chorusing was intense.

Frogs were sorted according to size to prevent cannibalism and predation and were transported to the laboratory for processing. Standard length (snout to the tip of the urostyle) was measured to the nearest 1 mm with calipers. An ink tattoo was applied to the ventral surface of each individual. Tattoos were applied using a 110-V AC Spaulding Special "A" veterinary tattoo gun (Spaulding & Rogers Mfg., Inc., Voorheesville, New York, U.S.A. 12186) fitted with a gang of 6 needles. All frogs marked with a tattoo were also marked by the excision of the distal 2 phalanges of the 4th digit of the right forefoot to allow evaluation of the permanency of tattoos.

Population Size

We estimated population size separately for each species in each area using Jolly's stochastic method (Krebs 1989). We selected this method because populations were geographically open—animals moved among the sampling areas. Differences in population size were considered significant ($P = 0.05$) when there was no overlap in 95% confidence intervals. Confidence limits of 95% were calculated by solving for N_L in the equation:

$$95\%\ \text{C.I.} = (-1.96 \leq (\ln N - \ln N_L)/(\text{var}(N)/N^2)^{-2} \leq 1.96),$$

in which var(N) = variance, N_L = confidence limit, and N = estimated population size (Skalski and Robson 1992).

We omitted transforming individuals from the analysis because of the negative effect of their low recapture rate on the precision of the population size estimates. We decided that it was better to have a reliable estimate of a significant portion of the population than a less reliable estimate of the entire population.

When calculating the ratio of maximum population size to minimum population size, N_{max}/N_{min}, we used means of all population estimates that were not significantly different. We used means rather than the single highest and lowest population estimates in order to provide a conservative estimate of natural variation in the size of populations that are not under stress from harvesting or habitat loss. For individuals < 1 yr post-transformation (YPT), the ratio N_{max}/N_{min} was calculated from the standardized (percent of total) number of individuals marked.

Age Distribution

Because of the lack of data on age-specific abundance, we based our discussion of fluctuations in age structure on variation in the number of transforming individuals captured in a given year, relative to the number of older individuals captured, and on yearly fluctuations in the mean and maximum size of females captured. Transforming individuals (< 1 YPT) were identified by the presence of the tail rudiment.

To test the *ad hoc* hypothesis that there were periods of increased mortality before 1985 and between 1987 and 1991, we used 1-tailed *t*-tests to test whether mean size of mature females decreased between 1987 and 1991, and increased between 1985 and 1986, 1986 and 1987, 1991 and 1992, and 1992 and 1993.

Results

We marked 1244 *R. catesbeiana*, 628 *R. clamitans*, and 959 *R. septentrionalis* between May, 1985 and October, 1987, of which 567 *R. catesbeiana*, 65 *R. clamitans*, and 72 *R. septentrionalis* were recaptured at least once. Between May 1991 and October 1993, we marked 996 *R. catesbeiana*, 486 *R. clamitans*, and 644 *R. septentrionalis*, of which 367 *R. catesbeiana*, 94 *R. clamitans*, and 100 *R. septentrionalis* were recaptured at least once. Only 1 frog (a *R. catesbeiana*) marked in the 1985 to 1987 period was recaptured in the 1991 to 1993 period.

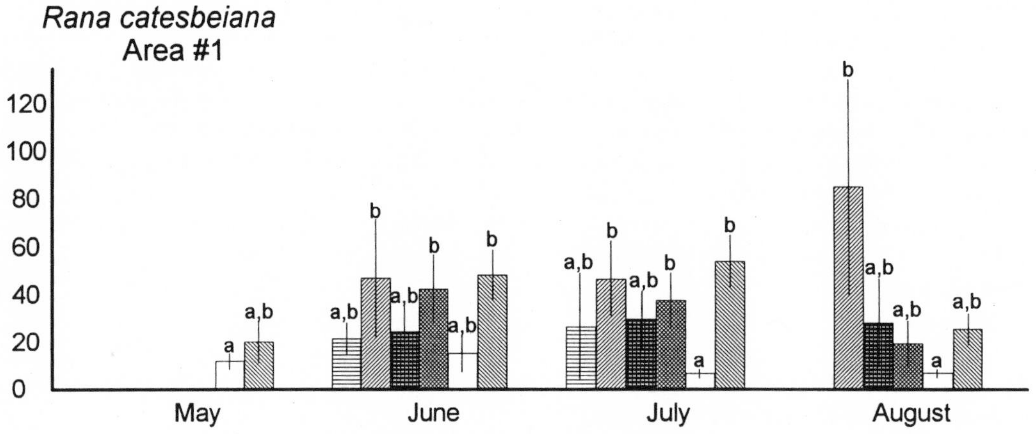

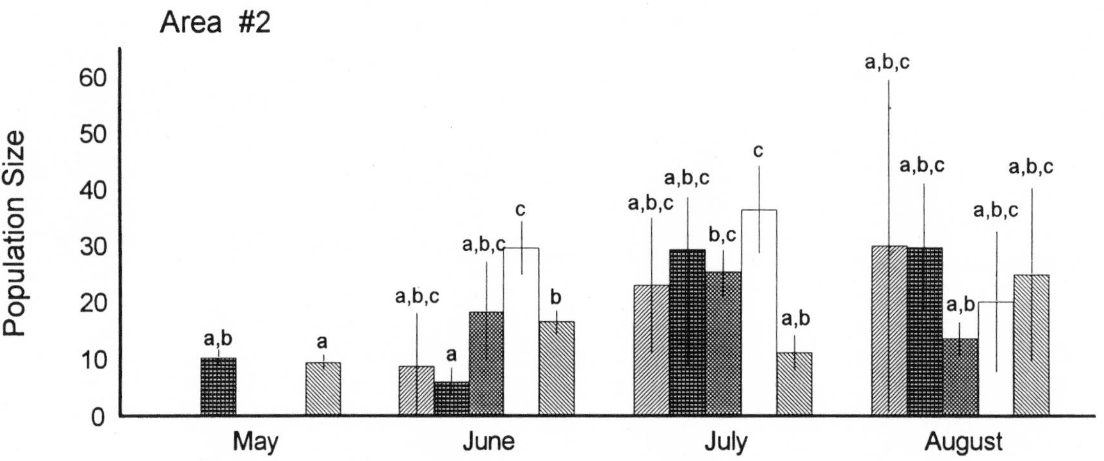

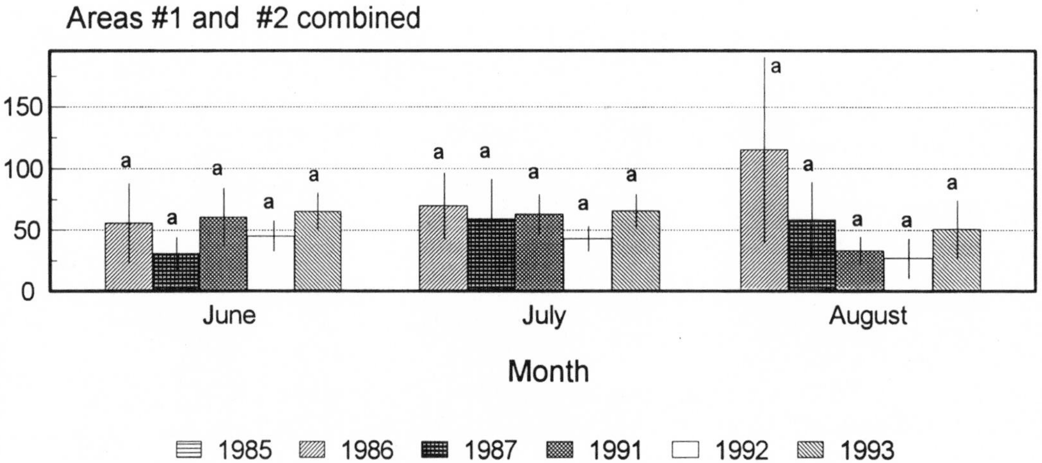

Figure 2. *Estimated size of* Rana catesbeiana *populations at Lake Sasajewun. Bars indicate SE. Within each Area, estimates sharing the same letter are not different at* P = 0.05.

Population Size

We were unable to estimate the size of the populations in Area 3. Sampling efficiency and recapture rates were low, because animals could escape into deep water when we were wading after them and into very shallow water or dense vegetation cover when we were attempting to capture them from the canoe. Low recapture rates also precluded estimates of population size of *R. clamitans* in Area 1 in 1991 and 1992, and in Area 2 in 1985, 1987, and 1991, and of *R. septentrionalis* in Area 1 in 1991. *R. septentrionalis* were rarely found in Area 2. Estimates of population size were consistent among months within each summer in all years sampled for *R. clamitans* and *R. septentrionalis*. Therefore, we present a single population size estimate for each year for these species.

Month-to-month comparisons indicate that the estimated size of the *R. catesbeiana* population was constant among months within the summers of all years in both study areas, except in Area 2 in 1993. There, the population was significantly smaller in May than in June but was not significantly smaller than in July or August (Fig. 2). Among years, there was no apparent trend in population size in either area. There were differences in population size among some years in each area but years that produced a high population in one area tended to produce a low population in the other. When estimates for the 2 areas were combined, there were no significant differences in population size either within each year or among different years (Fig. 2).

There was no apparent trend in the size of the *R. clamitans* population of either area. The size of the population in Area 1 was constant among the 4 yr for which estimation was possible (Fig. 3). There was a significant increase in the size of the population from 1992 to 1993 in Area 2. However, the size of the population in 1993 was not significantly greater than it had been in 1986 (Fig. 3). The size of the *R. septentrionalis* population in Area 1 was significantly greater in 1993 than in 1985 (Fig. 4).

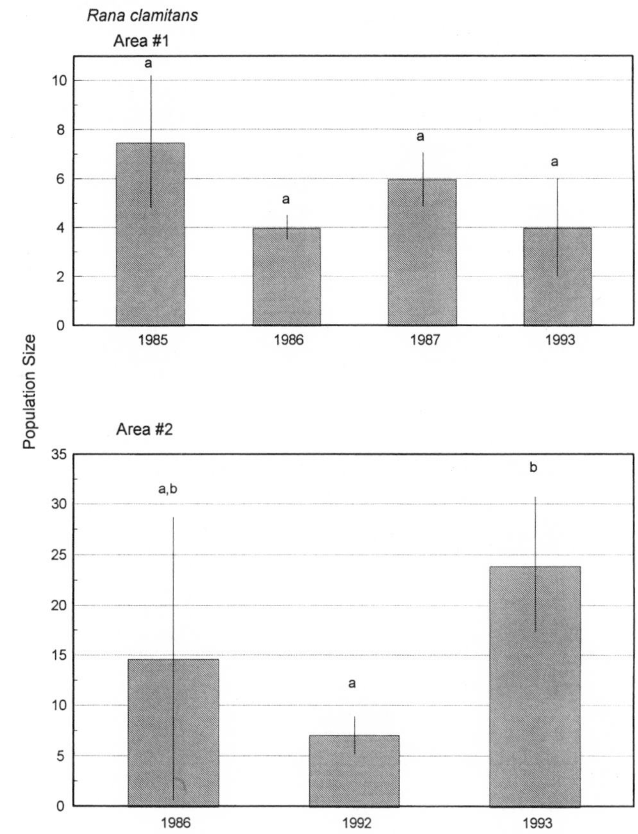

Figure 3. *Estimated size of* Rana clamitans *populations at Lake Sasajewun. Symbols are as in Fig. 2.*

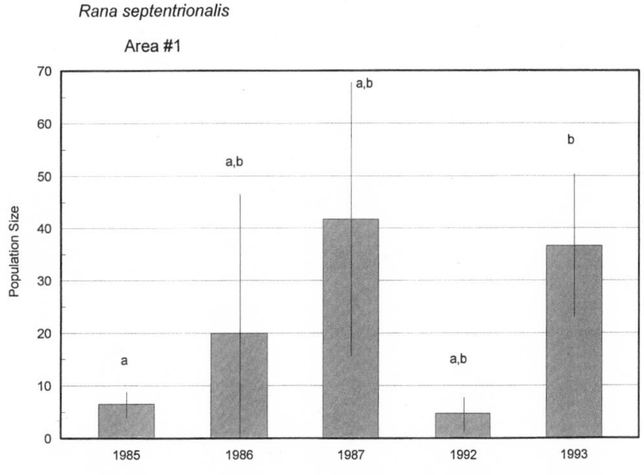

Figure 4. *Estimated size of* Rana septentrionalis *population at Lake Sasajewun, Area 1. Symbols are as in Fig. 2.*

The ratio of N_{max}/N_{min} for *R. catesbeiana* of age ≥ 1 YPT was 1.8 in Area 1 and 1.4 in Area 2. For *R. clamitans* the ratio was 1.0 in Area 1 and 1.8 in Area 2 and for *R. septentrionalis* the ratio was 1.5 in Area 1. The ratio N_{max}/N_{min} for transforming individuals was 1.6 (*R. catesbeiana*), 2.6 (*R. clamitans*), and 2.8 (*R. septentrionalis*).

Age Distribution

The percent of transforming individuals varied among years from 27.5 to 44.0% ($\bar{x}$ = 38.3, SD = 5.9, n = 6) in *R. catesbeiana*; 32.5 to 85.3% ($\bar{x}$ = 57.4, SD = 17.4, n = 6) in *R. clamitans*; 21.6 to 60.2% ($\bar{x}$ = 38.4, SD = 17.2, n = 6) in *R. septentrionalis* (Table 1).

Table 1. *Total number of individual frogs marked in all 3 areas combined. Frogs < 1 YPT (years post transformation) are transforming individuals. Percent equals the number of individuals < 1 YPT divided total number captured times 100.*

	1985	1986	1987	1991	1992	1993
Rana catesbeiana						
≥ 1 YPT	170	266	306	249	203	172
< 1 YPT	125	182	195	196	77	99
%	42.4	40.6	38.9	44.0	27.5	36.5
Rana clamitans						
≥ 1 YPT	102	89	94	28	51	106
< 1 YPT	106	137	100	163	87	51
%	51.0	60.6	51.5	85.3	63.0	32.5
Rana septentrionalis						
≥ 1 YPT	188	130	259	100	125	167
< 1 YPT	67	194	121	151	55	46
%	26.3	59.9	31.8	60.2	30.6	21.6

The maximum size of mature female *R. catesbeiana* increased by approximately 8 mm between adjacent years (Table 2). Eight mm corresponds to approximately 1 yr's growth (Shirose et al. 1993). The mean size of mature females increased between 1985 and 1986 (t = 3.48, 139 df, P < 0.001), decreased between 1987 and 1991 (t = –10.10, 138 df, P < 0.001), and increased between 1992 and 1993 (t = 2.18, 230 df, P = 0.02) (Table 2). There was no difference in the mean size of mature females between 1986 and 1987 or 1991 and 1992.

Table 2. *Standard lengths (mm) of mature female frogs.*

	1985	1986	1987	1991	1992	1993
Rana catesbeiana						
$\bar{x}$	108	116	113	93	94	98
SD	13.7	15.6	15.2	8.9	8.5	15.3
n	74	67	85	60	110	143
Max	142	149	157	118	126	135
Rana clamitans						
$\bar{x}$	69	73	80	63	72	78
SD	7.9	8.2	7.4	7.2	8.0	7.2
n	32	28	35	5	33	49
Max	91	92	98	77	88	89
Rana septentrionalis						
$\bar{x}$	59	58	60	58	58	59
SD	6.8	5.7	5.3	4.6	4.9	5.1
n	76	93	105	24	34	76
Max	76	73	76	67	66	71

The maximum size of mature female *R. clamitans* displayed less variation than in *R. catesbeiana* but followed the same pattern of increase (Table 2). The mean size of mature females increased between 1985 and 1986 (t = 1.73, 58 df, P = 0.046) and 1986 and 1987 (t = 3.66, 61 df, P < 0.001). There was no difference in the mean size of mature females between 1992 and 1993. Sample sizes were too small to allow comparison between 1987 and 1991 or 1991 and 1992.

The maximum size of mature female *R. septentrionalis* displayed little variation and no clear pattern (Table 2). The mean size of mature females increased between 1986 and 1987 (t = 2.17, 196 df, P = 0.02), and decreased between 1987 and 1991 (t = –3.13, 141 df, P < 0.001) (Table 2). There was no difference in the mean size of mature females between 1985 and 1986, 1991 and 1992, or 1992 and 1993. The maximum size of mature females of all 3 species in each year from 1991 to 1993 was less than the maximum size in any year from 1985 to 1987.

Discussion

Population Size

We found no evidence of a long-term decline in the population size of any of the species monitored. However, we found marked short-term fluctuations in all 3 species. The size of the *R. septentrionalis* population in Area 1 was significantly greater in 1993 than in 1985. However, this apparent increase in population size may be the result of a recovery following a population decline in 1988 to 1992 rather than the result of a gradual increase.

Because we did not capture every frog in the study area, we expected the estimates of population size to exceed the number of individuals marked in any year. However, population estimates for all species were lower than expected on the basis of the total number of individuals marked in each year. The assumption of equal catchability may have been violated and this would have caused the population size to be underestimated, with the magnitude of the bias being proportional to the magnitude of the variation in catchability (Begon 1979). Differences in catchability among *R. catesbeiana* may be attributed to intersexual differences in behavior and habitat preference (Shirose 1990) as well as to size-related (and presumably age-related) differences in rates of dispersal. Large individuals are more likely to occupy established home ranges (Durham and Bennett 1963) and are therefore more likely to be recaptured, whereas small individuals may be more likely to disperse from the point of their original capture (Willis et al. 1956; Shirose 1990) and are less likely to be recaptured. Size-related differences in catchability among green frogs have been attributed to differences in circadian patterns of behavior (Martof 1956). Because sampling efficiency was low in Area 3 and estimates of population size are not available, the size of the entire population of the lake is unknown. We know that there was movement of individuals among areas and that Area 3 was a source of frogs emigrating to Areas 1 and 2, and vice versa.

The pattern of population fluctuations in *R. catesbeiana* was not consistent between the 2 study sites. For example, in Area 1 the size of the population in July was greater in 1993 than in 1992, but in Area 2 the size of the population in July was greater in 1992 than in 1993. If the population size estimates for the 2 areas are combined, there is no difference between 1992 and 1993. It is possible that much of the fluctuation in population size in these years was the result of fluctuation in habitat preference (possibly related to climatic differences) rather than fluctuations in mortality and recruitment.

The ratio N_{max}/N_{min} indicates that population size may fluctuate by as much as 50 to 80%, even in the absence of long-term trends in population size. Population fluctuations are likely to be even more spectacular than this ratio suggests because fluctuations in the size of the transforming age-class, comprising approximately 38% of the *R. catesbeiana* and *R. septentrionalis* captured and 57% of the *R. clamitans* captured, appear to be at least as great as fluctuations in the older age classes.

Age Distribution

In the absence of perturbations we may expect a population to approach a state in which the age structure and rate of population growth are constant, with these parameters being determined by age-specific schedules of mortality and fecundity (Charlesworth 1980). Environmental factors that increase the mortality rate often affect all ages equally and may decrease the size of the population without greatly affecting the age structure. However, subsequent population growth will result in a relative increase in the younger age classes (Ricklefs 1983) which will diminish as the population returns to a stable age structure. An increase in the proportion of older individuals may therefore indicate recovery from a period of high mortality. However, it is possible for a population in decline to exhibit a relative increase in the proportion of older individuals during a period of poor recruitment due to high mortality in embryos or larvae.

The regular increase in maximum size of mature female *R. catesbeiana* by approximately one yr's growth per yr from 1985 to 1987, and from 1991 to 1993 suggests a population recovering from a period of increased mortality prior to 1985 and another between 1987 and 1991. The decreasing trend in the proportion of transforming individuals relative to older individuals, without a concomitant collapse in recruitment, and the pattern in the mean size of mature females over the same periods are generally consistent with the hypothesis of periods of increased mortality prior to 1985 and between 1987 and 1991. The dearth of recaptures in 1991 to 1993 of *R. catesbeiana* marked in the 1980s is also consistent with a period of increased mortality between these periods.

The evidence is less clear in the case of *R. clamitans*. The maximum size of mature females was close to the maximum size for the species in all years except 1991 in which the small sample size raises serious questions about the validity of the estimate. The oldest frogs may have been older in successive years, but the evidence is not conclusive. The trend in the proportion of transforming individuals relative to older individuals is consistent with a period of increased mortality between 1987 and 1991. The trend in the mean size of mature females is consistent with a period of increased mortality prior to 1985.

The trend in mean size of mature female *R. septentrionalis* is consistent with a period of increased mortality prior to 1985, and between 1987 and 1991. The trend in the proportion of transforming individuals relative to older individuals is consistent with a period of increased mortality between 1987 and 1991.

If the strength of the evidence of periods of increased mortality is proportional to the actual increase in mortality, then it appears that *R. catesbeiana* has been most affected and *R. septentrionalis* least affected. This purported gradient follows the gradients in body size and clutch size as well as the inverse of the latitudinal gradient in the northern limit of range for these species (Cook 1984). The precise causes and relative effects of the increased mortality on individuals of different species, age and sex are beyond the scope of this study. However, it is perhaps worth noting that in the winters of 1987 and 1988, the population of snapping turtles (*Chelydra serpentina*) in Lake Sasajewun endured high mortality from predation by otters (*Lutra canadensis*; Brooks et al. 1991). The turtles were killed while hibernating on the lake bottom in shallow waters, a habitat also used by hibernating frogs. Because anurans are an important item in the diet of some otter populations in both summer and winter (Lagler and Ostenson 1942; Hamilton 1961; Sheldon and Toll 1964), it is possible otters also killed a significant number of overwintering frogs during these years which might then account for the demographic and size changes we observed.

Periodic reassessment is adequate for monitoring long-term trends in population size but is not sufficient for revealing patterns of cyclic population change or for differentiating between cyclic and episodic increases in mortality. The intricacies of population dynamics in long-lived anurans can be illuminated only by monitoring populations for periods of at least the generation time of the species being studied (about 7 or 8 yr for *R. catesbeiana*). Periodic reassessment will yield useful information on temporal variation in rate of growth, longevity, survivorship, sex ratio, size, and age at maturity, but continuous long-term studies are required to determine whether variation in these parameters is cyclic and predictable enough to be useful in designing conservation strategies. Long-term studies are also essential to understand the effect of the stochastic nature of extrinsic influences such as predation, food abundance, and climate (Gibbons 1990; Pechmann et al. 1991). This is particularly true for species like *R. catesbeiana* whose long generation time is likely to produce a time lag between stimulus and reaction.

ACKNOWLEDGMENTS

We thank S.S. Desser and J.R. Barta for their help in starting this project. We also thank T. Gage, J. Siegrist, L. Standing, and L. Wilkinson for their assistance in field work over various years of this study. Our thanks to D. Strickland and the Ontario Ministry of Natural Resources for permission to conduct research in Algonquin Provincial Park. We are grateful to the User Committee of the Wildlife Research Station for permission to use the facility. We especially acknowledge the support of J. Baker and M. McLaren in helping to fund this research. This research was supported by grants from the

Ontario Ministry of Natural Resources and the Natural Sciences and Engineering Research Council of Canada (Number 5990 to RJB).

Literature Cited

Baker RH. 1942. The bullfrog. A Texas wildlife resource. Texas Game, Fish, and Oyster Commission Bulletin 23:1–7.

Barinaga M. 1990. Where have all the froggies gone? Science 247:1033–1034.

Begon M. 1979. Investigating animal abundance. Baltimore: University Park Press.

Berrill M, Bertram S, Toswill P, Campbell V. 1992. Is there a bullfrog decline in Ontario? In: Bishop CA, Pettit KE (editors). Declines in Canadian amphibian populations: designing a national monitoring strategy. Ottawa: Environment Canada. Canadian Wildlife Service Occasional Paper 76. p 32–36.

Bishop CA. 1990. Declining amphibian populations—a Canadian perspective. Ottawa: Canadian Wildlife Service.

Bishop CA, Brooks RJ, Carey JH, Ng P, Norstrom RJ, Lean DRS. 1991. The case for a cause-effect linkage between environmental contamination and development in eggs of the common snapping turtle (*Chelydra s. serpentina*) from Ontario, Canada. Journal of Toxicology and Environmental Health 33:521–547.

Brooks RJ. 1992a. Monitoring wildlife populations in long-term studies. In: Bishop CA, Pettit KE (editors). Declines in Canadian amphibian populations: designing a national monitoring strategy. Ottawa: Environment Canada. Canadian Wildlife Service Occasional Paper 76. p 94–97.

Brooks RJ. 1992b. Intensive monitoring: biology of amphibians in Canada. In: Bishop CA, Pettit KE (editors). Declines in Canadian amphibian populations: designing a national monitoring strategy. Ottawa: Environment Canada. Canadian Wildlife Service Occasional Paper 76. p 106–108.

Brooks RJ, Brown GP, Galbraith DA. 1991. Effects of a sudden increase in natural mortality of adults on a population of the common snapping turtle (*Chelydra serpentina*). Canadian Journal of Zoology 69:1314–1320.

Burton TM, Likens GE. 1975. Salamander populations and biomass in the Hubbard Brook Experimental Forest, New Hampshire. Copeia 1975:541–546.

Charlesworth, B. 1980. Evolution in age-structured populations. Cambridge, UK: Cambridge University Press.

Cook FR. 1984. Introduction to Canadian amphibians and reptiles. Ottawa: National Museums of Canada.

Culley DDJr, Gravois CT. 1971. Recent developments in frog culture. Proceedings Conference of the Southeast Association of Game and Fish Commissions 25:583–597.

Durham L, Bennett GW. 1963. Age, growth, and homing in the bullfrog. Journal of Wildlife Management 27:107–123.

Gibbons JW. (editor). 1990. The ecology and life history of the slider turtle. Washington, DC: Smithsonian Institution Press.

Gibbs EL, Nace GW, Emmons MB. 1971. The live frog is almost dead. BioScience 21:1027–1034.

Hamilton WJJr, 1961. Late fall, winter, and early spring foods of 141 otters from New York. New York Fish and Game Journal 8:106–109.

Krebs CJ. 1989. Ecological methodology. New York: Harper and Row.

Lagler KF, Ostenson BT. 1942. Early spring food of the otter in Michigan. Journal of Wildlife Management 6:244–254.

McCarrahar DB. 1977. Nebraska's Sandhills Lakes. Lincoln, NE: Nebraska Game and Parks Commission.

Martof BS. 1956. Factors influencing the size and composition of populations of *Rana clamitans*. American Midland Naturalist 56:224–245.

Obbard ME. 1977. A radio-telemetry and tagging study of activity in the common snapping turtle, *Chelydra serpentina* [thesis]. Guelph, ON: University of Guelph.

Obbard ME. 1983. Population ecology of the common snapping turtle, *Chelydra serpentina*, in north-central Ontario [dissertation]. Guelph, ON: University of Guelph.

Pechman JHK, Scott DE, Semlitsch RD, Caldwell JP, Vitt LJ, Gibbons JW. 1991. Declining amphibian populations: the problem of separating human impacts from natural fluctuations. Science 253:825–940.

Ricklefs RE. 1983. The economy of nature. New York: Chiron Press.

Rose FL, Armentrout D. 1974. Population estimates of *Ambystoma tigrinum* inhabiting playa lakes. Journal of Animal Ecology 43:674–679.

Sheldon WG, Toll WG. 1964. Feeding habits of the river otter in a reservoir in central Massachusetts. Journal of Mammalogy 45:449–455.

Shirose LJ. 1990. Population ecology of the postmetamorphic bullfrog (*Rana catesbeiana* Shaw) in Algonquin Provincial Park, Ontario [thesis]. Guelph, ON: University of Guelph.

Shirose LJ, Brooks RJ, Barta JR, Desser SS. 1993. Intersexual differences in growth, mortality and size at maturity in bullfrogs in Central Ontario. Canadian Journal of Zoology 71:2363–2369.

Skalski JR, Robson DS. 1992. Techniques for wildlife investigations: design and analysis of capture data. San Diego, CA: Academic Press, Inc.

Treanor RR. 1975. Management of the bullfrog (*Rana catesbeiana*) resource in California. California Department of Fish and Game, Inland Fisheries Administration Report 75-1:1–30.

Treanor RR, Nichola SJ. 1972. A preliminary study of the commercial and sporting utilization of the bullfrog, *R. catesbeiana* Shaw in California.California Department of Fish and Game, Inland Fisheries Administration Report 72-4:1–23.

Vogt RC. 1981. Natural history of amphibians and reptiles of Wisconsin. Milwaukee, WI: The Milwaukee Public Museum.

Wake DB. 1991. Declining amphibian populations. Science 253:860.

Wake DB, Morowitz HJ. 1990. Declining amphibian populations—a global phenomenon? Irvine, CA: National Research Council Board on Biology.

Willis YL, Moyle DL, Baskett TS. 1956. Emergence, breeding, hibernation, movements, and transformation of the bullfrog, *Rana catesbeiana*, in Missouri. Copeia 1956:30–40.

Wyman RL. 1990. What's happening to the amphibians? Conservation Biology 4:350–352.

Amphibians in decline: Canadian studies of a global problem. David M. Green, editor.
Herpetological Conservation 1:27–36.

Chapter 3

POPULATION CHARACTERISTICS OF THE RED-SPOTTED NEWT, *NOTOPHTHALMUS VIRIDESCENS*, AT THE MASTIGOUCHE RESERVE, QUÉBEC

RAYMOND LECLAIR JR.

Département de Chimie-Biologie, Université du Québec à Trois-Rivières, C.P.500, Trois-Rivières, Québec G9A 5H7, Canada

MARIA HELENA CAETANO

Faculdade de Ciências, Departamento de Zoologia e Antropologia, C2 Campo Grande, 1700 Lisboa, Portugal

ABSTRACT.—Adults of the red-spotted newt, *Notophthalmus viridescens*, were sampled in 1992 with minnow traps and dip-nets in 5 permanent oligotrophic lakes in the Mastigouche Reserve, Québec, to assess size and age structures. Ages were determined by skeletochronology. Snout-vent length (SVL) of 181 adult specimens ranged between 35 and 54 mm, with similar mean size in males (43.5 mm) and in females (44.1 mm). Overall sex-ratio was 1.5 males per female, with males strongly predominating in early June. Age correlated with SVL of adults and ranged from 2 to 13 yr. Depending on the lake, modal age frequencies were between 4 and 7 yr and maturity was reached between 2 and 4 yr. Growth was similar in males and females according to von Bertalanffy growth models. Metamorphs caught with drift fences leaving 1 lake in early September of 1994 averaged 18.3 mm SVL. Many smaller larvae were still in the lake at this time. The species is frequently encountered in lakes of the Mastigouche Reserve. Predation by fish and winter mortality of young terrestrial efts are potential causes of low abundance. Structure and dynamics of red-eft populations remain unknown.

RÉSUMÉ.—Des tritons verts adultes, *Notophthalmus viridescens*, ont été échantillonnés avec des nasses à ménés et au filet troubleau dans 5 lacs permanents oligotrophes de la réserve Mastigouche (Québec) en vue d'une évaluation des structures de taille et d'âge. L'âge a été déterminé par squelettochronologie. La longueur museau-cloaque (LMC) de 181 spécimens adultes était distribuée entre 35 et 54 mm avec des tailles moyennes similaires chez les mâles (43.5 mm) et les femelles (44.1 mm). Le rapport des sexes de l'ensemble était de 1.5 mâles par femelle, les mâles prédominant nettement en début juin. L'âge, corrélé à la taille chez les adultes, variait entre 2 et 13 ans. Selon les lacs, les fréquences modales d'âge variaient entre 4 et 7 ans et la maturité était atteinte entre 2 et 4 ans. La croissance était similaire chez les mâles et les femelles selon les modèles de croissance de von Bertalanffy. Des spécimens nouvellement métamorphosés quittant 1 lac en 1994, et capturés à l'aide de clôtures de dérive, avaient une LMC moyenne de 18.3mm. Plusieurs larves avec des taille plus réduites étaient encore dans le lac à ce moment. L'espèce est fréquemment rencontrée dans les lacs de la réserve Mastigouche. La prédation par les poissons et la mortalité hivernale des jeunes elfes en milieu terrestre sont des causes potentielles de la faible abondance. La structure et la dynamique des populations d'elfes rouges demeurent toutefois inconnues.

Canadian populations of the red-spotted newt, *Notophthalmus viridescens*, represent postglacial immigrants that invaded as the ice-sheets were retreating (Bleakney 1958). In Québec, the northernmost locality is Sept-Iles (50°13'N) on the lower North Shore of the St. Lawrence River (Bider and Matte 1991). *Notophthalmus viridescens* is accorded common status (Leclair 1985) in the Gaspé Peninsula, on Prince Edward Island, and in Nova Scotia (Bleakney 1958; Cook 1984) and is the 2nd most commonly reported amphibian species in Québec (Bider and Matte 1991). It is particularly prevalent in this region in lakes and marshes devoid of fish.

Red-spotted newt, Notophthalmus viridescens. *Photo by John Mitchell.*

However, the recorded frequency of a species does not tell very much about its abundance. Québec populations of *N. viridescens* are at the northern limit of the species distribution and most species of amphibians and reptiles become increasingly rare at the periphery of their range (Bleakney 1958). Apart from museum information on morphological characteristics of red-spotted newts (Cook 1967; Gilhen 1984) and general life history accounts (Logier 1952; Melançon 1961; Gorham 1970; Froom 1982; Cook 1984), no precise demographic studies have been published on any Canadian populations of this species. Detailed information on population dynamics mostly comes from studies made in northeastern United States (Mecham 1967; Hurlbert 1969, 1970; Healy 1970, 1973, 1974, 1975a,b, and Gill 1978a,b, 1979).

Smith and Pfingsten (1989) have summarized the variable and complicated life history of *N. viridescens*. Most commonly, eggs are deposited in water in spring or early summer, the aquatic larval stage lasts 4 to 6 mo, followed by a terrestrial eft stage of 2 to 7 yr. Adults return to the pond for a permanent aquatic life but may also return again to land in case of temporarily inappropriate aquatic conditions. Hibernation occurs either on land or in water depending on the quality (temperature, oxygen, food) and stability (drought) of the aquatic habitat. Populations omitting the terrestrial eft stage completely and populations containing paedomorphic adults also exist as alternate life histories. Factors responsible for such different life histories in urodeles are poorly understood and actively debated (Breuil 1992; Whiteman 1994).

Most ecological studies on *N. viridescens* have been conducted in small semi-permanent or temporary ponds. In this paper, we analyse size and age structures of populations of *N. viridescens* from 5 large permanent lakes characteristic of the Laurentian Shield, in Québec. Age of *N. viridescens*, an important demographic parameter, has usually been obtained through body size distribution analysis or capture-recapture methods (Hurlbert 1969; Healy 1974; Gill 1978a), but these methods are imprecise (Halliday and Verrell 1988). The use of growth marks in bones (skeletochronology) is preferable, because the method has been proven to be effective both for European newts, *Triturus* spp. (Smirina and Rocek 1976; Caetano et al. 1985; Miaud 1992) and for *N. viridescens* (Forester and Lykens 1991).

Materials and Methods

We initiated study of newt populations in 1992 in permanent oligotrophic lakes of the Mastigouche Reserve (46°38'N, 73°15'W) in the Laurentian Shield in Québec. The 5 lakes studied in 1992 (Mastigou, Lafond, Vautour, Simpson and Bouteille) varied in area from 7 to 254 ha and were 3.3 to 18 m deep. Similar climatic conditions prevailed in the study lakes, because they were all within a 35-km radius. Physico-chemical characteristics of the lakes (conductivity, dissolved oxygen, thermal stratification, and transparency) are typical of small, oligotrophic, temperate zone lakes (Magnan 1988).

The specimens were caught from late spring to mid-summer using dip-nets and unbaited minnow traps deposited at the vegetated margin of lakes in shallow water less than 1-m deep. We distributed 30 minnow traps in 1 lake at a time during daylight hours. These were left overnight and then checked and removed the following day. Traps were then either redistributed in the same lake or transfered to another lake (Table 1), with the objective of getting a minimum of 30 specimens per lake. Attempts to catch newts with minnow traps in deeper water proved to be unsuccessful and were abandoned.

The animals caught were killed by freezing. We recorded skin color, tail shape, condition of gonads, secondary sexual characters such as swollen cloaca and/or horny excrescences in males, body mass to the nearest 0.01 g, total length (TL, mm), and snout-vent length (SVL, mm, from the tip of the snout to the posterior end of the cloaca).

Table 1. *Sites, dates, and sample sizes of* Notophthalmus viridescens *caught in the Mastigouche Reserve in 1992.*

Lakes	Date	Males	Females
Boutielle	5 June	25	0
Mastigou	9–10 June	22	1
	16 June	12	5
	25 June	2	5
	28 June	0	4
	15 July	0	5
Lafond	12 June	12	20
	12 June[a]	6	8
	13 August	2	0
Vautour	13 June	18	3
	24 June	0	2
	1 July	1	6
Simpson	17 June	1	5
	23 June	3	5
	26 June	8	10

[a]Dip-net sampling

Weekly, from early June to the end of August, 1994, Lake Lafond (43.6 ha in area, 7.9 m maximum depth) was sampled again for 24 hr in the same manner as described above. However, the minnow traps were restricted in distribution to one of the most favorable bays over about 200 m of shoreline and each animal was toe-clipped and released alive at the site of capture. The procedure was initiated to follow the presence of newts in one specific zone of reproduction. We sampled the same area in late August and early September using dip-nets to catch larvae. We installed two 20-m long drift fences, with 10 pitfall traps on each side, 16 m from the shore. We trapped from the beginning of June to the 1st week of October, allowing the capture of red efts going in or out of the lake.

We assessed the age of each animal from the 1992 sample and some red efts using skeletochronology of growth marks in humeri (Flageole and Leclair 1992; Caetano and Castanet 1993). Each humerus was decalcified for 3 hr in 3% nitric acid, cross-sectioned (20 μm) at the midshaft diaphysis using a freezing microtome, and stained 15 min with Ehrlich's hematoxylin to reveal lines of arrested growth (LAG). Fifteen to 20 representative cross-sections per animal were mounted on slides in aquamount and protected with a cover-slip. Ages were determined independently by the 2 authors from observations of LAGs under the microscope. Growth curves were constructed from the von Bertalanffy (1938) model modified by Walford (1946) and Gullan (1968).

Results

Capture Phenology

Captures of newts for 1992 were initiated on 5 June in Lake Bouteille and ended on 13 August in Lake Lafond (Table 1). The animals were restricted to shallow narrow littoral zone habitats with floating and emergent vegetation. We caught 181 specimens with an overall adult sex-ratio of 1.5 males for every female. In only 1 lake (Lafond) did females outnumber males. In 3 lakes, males predominated in the first sampling dates (Table 1). The number of females caught thereafter gradually increased. Early in the season, when most of the animals were collected, capture rate averaged 0.66 newts/day/trap, with a maximum of 5 newts caught in a trap in 1 day.

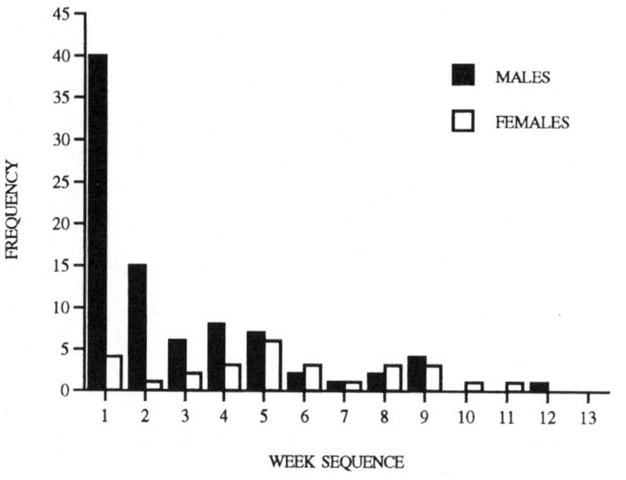

Figure 1. *Weekly captures of adult* Notophthalmus viridescens *in Lake Lafond from 7 June to 30 August 1994.*

The weekly captures in 1994 of adult newts in Lake Lafond yielded 86 males and 28 females (sex ratio 3:1), with nearly half of the males being caught on June 7, the 1st sampling date (Fig. 1). No recaptures were made thus all the specimens were different animals. Captures were less numerous after mid-July, and by mid-August almost no specimens were collected. For the first 5 wk of sampling, capture rate averaged 0.61 newts /day/trap.

Secondary sexual characteristics were evident in males caught in June and early July. The few males caught in August had lost those characteristics. Most females captured in June were gravid.

During the first 2 wk of September 1994, we caught 23 newt larvae ranging from 8.5 to 19 mm SVL ($\bar{x}$ = 14.8) in Lake Mastigou, and 14 others ranging from 15 to 19 mm ($\bar{x}$ = 16.7) in Lake Lafond. On 1 September the previous year (1993), we had also caught 15 larvae in Lake Lafond having a similar range in SVL (15 to 17.5 mm), that metamorphosed within few days in the laboratory.

In 1994, the drift fences started yielding postlarval metamorphs on 30 August, but massive migration towards the forest occurred during the last 3 wk of September when we collected 1039 metamorphs, most of which still showed gill stubs. A subsample of 72 specimens taken on 23 September had a mean SVL of 18.3 ± 1.1 mm (SD).

Red efts were rarely seen active on the surface during the study, even during warm summer rains. However, the drift fences yielded 147 red efts in migration toward the lake during June to August, 1994. The 128 that were measured averaged 32.3 ± 2.7 mm (range = 26 to 39 mm). We never captured large specimens with adult characteristics in pitfall traps, either leaving or entering the lake.

Size Structure

Snout-vent length in adult specimens (Fig. 2) varied from 35 to 54 mm, ($\bar{x}$ = 44.1 ± 3.6 mm in females, n = 68; $\bar{x}$ = 43.5 ± 3.72 mm in males, n = 106). Total length varied from 70 to 126 mm with mean values of 93.2 ± 11.3 mm in females and 94.2 ± 10.3 mm in males. The recorded bimodal frequency distribution for male SVL is partly due to the sample taken from Lake

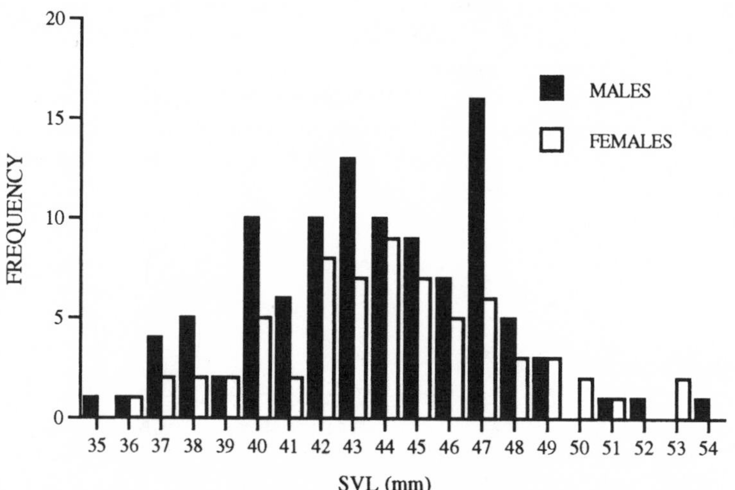

Figure 2. *Snout-vent lengths of adult* Notophthalmus virides-cens *from 5 permanent oligotrophic lakes of the Mastigouche Reserve, Québec in 1992.*

Bouteille, which contained many specimens larger than the general mean (Fig. 2). Thirty-two percent of the male newts from this lake had an SVL of 47 mm. Other interlake differences in mean SVL were found (Fig. 3). Our data are compared with values from other populations in Table 2.

Age Structure

In all humeri examined, clear lines of arrested growth (LAGs) were distinguished in the periosteal bone. The first (deepest) LAG was generally more hematoxylinophilic than all the others and corresponded to the first winter encountered by the post-metamorphic animals. Although this LAG may sometimes be partially destroyed by endosteal resorption that occurs around the marrow cavity, this phenomenon did not affecte the accuracy of age determination. For this study, we made the supposition that the LAGs increase in number by only 1 each yr, as was demonstrated in *Triturus cristatus* (Francillon 1979) and *T. marmoratus* (Caetano and Castanet 1987). Double marks were uncommon in *N. viridescens* and readily recognized. So, in

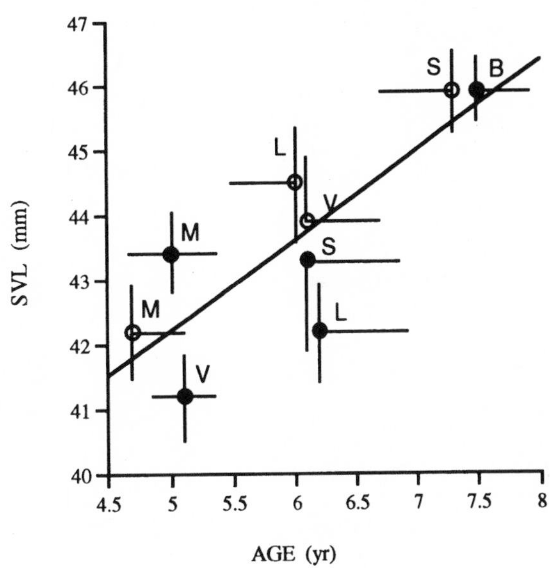

Figure 3. *Mean age and size of male (black symbols) and female (white symbols) newts from 5 lakes in 1992. M = L. Mastigou, V = L. Vautour, L = L. Lafond, S = L. Simpson, and B = L. Bouteille.*

Table 2. *Sizes (mean and/or range) of* Notophthalmus viridescens *from different regions.*

	Snout-vent length (mm)		Total length (mm)		
Region	Male	Female	Male	Female	Reference
Prince Edward Island	50.3	52.1	106.3	109.8	Cook 1967
	43–59	42–64	86–131	81–137	
Nova Scotia	45.2	46.9	91.3	94.5	Gilhen 1984
	36–56	37–57	73–106	70–119	
Québec	43.5	44.1	94.2	93.2	This study
	34–54	36–53	70–126	72–116	
Ontario			63–123	63–123	Logier 1952
New York			81–98	83–87	Bishop 1941
New York	38.9[a]	38.9[a]	86.5[b]	84.4[b]	Hurlbert 1969
	33–48[a]	33–48[a]	104.1[c]	100.5[c]	
Ohio	42.1	41.4	93.8	87.6	Smith et al. 1989
North Carolina	45.1	45.1	92.1	91.1	Chadwick 1944
	40–51	40–51	81–108	81–99	

[a]terrestrial adult migrants

[b]mean from 3 lakes

[c]mean from 2 lakes

most of the cases, age could be simply determined by counting the number of LAGs in the periosteal bone.

Estimated age varied between 2 and 13 yr, with modal frequencies at 4 and 7 yr for both males and females (Fig. 4). The pattern observed in age frequency distribution was mainly due to interpopulational differences. Mean ages varied from 5.0 to 7.5 yr in males, and from 4.7 to 7.3 yr in females (Fig. 3). Differences in cohort strength within single populations may also have contributed to the observed pattern of age distribution.

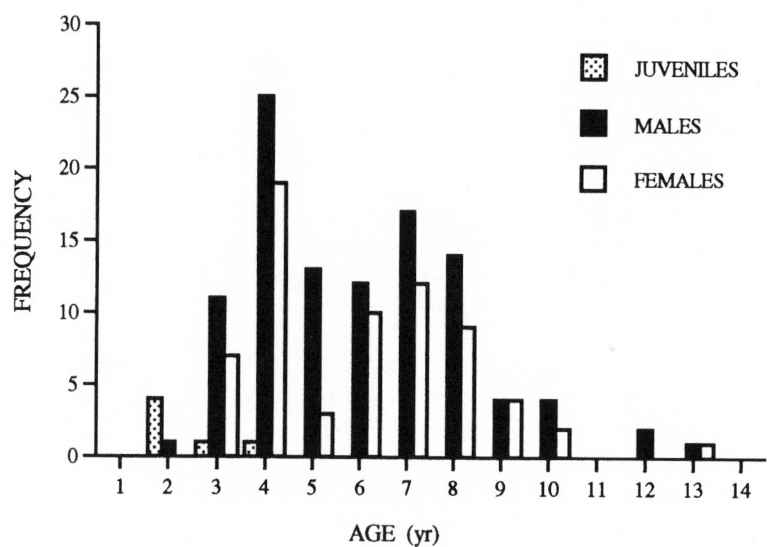

Figure 4. *Age distribution of* Notophthalmus viridescens *caught in 5 permanent oligotrophic lakes of the Mastigouche Reserve, Québec in 1992.*

Minimum age of maturity was commonly 3 yr for both sexes, except for males from Lake Bouteille and females from Lake Simpson, where 4 yr was the minimum. A single 2-yr old male from Lake Mastigou was found with well developed gonads.

Age and SVL were correlated in both sexes in all populations ($r = 0.78$ for all males combined, and 0.80 for all females combined). However, because the age/size classes of adults overlapped, SVL alone was not adequate to discriminate age classes. Males and females showed similar growth curves although the asymptotic size determined by the von Bertalanffy model (Fig. 5) was larger in females (52.2 mm) than in males (50.2 mm).

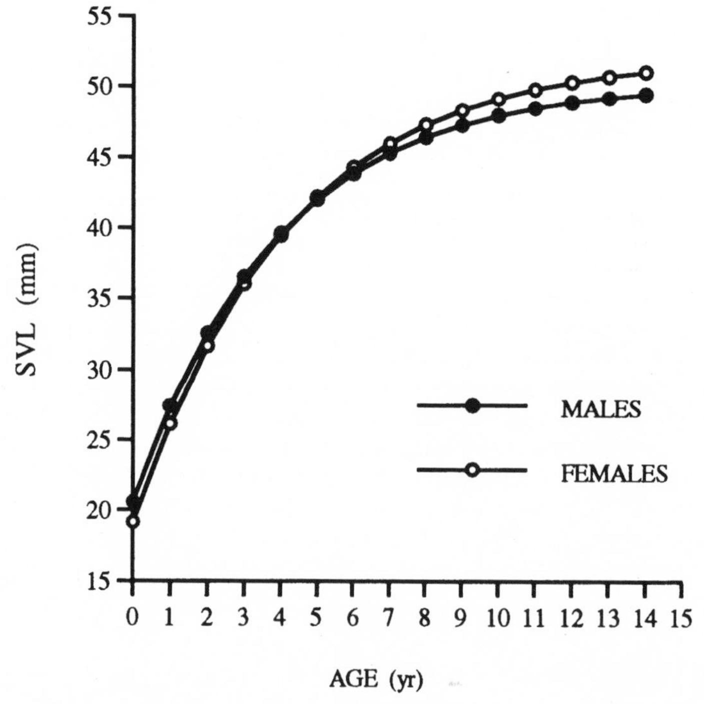

Figure 5. *von Bertalanffy growth curves for* Notophthalmus viridescens *from 5 lakes of the Mastigouche Reserve. For males,* $L_t = 50.26(1 - e^{-0.25(t + 2.12)})$, *and for females,* $L_t = 52.17(1 - e^{-0.24(t + 1.92)})$.

Discussion

Abundance

Between 1988 and 1992, 15 lakes in the Mastigouche Reserve with experimental fishing activities, including lake Vautour, were recorded as sustaining populations of *N. viridescens* (J. Archambault, pers. com.) but nowhere were the newt populations noted as abundant. Our own data add Lakes Mastigou, Lafond, Simpson, Bouteille, Jones, Chamberlain, and Régis to the list. *Notophthalmus viridescens* appears to be well distributed throughout the study ecosystem but abundance is low compared to what has been reported in many studies conducted in United States where, although methods of capture differ, thousands of specimens can be caught (Hurlbert 1969; Healy 1973; Burton 1977; Gill1978b).

Adult newts usually live in water no deeper than 1 m, in areas with abundant vegetation (Smith and Pfingston 1989), and in Canada, concentrate their reproductive activities between April and June (Logier 1952). Sampling with dip-nets in such "ideal" circumstances in the Mastigouche Reserve yielded only few specimens. In Lake Deux-Étapes, where red-spotted newts had already been recorded, a 1-day sampling with minnow traps in June 1992 yielded no specimens. In Lake Vernez during June 1994, only 1 newt was captured by sampling with 30 minnow traps on 2 occasions. With the same number of traps, 3 different sampling dates for Lake Jones in 1993, 7 for Lake Régis, and 8 for Lake Chamberlain in 1994 were required to procure 49, 38, and 39 specimens respectively. The 114 adult newts caught and released in 1994 in Lake Lafond were sampled on 12 different dates (Fig. 1). This suggests that adult newts do not maintain very high abundance in the studied lakes.

Factors responsible for the relatively low abundance observed in the Mastigouche Reserve may be related to habitat preferences (Gates and Thompson 1982), low productivity of the lakes, and/or predation or competition pressure from fish. Brook trout (*Salvelinus fontinalis*), redbelly dace (*Phoxinus eos*), white sucker (*Catostomus commersoni*), pearl dace (*Semotilus margarita*) and creek chub (*Semotilus atromaculatus*) are present in varying numbers in the study lakes (Magnan 1988; Lacasse and Magnan 1992).

Sex Ratio

Strongly biased sex ratios may be detrimental to population stability. However, the values observed in this study (1.5 males : 1 female for five study lakes in 1992, or 3:1 in Lake Lafond for 1994) do not represent a very strong bias and unbalanced sex-ratios in favor of males seem rather common in populations of *N. viridescens*. Healy (1974) in Massachusetts, Gill (1978b) in Virginia, and, to a lesser extent, Hurlbert (1969) in New York made similar observations in newt populations. Such results can be explained by greater catchability of males, which are actively searching for females during the reproductive period. As shown by Healy (1974), and by our capture data, males also arrive first in the breeding sites and outnumber females at that time. Early maturity in males can also contribute to unbalanced sex-ratios in adult urodeles (Flageole and Leclair 1992), but this factor appears insignificant here, because the number of males in the youngest adult age classes (2-to 3-yr old) was small. Better survivorship in males, as noted by Gill (1978b, 1985), was not observed here. Females averaged 1-yr older than males in Lakes Vautour and Simpson. On the other hand, equal numbers of males and females have been observed by Gill (1978b) among young recruits leaving the ponds, and by Pitkin and Tilly (1982) among large winter aggregations of adults in cold water.

Size and Age Structures

The mean values and ranges of body sizes (both SVL and TL) of newts from the Mastigouche Reserve were well within the reported values for the species in northeastern United States and Canada (Table 2). There was no apparent geographic trend in body size despite the large sizes of specimens from Prince Edward Island (Cook 1967). Age structure in Mastigouche *N. viridescens* also did not deviate from other reported values. On the basis of age-specific survivorship curves, Gill (1985) suggested that longevity of red-spotted newts could reach 12 yr in females and 15 yr in males. By skeletochronological methods, Forester and Lykens (1991) found a maximum age of 9 yr in a population of *N.*

viridescens from the Allegheny Plateau of Maryland. Maximum ages observed in our samples ranged between 9 yr, in Lake Vautour, to 13 yr in Lakes Bouteille and Simpson.

Variation in body size of salamanders among lakes corresponded to variation in age (Fig. 3) largely because the 2 parameters were correlated (Fig. 5). Because the newts breed for the 1st time between ages 2 and 4, and growth continues for varying numbers of years after the attainment of maturity depending upon the population (Fig. 3), the mean number of opportunities for reproduction, and consequently the whole population dynamics, may vary greatly from lake to lake. Both the level of these variations and reasons for the interlake differences need investigation.

Metamorphs

Metamorphosing larvae of *N. viridescens* were observed in mid-June in lowland Maryland by Worthington (1968) and between late August and mid-September at an elevation of 2200 m in Highlands, North Carolina, by Chadwick (1950). Metamorphosis occurs in the fall in Massachusetts according to Healy (1974, 1975a). The 3 authors reported similar size at metamorphosis of about 20-mm SVL. Gilhen (1984) reported the size of newly transformed juveniles in Nova Scotia to range from 18.5 to 24.5 mm ($\bar{x}$ = 22.6).Larvae from

Larvae from the Mastigouche Reserve metamorphosed at a SVL between 15 and 20 mm during September. The mean SVL of 18.3 mm in 1994 was for small metamorphs that had already travelled at least 16 m overland. Many larvae as small as 8 mm were still in the water in mid-September. This meant that the young terrestrial efts were still quite small some weeks before the arrival of freezing air temperatures. If these represent the characteristic body size and phenology of the newts in the Mastigouche region, recruits may be at risk of high winter mortality compared to populations further south. this could partly explain the relatively low number of maturing efts that were caught in migration toward the lake compared to the abundance of metamorphs leaving the lake.

Acknowledgments

We are grateful to Daniel Courtois, Jean Leclerc, Patrick Carrier, and Emmanuel Milot for their participation in newt sampling, and to the Ministère québécois du Loisir, de la Chasse et de la Pêche for providing logistical help at the Mastigouche Reserve and legal authorization to catch the animals. Financial support was provided by the Fond Institutionnel de Recherche of Université du Québec à Trois-Rivières.

Literature Cited

Bider JR, Matte S. 1991. Atlas des amphibiens et des reptiles du Québec. Ste.-Anne-de-Bellvue, QE: Société d'Histoire Naturelle de la Vallée du St.-Laurent.

Bleakney JS. 1958. A zoogeographical study of the amphibians and reptiles of eastern Canada. Ottawa: National Museum of Canada Bulletin 155 (Biological Series 54). 119 p.

Breuil M. 1992. La néoténie dans le genre *Triturus:* mythes et réalités. Bulletin Societé Herpétologique de France 61:11–44.

Burton TM. 1977. Population estimates, feeding habits and nutrient and energy relationships of *Notophthalmus v. viridescens*, in Mirror Lake, New Hampshire. Copeia 1977:139–143.

Caetano MH, Castanet J. 1987. Experimental data on bone growth in *Triturus marmoratus* (Amphibia, Urodela). In: Van Gelder JJ, Strijbosch H, Bergers PJM editors. Proceedings of the General Meeting of the Societas Europaea Herpetologica. Nijmegen, Netherlands: Faculty of Sciences. p 87–90.

Caetano MH, Castanet J. 1993. Variability and microevolutionary patterns in *Triturus marmoratus* from Portugal: age, size, longevity and individual growth. Amphibia-Reptilia 14:117–129.

Caetano MH, Castanet J, Francillon H. 1985. Détermination de l'âge de *Triturus marmoratus mamoratus* (Latreille 1800) du Parc National de Peneda Gerês (Portugal) par squelettochronologie. Amphibia-Reptilia 6:117–132.

Chadwick CS. 1950. Observations on behavior of the larvae of the common American newt during metamorphosis. American Midland Naturalist 43:392–398.

Cook FR. 1967. An analysis of the herpetofauna of Prince Edward Island. National Museum of Canada Bulletin 212:1–60.

Cook FR. 1984. Introduction aux amphibiens et reptiles du Canada. Ottawa: National Museum of Natural Sciences, National Museums of Canada.

Flageole S, Leclair RJr. 1992. Étude démographique d'une population de salamandres (*Ambystoma maculatum*) à l'aide de la méthode squeletto-chronologique. Canadian Journal of Zoology 70:740–749.

Forester DC, Lykens DV. 1991. Age structure in a population of red-spotted newts from the Allegheny Plateau of Maryland. Journal of Herpetology 25:373–376.

Francillon H. 1979. Étude expérimentale des marques de croissance sur les humérus et les fémurs de Triton crêtés (*Triturus cristatus cristatus* Laurenti) en relation avec la détermination de l' âge individuel. Acta Zoologica 60:223–232.

Froom B. 1982. Amphibians of Canada. Toronto: McClelland and Stewart.

Gates JE, Thompson EL. 1982. Small pool habitat selection by red-spotted newts in Western Maryland. Journal of Herpetology 16:7–15.

Gilhen J. 1984. The amphibians and reptiles of Nova Scotia. Halifax, NS: Nova Scotia Museum.

Gill DE. 1978a. Effective population size and interdemic migration rates in a metapopulation of the red-spotted newt, *Notophthalmus viridescens* (Rafinesque). Evolution 32:839–849.

Gill DE. 1978b. The metapopulation ecology of red-spotted newt, *Notophthalmus viridescens* (Rafinesque). Ecological Monographs 48:145–166.

Gill DE. 1979. Density dependence and homing behavior in adult red-spotted newts *Notophthalmus viridescens* (Rafinesque). Ecology 60:800–813.

Gill DE. 1985. Interpreting breeding patterns from census data: a solution to the Husting dilemma. Ecology 66:344–354.

Gorham SW. 1970. The amphibians and reptiles of New Brunswick. Saint John, NB: New Brunswick Museum.

Gullan JA. 1968. Manual of methods for fish assessment. I. Fish population analysis. Rome: United Nations Food and Agriculture Organization. (FAO Fisheries Technical Paper 40, Rev. 2.).

Halliday TR, Verrell PA. 1988. Body size and age in amphibians and reptiles. Journal of Herpetology 22:253–265.

Healy WR. 1970. Reduction of neoteny in Massachusetts populations of *Notophthalmus viridescens*. Copeia 1970:578–581.

Healy WR. 1973. Life history variation and growth of juveniles *Notophthalmus viridescens* from Massachusetts. Copeia 1973:641–647.

Healy WR. 1974. Population consequences of alternative life histories in *Notophthalmus v. viridescens*. Copeia 1974:221–229.

Healy WR. 1975a. Breeding and postlarval migration of the red-spotted newt, *Notophthalmus viridescens*, in Massachusetts. Ecology 56:673–680.

Healy WR. 1975b. Terrestrial activity and home range in efts of *Notophthalmus viridescens*. American Midland Naturalist 93:131–138.

Hurlbert SH. 1969. The breeding migrations and interhabitat wandering of the vermilion-spotted newt *Notophthalmus viridescens*. (Rafinesque). Ecological Monographs 39:465–488.

Hurlbert SH. 1970. The post-larval migration of the red-spotted newt, *Notophthalmus viridescens* (Rafinesque). Copeia 1970:515–528.

Lacasse S, Magnan P. 1992. Biotic and abiotic determinants of the diet of brook trout, *Salvelinus fontinalis,* in lakes of the Laurentian shield. Canadian Journal of Fisheries and Aquatic Sciences 49:1001–1009.

Leclair RJr. 1985. Les Amphibiens du Québec: Biologie des espèces et problématique de conservation des habitats. Québec: Bibliothèque Nationale du Québec.

Logier EBS. 1952. The frogs, toads and salamanders of eastern Canada. Toronto: Clarke, Irwin, and Company.

Magnan P. 1988. Interactions between brook charr, *Salvelinus fontinalis*, and nonsalmonid species: ecological shift, morphological shift and their impact on zooplankton communities. Canadian Journal of Fisheries and Aquatic Sciences 45:999–1009.

Mecham JS. 1967. *Notophthalmus viridescens*. Catalogue of American Amphibians and Reptiles 53:1–4.

Melançon C. 1961. Inconnus et méconnus: amphibiens et reptiles de la province de Québec. 2e ed., Orsainville, QE: La Société Zoologique de Québec Inc.

Miaud C. 1992. La squelettochronologie chez les *Triturus* (Amphibiens, Urodèles) à partir d'une étude de *T. alpestris, T. helveticus* et *T. cristatus* du sud-est de la France. In: Baglinière JL, Castanet J, Conand F, Meunier FJ, editors. Tissus Durs et Âge Individuel des Vertébrés. Bondy, France: Colloque et Séminaires ORSTOM-INRA. p 363–384.

Pitkin RB, Tilley SG. 1982. An unusual aggregation of adult *Notophthalmus viridescens*. Copeia 1982:185–186.

Smirina EM, Rocek Z. 1976. On the possibilty of using annual bone layers of alpine newts, *Triturus alpestris* (Amphibia: Urodela), for their age determination. Vestnik Československe Spolecnosti Zoologicke 40:232–237.

Smith CJ, Pfingsten RA. 1989. *Notophthalmus viridescens* (Rafinesque), eastern newt. In: Pfingsten RA, Downs FL, editors. Salamanders of Ohio. Columbus, OH: The Ohio State University. p 79–87. (Ohio Biological Survey, New Series 7).

von Bertalanffy L. 1938. A quantitative theory of organic growth. Human Biology 10:181–213.

Walford LA. 1946. A new graphic method of describing the growth of animals. Biological Bulletin 90:141–147.

Whiteman HH. 1994. Evolution of facultative paedomorphosis in salamanders. Quarterly Review of Biology 69:205–221.

Worthington RD. 1968. Observations of the relatve sizes of three species of salamander larvae in a Maryland pond. Herpetologica 24:242–246.

©1997 by the Society for the Study of Amphibians and Reptiles
Amphibians in decline: Canadian studies of a global problem. David M. Green, editor.
Herpetological Conservation 1:37–44.

Chapter 4

POPULATION BIOLOGY OF THE LONG-TOED SALAMANDER, *AMBYSTOMA MACRODACTYLUM*, IN THE FRONT RANGE OF ALBERTA

G. LAWRENCE POWELL, ANTHONY P. RUSSELL, JANICE D. JAMES, STEPHEN J. NELSON, AND SHERI M. WATSON

Vertebrate Morphology Research Group, Department of Biological Sciences and Kananaskis Research Stations, University of Calgary, 2500 University Dr. NW, Calgary, Alberta T2N 1N4, Canada

ABSTRACT.—A 2-yr capture-recapture study of 2 populations of the long-toed salamander (*Ambystoma macrodactylum*) in the Front Range of western Alberta yielded large total capture numbers for 1992 and 1993 (881 salamanders at the Lafarge Borrow Pit and 401 at the Quarry Ponds) and relatively few recaptures (60 at the Lafarge Borrow Pit and 91 at the Quarry Ponds). Populations thus appear to be large at these 2 localities. The Lafarge Borrow Pit sample represented all terres trial age classes except the youngest. A shift in the peak of the snout-vent length distribution between years suggests the existence of a large cohort. At the Quarry Ponds, the sample represented breeding adults and emigrating metamorphs almost entirely. Breeding adults moved a significant distance between terrestrial and breeding habitats at the Quarry Ponds but not at the Lafarge Borrow Pit.

RÉSUMÉ.—Une étude de 2 ans sur 2 populations de salamandres à longs doigts (*Ambystoma macrodactylum*) dans le Front Range de l'ouest de l'Alberta s'est soldée par un nombre important de captures en 1992 et 1993 (881 salamandres à Lafarge Borrow Pit et 401 à Quarry Ponds) et relativement peu de reprises (60 à Lafarge Borrow Pit et 91 à Quarry Ponds). Les populations sont par conséquent importantes dans ces 2 localités. L'échantillon de Lafarge Borrow Pit représentait toutes les catégories d'âges terrestres à l'exception des plus jeunes. Un glissement dans la répartition de la longueur maximale museau-anus entre les différentes années donne à penser qu'il existe une importante cohorte. À Quarry Ponds, l'échantillon représentait des adultes reproducteurs et presque tous les stades de la métamorphose. Les adultes reproducteurs parcouraient des distances plus importantes entre les habitats terrestres et les habitats de reproduction à Quarry Ponds, qu'à Lafarge Borrow Pit.

Ambystoma macrodactylum, the long-toed salamander, ranges from central California north through the American Pacific Northwest and British Columbia to the Alaska panhandle (Bishop 1947; Ferguson 1961; Nussbaum et al. 1983; Cook 1984). It is found in a broad array of habitat types (Bishop 1947; Ferguson 1961; Nussbaum et al. 1983), and frequently maintains high population densities (Ferguson 1961; Nussbaum et al. 1983). As in other *Ambystoma*, transformed individuals are rarely seen outside of the breeding season, leading largely subterranean lives for the rest of the year (Ferguson 1961; Anderson 1967; Nussbaum et al. 1983). Life history traits vary with habitat, climate, and elevation. Breeding takes place early in the spring or in mid-winter (Anderson 1967; Nussbaum et al. 1983; Beneski et al. 1986). Russell and Bauer (1993) recorded 22 populations of *A. macrodactylum* in Alberta. The species has been placed on the Red List in the Status Evaluation System of the Alberta Wildlife Management Branch (Anonymous 1991).

Here, we present the results of a 2-yr study of two populations of the eastern long-toed salamander (*Ambystoma m. krausei*) in the upper Bow Valley of southwestern Alberta. These results are from an ongoing long-term intensive monitoring project intended to provide baseline data on population size and fluctuations over time as part of the global effort to assess the widely-observed phenomenon of declining amphibian populations (Vial and Saylor 1993).

MATERIALS AND METHODS

Study Sites

The Lafarge Borrow Pit (Lafarge; 51°5'20' N, 115°4' W; elevation 1290 m) is located on a tilly terrace, in an extensive oval, flat-bottomed excavation that is approximately 400 m along its northeast-southwest axis and some 200-m wide. This borrow pit broaches the water table along much of its north side and thus contains shallow standing water of fluctuating depth throughout the salamanders' active season. The bottom of the pit is underlain with large cobbles which are always close to, or comprise, the substrate surface. The pond bottoms are muddy, with little aquatic vegetation. Adult and metamorph salamanders are to be encountered over all of the pit bottom, an area of approximately 2.5 ha.

The Quarry Ponds (Quarry; 51°4'33" N, 115°22'30" W; elevation 1375 m) are in a reclaimed open-pit coal mine consisting of rolling meadows underlain by unsorted mine spill on the lower north slope of Chinaman's Peak. The eastern half of the meadow contains the 2 small spring-fed Quarry Ponds. Breeding occurred only in the smaller Upper Quarry Pond, which has a 150-m circumference, a maximum depth of 1.5 m, and an uncomplicated muddy bottom. The main terrestrial habitat is the woods (mixed *Populus* sp./*Pinus contorta*) on the south slope of Chinaman's Peak. The nearest margin of forest outlier is some 60 m away from the pond, while the main body is 175 m away.

Field Methods

Field work in 1992 was conducted from 1 May to 28 August and 20 to 30 September. Field work in 1993 was conducted 3 May to 18 August. We installed a T-shaped, 12 × 6-m drift fence in a wooded gully just off the main borrow pit at Lafarge on 11 June 1992. Thirteen 1-l pitfall traps, capped with slate roofs and paired across the fence, were installed at 3-m intervals. In 1992, the fence consisted of 22-cm wide polyethylene sheeting, stapled to posts, but in 1993 it was constructed from 20-cm wide corrugated plastic lawn edging. The bottom 2 cm of the fence was buried. An 80-m drift fence was installed along the southeast shore of Upper Quarry Pond at the beginning of May, 1993, using the same materials and with the pitfall traps paired at the same intervals. Both drift fences were taken down at the end of the 1993 field season.

Drift fences were checked daily and captures immediately processed and released on the opposite side under the nearest available cover. Salamanders were also captured during nocturnal jacklighting forays, processed on the spot, and then released. Each salamander was weighed to the nearest 0.1 g, and snout-vent length (SVL) to the nearest mm was measured with a ruler. Injuries and unusual marks were also noted at capture. Each individual was given a unique toe-clip for that locality (Ferner 1979). No more than 1 toe was taken from each foot, except at Lafarge, where capture numbers became very high in late 1993. Sheppard (1977) indicated that clipped toes of *Ambystoma macrodactylum* did not regenerate appreciably within 1 yr, and the technique has been found useful over several years in other species of salamanders (Husting 1965; Hairston 1987). Toes collected in both years from 170 individuals were preserved in 10% neutral buffered formalin and

Long-toed salamander, Amystoma macrodactylum. *Photo by David M. Green.*

retained for age determination using skeletochronological analysis (Russell et al. 1996). Sex was recorded from salamanders > 3 g in the breeding season, but the secondary sexual characteristics of adult animals became too ambiguous after June for sex to be assigned. The sex of subadults could not be externally determined.

Analysis

Our data likely violated the assumption of equal probability of capture or recapture required by methods using mark-recapture data for animal population size estimates (Begon 1979). The assumption of equal probability of death (Begon 1979) was shown by our skeletochronological data (Russell et al. 1996) to be invalid. Removal methods are inappropriate for amphibian species which disperse widely from breeding locales (Hayek 1994), as is the case at Quarry. Although the Lafarge population may be geographically closed, the size of the site would have required considerably more effort if removal methods had been attempted. Capture and recapture numbers are therefore presented as tallies for each locality for each year.

Population structures were examined through site-specific SVL histograms for each year's data. Median SVL was compared between sites for each year, and between years for each site, by Mann-Whitney U tests, and distributions compared by Kolmogorov-Smirnov 2-sample tests. Acceptance levels were fixed through stepwise Bonferroni adjustments (Rice 1989) for the Kolmogorov-Smirnov 2-sample tests. All statistical analyses were performed with SYSTAT (Wilkinson 1990). Skeletochronological analysis indicated that the relationship between size and age was too indeterminate to allow the statistical dissections of these SVL distributions into size-age classes (Russell et al. 1996).

RESULTS

Habitat Use

We encountered adult salamanders at Lafarge only in the breeding pools during May and June. Salamander surface activity was high over the wetter portions of the pit bottom throughout the study period. At Quarry, adults were also only found around the breeding pond during the early part of the active year. Most hand captures made in the terrestrial habitat were made just within the woods, but most of the terrestrial habitat was extremely difficult to search for salamanders. Captures were generally made within a few metres of the woods' edge. Newly-metamorphosed individuals were only trapped as they left the breeding pond and salamanders < 60-mm SVL were otherwise seldom encountered.

Number of Captures

There were considerable differences in numbers captured and recaptured between study sites, and between years within study sites. Absolute capture number was larger at Lafarge (881 individuals), with a smaller recapture number (60 recaptures) than that at Quarry (401 individuals and 91 recaptures). However, capture number diminished markedly at Lafarge between years (546 captures in 1992 and 335 in 1993) while the recapture number remained more or less constant (32 recaptures in 1992 and 28 in 1993). Rising groundwater levels steadily flooded our drift fence through the second half of the 1993 field season and also, but to lesser effect, in 1992. This contributed to the drop in capture number. The rising water levels through the summer of 1993 also greatly diminished the area of the terrestrial habitat, restricting the potential for hand captures. Capture and recapture numbers at Quarry rose between years (51 captures, 12 recaptures in 1992 and 350 captures, 79 recaptures in 1993). The 1992 total capture numbers for Quarry were partially the result of an early spring. Much of the breeding adult movement was finished by the time we started our field work that year. The onset of breeding was later in 1993, contributing to the great increase in capture and recapture numbers between the 2 yr at that site.

Daily fluctuations in new capture number at Lafarge were more marked in 1992 than in 1993 (Fig. 1). Capture numbers peaked in the middle of 1992, declined steeply and then increased near the end of

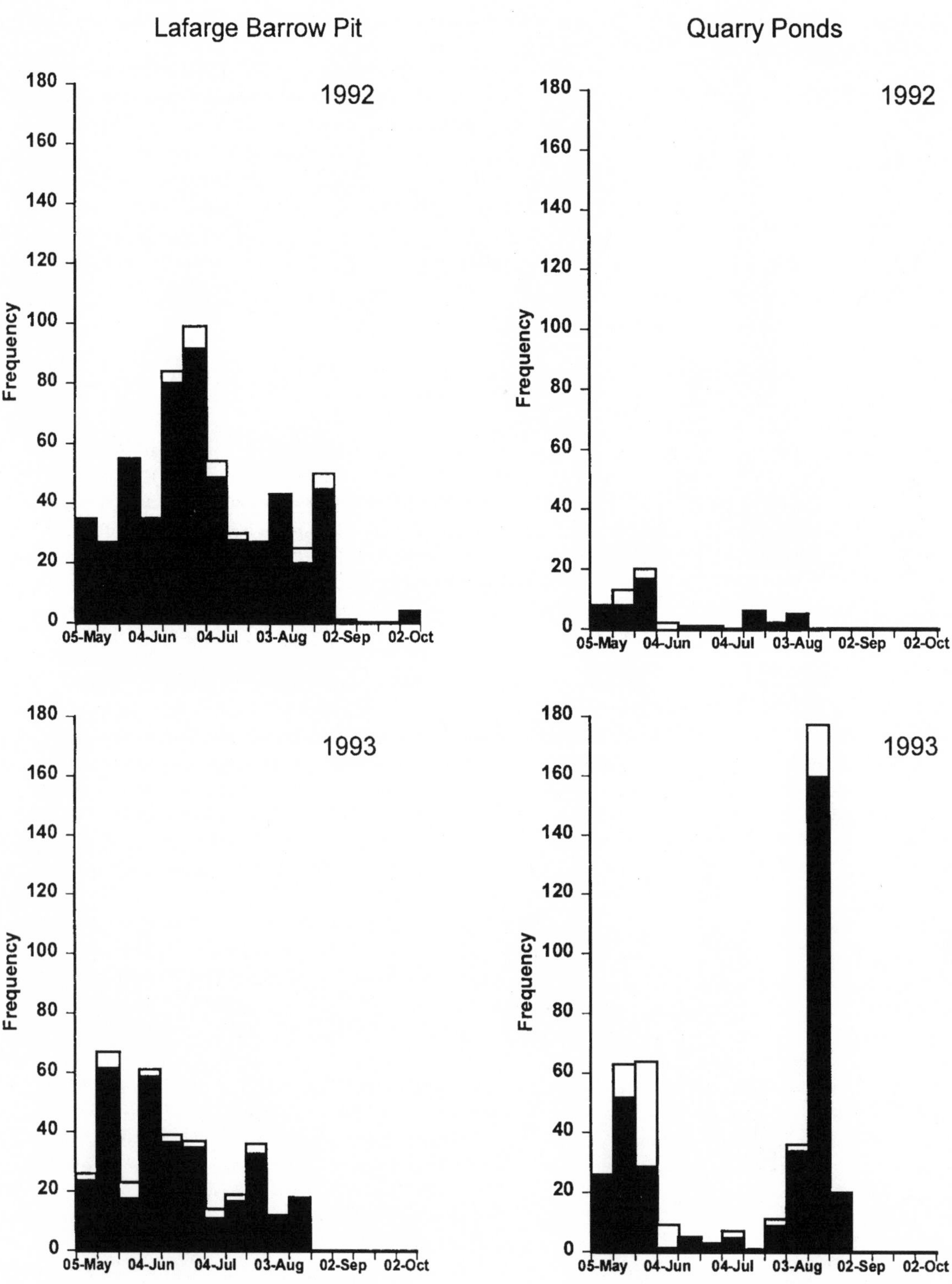

Figure 1. *Numbers of newly captured (shaded bars) and recaptured (unshaded bars)* Ambystoma macrodactylum *captured at Lafarge Barrow Pit and Quarry Ponds in 1992 and 1993.*

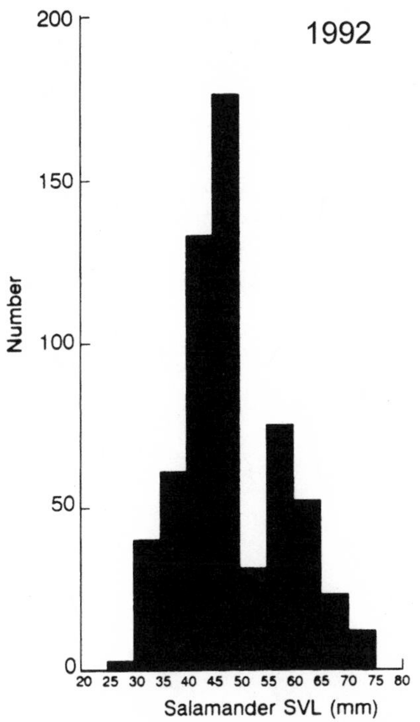

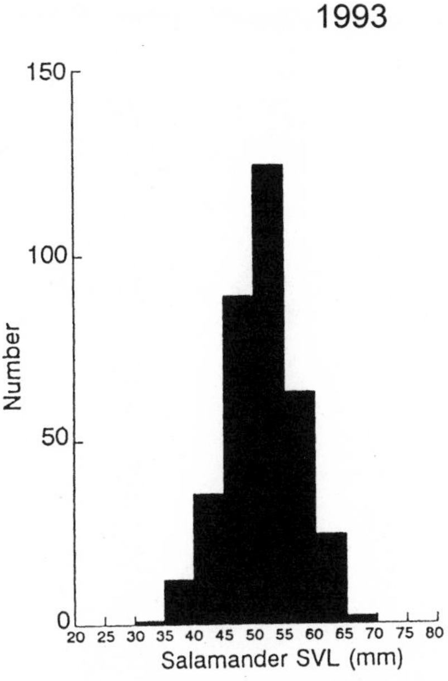

Figure 2. *Snout-vent length distributions of salamanders caught at Lafarge Barrow Pit in 1992 and 1993.*

August. New captures at Lafarge generally declined during 1993. Numbers of daily recaptures were very low relative to new captures in both years at this site.

The SVL distribution of the 1992 captures at Lafarge was wider and had more peaks than that of the 1993 captures at the same locality (Fig. 2). There were significant differences in central tendency ($U = 78399$; $P < 0.001$) and distribution (Kolmogorov-Smirnov 2-sample statistic = 0.332; $P < 0.001$) in SVL between the two years. Small individuals were under-represented in both years. The 1993 SVL distribution had a larger modal value than the 1992 distribution, but the largest individuals were caught in 1992.

Numbers of new daily captures at Quarry showed the same general pattern in 1992 and 1993 (Fig. 1). The late peak in 1993 was not found in 1992, because it represented emigrating metamorphs that were not sampled the 1st year. The May peaks of captures and recaptures in both years at Quarry represented breeding adults moving into and out of the breeding pond. Captures between the 2 peak periods in 1993 (Fig. 1) included a few adults around the breeding pond, but were mostly individuals captured in the terrestrial habitat. Recaptures were more frequent early in both years of the study. Breeding adults were the most frequent recaptures.

The SVL distribution of the 1993 Quarry sample (Fig. 3) was strongly bimodal, metamorphs and breeding adults being the principal fractions of the population represented. This sample was significantly different in distribution and central tendency from the Lafarge samples of 1992 (Kolmogorov-Smirnov 2-sample statistic = 0.478; $P < 0.001$; $U = 133644$; $P < 0.001$) and 1993 (Kolmogorov-Smirnov 2-sample statistic = 0.557; $P < 0.001$; $U = 77625$; $P < 0.001$). Representation of all age classes (excepting year 0 individuals and breeding adults) was much better for the Lafarge population than for the Quarry population.

Age in these populations of *A. macrodactylum* cannot be equated precisely with SVL, because same-aged individuals vary greatly in size (Russell et al. 1996). Therefore, we have no way of directly estimating survivorship from our SVL distributions. The SVL distribution for Quarry in 1993 (Fig. 3) applied mostly to individuals from the vicinity of the breeding pond and could not be representative of the true SVL distribution of the population as a whole. Those for Lafarge in 1992 and 1993 (Fig. 2) evidently under-represent the year 0 cohort.

Discussion

Possible Sources of Error

We carefully examined each capture's toes for evidence of regeneration. If a toe showed signs of regeneration in a recapture it was re-cut. On a few occasions, natural amputations caused mistaken classifications of new captures as recaptures, when the original clip involved two or fewer digits.

Breeding ponds at Lafarge are large and surrounded by terrain very poorly suited to the construction of drift fences, and the terrestrial habitat is immediately at hand. Thus emerging newly transformed individuals are difficult to capture and are likely under-represented in our samples. Conversely, the subadult terrestrial component of the Quarry population remains largely uncensused, because it lives in terrain not lending itself to drift fence construction or successful jacklighting.

Habitat Use

Adult *A. macrodactylum* are almost entirely nocturnal, tend to be found below ground during the day, and are rarely active at the surface except during the breeding season (Anderson 1967; Nussbaum et al. 1983; Beneski et al. 1986; Leonard et al. 1993; Russell and Bauer 1993). We found that nocturnal above-surface activity was common throughout the active seasons in both years. Uncharacteristically rainy springs and summers may have accounted for a great deal of this surface activity. More normally dry conditions could partially justify the Red List status of this species in Alberta. Data on the Lake Linnet breeding population in Waterton National Park, Alberta (Fukumoto 1995) indicate that even when the population is large, this species is not commonly encountered at the surface outside of the breeding season.

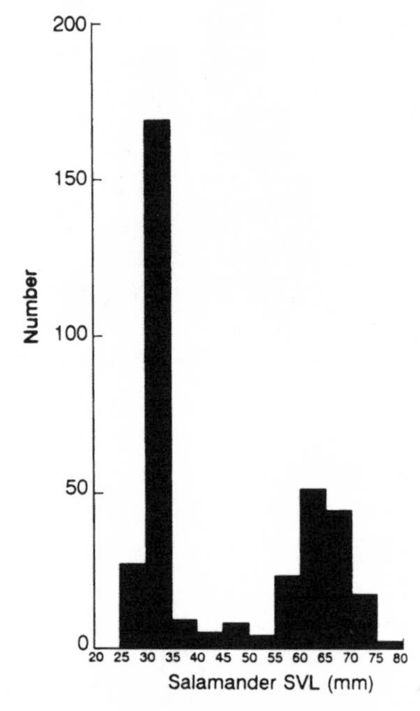

Figure 3. *Snout-vent length distributions of salamanders caught at Quarry Ponds in 1993.*

Population Sizes

The 881 captures at Lafarge and 401 captures at Quarry indicate large populations. Toe clips on most if not all individuals should still have been readable at the end of the study period (Sheppard 1977) and the 6 to 10 yr lifespan in these populations is long enough (Russell et al. 1996) that normal mortality would not have accounted for the paucity of recaptures. The number of recaptures at Lafarge was small in 1992 and, though more evenly distributed, decreased in 1993 (Fig. 1). This also suggests a very large population, even if all individuals had equal probabilities of being captured and recaptured.

The interstices between the cobbles lining the Lafarge Borrow Pit afford a potentially very large and easily accessible subsurface habitat for salamanders, and salamanders were easily found beneath the rocks. Fluctuations in capture number within years could represent movement in and out of this refugium. The diminution in the capture number between 1992 and 1993 may be real but was more likely attributable to rising groundwater levels at the site in 1993.

The increase in body size of salamanders at Lafarge between 1992 and 1993 may represent growth in the individuals of a particular, numerically large cohort. This is most easily imagined as the product of one unusually successful recruitment and implies that the size of this population can fluctuate as a result of such events. This sort of fluctuation is consistent with the estimated generation time (from skeletochronology of this species in the upper Bow Valley (Russell et al. 1996), and is to be expected in a salamander species with an aquatic larval stage (Semlitsch 1983; Hairston 1987; Pechman et al. 1991), particularly at localities, such as this, where the breeding ponds are shallow and water levels vary unpredictably but in synchrony with the water table.

Post-metamorphic salamanders at Lafarge live immediately about the breeding habitat, and all size classes, with the possible exception of the very recent metamorphs, are readily captured on the surface throughout the active season. Anderson (1967) found that recently metamorphosed *A. macrodactylum* stayed close to the breeding pond for some time, moving away with the summer rains, but this seems to be due to the dryness of the surrounding habitat after emergence. The terrestrial habitat at Quarry is at least 60 m from the breeding habitat and affords good surface cover. Thus, only breeding adults and new metamorphs are generally encountered.

Beneski et al. (1986) found that immigration into a breeding pond by adult *A. macrodactylum* was brief and concerted, but that emigration was episodic. We also noted that numbers of adults around the Quarry Ponds fluctuated strongly during the breeding season. Beneski et al. (1986) captured 2030 individuals and estimated a breeding population of 3141 salamanders by correcting for fence trespass. We captured fewer breeding adults at Quarry, but we used an incomplete fence. Fukumoto (1995) captured 655 *A. macrodactylum* in 1994 at Lake Linnet in Waterton Lakes National Park, Alberta, and also recorded a low recapture rate of 36 out of 619 adults (5.0%). A Schumacher-Eschmayer estimate of the breeding adult population size for the locality for that year was 3856 individuals (95% confidence interval = 3274 to 4690; Fukumoto 1995). This population is at a locality similar to Quarry, with a breeding pond widely separated from the terrestrial habitat.

The Status of *Ambystoma macrodactylum* in Alberta

The 2 populations of *A. macrodactylum* we have studied are large, and there is evidence of at least 1 other large population of this species in Alberta (Fukumoto 1995). *Ambystoma macrodactylum* is spottily distributed along the Front Range in western Alberta (Russell and Bauer 1993) and generally considered to be rare (Anonymous 1991). This impression is likely due to its retiring habits and small size (Bishop 1947; Ferguson 1961; Anderson 1967; Nussbaum et al. 1983; Cook 1984; Leonard et al. 1993; Russell and Bauer 1993) as well as its localized distribution (Russell and Bauer 1993). Other large populations may be present in the southern part of the Alberta range. Its persistence in disturbed habitats suggests that it will recover rapidly from anthropogenic disturbances.

Long-term Monitoring

Our study illustrates the importance of intensive long-term population studies to the assessment of status for rare and/or endangered species, particularly when the species is cryptic. Cursory sampling might not have revealed the large population at Lafarge. Similarly, any size estimate based on our 1992 data from Quarry would have suggested a smaller population than actually present. Pechmann et al. (1991) demonstrated that data on population sizes accrued over long time bases are necessary to address the question of whether or not an amphibian population is declining. Considerably more field work at these 2 sites is necessary before any question of trends can be addressed, although we can now say that the species is locally common in the upper Bow River Valley.

Acknowledgments

This study was funded partly through a contract awarded by Alberta Lands, Forestry and Wildlife, Fish and Wildlife Division to GLP (divisional contract number 8-119-1), partly through two Recreation, Parks, and Wildlife Foundation grants to APR, and partly through an NSERC operating grant to APR. SMW was funded partially through an NSERC summer studentship. We are grateful to the management of Lafarge Canada Inc.'s Exshaw plant (particularly Paul Fitzpatrick, Didier Catillon, and Dave McKinnon) for providing funding for a field vehicle for both years, and to Frank Staubitz and Bill Marshall for permission to park our field headquarters in the Willow Rock campground for the summer of 1992. Jon Jorgenson of the Alberta Fish and Wildlife Division's Calgary Office must be thanked for the use of the trailer in 1992. Francis Cook of the National Museum of Nature kindly sent us a copy of his report on the Lafarge Borrow Pit. Dean Hall shared his research on Long-toed salamander skeletochronology with us. We are obliged to the Kananaskis Field Station and its staff for accommodations and facilities in 1993. Comments by two anonymous reviewers and by David Green upon an earlier draft of this paper greatly improved the final product.

Literature Cited

Anonymous. 1991. The status of Alberta wildlife. Wildlife Management Branch Status Report. Edmonton, AB: Alberta Forestry, Lands and Wildlife. Publication Number I/413.

Anderson JD. 1967. A comparison of the life histories of coastal and montane populations of *Ambystoma macrodactylum* in California. American Midland Naturalist 77:323–355.

Begon M. 1979. Investigating animal abundance. Capture-recapture for biologists. London: Edward Arnold Limited.

Beneski JT Jr, Zalisko EJ, Larsen JH Jr. 1986. Demography and migratory patterns of the eastern long-toed salamander, *Ambystoma macrodactylum columbianum*. Copeia 1986:398–408.

Bishop SC. 1947. Handbook of Salamanders. Ithaca, NY: Comstock.

Cook FR. 1984. Introduction to Canadian amphibians and reptiles. Ottawa: National Museum of Natural Sciences.

Ferner JW. 1979. A review of marking techniques for amphibians and reptiles. Society for the Study of Amphibians and Reptiles Herpetological Circular 9:1–41.

Ferguson DE. 1961. The geographic variation of *Ambystoma macrodactylum* Baird, with the description of two new subspecies. American Midland Naturalist 65:311–338.

Fukumoto JM. 1995. Long-toed salamander (*Ambystoma macrodactylum*) ecology and management in the Waterton Lakes National Park [thesis]. Calgary, AB: University of Calgary.

Hairston NG Sr. 1987. Community ecology and salamander guilds. Cambridge, UK: Cambridge University Press.

Hayek L-AC. 1994. Removal sampling. In: Heyer WR, Donnelly MA, McDiarmid RW, Hayek L-AC, Foster MS, editors. Measuring and monitoring biological diversity. Standard methods for amphibians, Washington: Smithsonian Institution Press. p 201–205.

Husting EL. 1965. Survival and breeding structure in a population of *Ambystoma maculatum*. Copeia 1965:352–362.

Leonard WP, Brown HA, Jones LLC, McAllister KR, Storm RM. 1993. Amphibians of Washington and Oregon. Seattle: Seattle Audubon Society.

Nussbaum RA, Brodie ED Jr, Storm RM. 1983. Amphibians and reptiles of the Pacific Northwest. Moscow, ID: University of Idaho Press.

Pechmann JHK, Scott DE, Semlitsch RD, Caldwell JP, Vitt LJ, Gibbons JW. 1991. Declining amphibian populations: the problem of separating human impacts from natural fluctuations. Science 253:892–895.

Rice WR. 1989. Analyzing tables of statistical tests. Evolution 43:223–225.

Russell AP, Bauer AM. 1993. The amphibians and reptiles of Alberta: a field guide and primer of boreal herpetology. Calgary, AB and Edmonton, AB: University of Calgary Press and University of Alberta Press.

Russell AP, Bauer AM, Powell GL, Hall DR. 1996. Growth and age in Alberta long-toed salamanders (*Ambystoma macrodactylum krausei*): a comparison of two methods of estimation. Canadian Journal of Zoology 74:397–412.

Semlitsch RD. 1983. Structure and dynamics of two breeding populations of the eastern tiger salamander, *Ambystoma tigrinum*. Copeia 1983:608–618.

Sheppard RF. 1977. The ecology and home range movements of *Ambystoma macrodactylum krausei* (Amphibia: Urodela) [thesis]. Calgary, AB: University of Calgary.

Vial JL, Saylor L. 1993. The status of amphibian populations: a compiliation and analysis. Corvallis, OR: IUCN—The World Conservation Union/Species Survival Commission Declining Amphibian Populations Task Force. Working Document 1. 98 p.

Wilkinson L. 1990. SYSTAT: the system for statistics. Evanston, IL: SYSTAT Incorporated.

©1997 by the Society for the Study of Amphibians and Reptiles
Amphibians in decline: Canadian studies of a global problem. David M. Green, editor.
Herpetological Conservation 1:45–56.

Chapter 5

TEMPORAL VARIATION IN ABUNDANCE AND AGE STRUCTURE IN FOWLER'S TOADS, *BUFO FOWLERI*, AT LONG POINT, ONTARIO

DAVID M. GREEN

Redpath Museum, McGill University, 859 Sherbrooke St. W., Montréal, Québec H3A 2K6, Canada

ABSTRACT.—Fowler's toad (*Bufo fowleri*) abundance was surveyed during 7 consecutive breeding seasons (1988 to 1994) at Long Point, Ontario by tracking and marking all calling males within a study area along 10 km of Lake Erie shoreline. Reliable estimates of total numbers of toads, $\hat{N}$, were based upon continuous capture-recapture data (removal sampling) throughout the breeding season. The number of captured and marked males rose markedly from 12 toads in 1988 ($\hat{N}$ = 12 ± 0.0) to 294 in 1991 ($\hat{N}$ = 487 ± 48.8) then declined to 83 ($\hat{N}$ = 115 ± 12.1) in 1994. Males in 1994 (mean SVL = 54.3 mm) and 1989 (mean SVL = 56.6 mm) were significantly larger than males in other years. In 1992, when toads were abundant, 13 of 54 adult males were 1-yr old but in 1994 only 4 of 81 males were 1-yr old. This evidence of fewer but larger and older toads in 1994 implies that a poor summer growth rate in 1993, which would diminish the number of yearling males reaching breeding condition, may have determined the decline in calling males. However, this cannot be correlated with total abundance of post-metamorphic toads. Because age structure among adults is evidently variable and chorus behaviour is weather-dependant, chorus size is not directly indicative of population size. Spring field studies and reference to weather records failed to identify factors correlated with the either the increase in adult males beginning in 1990 to 1991 or the decline of 1994.

RÉSUMÉ.—L'abondance des crapauds de Fowler (*Bufo fowleri*) a été évaluée pendant 7 saisons consécutives de reproduction (entre 1988 et 1994), à Long Point (Ontario), en retraçant et marquant tous les mâles qui coassaient dans un rayon de 10 km. Des évaluations fiables du nombre total de crapauds, $\hat{N}$, ont pu être faites au moyen de données continues de marquage/recapture (échantillon d'enlèvement), pendant toute la saison de reproduction. Le nombre de mâles capturés et marqués a progressé de façon spectaculaire, passant de 12 crapauds en 1988 ($\hat{N}$ = 12 ± 0,0) à 294 en 1991 ($\hat{N}$ = 487 ± 48,8) pour revenir à 83 ($\hat{N}$ = 115 ± 12,1) en 1994. Les mâles de 1994 (LMC moyen = 54,3 mm) et de 1989 (LMC moyen = 56,6 mm) étaient beaucoup plus gros que les mâles des autres années. En 1992, année où les crapauds étaient particulièrement abondants, 13 des 54 mâles adultes avaient 1 an, contre seulement 4 sur 81 mâles en 1994. Le moins grand nombre de crapauds plus âgés et plus gros en 1994 traduit un taux de croissance estival médiocre en 1993 qui a contribué à faire décroître le nombre de mâles d'un an qui atteignent leur maturité sexuelle, ce qui peut avoir joué un rôle clé dans le déclin des populations de mâles reproducteurs. Toutefois, ce phénomène ne peut être corrélé avec l'abondance totale de crapauds postmétamorphiques. Dans la mesure où la répartition par âge des adultes est variable et que les coassements dépendent des conditions météorologiques, le nombre de crapauds qui coassent n'est pas directement révélateur de la taille de la population. Des études sur le terrain menées au printemps et l'étude des relevés météorologiques n'ont pas permis d'identifier de facteurs ayant un rapport avec l'augmentation du nombre de mâles adultes à partir de 1990 et 1991 ou avec leur déclin en 1994.

Fowler's toads (*Bufo fowleri*) are at the northern limit of their range along the Canadian shore of Lake Erie. Once distributed along the whole northern shoreline, they have not been recorded at Pt. Pelée or on Pelée Island in western Lake Erie since the 1940s (Green 1989). The largest remaining population is at Long Point, a 35 km sand spit extending eastward into the lake (Green 1989). Museum and field records for Long Point, although patchy and largely anecdotal, show that *B. fowleri* appeared to be reasonably abundant in the late 1970s and early 1980s but their numbers precipitously declined around 1986 (Green 1989, 1992; Oldham 1988; Oldham and Sutherland 1986).

Deciphering population fluctuations from declines in amphibian abundance is a critical problem (Wake 1991; Pechmann and Wilbur 1994; Pechmann et al 1991; Blaustein 1994; Green, this volume). It has been said that before we may understand the reasons behind declines, we need to understand the normal parameters of population abundance (Vial and Saylor 1993) and measure them effectively (Heyer et al. 1994). This is a formidable problem. Demographic parameters such as sex-ratio and age-, size-, or sex-specific survivorship rates may differ significantly for different species, for different populations of the same species, or even at different times within a single population.

At the simplest level of analysis, the size (N) of an animal population at some time (t) is a balance between birth (B) and immigration (I) of new individuals versus their removal by death (D) and emigration (E):

$$N_t = N_{t-1} + (B + I) - (D + E).$$

Although charting the change in size over time (ie. ΔN) is a phenomenological approach to addressing whether or not a population is in decline (Pechmann and Wilbur 1994), estimating N is not easy. Like many North Temperate anurans, *Bufo fowleri* is a short-lived species with high fecundity (B) and a high mortality rate (D). The extremely low probability of surviving to maturity is counteracted by the laying of thousands of eggs by every female that does manage to survive (Clarke 1977). Breden (1988) calculated that 0.1% of all eggs laid survive to age of 1st reproduction in a population of *B. fowleri* in northern Indiana. With this Type III survivorship curve, the actual number of adult individuals can be profoundly affected by changes in B and D due to unpredictable events.

The accurate estimation of N, particularly of male numbers in a breeding population, is crucial for the validity of certain amphibian monitoring methods. Road call-count surveys that rely upon observations of calling intensity (Lepage et al, this volume; Bishop et al., this volume) necessarily rely exclusively on data from males. Chorus intensity is frequently estimated subjectively on a scale of 0 (silence) to 3 (full chorus of overlapping calls) by listeners. In order to compare estimates among populations and years, the analysis of this sort of survey data must assume a constant relationship between the number of callers and the total number of individuals within the population. Estimation of N by call-count surveys is therefore subjective and laden with untested assumptions. Furthermore, the demographic characteristics of all populations, such as age structure, are implicitly assumed to be the same and stable from year to year. If age structure among the callers is not constant then they will not represent a constant proportion of the overall population.

*Fowler's toad (*Bufo fowleri*). Photo by David M. Green.*

I have annually surveyed the population of *B. fowleri* at Long Point from 1988 to 1994, inclusive, seeking to determine population sizes and to chronicle fluctuations in their abundance (Laurin and Green 1990; Green 1992). My students and I searched for every calling male Fowler's toad in a predefined study area along 10 km of Lake Erie shoreline. This was feasible with intensive effort, because males are easily visible at breeding sites and can be tracked by their persistent calling. They also return to the same general calling site on successive nights (Laurin and Green 1990). Because females are cryptic in behaviour and spend less time at the breeding sites, they cannot be censused as intensively or thoroughly. Nevertheless, our capture-recapture data from females show no evidence for a departure from a 1:1 sex ratio among breeding adults. This time-intensive approach has enabled evaluations of numbers of males from year to year. I present the evidence of significant year to year changes in the overall size distribution and age structure of the Long Point population of *B. fowleri*.

MATERIALS AND METHODS

The study area was a region at the western base of Long Point, Ontario (42°35′ N, 80°25′ W). It extended from the western end of Hastings Drive (Hahn Beach), east 10 km to the Thoroughfare Point Management Unit of the Canadian Wildlife Service. Strips of dune and marshland run parallel to the lakeshore. Along much of the dune area are cottages and roads. The area of settlement extends north in some places into regions that were previously marshland, however most of the marsh and much of the dune and foreshore is protected land. Nowhere is the region more than about 500-m wide. The individual study sites where toad choruses were encountered were as I previously reported (Green 1992). Some of these chorus sites were in the marshes amid bulrushes, cattails, or emergent sedges while others were at ponds formed in shallow sandy depressions.

The methods used to survey the toads, though steadily refined over the years from 1988 through 1994, did not differ substantially from those described by Laurin and Green (1990) and Green (1992). Each spring I employed the help of students and volunteers to find, hand-capture, and mark every breeding male toad in the study area. The area was traversed by car, by bicycle, and on foot. All study sites were inspected nightly throughout the entire breeding season in May and early June (Table 1). We measured snout-vent length (SVL) of all adult toads with dial calipers and marked them by toe-clipping (Green 1992). We employed an additive numbering scheme which relied on the toes of the hind feet and refrained from cutting the innermost front toe (the thumb) which is important to males in amplexus. We could number toads consecutively from 101 to 1599 while cutting no more than 2 toes on each hind foot and 1 toe on each front foot. This provided enough numbers for several years' study (longer than the life-span of the toads, as we discovered) before having to re-use numbers. Toads' body temperatures were taken with a quick-reading mercury thermometer (Miller and Weber Co.) or an electronic thermometer with a T-type thermistor (Omega Corp). Daily air temperatures and rainfall were recorded throughout each study period using a drum-type recording thermometer (Taylor Weather Hawk) and a tipping-bucket electronic rain gauge (Rainwise). Weather records for the whole year were obtained for the nearby St. Williams weather station from Environment Canada.

Table 1. *Survey dates, numbers of individuals captured and identified, recaptures, and mean snout-vent lengths (SVL) of male* Bufo fowleri *at Long Point, Ontario, 1988 to 1994.*

	1988	1989	1990	1991	1992	1993	1994
Survey start date	2 May	3 May	3 May	4 May	3 May	5 May	1 May
Survey end date	2 June	20 June	6 June	12 June	10 June	9 June	7 June
Emergence of toads	15 May	17 May	9 May	10 May	10 May	9 May	20 May
Number of activity nights	13	21	9	25	18	21	13
Number of males captured	12	46	61	294	229	265	83
$\hat{N}_{males}$(SE)	12 (0.0)	54 (6.5)	248 (82.4)	487 (48.8)	367 (32.7)	515 (61.9)	115 (12.1)
Previous year recaptures	N/A	1	2	1	8	13	1
Yearly recapture rate	N/A	0.083	0.043	0.016	0.027	0.057	0.004
Mean SVL (SD)	52.5 (3.6)	56.6 (3.4)*	51.9 (3.6)	51.7 (2.7)	52.4 (2.8)	50.9 (3.0)	54.3 (2.2)*

*Toads were larger than in other years ($P < 0.001$)

Estimation of population size during each breeding season used the nightly individual capture/recapture data (Donnelly and Guyer 1994). Computation was done using the 1994 update of the program CAPTURE (White et al. 1982; Rexstad and Burnham 1991). The assumptions that capture probabilities varied with time and with individual, the M_{th} model of Chao et al. (1992), were used for all years' data.

Ages were determined for samples of male toads from 1992, a year of high abundance, and from 1994, a year of declined numbers, using skeletochronology (Acker 1993; Kellner and Green 1995; LeClair and Castanet 1987). Toes clipped for marking the animals were collected and stored in 10% buffered formalin in individually marked microfuge tubes. After storage for 3 mo, these toes were stripped, decalcified, sectioned, and stained with hematoxylin in order to visualize lines of arrested growth (LAGs) in the bones (Kellner and Green 1995). Each LAG corresponded to a period of winter hibernation. Age of the animal could thus be read directly from the number of LAGs (Leclair and Castanet 1987). Fifty-four males from the 1992 season and 81 males from the 1994 season were aged in this way (N. Pelzer and D.M. Green, unpublished). Statistical procedures to detect significant differences is size and age structure (ANOVA) used SYSTAT 5.02 software.

Results

The timing of emergence from hibernation, and the dates when the toads chorused varied considerably from year to year (Table 1, Figs. 1 and 2). The toads were never observed to call below a body temperature of 14° C. The earliest dates when full chorusing commenced were 9 May in 1990 and 1993, although I had reports from local observers that toads were calling 27 April in 1990. Onset of chorus-

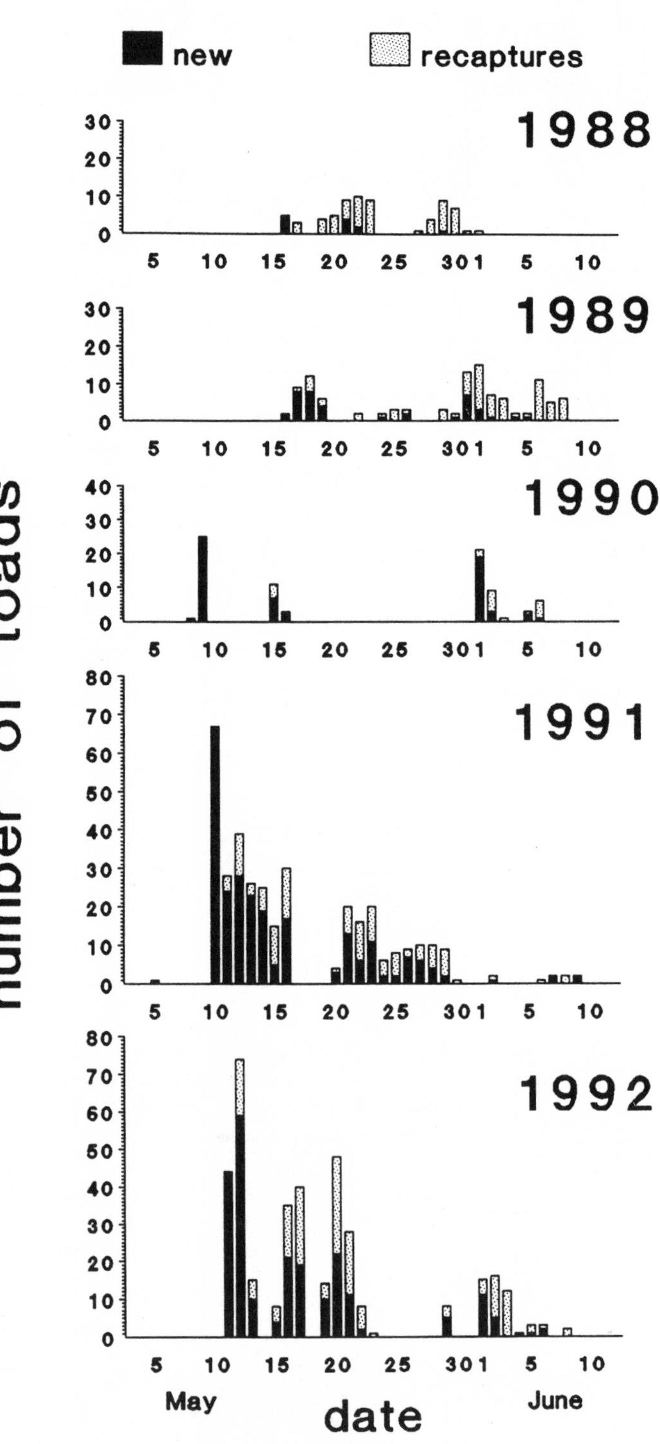

Figure 1. *Daily captures (black bars) and recaptures (stippled bars) of adult male* B. fowleri *at Long Point, Ontario in 1988 to 1992.*

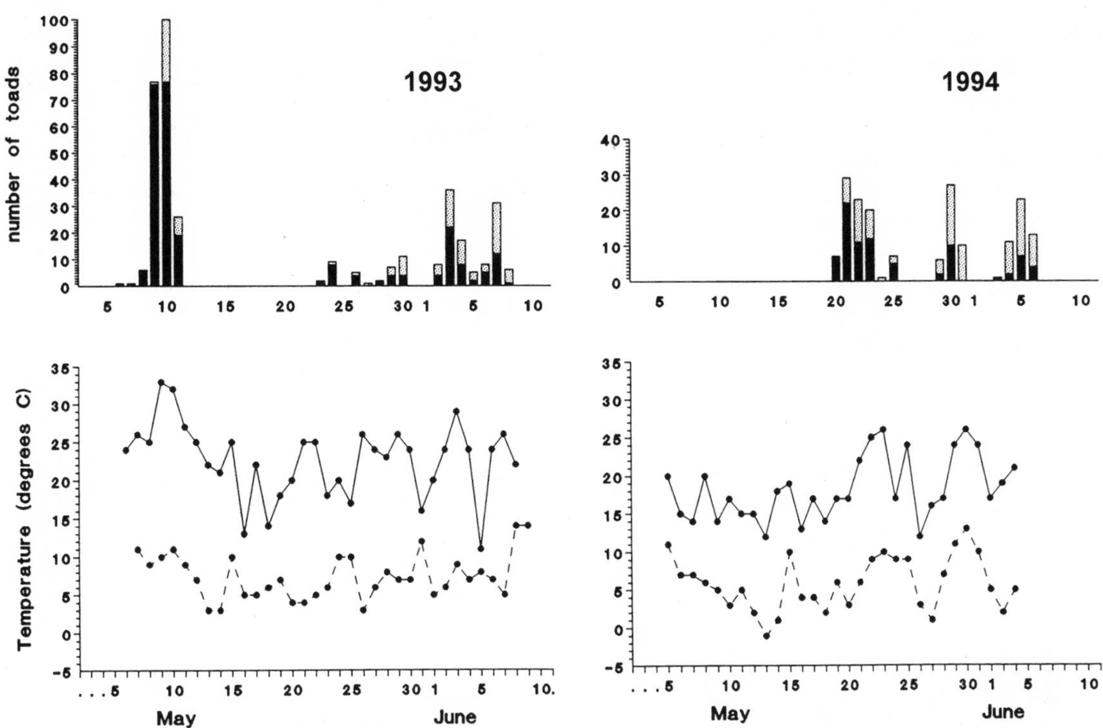

Figure 2. *Daily captures (black bars) and recaptures (stippled bars) of adult male* B. fowleri *in 1993 and 1994 in comparison to daily maximum (solid line) and minimum (dashed line) air temperatures.*

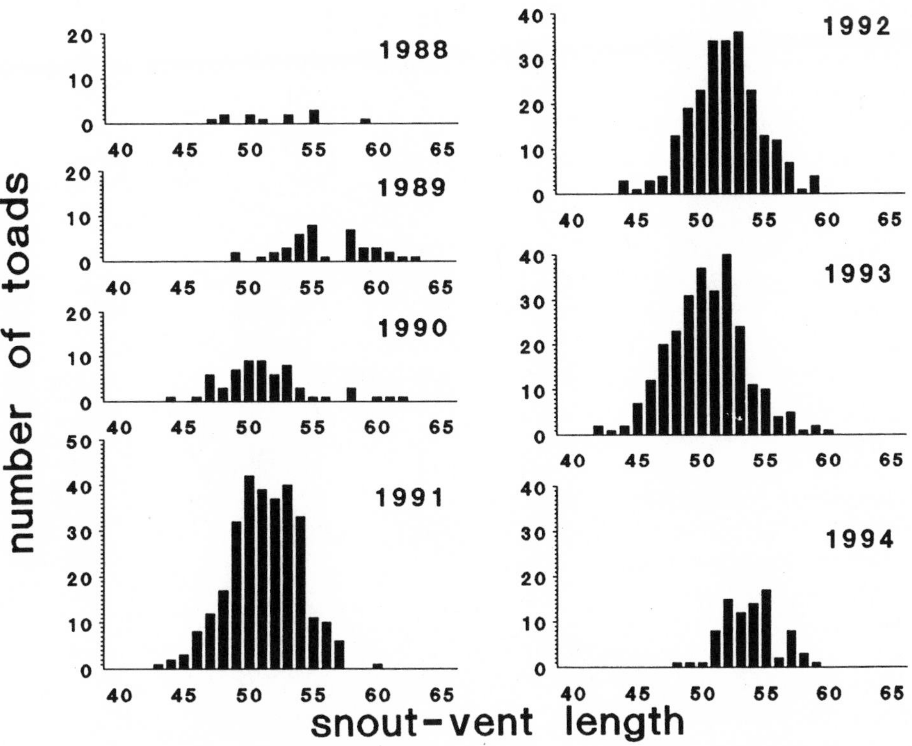

Figure 3. *Size distributions of male* B. fowleri *captured each year from 1988 to 1994. Toads in 1989 and 1994 were significantly larger than toads in other years (*P *< 0.001).*

ing was delayed until 20 May in 1994. This variation in timing was entirely due to the weather. Cold weather spells in every year also interrupted breeding activity after it had commenced. Notably long interruptions in toad activity occurred from 16 to 31 May 1990, and 11 to 22 May 1993, resulting in 2 clearly separated waves of chorusing activity. The number of nights that toads were active during the breeding season varied considerably, from 9 nights in 1990 to 25 nights in 1991 (Table 1). However, there was no strong correlation between number of active nights and numbers of toads caught (r = 0.672, P = 0.098). Capture-recapture rates during the breeding season showed that we were able to catch and mark most breeding male toads (Figs. 1 and 2) by the end of the breeding season in most years. The total numbers of males captured and marked ranged from the low of 12 in 1988 to a high of 248 in 1991 (Table 1). Low numbers prevailed from 1988 to 1990, there was a period of relatively high numbers during 1991 to 1993, and a decline in 1994 (Fig. 3). Population size estimates (Fig. 4) closely paralleled the total number of captures each year.

Year-to-year recapture data were unencouraging. Although marks could easily be read throughout a breeding season, toes frequently regrew after an intervening year. Frequently, marks from a previous year were difficult to read beyond noting that a particular toad had apparently been captured before. Year-to-year recaptures (Table 1) were too rare to evaluate their significance. The rate of return of marked toads, i.e. their survivorship, s_x, derived in this way was at most 8.3% (1989), and at the least 0.04% (1994).

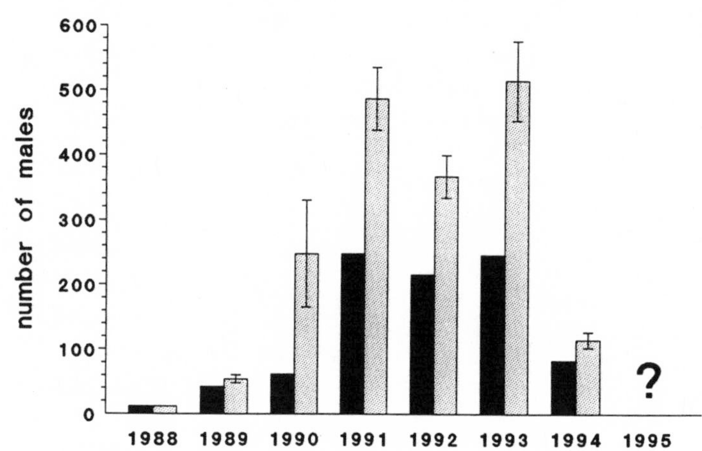

Figure 4. *Numbers of male* B. fowleri *captured (black bars) and estimated (stippled bars) in the breeding population at Long Point, Ontario, 1988 to 1994.*

Among all male toads caught and measured from 1988 through 1994 (n = 984), the mean snout-vent length was 52.2 ± 3.1 mm (SD) but that average varied significantly from year to year (Fig. 3; Table 1), over a range of 5.7 mm. The largest toads overall were those of 1989, averaging 56.6 ± 3.4 mm. They were followed in size by those of 1994, which averaged 54.3 ± 2.2 mm. Toads in both of these years were significantly larger than in other years ($P < 0.001$). The smallest average sizes of toads were found in 1993 ($\bar{x}$ = 50.9 ± 3.0 mm) and 1991 ($\bar{x}$ = 51.7 ± 2.7 mm). A significant difference in size was even detected between the large samples of 1993 and 1992 ($P < 0.001$), 1992 having the larger toads. A regression between yearly population size and the yearly mean SVL of the toads yielded a significant correlation coefficient: r = 0.58.

Age structures determined by skeletochronology for the 1992 and 1994 adult male cohorts were markedly different (Fig. 5). As presented by Kellner and Green (1995), 13 of 54 adult male toads (24%) were 1-yr old in 1992, 29 were 2-yr old (54%) and 12 were 3-yr old (22%). There was one 4-yr-old toad. But in 1994, there were only four 1-yr-old adults out of 81 toads (5%). The great majority were 2- or 3-yr old and only one 4-yr old was found. There were significant differences in size between 1-, 2-, and 3-yr-old toads in 1992 but not in 1994. That the toads in 1994, on the whole, were larger than in 1992, was due mainly to the significantly smaller sizes of the 2-yr olds of 1992 compared to the 2-yr olds of 1994 ($P < 0.001$), as well as to the smaller sizes of the significant numbers of yearling adults in 1992.

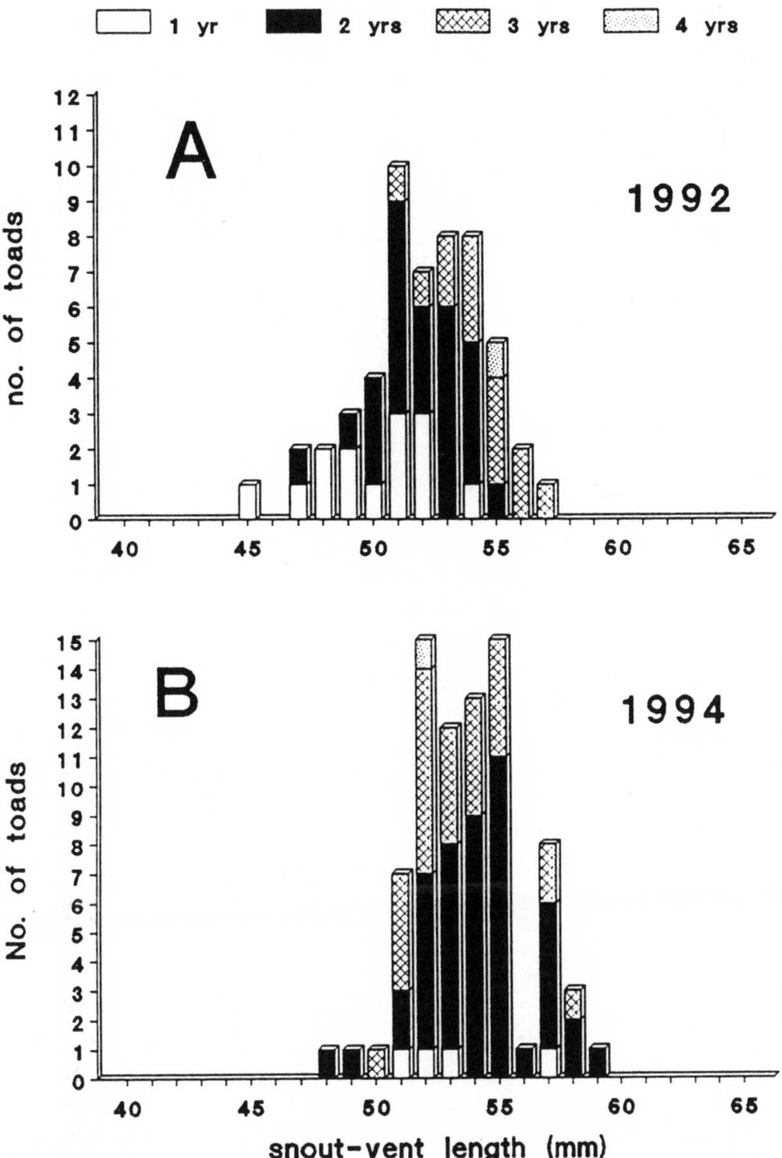

Figure 5. *Age/size distributions of adult male* B. fowleri. *Yearling adults comprised 25% of the sample in 1992 (*n *= 53; Kellner and Green 1995) and 5% of the sample in 1994 (*n *= 81; N. Pelzer and D. Green, unpublished data).*

Discussion

The *B. fowleri* at Long Point may reach sexual maturity at 1-yr old (Kellner and Green 1995). This is younger than Breden (1987, 1988) could discern for Fowler's toads at Indiana Dunes on Lake Michigan using only capture-recapture methods, and younger than generally acknowledged for *Bufo* (Blair 1953; Clarke 1975). Breden (1988) calculated a life table assuming a stable age structure, but this assumption does not hold at Long Point. Anuran populations are known to vary in the age of attaining sexual maturity (Turner 1960; Augert and Joly 1993) and demonstrate considerably individual variation in growth rates (Halliday and Verrell 1988). Variations in age at maturation in toads that have been related to latitude and altitude of populations (Acker, et al. 1986; Hemelaar 1988; Verrell and Francillon 1986) may ultimately be related to relative growing conditions. Male *B. bufo* were reported to achieve sexual maturity at 2- to 3-yr of age in the Netherlands but at 4 to 7 yr in Norway, and 6 to 9 yr in Switzerland, where the growing seasons are shorter (Hemelaar 1988). My evidence shows that this variability may occur also between years within a single population and that toads clearly do not necessarily mature at a particular age.

Adult size in toads also has considerable variability. The mean size of adult male toads at Long Point over the 7 yr of study was 52.2 mm SVL, but varied over a range of nearly 6 mm. Breden (1988) derived an average SVL of 58.9 mm for male *B. fowleri* at Indiana Dunes while Hranitz et al. (1993) found average SVL among males varied from 38.3 mm to 57.2 mm for different populations in 2 different yr in Virginia. As with the toads I have studied, Hranitz at al. (1993) reported significant, negatively correlated differences in average adult size vs. population size and the same phenomenon has been seen among *B. bufo* populations in southern Bavaria (J. Kuhn, pers. comm.).

Anurans grow quickly until sexual maturity is reached and then slow or even stop growth, depending upon conditions (Blair 1953; Brown and Acala 1970; Durham and Bennett 1963; Turner 1960; Underhill 1960). Lack of a correlation between age and size in amphibians mostly reflects the variance in postmetamorphic growth rates (Halliday and Verrell 1988; Hemelaar 1988). Growth is very rapid, and variable, in *B. fowleri* immediately post-transformation. Labanick and Schleuter (1976) found that toadlets grew from 9 mm to an average of 47 mm after 2 mo, a rate of 0.36 mm/day. Clarke (1975) found a postmetamorphic growth rate 0.42 mm/day and Breden (1988) found that toadlets grew at an average rate of 0.61 mm/day over 2 mo and virtually ceased growing upon attaining sexual maturity. At Long Point, the natural growth rate in 1993 averaged 0.37 mm/day from 19 June to 25 September (D.M Green and H.-S. Chan-Tang, unpublished).

Because the variance in size among adults depends mainly upon early growth (Halliday and Verrell 1988), cohorts resulting from eggs laid at different times may have very different maturation rates and ages of maturity (Turner 1960). In years such as 1990 and 1993 at Long Point, when there were particularly distinct waves of early and late breeding, cohorts of toadlets were separated in age by several weeks. This may profoundly affect when individuals may attain sexual maturity during a growing season of unpredictable length. The age at which sexual maturity is reached will significantly affect final adult size and seasonal growing conditions will determine the proportions of animals maturing at particular ages. Individuals that grow sufficiently by their 1st fall will be observed to be sexually active the following spring as yearlings. Members of the same year class which do not reach sufficient size in 1 yr will have a 2nd yr of growth available before sexual maturity and so should be larger. Therefore, under good conditions at Long Point, there should be more, smaller animals in a chorus and a positive correlation between size and age, as observed in 1992. Conversely, under less optimal growing conditions, more toads will suffer delayed maturation. This variation in age at 1st reproduction might conceivably help protect it from bad years via the "storage effect" (Warner and Chesson 1985) but these late-maturing toads will also be subject to an additional year's mortality risk before joining the adult chorus. The result will be a smaller chorus of larger toads and no correlation between age and size, which is precisely what we observed in the 1994 chorus. Thus the size distribution and age structure of the breeding chorus can be read forensically to give evidence of the type of growing conditions prevailing the previous year. However, I could find no relationship between weather records and the inferred relative growth rates or population sizes of the toads. Although we may be sure of the date of first emergence over the 7 yr, the end of the growing season is indefinable. Nor is there present information on total activity time of the toads during the season.

Estimating the number of breeding animals in an anuran chorus is not a trivial undertaking. Jolly-Seber methods (Pollock et al. 1990) estimate abundance from 1 capture period to the next and so are appropriate for year-to-year estimations where recapture rates are high. This implies a reasonable longevity among individuals, reasonably stable age structure, and reasonably high survivorship of marked animals, none of which apply to a small anuran. Capture-recapture and removal methods embodied in the program CAPTURE are more appropriate for a breeding chorus of toads and can make use of detailed, nightly capture information for marked individuals. We can assume that the population of adult toads is closed, a key assumption of the method, due to their high site fidelity over the course of the breeding season. But we cannot assume equal catchability per individual. Some animals are more persistent than others and may call on many more active nights than others. Nor may we assume equal catchability over time. Toads do not emerge synchronously and their activity, even in the heart of the breeding season, is limited by weather conditions. The numbers of active toads was consistently highest during the 1st few days of the breeding season, so long as the toads could maintain an active body temperature above 14° C (Figs. 1 and 2). Yet every year by mid-June, even at evening temperatures over 20° C range, toad chorus activity had declined and virtually ceased. We do not have evidence of any variable behavioural responses to capture such as "trap-happiness" or increased shyness among previously captured animals.

It is clear that the number of calling males and the total number of post-metamorphic toads in a population are 2 different things. Percentages of yearlings among calling males differed in 1992 and 1994. A rise in the numbers of calling males may not necessarily reflect an increase in the total number of postmetamorphic individuals since it may be due simply to a large cohort of precociously mature yearlings. Increased breeding population size may have eventual repercussions but because of stochastic variability in the whole system, the downstream effect on total population size is not predictable.

Even the intensive, daily sampling, such as we have done, cannot absolutely be relied upon to give unambiguous abundance estimates without intensive analysis. They depend upon conformity to various assumptions, especially population closure and aspects of the catchability of animals. It seems unlikely, therefore, that any extrapolations of abundance from the crude chorus intensity scores possible from call-count surveys will be of much real value. Age structure inconsistency poses still a further problem for such data to estimate population size variation, and therefore to monitor trends. Call-count surveys implicitly assume stable age structure. But age structure in anuran populations, particularly among short-lived species such as *Bufo fowleri*, *Pseudacris crucifer*, *P. triseriata*, and *Rana sylvatica*, is unlikely to remain constant and therefore even if the data were otherwise rigorous and consistent they may still not accurately reflect population size from year to year. Unless the age structure of populations is also known, call surveys may only be reliable to produce presence/absence information. This is a problem beyond the inherent difficulty of obtaining consistent data on chorus intensity from volunteers or the inability to discern variable numbers of individuals in full choruses due to the saturation of chorus intensity beyond a certain density of individuals (Arak 1983).

Year to year variability in age structure, growth rate, age-specific mortality, and age at maturity mean that there is no single, stable life history to characterize a population such as these toads. There can be no one life table for a fluctuating population (Shirose and Brooks, this volume). This population, and many other North Temperate anuran populations, likely never achieves stability since it may never approach the carrying capacity of the environment. For the toads at Long Point, there are no evident limitations on breeding sites, hibernation sites, or food. Yet numbers of toads careen from one unpredictable level to the next. They are genetically depauperate compared to sympatric *Bufo americanus* or to *B. fowleri* south of Lake Erie (Green 1984) and they may have been wiped out completely more than once at Long Point, as almost happened in the mid-1980s.

Although the yearly breeding choruses may be considered closed for intensive sampling, the population as a whole in my study area is not closed. If the low year-to-year recapture rate truly reflected survivorship of about 5% in a closed population, then we would expect to see a 1:20 ratio of 3-yr-old to 2-yr-old toads. The true ratio, for both 1992 and 1994 was closer to 1:2. Marked animals must be leaving and new animals arriving each year. This cannot be discounted as a factor in population fluctuations (Breden 1987, 1988; Clarke 1974). In my study area, immigration would be dependent upon the numbers of toads in the plausible source population immediately further east along the point. Toads on the Courtright Ridge east of the study area were observed in much greater densities than at our study sites in 1989 (D.M. Green and D. Cantin, unpublished). Immigrating toads did not recolonize Hahn Beach at the western end of the study area until 1991.

Consideration of amphibian population declines is complicated by natural fluctuations in abundance which are largely unpredictable (Pechman and Wilbur 1994; McCoy 1994). This population of *Bufo fowleri* has clearly risen and fallen over 7 yr. There may be several plausible reasons, including variable effects of temperature, predation, and habitat alterations due to succession. *Bufo fowleri* in Ontario are at the northern limit of their range and are thus approaching the limits of their habitat tolerance. Degradation of habitat is often justifiably implicated in population declines (Blaustein et al. 1994; Johnson 1992). However, human disturbance does not appear to have dramatically changed at Long Point over the past decade. Cottages are confined to a limited area at the base of the point and most of the point is natural preserve.

Presumably normal fluctuations in abundance may be driven both by catastrophic conditions, such as severe storms, and by the more usual range of biotic and abiotic variation. One unpredicted event in particular profoundly influenced the population size of *B. fowleri* at Long Point. Calamitous decline of the population immediately followed the winter of 1985 and its severe storms. An unusually thick ice build-up on Lake Erie, lifted by high water and driven by strong storm winds on 2 December 1985 destroyed and inundated large sections of dune and foreshore along Long Point. The beach along Hastings Drive was devastated; cottages were simply pushed off their foundations and blown into the marsh. Dune blow-outs created gaps in the beach face and covered the marsh with tracts of new sand. Although this occurred before the toads were surveyed on an intensive, annual basis, my own field observations for 1979 through to 1981 showed them to be abundant previous to the storm.

Yet the destructive 1985 storm at Long Point nevertheless created new, open habitat which could be occupied by the surviving toads, contributing to the subsequent rise in their numbers. These early successional stages may be necessary for certain species and the oligotrophic, sandy-bottomed, and vegetation-poor ponds were clearly favoured by the toads. Since 1985, there has be no comparable disturbance of the dunes at Long Point and much of the open habitat the toads require has been overgrown. This reliance upon early successional habitats can further contribute to the instability of the toad population.

Predator-prey interactions between the toads and the garter snakes, water snakes, and raccoons which prey upon them (Green 1989) may be significant but the impact upon toad population size is undocumented. Since the diet of the snakes, in particular, may be tied to the abundance of anuran prey (Lagler and Salyer 1945), fluctuating predator-prey cycles might be expected. Snakes were rarely encountered in the early years of this study but have since become increasingly common.

Mortality is very high in toads, especially for tadpoles and newly metamorphosed toadlets (Blair 1953; Clarke 1977; Breden 1987, 1988). A critical stage in the life history, therefore, is the breeding adult, for it is upon them that the continuance of the population relies. *Bufo fowleri* is classed as Vulnerable on Canada's endangered species list, based upon my estimations of their numbers in the early 1980's (Green 1989). But how can we fix a status upon a population when populations are inherently not stable? Should a species be delisted because its numbers rise? Should there be heightened concern and action because its numbers fall? Periodic population lows increase the risk of stochastic extinction and greatly affect genetic effective population size and the magnitude of genetic drift. Yet factors that influence the numbers of toads, or frogs, or salamanders are ever present and unpredictable. Amphibians remain subject to the stochasticity of their local environments, which remains the prime determiner of their abundance.

Acknowledgments

Permission to collect and study toads at Long Point was granted by the Ontario Ministry of Natural Resources and the Canadian Wildlife Service. I thank (in alphabetical order) Evelynne Barten, Laurie Bierbrier, Joelle Boudrault, Danièlle Cantin, Hoi-Sing Chan-Tang, Alia El-Yassir, Li Fanglin, Mandy Kellner, Erika Krausz, Geneviève Laurin, Martin Ouellet, Noemi Pelzer, Sue Porebski, Cathy Saumure, Raymond Saumure, Tim Sharbel, Tanja Tajvassallo, Joe Tambasco, Amy Vallachovic and Cliff Zeyl for help in the field and in the lab. I also thank the Long Point Bird Observatory, Long Point

Provincial Park, and the Canadian Wildlife Service Big Creek Station for their valuable assistance. I thank David Galbraith and Henry Wilbur for their comments on the manuscript. The research was funded by operating grants from the World Wildlife Fund Endangered Species Recovery Fund and NSERC Canada.

LITERATURE CITED

Acker PM, Kruse KC, Krehbiel EB. 1986. Aging *Bufo americanus* by skeletochronology. Journal of Herpetology 20:570–574.

Arak A. 1983. Male-male competition and mate choice in anuran amphibians. In: Bateson PPG, editor. Mate choice. Cambridge, UK: Cambridge University Press. p 181–210.

Augert D, Joly P. 1993. Plasticity of age at maturity between two neighbouring populations of the common frog (*Rana temporaria* L.). Canadian Journal of Zoology 71:26–33.

Blair WF. 1953. Growth, dispersal and age at sexual maturity of the Mexican toad (*Bufo valliceps* Wiegmann). Copeia 1953:208–212.

Blaustein AR. 1994. Chicken little or Nero's fiddle? A perspective on declining amphibian populations. Herpetologica 30:85–97.

Blaustein AR, Wake DB, Sousa WP. 1994. Amphibian declines: judging stability, persistence, and susceptibility of populations to local and global extinctions. Conservation Biology 8:60–71.

Breden F. 1987. Population structure of Fowler's Toad, *Bufo woodhousei fowleri*. Copeia 1987:386–395.

Breden F. 1988. The natural history and ecology of Fowler's toad, *Bufo woodhousei fowleri* (Amphibia: Bufonidae) in the Indiana Dunes National Lakeshore. Fieldiana Zoology 49:1–16.

Brown WC, Acala AC. 1970. Population ecology of the frog *Rana erythraea* in southern Negros, Philippines. Sencken. Biol. 45:591–611.

Chao S, Lee M, Jeng SL. 1992. Estimation of population size for capture-recapture data when capture probabilities vary by time and individual animal. Biometrics 48:201–216.

Clarke RD. 1974. Activity and movement patterns in a population of Fowler's toad, *Bufo woodhousei fowleri*. American Midland Naturalist 92:257–274.

Clarke RD. 1975. Post-metamorphic growth rates in a natural population of Fowler's toad, *Bufo woodhousei fowleri*. Canadian Journal of Zoology 52:1489–1498.

Clarke RD. 1977. Postmetamorphic survivorship of Fowler's toad, *Bufo woodhousei fowleri*. Copeia 1977:1489–1498.

Donnelly MA, Guyer C. 1994. Mark-recapture. In: Heyer WR, Donnelly MA, McDiarmid RW, Hayek L-AC, Foster MS, editors. Measuring and monitoring biological diversity. Standard methods for amphibians. Washington: Smithsonian Institution Press. p 183–200.

Durham L, Bennett GW. 1963. Age, growth, and homing in the bullfrog. Journal of Wildlife Management 27:107–123.

Green DM. 1982. Mating call characteristics of hybrid toads (*Bufo americanus* × *Bufo fowleri*) at Long Point, Ontario. Canadian Journal of Zoology 60:3292–3297.

Green DM. 1984. Sympatric hybridization and allozyme variation in the toads *Bufo americanus* and *B. fowleri* in southern Ontario. Copeia 1984:18–26.

Green DM. 1989. Fowler's Toads, (*Bufo woodhousii fowleri*) in Canada: biology and population status. Canadian Field-Naturalist. 103:486–496.

Green DM. 1992. Fowler's toads, *Bufo woodhousii fowleri*, at Long Point, Ontario: changing abundance and implications for conservation. In: Bishop CA, Pettit KE, editors. Declines in Canadian amphibian populations: designing a national monitoring strategy. Ottawa: Canadian Wildlife Service. Occasional Publication Number 76. p 37–45.

Halliday TR, Verrell PA. 1988. Body size and age in amphibians and reptiles. Journal of Herpetology 22:253–265.

Hemelaar ASM 1988. Age, growth and other population characteristics of *Bufo bufo* from different latitudes and altitudes. Journal of Herpetology 22:369–388.

Heyer WR, Donnelly MA, McDiarmid RW, Hayek L-AC, Foster MS, editors. 1994. Measuring and monitoring biological diversity. Standard methods for amphibians. Washington: Smithsonian Institution Press.

Hranitz JM, Klinger TS, Hill FC, Sagar RG, Mencken T, Carr J. 1993. Morphometric variation between *Bufo woodhousii fowleri* Hinckley (Anura: Bufonidae) on Assateague Island, Virginia and the adjacent mainland. Brimleyana 19:65–75.

Johnson B. 1992. Habitat loss and declining amphibian populations. In: Bishop CA, Pettit KE, editors. Declines in Canadian amphibian populations: designing a national monitoring strategy. Ottawa: Canadian Wildlife Service. Occasional Publication Number 76. p 71–75.

Kellner A, Green DM. 1995. Age structure and age of maturity in Fowler's toads, *Bufo woodhousii fowleri*, at their northern range limit. Journal of Herpetology 29:485–489.

Labanick GM, Schleuter RA. 1976. Growth rates of recently transformed *Bufo woodhousei fowleri*. Copeia 1976:824–826.

Lagler KF, Salyer JG. 1945. Influence of availability of the feeding habits of the common garter snake. Copeia 1945:159–162.

Laurin G, Green DM. 1990. Spring emergence and male breeding behaviour of Fowler's toads, (*Bufo woodhousei fowleri*), at Long Point, Ontario. Canadian Field-Naturalist 104:429–434.

Leclair R Jr, Castanet J. 1987. A skeletochronological assessment of age and growth in the frog *Rana pipiens* Schreber (Amphibia, Anura) from Southwestern Quebec. Copeia 1987:361–369.

McCoy ED. 1994. "Amphibian decline": a scientific dilemma in more ways than one. Herpetologica 50:98–103.

Oldham MJ. 1988. 1985 Ontario herpetofaunal summary. Toronto: Ontario Ministry of Natural Resources.

Oldham MJ, Sutherland DA. 1986. 1984 Ontario herpetofaunal summary. Toronto: Essex Region Conservation Authority and World Wildlife Fund Canada.

Pechmann JHK, Scott DE, Semlitsch RD, Caldwell JP, Vitt LJ, Gibbons JW. 1991. Declining amphibian populations: the problem of separating human impacts from natural fluctuations. Science 253:892–895.

Pechmann JHK, Wilbur HM. 1994. Putting declining amphibian populations in perspective: natural fluctuations and human impacts. Herpetologica 50:65–84.

Pollock KH, Nichols JD, Brownie C, Hines JE. 1990. Statistical inference for capture-recapture experiments. Wildlife Monographs 107:1–97.

Rexstad E, Burnham K. 1991. User's guide for interactive program CAPTURE. Fort Collins, CO: Colorado Cooperative Fish and Wildlife Research Unit.

Turner FB. 1960 Postmetamorphic growth in anurans. American Midland Naturalist 64:327–338.

Underhill JC. 1960. Breeding and growth in Woodhouse's toad. Herpetologica 16:237–242.

Verrell PA, Francillon H. 1986. Body size, age, and reproduction in the smooth newt, *Triturus vulgaris*. Journal of Zoology (London) 210A:89–100.

Vial JL, Saylor L. 1993. The status of amphibian populations: a compilation and analysis. Corvallis, OR: IUCN—The World Conservation Union Species Survival Commission, Declining Amphibian Populations Task Force. Working Document 1.

Wake DB. 1991. Declining amphibian populations. Science 253:860

Warner RR, Chesson PL. 1985. Coexistence mediated by recruitment fluctuations: a field guide to the storage effect. American Naturalist 125:769–787.

White GC, Anderson DR, Burnham KP, Otis DL. 1982. Capture-recapture and removal methods for sampling closed populations. Los Alamos, NM: Los Alamos National Laboratory. Publication LA-8787-NERP.

Amphibians in decline: Canadian studies of a global problem. David M. Green, editor.
Herpetological Conservation 1:57–63.

Chapter 6

FLUCTUATIONS IN A NORTHERN POPULATION OF GRAY TREEFROGS, *HYLA VERSICOLOR*

SUSAN BERTRAM[1] AND MICHAEL BERRILL

Department of Biology, Trent University, Peterborough, Ontario K9J 7B8, Canada.

ABSTRACT.—For 3 breeding seasons (1991 to 1993), we studied a gray treefrog (*Hyla versicolor)* population in southern Ontario. We identified the number of vocalizing males every night of the breeding season and recorded mating events. Breeding seasons varied from 4 to 7 wk in length and number of nights suitable for calling each year ranged from 13 to 37. The total number of calling males varied from 30 to 77, and nightly chorus attendance varied greatly. Variation in number of males chorusing nightly was not explained through air temperature fluctuations, except that when the temperature fell below 8° C, males would not call. Relatively few males mated, resulting in an apparently skewed sex ratio. Number of males calling, breeding season length, number of nights males called, nor proportion of nights within a season suitable for calling were successful in predicting female presence at the pond. We combined our data with data from 2 other intensive field studies on *H. versicolor* populations to find that seasonal population size and number of males seasonally present were positively correlated with the proportion of nights suitable for chorusing (nights that air temperature at dusk was above 8° C and males chorused) in relation to breeding season length. Variation in numbers of males calling and mating in our 3-yr study suggests that establishing decline due to human impact will require long-term intensive study.

RÉSUMÉ.—Pendant 3 saisons de reproduction (1991 et 1993), nous avons étudié une population de rainettes versicolores (*Hyla versicolor*) dans le sud de l'Ontario. Nous avons identifié le nombre de mâles qui coassaient chaque nuit pendant la saison de reproduction et répertorié les accouplements. La saison de reproduction dure entre 4 et 7 semaines et le nombre de nuits propices aux coassements chaque année varie entre 13 et 37. Le nombre total de mâles qui coassent varie entre 30 et 77 et l'assistance nocturne au choeur des mâles varie considérablement. Les fluctuations de la température de l'air ne permettent pas d'expliquer les variations dans le nombre de mâles qui coassent chaque nuit, sauf lorsque celle-ci chutait en-deçà de 8° C puisqu' à cette température, les mâles ne vocalisaient pas. Relativement peu de mâles se sont accouplés, ce qui a donné un proportion mâle/femelle apparemment dyssymétrique. Le nombre de mâles qui coassent, la durée de la saison des amours, le nombre de nuits au cours desquelles les mâles coassent et la proportion de nuits dans la saison propice aux coassements n'ont pas permis de prédire la présence des femelles dans l'étang. Nous avons combiné nos données aux données de 2 autres études intensives sur le terrain correspondant à des populations de *H. versicolor* et avons constaté que la taille de la population saisonnière et le nombre de mâles présents à chaque saison étaient positivement corrélés à la proportion de nuits propices aux coassements (nuits au cours desquelles la température de l'air à la tombée de la nuit était supérieure à 8° C et au cours desquelles les mâles coassaient), par rapport à la durée de la saison de reproduction. La variation dans le nombre de mâles qui ont coassé et se sont accouplés pendant notre étude de 3 ans donne à penser que des études intensives à long terme s'imposent avant d'attribuer le déclin des populations de *H. versicolor* à l'influence humaine.

1 Present Address: *Department of Zoology, Arizona State University, Tempe, Arizona 85287-1501 USA*

Understanding factors that regulate anuran populations during the breeding season is hampered by the intensive human effort required to determine the number of individuals visiting and breeding in a pond on a seasonal basis. To properly assess amphibian population declines (Blaustein et al. 1994; Pechmann et al. 1991; Bishop and Pettit 1992) and future research directions, we need to know the extent of variation in numbers of vocalizing males, the variation in the proportion of males that mate, and what factors influence these numbers.

A number of studies have reported numbers of adult male and female anurans at temperate ponds during the breeding season (for example, Fellers 1979; Howard 1980; Berven 1990; Elmberg 1990; Sullivan and Hinshaw 1992). In these studies, the overall sex ratio (yearly, not operational) appeared skewed toward males in most years. Our study addressed variation in male and female attendance of gray treefrogs, *Hyla versicolor,* at a breeding pond at the northern edge of the species' range.

Hyla versicolor ranges from the southern United States northward to central Ontario. It breeds over 4 to 6 wk in late spring and early summer. Both males and females remain in trees surrounding the pond during the day and males move to the pond on each night suitable for chorus attendance. Males select calling sites and vocalize to attract females and maintain inter-male distances (Fellers 1979). Females enter the pond silently, approach a calling male, and induce amplexus, which lasts most of the night. The pair adhere their eggs to vegetation in small scattered clumps just before dawn. Tadpoles undergo metamorphosis by mid-summer but the age at which adult frogs enter a pond to breed for the 1st time is unknown.

*Gray treefrog (*Hyla versicolor*). Photo by John Mitchell.*

We monitored breeding behavior of a population every night of the breeding season over 3 yr. We tested 2 hypotheses. Because temperature is highly variable during the spring and anurans are ectothermic, the length of the breeding season, the number of evenings males chorus, and the number of males chorusing each evening, should vary. Second, because females are thought to be drawn to the chorus through male attraction vocalizations (Wells 1977a), the number of females found mating is likely to be related to the number of males vocalizing.

Materials and Methods

Our study site was a small pond located on Trent University property, 6 km north of Peterborough, Ontario (44°22′ N, 78°17′ W). The pond is permanent, measuring 80 × 45 m in spring, and 20 × 15 m at summer's end, with a maximum depth of 1 m. The pond has adequate vegetation for use as calling perches by males and for females to adhere eggs, but not enough to hinder our movements. Although treefrog immigration into the pond is possible, we believe it is unlikely because the pond is unusually isolated, with the closest pond located 2 km away.

Beginning in the 1st week of May in 1991, 1992, and 1993, we arrived at the pond at dusk every night to listen for calling males. Once vocalizations began, we systematically circled the pond at least twice and located every calling male using both auditory and visual cues. One person would move into an area of the pond, identifying each individual and tagging its location, and searching for non-vocalizing or satellite males. A 2nd person followed, confirming frog identities and searching for frogs that were missed. We identified all vocalizing males by making accurate sketches their natural dorsal markings. These markings are complex and it is unlikely that 2 individuals would have the same patterns. Dorsal markings have also been used in field studies of other anurans (Wells 1977b; Tuckerman 1982). We reared 1 *H. versicolor* in a lab for 3 yr and the frog was re-sketched annually without access to the previous sketches; its back pattern remained unchanged through time. Furthermore, all field and lab sketches were completed by 1 person who also confirmed the identity of any frog identified by any other observer. The sketch of each individual new to the chorus was added to an album containing the sketches of all individuals' back patterns thus allowing us to determine whether each male was returning or new to the chorus. During the 1992 and 1993 field seasons, sketches from prior seasons were used to identify individuals returning for a 2nd or 3rd year. Therefore, we have assumed that our technique located all males present, clearly identified each male, and that each individual's back pattern remained unchanged through time.

Due to the silent nature of female treefrogs, our attempts to identify females at the pond were limited to locating females already in amplexus. In 1991, while thoroughly searching the pond each night, we recorded all amplexed pairs found. In 1992 and 1993, upon locating and identifying a vocalizing male, we marked its vocalization site with fluorescent flagging tape, allowing rapid relocation. As each evening progressed, we returned to every individual's vocalization site to relocate the individual. If a male was no longer vocalizing, we thoroughly searched an area of at least 2 m in diameter from where the chorusing male was last seen to determine if he was in amplexus with a female. At the end of each chorus we remained at the pond until we had either located all mating pairs, or ensured ourselves that the males had left the chorus for the evening. As a result we have assumed that we observed all matings that occurred over the 3 yr. Our technique did not allow for the identification of females that may have been present at the chorus that did not mate.

Nightly maximum and minimum air temperatures were obtained from the Trent University weather station, located within 1 km of the study pond. The weather station recorded air temperature every minute, giving hourly means. From the hourly means, we determined the maximum and minimum air temperatures for each night during the 3 breeding seasons. To determine whether there was a temperature switch (an air temperature indicative of non-chorusing male behavior), we completed a Chi-square analysis for temperatures from 5 to 15° C and identified the temperature that had the highest significant level indicative of chorusing male presence. All other analyses were completed using regressions.

RESULTS

Males started vocalizing at dusk during each breeding seasons. Calling individuals moved from the trees to the floating mats of vegetation on the pond or into bushes surrounding the pond. Although some males moved around the pond early in the evening, most remained at the same calling site from their arrival at the pond until chorus end (1930 to 0300 hr). All observed females that entered the chorus initiated contact with males and the amplexed pair then remained in the same general area of the male's calling site for several hours. They laid their fertilized eggs just prior to dawn.

Chorusing activity started between 5 and 10 May in all 3 seasons. Number of males vocalizing on any particular evening was highly variable both within and between breeding seasons. On nights when males called, the mean number of individuals vocalizing was 22 in 1991 (range = 1 to 42), 10 in 1992 (1 to 19), and 12 in 1993 (1 to 37). Although the number of males calling in an evening was positively correlated with the minimum, maximum, and average temperatures recorded for that evening, less than 15% of the variation was explained by temperature (Table 1). However, low air temperatures did hinder chorusing behavior. When the air temperature dropped below 8° C by dusk, males seldom vocalized ($P = 0.001$).

On nights that females were present and mating, chorus size ranged from 5 to 40 males. Maximum, minimum and average temperatures were not correlated with presence of females (Table 1). Although number of females in amplexus each evening was positively correlated with number of males vocalizing ($P = 0.001$, $r^2 = 0.140$), only 13% of the variation in female presence was explained by male presence (Table 1).

Table 1. *Influence of temperature on presence of* Hyla versicolor *at breeding choruses.*

		Nightly temperature		
		Minimum	Maximum	Mean
Total frogs/night	r^2	0.093	0.089	0.135
	F	8.632	8.265	12.564
	P	0.004	0.005	0.001
Males/night	r^2	0.095	0.090	0.141
	F	8.737	8.915	13.172
	P	0.004	0.004	0.001
Females/night	r^2	0.000	0.000	0.000
	F	0.958	0.008	0.257
	P	0.311	0.930	0.614

Table 2. *Numbers of adult male and female* H. versicolor *present during the breeding season.*

Year	Males	Females	Duration of breeding (days)	Chorus nights (proportion)
1978[1]	35	9	50	33 (0.66)
1988[2]	26	6	42	12 (0.28)
1989[2]	36	39	37	23 (0.62)
1990[2]	19	10	30	15 (0.50)
1991[3]	77	10	30	24 (0.80)
1992[3]	30	4	29	13 (0.45)
1993[3]	64	19	54	37 (0.69)

[1]Fellers (1979)
[2]Sullivan and Hinshaw (1992)
[3]This study

Chorusing activity ceased in early June in 1991 and 1992 (7 and 8 June), but chorusing continued until 27 June in 1993 (Fig. 1). There did not appear to be a trend between seasonal air temperatures and length of breeding seasons. Air temperatures during 1992 were cooler than those of 1991 yet the breeding season lengths were similar (Fig. 1). Conversely, daily maximum and minimum air temperatures during 1991 and 1993 were similar, but the 1993 breeding season was three weeks longer than that of 1991 (Fig. 1). The number of nights suitable for male chorusing, i.e. nights when the air temperature at dusk was above 8° C and males were found vocalizing, varied among seasons, with 24 nights in 1991, 13 in 1992, and 37 in 1993.

Numbers of breeding adults fluctuated considerably. In 1991, 77 different males vocalized at the pond, and ten matings occurred. In 1992, only 30 different males attended the chorus and only four matings occurred. During the longer season of 1993, 68 different males joined the chorus and nineteen matings occurred. Thus the sex ratios, based on number of males vocalizing and number of females found in amplexus, were all male biased and varied from 7.7:1 in 1991, to 7.5:1 in 1992, and 3.4:1 in 1993.

New males were the most abundant class in the chorus each year. Of the 77 males at the pond in 1991 only 16 (21%) returned in 1992. In 1993, 9 (33%) of the 1992 chorus returned, including 6 from the 1991 chorus. Because the population in 1993 was so much larger than that of 1992, 83% of the males were there for the first time.

Table 3. *Numbers of male and female* H. versicolor *from 3 breeding populations related to number of nights males chorused and duration of season. Data from this study, Fellers (1979) and Sullivan and Hinshaw (1992) are combined.*

		Number of chorus nights	Duration of breeding season	Proportion of chorus nights to total
Total frogs	r^2	0.228	0.000	0.543
	F	2.768	0.159	8.128
	P	0.157	0.706	0.036
Total males	r^2	0.305	0.000	0.587
	F	3.629	0.225	9.534
	P	0.115	0.635	0.027
Total females	r^2	0.110	0.000	0.000
	F	0.617	0.132	0.657
	P	0.468	0.731	0.657

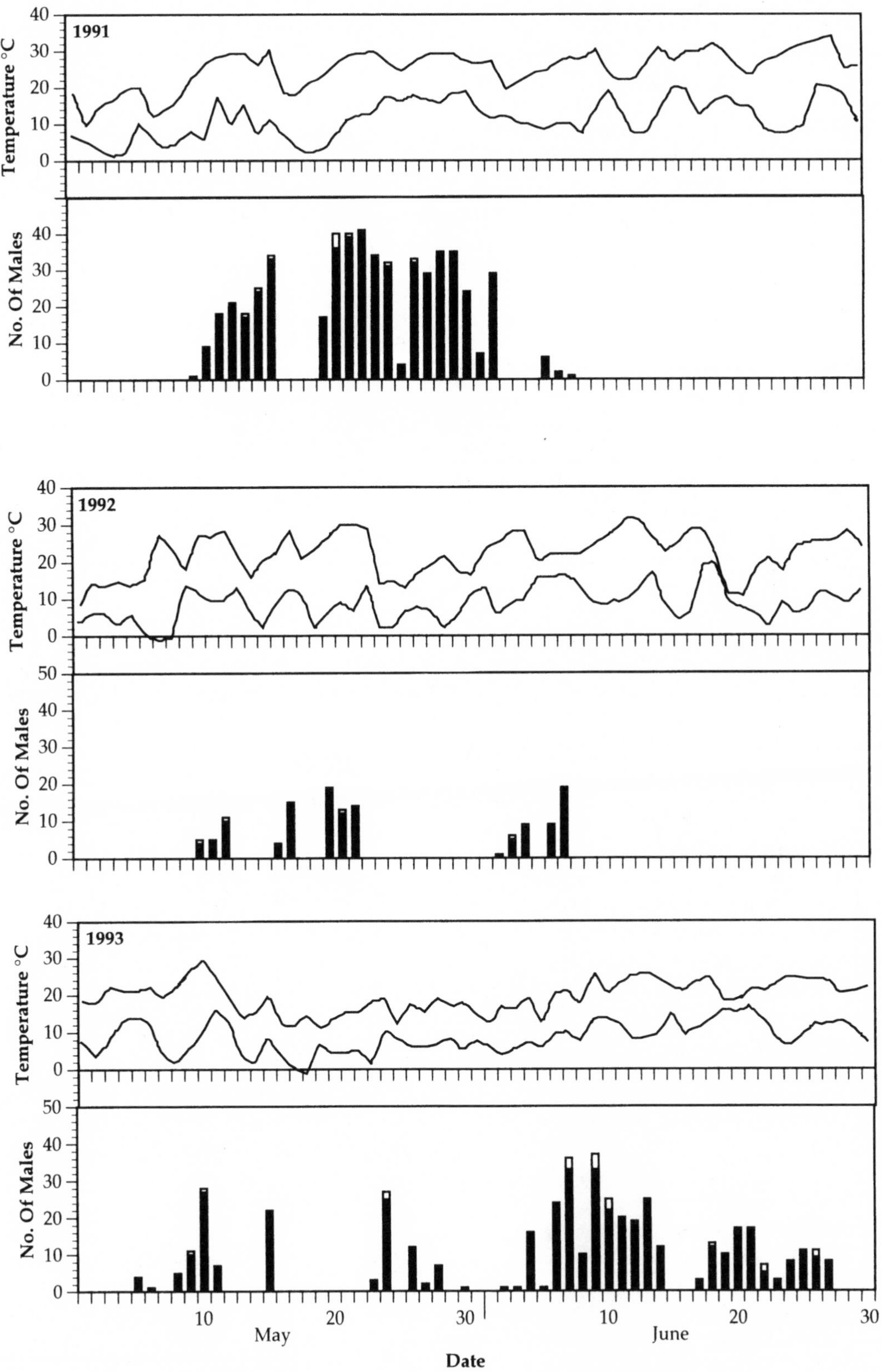

Figure 1. *Comparison of numbers of males vacalizing (black bars) and mating (clear bars) each night, 1991 to 1993. Minimum and maximum air temperatures indicate a relationship between low minimum temperature (< 8°) and nights without any calling males.*

Discussion

The degree of nightly and seasonal variability in the *H. versicolor* breeding system is immense. Breeding seasons differed in duration as well as in numbers of males and females in attendance. Trends indicate we cannot use overall breeding season air temperatures to estimate breeding season length or breeding population numbers and structure. We are also unable to use nightly air temperatures to predict number of males or number of females present. However, we are able to make predictions on how nightly air temperature fluctuations influence the presence or absence of a chorus. When the temperature fell below 8° C males never chorused. In more southern populations this temperature switch appears to occur at a higher temperature (Fellers 1979), indicating that adaptation to local conditions may occur throughout the range of the species.

We combined our 3 yr of data on population size, length of breeding season, number of nights males vocalized during a season, and the proportion of suitable calling nights (suitable calling nights per breeding season length), with 4 other yr of complete breeding season data from populations of vocalizing gray treefrogs (Table 2; Fellers 1979; Sullivan and Hinshaw 1992). Although sample size is only 7 breeding seasons, there is a positive correlation between total number of males (with or without females) identified at the pond each year and proportion of suitable calling nights (Table 3, Fig. 2). However, we found no such relationships between yearly population size and length of breeding seasons or number of nights males vocalized, nor between number of males present and season length or number of nights males vocalized (Table 3). Female presence at the pond was not correlated with number of nights of male chorusing, breeding season duration, or proportion of nights males chorused (Table 3). It was also not correlated with number of males chorusing (P = 0.788).

Over-wintering conditions are likely to influence population sizes and sex ratios. All 3 intensive studies of *H. versicolor* (Table 2) were conducted in areas where over-wintering treefrogs have the possibility of being exposed to harsh environmental conditions. Our data indicate that relatively few males return to the breeding pond for a 2nd or a 3rd season. Sullivan and Hinshaw (1992) reported a similar low rate of return (23% in 1989; 22% in 1990) in a central Maine population. It is possible that non-returning males have migrated elsewhere, but the isolation of our study pond and the evidence that other species return to the same location to breed (Oldham 1966; Gill 1978; Berven and Grudzien 1990) suggests that non-returning males may not survive the winter. Winter survival could be greater in more southern parts of the species range and populations there should provide a means of testing this idea.

Female anurans are notoriously difficult to find at breeding sites and, as a result, male-biased sex ratios are frequently reported (Berven 1990; Pechman et al. 1991; Sullivan and Hinshaw 1992). However, the combination of the ease of accessibility to all parts of our pond, the silence of a male in amplexus, the prolonged amplexus of a mating pair, and our searching technique, which compelled us to scour the area for any male that was identified previously in the evening as vocalizing, gives us confidence that we saw all matings that occurred. Therefore, we believe that the male-biased sex ratios we report for *H. versicolor* are real. Over-wintering mortality may also account in part for such a highly skewed sex ratio years. Since female anurans generally take 1-yr longer to reach sexual maturity than males (Berven 1990), harsh winter conditions occurring for consecutive years may make it difficult for many females to reach sexual maturity before dying.

We conclude that high variability in nightly and seasonal chorus activity, coupled with the low probability of males mating, characterizes *Hyla versicolor* populations. If this treefrog system is indicative of other anuran species, we believe that long term and intensive studies will be necessary to establish whether decline is occurring.

Acknowledgments

We would like to thank Victoria Campbell and Andrea Smith for their dedicated assistance in the field. Thanks also goes to Dr. Brian Sullivan for supplying us with more detailed data on his and Dr. Hinshaw's 3 yr of treefrog research.

Literature Cited

Berven KA. 1990. Factors affecting population fluctuations in larval and adult stages of the wood frog (*Rana sylvatica*). Ecology 71:1599–1608.

Berven KA, Grudzien TA. 1990. Dispersal in the wood frog (*Rana sylvatica*): implications for genetic population structure. Evolution 44:2047–2056.

Bishop CA, Pettit KE, editors. 1992. Declines in Canadian amphibian populations: designing a national monitoring strategy. Ottawa: Canadian Wildlife Service. Occasional Paper 76.

Blaustein AR, Wake DB, Sousa WP. 1994. Amphibian declines: judging stability, pesistence, and susceptibility of populations to local and global extinctions. Conservation Biology 8:60–71.

Elmberg J. 1990. Long-term survival, length of breeding season, and operational sex ratio in a boreal population of common frogs, *Rana temporaria*. Canadian Journal of Zoology 68:1271–1277

Fellers GM. 1979. Mate selection in gray treefrogs *Hyla versicolor*. Copeia 1979:286–290.

Gill DE. 1978. Effective population size and interdemic migration rates in a metapopulation of the red-spotted newt, *Notophthalmus viridescens* (Rafinesque). Evolution 32:839–849.

Howard RD. 1980. Mating behavior and mating success in woodfrogs, *Rana sylvatica*. Animal Behavior 28:705–716.

Oldham RS. 1966. Spring movements in the american toad, *Bufo americanus*. Canadian Journal of Zoology 44:63–100.

Pechmann JH, Scott DE, Semlitsch RD, Caldwell JP, Vitt LJ, Gibbons JW. 1991. Declining amphibian populations: the problem of separating human impact from natural fluctuations. Science 253:892–895.

Sullivan BK, Hinshaw SH. 1992. Female choice and selection on male calling behavior in the gray treefrog *Hyla versicolor*. Animal Behavior 44:733–744.

Tuckerman RD. 1982. The breeding behavior and life history of a northern population of the spring peeper [thesis]. Peterborough, ON: Trent University.

Wells KD. 1977a. The social behavior of anuran amphibians. Animal Behavior. 25:666–693.

Wells KD. 1977b. Territoriality and male mating success in green frogs (*Rana clamitans*). Ecology 58:750–762.

©1997 by the Society for the Study of Amphibians and Reptiles
Amphibians in decline: Canadian studies of a global problem. David M. Green, editor.
Herpetological Conservation 1:64–72.

Chapter 7

NORTHERN LEOPARD FROG (*RANA PIPIENS*) DISPERSAL IN RELATION TO HABITAT

CAROLYN N.L. SEBURN[1]

Department of Zoology University of Alberta, Edmonton, Alberta T6G 2E9, Canada

DAVID C. SEBURN[1]

Department of Geography, University of Alberta, Edmonton, Alberta T6G 2E9, Canada

CYNTHIA A. PASZKOWSKI

Department of Zoology University of Alberta, Edmonton, Alberta T6G 2E9, Canada

ABSTRACT.—We marked 938 young-of-the-year northern leopard frogs, *Rana pipiens*, at a central source pond in southern Alberta. Fifty-two frogs (4 marked at the source pond) were caught in traps up to 2 km downstream, confirming that streams are used for dispersal. During 3 censuses, conducted at 23 upstream, downstream or seepage ponds within 4 km of the source pond, 104 frogs marked at the source pond were recaptured. Frogs successfully dispersed to downstream ponds 2.1 km from the source site, upstream 1 km and overland only 0.4 km. Dispersal of hundreds of young frogs to ponds within 1 km occurred within 3 wk of metamorphosis. More distant ponds were reached within 6 wk. Ponds that receive dispersers can easily be confused with successful breeding sites unless tadpoles are monitored closely and dispersers marked. Some historic sites may be recolonized if intervening dispersal routes are protected.

RÉSUMÉ.—Un total de 938 grenouilles léopard âgées d'un an (*Rana pipiens*) ont été marquées dans un étang central du sud de l'Alberta. Cinquante-deux grenouilles (4 marquées à l'étang source) ont été capturées dans des pièges jusqu'à 2 km en aval, ce qui confirme que le courant est utilisé pour la dispersion. À l'occasion de 3 recensements entrepris dans 23 sites en amont, en aval ou dans des étangs situés dans un rayon de 4 km de l'étang source, 104 grenouilles marquées à l'étang source ont été recapturées. Les grenouilles avaient réussi à coloniser des étangs situés à 2,1 km en aval de leur étang source, à 1 km en amont et à seulement 0,4 km à la périphérie. La dispersion de centaines de jeunes grenouilles dans les étangs situés dans un rayon de 1 km intervient dans les 3 premières semaines de la métamorphose. Les grenouilles atteignent les étangs plus éloignés dans les 6 semaines qui suivent. Les étangs qui accueillent ces grenouilles peuvent facilement être pris pour des sites de reproduction, sauf si les têtards sont étroitement surveillés et que les grenouilles sont marquées. Certains sites historiques pourraient ainsi être recolonisés si les trajets que les grenouilles suivent pour se disperser étaient protégés.

---

1 Present address: *Seburn Ecological Services, 920 Mussell Road, RR #1, Oxford Mills, Ontario K0G 1S0, Canada.*

Local amphibian populations tend to fluctuate widely because of their susceptibility to vagaries of the weather and their dependence on seasonal habitats. Dispersal can stabilize populations on a larger metapopulation scale (Roff 1974; Gill 1978; Sjögren 1988, 1991). Studies have been undertaken on amphibians (Dole 1971; Van Gelder et al. 1986; Berven and Grudzien 1990; Sinsch 1991) but these are few because of logistical difficulties. The northern leopard frog, *Rana pipiens* Schreber, has vanished from much of its historic range in Alberta in recent years (Roberts 1981) and there are currently only 13 known breeding populations scattered throughout southeastern Alberta (Seburn 1992). The stability of the provincial populations may depend, in part, on the species' ability to recolonize vacated sites and maintain connections among extant populations. In eastern North America, terrestrial dispersal usually occurs on warm rainy nights and frogs disperse overland (Dole 1971). Warm rains are less common in Alberta than in the east and hence dispersal may be triggered by other factors. Overland movements may not be as important.

*Northern leopard frog (*Rana pipiens*). Photo by Martin Ouellet.*

An understanding of dispersal is important because if natural recolonization is insufficient, reintroductions may be necessary to maintain populations. In this study we examine the importance of streams to the dispersal route and the distance moved by young-of-the-year (YOY) *R. pipiens* from a population in southern Alberta. Because this study only follows individuals from transformation to first hibernation, no effort is made to distinguish between dispersal and migration.

Materials and Methods

Study Site

Research was conducted on the southern slopes of the Cypress Hills in southern Alberta, Canada (49°35'N, 110°20'W). The study area consisted of a circle of 4-km radius centred on a large, 2 ha marsh in the Sexton-Lodge Creek Watershed (Fig. 1) at which *R. pipiens* had successfully bred for several years prior to the study. In May and June of 1993, 24 study ponds were chosen, including a central "source" pond, 11 ponds less than 1 km away, and 12 ponds 1 to 4 km away (Fig. 1). Ponds were classified as being upstream (5 ponds), downstream (10 ponds), on parallel streams (5 ponds), or isolated (3 ponds) relative to the source pond. Ponds on parallel streams were located on Read and Thelma Creeks, which feed into Lodge Creek downstream of the study area. Isolated ponds were seepage ponds not connected to any stream. Data collected from each pond included surface area, shortest linear distance from the source pond as measured on aerial photos, and change in elevation relative to the source pond. Pond area ranged from 0.01 to 2 ha ($\bar{x} = 0.2$ ha). The source pond was at an elevation of approximately 1200 m above sea level, while the other ponds ranged from 1100 to 1350 m.

In general, downstream ponds within 0.5 km of the source pond were in white spruce (*Picea glauca*) forest and more distant ponds were in prairie. Upstream ponds were either in, or very close, to forested habitat, whereas parallel stream ponds were on elevated prairie. Two of the isolated ponds were in prairie habitat, 1 was in forest.

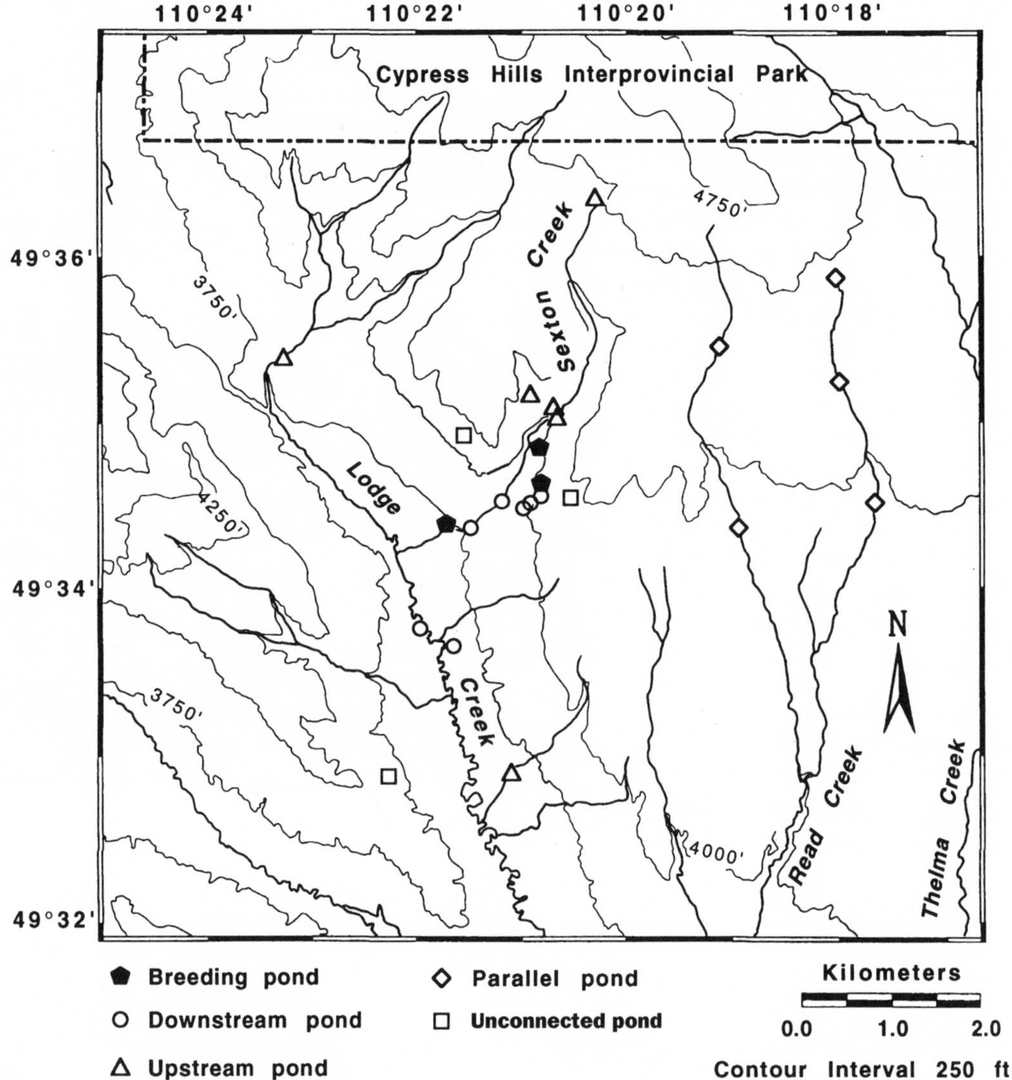

Figure 1. *Location of ponds surveyed at Cypress Hills study area, Alberta, Canada.*

Trapping

To determine whether frogs found within the Sexton-Lodge watershed actually moved through the streams, drift fencing was used to channel swimming frogs into pitfall traps. Traps were placed at culverts 0.4, 1.8, and 3.3 km downstream from the source pond (Fig. 2). Traps 1 and 2 were located at a small private gravel road in a wooded area directly downslope of the source pond and adjacent to another pond. Traps 3 and 4 were set at larger culverts crossing a gravel county road in prairie habitat.

Traps were checked twice daily (0800 and 2000 hr) from 18 August to 6 September. Captured frogs were toe-clipped and released downstream of the trap. The ratio of day to evening captures was compared to an expected 1:1 ratio using a *G*-test (Sokal and Rohlf 1981). Data from the 4 traps were pooled as daily morning and evening captures and compared to meteorological data collected at Cypress Hills Provincial Park. Relative humidity, temperature and accumulated rainfall were measured daily at 0800 and 1300 hr. Maximum and minimum temperatures for the past 24 hr were recorded at 0800 hr.

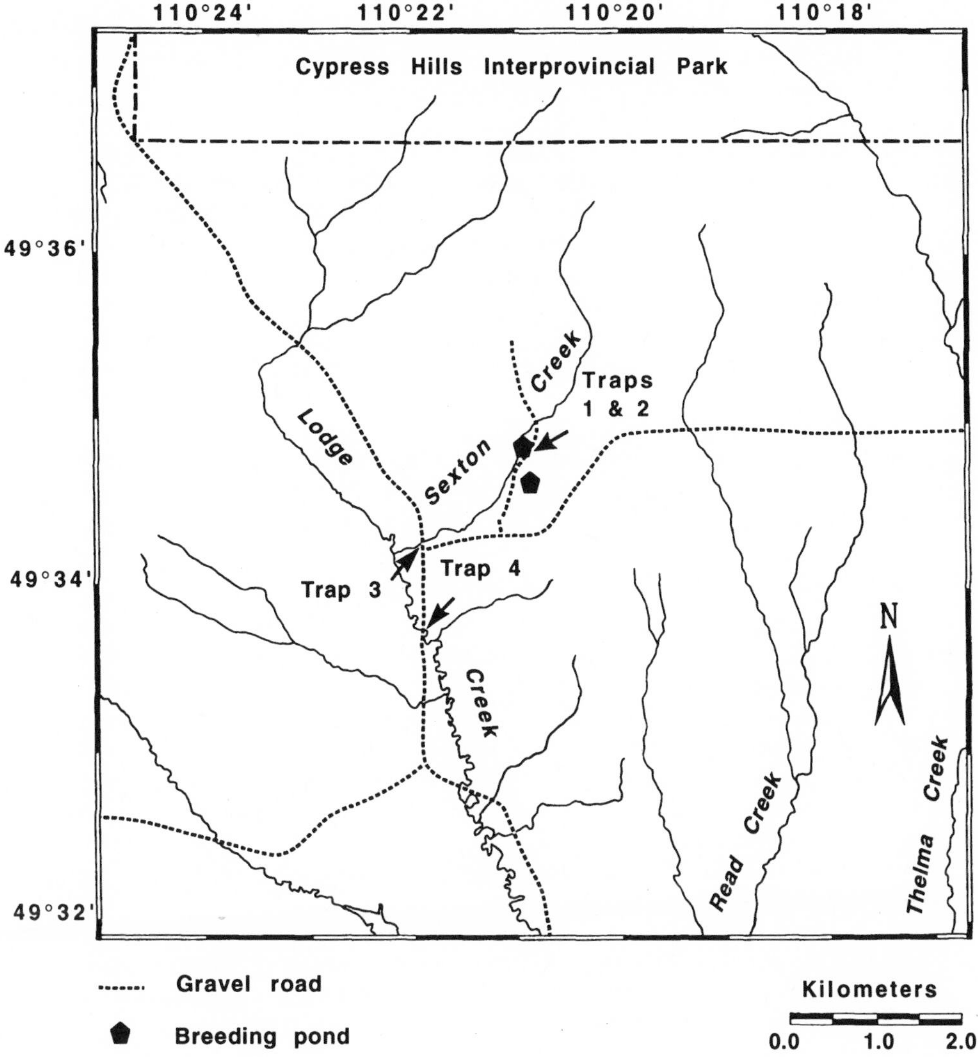

Figure 2. *Location of traps at Cypress Hills study area, Alberta, Canada.*

Census Data

To attempt to determine the number of frogs dispersing from the source pond, 791 YOY were marked by toe-clipping at the source pond from 5 to 15 August 1993. From 23 August to 15 September, 3 censuses were conducted at all the ponds for YOY frogs. The order of ponds censused was randomized with the exception that the source pond was always censused last. Unmarked frogs were marked according to the type of pond at which they were captured. Marked frogs were not remarked. Each census took from 4 to 6 days. Frogs were captured by hand and dip net and stored in buckets until either remaining frogs were too wary to be caught or over 1 hr had been spent capturing frogs.

Cypress Hills *R. pipiens* are dimorphic for dorsal colour, being either green or brown (Fogleman et al. 1980), so the proportion of each colour morph among the frogs collected was recorded at each census. The ratio of frogs marked at the source pond (MSP), compared with unmarked frogs and the ratio of green to brown frogs captured at each pond was determined. These ratios were compared to results at the source pond using a *G*-test to determine whether unmarked frogs at each pond were likely to have come from the source pond.

The number of frogs captured at each pond during each census was sensitive to the time of day and the prevailing weather. Therefore, data from all 3 censuses were pooled by calculating the total minimum number of frogs at each pond. This value was defined as the total number of unmarked frogs captured during all censuses plus the maximum number of MSP frogs caught in 1 census.

Stepwise regression was used to determine if both area and distance from the source pond were required to predict minimum number of frogs. Data were only used from ponds at which at least 1 MSP frog was found. Analysis of variance (ANOVA) was performed using standardized residuals from the regression of area on minimum number of frogs, comparing ponds upslope to ponds downslope of the source pond. The upslope ponds included all upstream ponds and the 1 isolated pond at which MSP frogs were found. The downslope ponds included all downstream ponds at which MSP frogs were found.

Results

In 1993, *R. pipiens* attempted to breed in 3 of the 24 ponds and were only successful at the source pond. In the other 2 ponds, tadpoles were produced but growth and development were retarded by 4 to 8 wk relative to the source pond. It is unlikely that those frogs which did transform had sufficient time to disperse before hibernating or that they could survive the winter.

Trap Data

Fifty-two YOY frogs were captured in all 4 traps. In traps 1 and 2, 21 frogs (2 MSP, 19 unmarked) were captured starting 18 August. Twenty-four frogs (2 MSP, 22 unmarked) were captured in trap 3 starting 21 August. Three unmarked frogs were captured at trap 4 starting 2 September. Of the 52 frogs trapped, 24 were found in traps at 0800 hr and 28 were found at 2000 hr ($G = 0.15$, $P > 0.5$; Fig.3). There was no correlation between relative humidity and captures in either the morning ($r = 0.38$) or evening ($r = -0.10$). Frogs were only caught when the relative humidity was > 65% (at 0800 hr) or 50% (at 1300 hr). However, at both times, there were only 2 days when the relative humidity was lower. The number of captures was not correlated with daily maximum temperature (morning: $r = -0.32$, evening: $r = 0.18$). Dispersal was not triggered by rainfall, number of morning captures was weakly correlated with rainfall ($r = 0.44$) suggesting that it may influence the number of dispersers.

Dispersal occurred in 2 pulses, the 1st from 16 to 23 August and the 2nd from 31 August to 5 September. The 1st pulse ended when the minimum temperature dropped from 6° to 3° C. The 2nd pulse began when the minimum temperature rose to 5° and ended when it fell from 4° to 2°. The 2nd pulse was larger than the 1st.

Census Data

By the end of the 3 censuses, we had marked 938 MSP frogs at the source pond and made 104 recaptures at the other ponds. An additional 698 unmarked frogs were marked at the other ponds. The total population of YOY frogs was estimated to be 5909 (4604 to 8247) using the Schnabel method (Krebs 1989). About 28% of the population was marked at some time.

The furthest pond at which MSP frogs were recaptured was 2.1 km from the source pond. During censuses, the proportion of MSP frogs to unmarked frogs did not differ significantly between the source pond and other ponds ≤ 2.1 km away (12.3% at source; 14.9% elsewhere: $G = 0.38$, $P > 0.5$) nor did the proportion of green to brown frogs differ (45% at both). Beyond 2.1 km, 37 YOY frogs were found at 4 ponds. Of these frogs, only 1 was brown and none were MSP. The ratio of green to brown frogs ($G = 45.34$, $P < 0.0001$) and the proportion of MSP frogs ($G = 37.49$, $P < 0.0001$) differed significantly from the source pond. These results suggest that all frogs captured within a 2.1 km radius of the source pond came from that pond and frogs captured beyond 2.1 km came from a different, unknown breeding pond(s). This assumption is used throughout the rest of the analyses.

MSP frogs were found up to 2.1 km downstream from the source pond (Fig. 4). No YOY were found greater than 1 km upstream. The furthest isolated pond with MSP (or any other) frogs was 0.4 km from the source pond and no frogs were found in the parallel stream ponds during censuses, although yearlings and adults were seen at some of these ponds earlier in the season. Thus the pres-

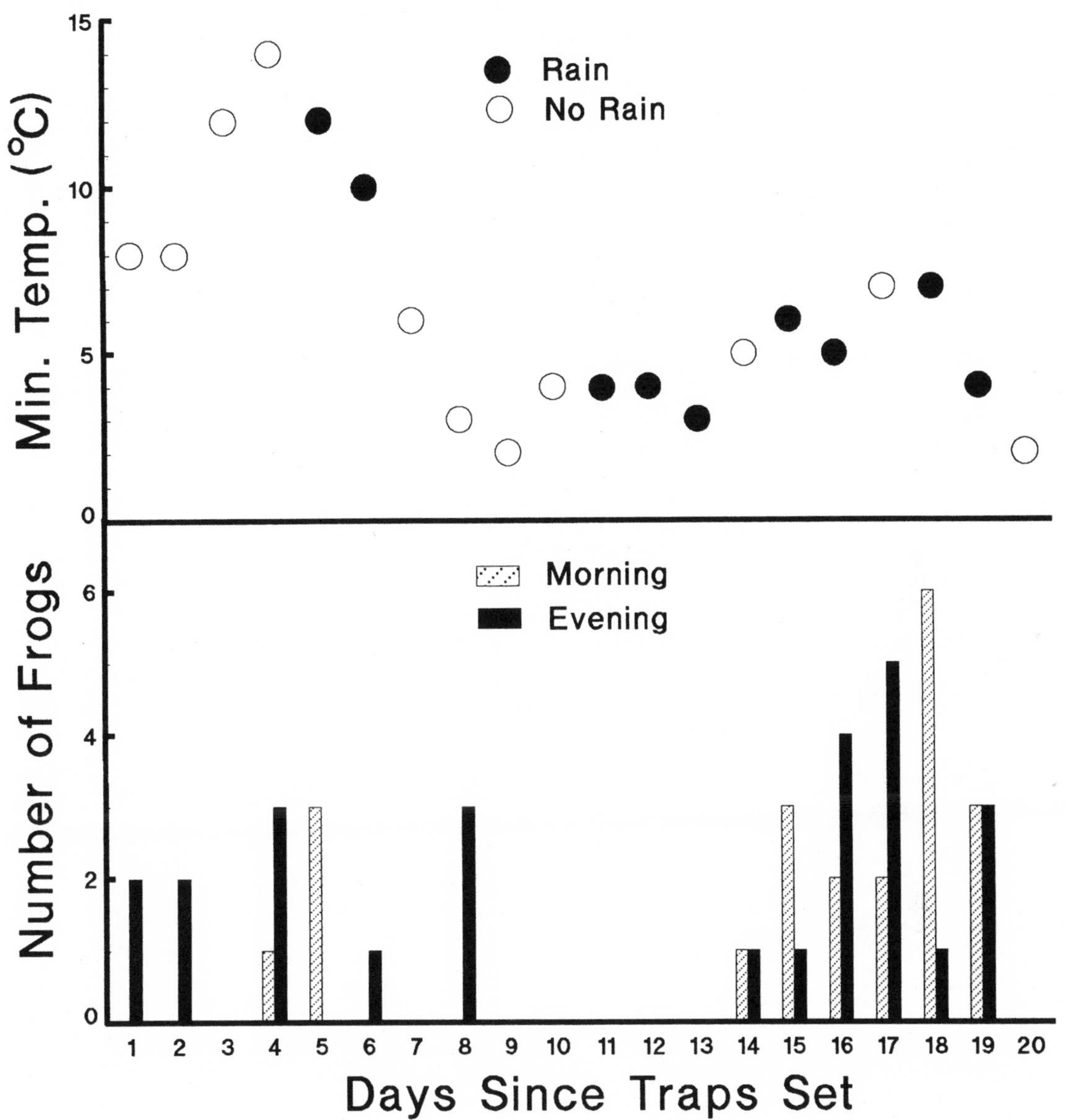

Figure 3. *Relationship between daily minimum temperature (at 0800 hr) and the number of frogs recovered from traos at morning (0800 hr) and evening (2000 hr). Trapping began on 18 August. Days designated "rain" had > trace precipitation.*

ence of an aquatic connection appears to increase the distance YOY frogs successfully disperse and the direction of water flow may also affect distance moved, although it should be noted that no upstream ponds between 1 to 3.5 km from the source pond were sampled due to inaccessibility.

There was no difference in the minimum number of frogs caught at upslope and downslope ponds (F = 1.03, $P > 0.10$). The minimum number of frogs was positively correlated with pond area (F = 23.96, r = 0.828, $P < 0.001$) and negatively correlated with distance from the source pond (F = 4.80, r = –0.534, $P < 0.05$; Fig. 4). Adding distance to area did not substantially improve the fit of the data ($F_e = 2.05$, $P > 0.05$).

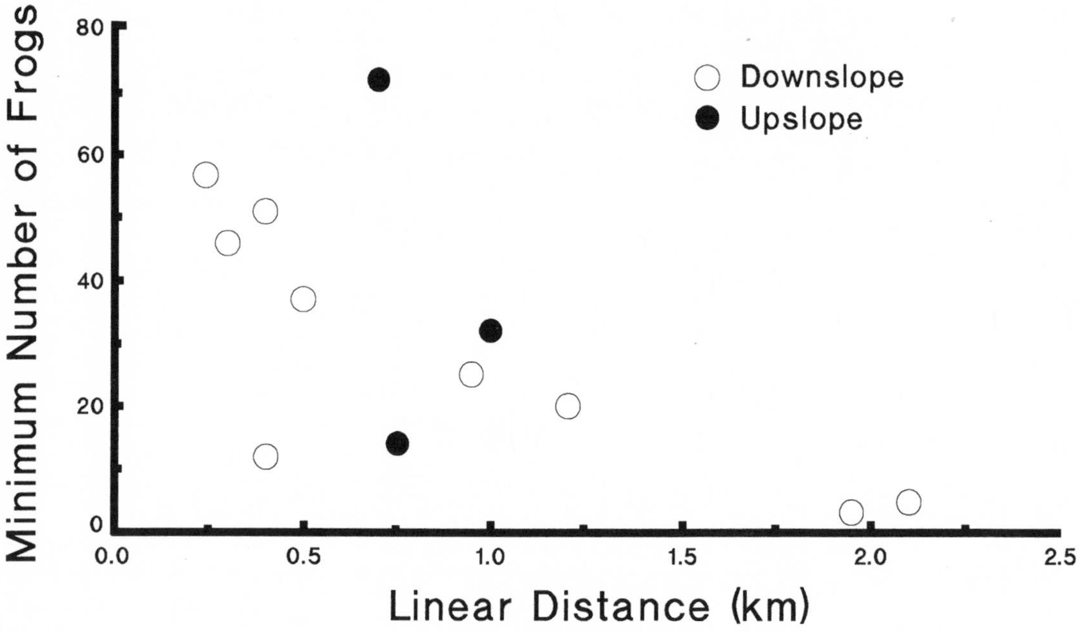

Figure 4. *Relationship between the minimum number of frogs captured in censuses and distance from the source pond for downslope and upslope ponds, at which at least 1 frog marked at the source pond was recovered.*

Discussion

Given the proximity of Traps 1 and 2 to the source pond, our data suggest that they captured only a small proportion of dispersers in the area whereas Traps 3 and 4 captured a larger proportion (Table 1). Traps 1 and 2 were set in a forested environment within 0.4 km of the source pond. MSP frogs were found at the adjacent pond before the traps were even installed and YOY frogs were commonly seen crossing the road above the culverts. The habitat connecting the source pond and this area consisted of a large number of small outflows running downhill through moist forest, obscuring the distinction between aquatic and terrestrial habitats. Traps 3 and 4, in contrast, were set in prairie habitat next to a wider, drier gravel road. Although 3 YOY frogs were found in the grass next to Trap 3 without having been captured in the trap, it appears that a high proportion of dispersers used the aquatic route. At Trap 4, after 31 August, a recurring problem with muskrat (*Ondatra zibethicus*) chewing the drift nets reduced the effectiveness of the trap. Nonetheless, both the trap and census data suggest that few dispersers got that far and that they did not arrive until much later than at the other ponds. Unfortunately, none of the frogs marked at Traps 3 and 4 were recaptured during the censuses, so there is no way to determine the proportion of dispersers captured in the traps. These results suggest that aquatic habitat may play a more important role in dispersal in a prairie environment than in a moist forested environment. However, it is important to recall that habitat is confounded with the distance from the source pond and the effect of these factors can not be separated.

Table 1. *Comparison of trap and census data. Trap data include all frogs trapped during 19 days. Census data include the minimum number of frogs captured in 3 censuses in the pond immediately downstream of the trap.*

Trap number	Trap data	Census data
1 and 2	21	112
3	24	9
4	3	5

Most research on movements in amphibians has focused on breeding migrations (for example, Van Gelder et al. 1986; Sinsch 1987). Data on post-metamorphic movements are difficult to gather and generally emphasize factors initiating dispersal. It is commonly believed that amphibians preferentially disperse in association with rainfall (Gibbons and Bennett 1974). Dole (1971) found that the largest emigrations by YOY *R. pipiens* were on rainy nights or in conjunction with rapid drops in barometric pressure and almost all movements were at night. Number of frogs captured was negatively correlated with minimum temperature. In our study, rainfall had a small effect on number of frogs trapped and minimum daily temperature appeared to have a threshold effect such that dispersal stopped when temperatures dropped too low (Fig. 3). Approximately half of the frogs were captured during the day. These differences may be a result of cool nighttime temperatures in our study (2° to 14°) compared to Dole's (9° to 18°) or the fact that dispersers were moving along aquatic corridors. Barometric pressure data were unavailable. Dole (1971) found YOY leopard frogs dispersed up to 800 m per night and recaptured 2 as adults 5 km from their natal pond, a distance that could only have been travelled overland. In our study, YOY *R. pipiens* moved up to 2.1 km in less than 6 wk after metamorphosis. In spring 1994, 1 frog that had been marked YOY at the source pond was recaptured 8 km away (L. Yaremko and C. Paszkowski, unpublished data).

Bovbjerg (1965), comparing the timing of emigration from a natural pond with a simulated indoor pond, found that the peak time for both coincided. Bovberg concluded that the timing of emigration was determined developmentally rather than by the weather, although he could not exclude barometric pressure. Rainfall or humidity may be an important factor in dispersal because odours carry further in moist air making it easier for individuals to orient to ponds (Sinsch 1991). It may also be important in reducing dehydration. Whether it is orientation or hydration requirements that are most important, an aquatic route may provide the required cues and moisture and may therefore increase the number of days during which dispersal can occur. Freeland and Martin (1985) found that in the early 1980's the cane toad (*Bufo marinus*) expanded its range in Queensland, Australia, at an average rate of 27 km/yr, but up to 110 km/yr within watersheds. It was not determined if the toads actually moved through the water or simply followed the riparian vegetation. However, Maynard (1934) reported that Amercian toads (*Bufo americanus*) in New York State used a stream to migrate to the vicinity of a breeding pond. Toads floated downstream at least 0.8 km and possibly as far as 4 km, over the course of several nights.

Results of the present study have important implications for monitoring breeding success in populations of *R. pipiens*. Because YOY frogs can disperse rapidly and in large numbers over a wide area, it may appear that successful reproduction has occurred in several ponds rather than at a single site. In fact, at 1 pond more than 1 km from the source site, adults were observed in the spring, tadpoles in July and August, and a large number of YOY in September. Nevertheless, we know that reproduction was unsuccessful at the pond because the local tadpoles never completed metamorphosis. We would not have known that reproduction failed at this location if we had not monitored the rate of development of the tadpoles or marked YOY frogs at the source pond. Casual historic observations of seemingly successful breeding sites may thus be called into doubt. Although the extent of dispersal varies among species and even among populations, it is possible that similar problems could occur in monitoring and assessing other populations of amphibians.

If successful breeding ponds are not accurately identified, then management of populations becomes difficult. Gill (1978) showed that red-spotted newts, *Notophthalmus viridescens*, have a source-and-sink metapopulation structure similar to that seen here. He postulated that source sites shifted as ponds were established, aged, and filled with organic matter. In most amphibians, reproductive success varies within each pond from year to year (Pechmann et al. 1991). A key factor for management, therefore, is to identify those ponds where reproduction is consistently successful over a long period. At our study site, *R. pipiens* could be extirpated from an area with at least a 2-km radius by the destruction of a single pond even though adults and YOY frogs are commonly seen in many of the other surrounding ponds.

Although the high dispersal capacity of some amphibian species complicates matters related to monitoring and management, it also offers hope of natural recolonization of areas where local populations have declined because of short-term factors such as weather or disease. We recommend that the natural dispersal ability of species be exploited in developing management and reintroduction strategies. For this to take place, we need to know a great deal more about the dispersal capacity of various species and the effect of habitats on distances dispersed. This style of management requires protection of dispersal routes and hibernation sites as well as identification of potential breeding sites. In addition, the relative reproductive success of dispersers and non-dispersers must be determined.

Acknowledgments

We thank L. Yaremko and F. Reintjes for assistance in the field, the staff of Cypress Hills Provincial Park for the use of facilities, the many land owners who allowed us access to their land and especially E. Mudie, J. and J. Harpell and the people of Eagle's Nest Ranch for their cooperation and kindness. M. Rankin (Canadian Museum of Nature) assisted with references and O. Stephen with graphics. S. Brechtel (Alberta Fish and Wildlife), S. Boutin, D. Schindler and J. Spence (University of Alberta) helped plan this research. This research was funded by the Alberta Ministry of Environmental Protection, Recreation, Parks and Wildlife Foundation, an NSERC operating grant to C. Paszkowski and an NSERC post graduate scholarship to C. Seburn. We thank two anonymous reviewers for their helpful suggestions.

Literature Cited

Bereven KA, Grudzien TA. 1990. Dispersal in the wood frog (*Rana sylvatica*): implications for genetic population structure. Evolution 44:2054–2056.

Bovbjerg RV. 1965. Experimental studies on the dispersal of the frog *Rana pipiens*. Proceedings of the Iowa Academy of Sciences 72:412–418.

Dole JW. 1971. Dispersal of recently metamorphosed leopard frogs, *Rana pipiens*. Copeia 1971:221–228.

Fogleman JC, Corn PS, Pettus D. 1980. The genetic basis of a dorsal color polymorphism in *Rana pipiens*. Journal of Heredity 71:439–440.

Freeland WS, Martin KC. 1985. The rate of range expansion by *Bufo marinus* in Northern Australia 1980–1985. Australian Wildlife Research 12:555–559.

Gibbons JW, Bennett DH. 1974. Determination of anuran terrestrial activity patterns by a drift fence method. Copeia 1974:236–243.

Gill DE. 1978. The metapopulation ecology of the Red-Spotted Newt, *Notophthalmus viridescens* (Rafinesque). Ecological Monographs 48:145–166.

Krebs CJ. 1989. Ecological methodology. New York: Harper and Row.

Maynard EA. 1934. The aquatic migration of the toad, *Bufo americanus* LeConte. Copeia 1934:174–177.

Pechmann JHK, Scott DE, Semlitsch RD, Caldwell JP, Vitt LJ, Gibbons JW. 1991. Declining amphibian populations: the problem of separating human impacts from natural fluctuations. Science 253:892–895.

Roff DA. 1974. The analysis of a population model demonstrating the importance of dispersal in a heterogeneous environment. Oecologia 15:259–275.

Roberts WE. 1981. What happened to the leopard frogs? Alberta Naturalist 11:1–4.

Seburn CNL. 1992. Leopard frog project: field report 1991. Edmonton, AB: Alberta Forestry, Lands and Wildlife, Fish and Wildlife Division.

Sinsch U. 1987. Orientation behaviour of toads (*Bufo bufo*) displaced from the breeding site. Journal of Comparative Physiology A. 161:715–727.

Sinsch U. 1991. Mini-review: The orientation behaviour of amphibians. Herpetological Journal 1:541–544.

Sjögren P. 1988. Metapopulation biology of *Rana lessonae* Camerano on the northern periphery of its range. Acta University Uppsaliensis Comprehensive Summaries of Uppsala Dissertations 157:1–35.

Sjögren P. 1991. Extinction and isolation gradients in metapopulations: the case of the pool frog (*Rana lessonae*). Biological Journal of the Linnean Society 42:135–147.

Sokal RR, Rohlf FJ. 1981. Biometry: the principles and practice of statistics in biological research. San Francisco: W.H. Freeman.

Van Gelder JJ, Aarts HMJ, Staal HWJM. 1986. Routes and speed of migrating toads (*Bufo bufo* L.): A telemetric study. Herpetological Journal 1:111–114.

©1997 by the Society for the Study of Amphibians and Reptiles
Amphibians in decline: Canadian studies of a global problem. David M. Green, editor.
Herpetological Conservation 1:73–77.

Chapter 8

DISTRIBUTION AND ABUNDANCE OF THE CHORUS FROG *PSEUDACRIS TRISERIATA* IN QUÉBEC

CLAUDE DAIGLE

Ministère de l'Environnement et de la Faune, Direction de la faune et des habitats, 150 boulevard René-Lévesque est, Québec, Québec G1R 4Y1, Canada

ABSTRACT.—During the spring in 1992 and 1993, a survey of the known distribution area of the chorus frog, *Pseudacris triseriata*, was carried out in Québec to verify its actual distribution and abundance. In the eastern part of the study area, near Montréal, the species was heard at only 10 of the 279 sites surveyed (3.6%). Most of these are threatened by human activities. *Pseudacris triseriata* was more widely distributed along the Ottawa river lowlands. It was heard at 62 of the 446 stations visited in this area (13.9%). Generally, the species is common where present. Comparison of these results with those of Bleakney (1959) indicates that *P. triseriata* has disappeared from a large area southeast of Montréal.

RÉSUMÉ.—Au cours des printemps 1992 et 1993, nous avons inventorié l'aire de distribution connue de la rainette faux-grillon de l'ouest, *Pseudacris triseriata*, au Québec. Dans la partie est de l'aire d'étude, près de Montréal, nous avons entendu l'espèce à seulement 10 des 279 sites inventoriés (3,6%). La plupart de ces endroits sont menacés par les activités humaines. *Pseudacris triseriata* est plus largement distribuée dans les basses terres le long de la rivière des Outaouais. Dans ce secteur, nous l'avons entendu à 62 des 446 stations visitées (13,9%). Cette rainette est généralement commune là où elle est présente. La comparaison de nos résultats avec ceux de Bleakney (1959) indique que *P. triseriata* est disparue d'une étendue relativement importante au sud-est de Montréal.

Although widespread in United States, the western chorus frog, *Pseudacris triseriata,* is at the northeastern limit of its range in Québec (Behler and King 1979; Conant and Collins 1991; Cook 1984). Bider and Matte (1991) report its presence in the Ottawa valley and the St. Lawrence lowlands surrounding Montréal. Favourable conditions for *P. triseriata* were created by deforestation of the St. Lawrence lowlands during European colonization and establishment of the species in Québec could have occurred during this period (Bonin and Galois 1994). *Pseudacris triseriata* is typically found in open areas such as prairies and meadows (Cook 1984; Conant and Collins 1991; Froom 1982) and breeding occurs in shallow water close to these sites where its calls may be heard both day and night (Froom 1982).

There have been few extensive surveys of anurans in Québec (Cook 1984). Although Bleakney (1959) found *P. triseriata* to be abundant in southeastern Québec in the mid-1950s, it is now considered to be the rarest anuran in Québec (Bider and Matte 1991). It is likely to be designated threatened or vulnerable by the Québec government (Beaulieu 1992). Government involvement with species so designated follows a 3-tiered approach. First, information for determining the species' status province-wide is acquired. Then, if needed, the species is legally designated threatened or vulnerable, and finally, measures ensuring its conservation are undertaken. As part of the 1st step, surveys were conducted during the spring in 1992 and 1993 to look at the distribution and abundance of *P. triseriata* in Québec.

Materials and Methods

The study area extended from the Canadian border south of Montréal, west to Chapeau along the Ottawa river (Fig. 1). It included most of the locations where *P. triseriata* had been previously reported (Bider and Matte 1991). Special efforts were made to cover roads surveyed by Bleakney (1959). Surveys were conducted mainly in the St. Lawrence lowlands (Ministère du Loisir, de la Chasse et de la Pêche 1986), extending to the foothills of the Adirondack, Appalachian, and Laurentian mountains. The study area was divided into 3 sectors: Montréal and Hull surroundings (sector A), St. Lawrence lowlands around Montréal (sector B), and St. Lawrence lowlands, along the Ottawa river, and the edge of the Adirondack and Appalachian mountains (sector C).

A survey of anurans based on calls of breeding males was made during the last week of April and the 1st week of May in 1992 and 1993. Known *P. triseriata* populations were visited first in order to ensure that the breeding season had in fact started. Both typical and non-typical habitats were considered in this study. During the day, teams of 2 persons explored the roads of the study area in search of potential sites for anuran reproduction. Marshes, lakes, ponds, fields and woodlands, wet or flooded, were located on maps (1:20,000). To identify other anuran species calling mostly after dusk, sites identified earlier in the day were surveyed between 1900 hr and midnight. Listening usually lasted less than 5 min. We recorded species that were heard calling and estimated their abundance by the following criteria: 0 = none heard, 1 = one male heard, 2 = most individuals can be counted, few calls overlapping, and 3 = full chorus, calls continuous and overlapping.

Western chorus frog, Pseudacris triseriata. *Photo by Jacques Brisson.*

Results

This 2-yr survey allowed us to visit 725 potential breeding sites for anurans. We found 446 sites along the Ottawa river lowlands in 2 wk. Four wk of research produced 279 potential breeding sites in the eastern part of the study area.

Pseudacris triseriata was heard at 72 stations, or about 10% of the total number of potential sites surveyed (Table 1). Abundance at these stations was moderate with 33 stations in class 1, 19 in class 2, and 20 in class 3. Most of the records came from the western part of the study area (Fig. 1). Five other species of anurans were heard during the surveys (Table 1). The northern spring peeper, *Pseudacris crucifer*, was the most widely distributed and abundant frog heard in the study area at that time of the year. We found it at 82% of the stations, of which most rated 3 on the abundance scale. Wood frogs, *Rana sylvatica*, were heard at 40.7% of potential sites

Table 1. *Frequency (number of sites and percentage of total) and abundance of anurans heard at 725 sites in southern Québec at the end of April and beginning of May, 1992 and 1993. Abundance classes are defined in the text.*

	Abundance class			
Species	0	1	2	3
Pseudacris triseriata	653 (90.0)	33 (4.6)	19 (2.6)	20 (2.8)
P. crucifer	128 (17.7)	43 (5.9)	53 (7.3)	501 (69.1)
Rana sylvatica	431 (59.4)	100 (13.8)	92 (12.7)	102 (14.1)
Bufo americanus	510 (70.3)	100 (13.8)	63 (8.7)	52 (7.2)
R. pipiens	571 (78.8)	93 (12.8)	33 (4.5)	28 (3.9)
Hyla versicolor	708 (97.6)	13 (1.8)	2 (0.3)	2 (0.3)

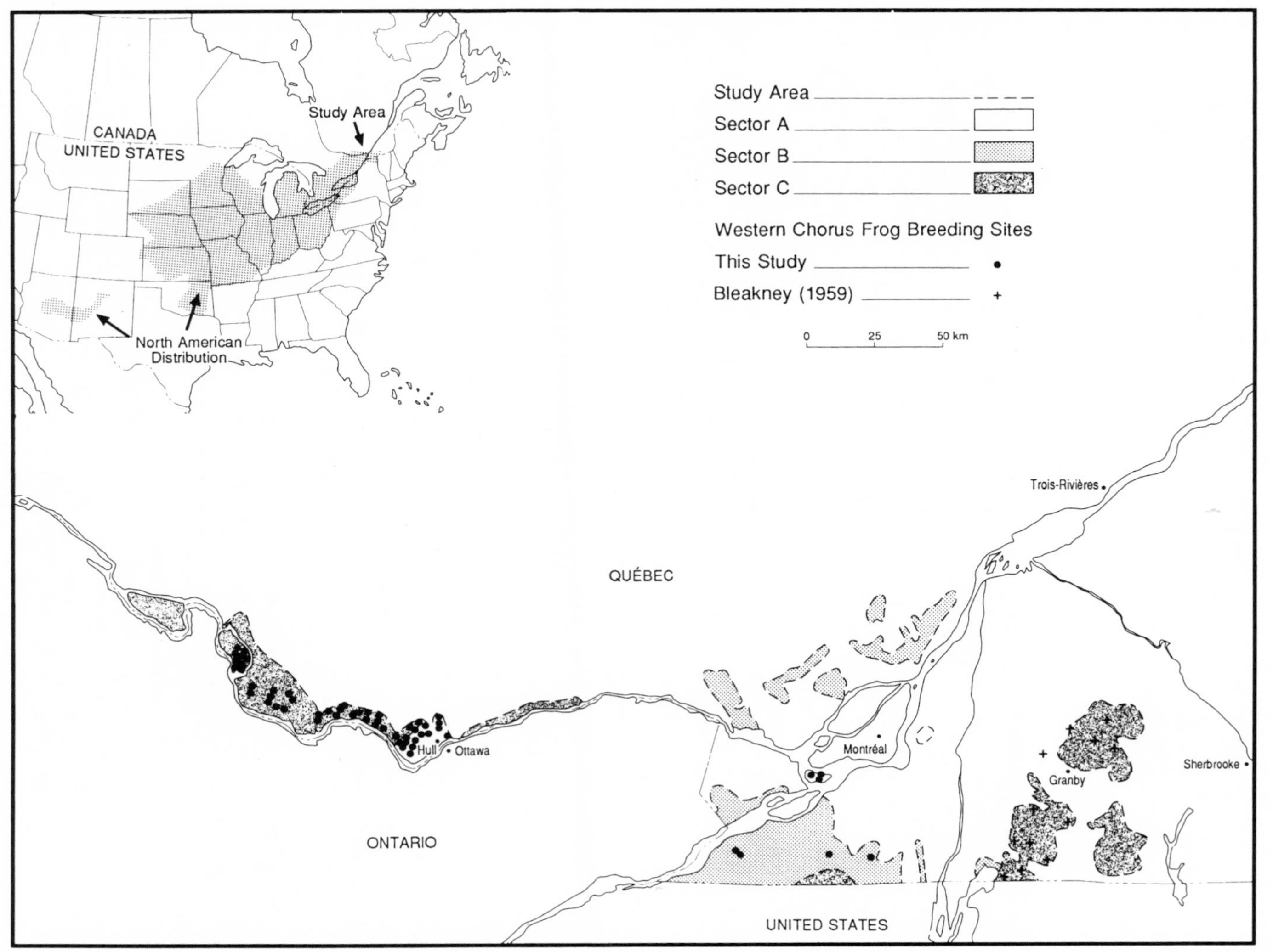

Figure 1. Study area indicating sectors of search and Pseudacris triseriata *breeding sites found during this study and by Bleakney (1959).*

surveyed and were moderately abundant, stations being equally distributed in class 1, 2 and 3. American toads, *Bufo americanus*, and northern leopard frogs, *Rana pipiens*, were common in the study area and they were found, respectively, at 29.8% and 21.1% of the 725 stations. We heard fewer *B. americanus* and *R. pipiens* at each station than *P. crucifer* or *R. sylvatica*. A few early gray treefrogs, *Hyla versicolor*, were heard at 17 stations (2.3%).

Discussion

The study region reports the warmest climate in Québec, with a mean annual temperature of 5° C and 3000 and 3200 degree-days per year (Wilson 1971). This is the most utilized area by both humans and amphibians in Québec. Considering human modifications to the landscape, the study area presented three types of environment. Montréal and Hull surroundings (sector A) are mainly open areas that are still undergoing heavy development for residential and industrial purposes. The St. Lawrence lowlands around Montréal (sector B) are heavily used for agriculture. Extensive drainage programs have been conducted in this part of Québec (Labrecque 1987), resulting in a loss of wetlands. In the rest of the St. Lawrence lowlands, the Ottawa river basin, and the river junction at the edge of the Adirondack and Appalachian mountains (sector C) drainage is less intensive and wetlands are still present. Potential breeding sites for anurans were more numerous in the Ottawa river lowlands than in the Montréal region. *Pseudacris triseriata* populations are more abundant and less endangered along the Ottawa river. Only a few sites in the Hull surroundings are endangered by suburban development. Human activities do not appear to threaten the others, at least not in the short term.

Because the Montréal region is more heavily used by humans, greater habitat losses have resulted over the last decades. In the Montréal region, we found only 10 *P. triseriata* breeding sites and most of them were endangered either by suburban, industrial, or drainage development. *Pseudacris triseriata* was abundant in the southeastern part of the Montréal region during the mid-1950s (Bleakney 1959). In our survey of the same area in 1992 and 1993, none was heard. Reasons for this decline are not evident.

Marginal populations, such as Québec *P. triseriata* at the northeast limit of their range, are known to be particularly vulnerable to natural fluctuations (Blaustein and Wake 1990). Extinctions and large, short-term fluctuations in population size are often related to annual climatic variations such as drought (Corn and Fogleman 1984; Pechmann et al. 1991). Anthropogenic causes may generate long-term variations in population distribution and abundance. Habitat loss and degradation are the most important and best understood causes of human-related species decline (Johnson 1992). Suitable habitat appeared to remain in the area visited both by Bleakney (1959) and for this study. In fact, in terms of human modifications and landscape, the area seems comparable (both sector C) to the western part of the study area where the species is present.

There are various explanations for amphibian declines in areas where human influence is not evident. Interspecific hybridization (Bogart 1992), disease (Crawshaw 1992, this volume), acidic deposition (Clark 1992), and pesticides (Bishop 1992; Berrill et al., this volume) have been identified as factors affecting amphibian survival. There may be synergistic effects between those and more global phenomena such as increased ultraviolet radiation and higher temperatures (Blaustein and Wake 1990; Ovaska, this volume). However, this particular study was not designed specifically to verify change in *P. triseriata* populations in Québec, and even less to explain eventual decline. In many cases, it is difficult to distinguish declines resulting from human activities from natural population fluctuations without long-term data on both populations and habitats (Pechmann et al. 1991). Data on environmental conditions prevailing at the time of the surveys were not collected either by us or Bleakney (1959). Changes in distribution and abundance are thus difficult to explain.

Acknowledgments

Many persons have contributed to this study. Joël Bonin and Alain Desrosiers helped with planning. Lyne Bouthillier, Alain Desrosiers, Cécile Dubé, Michel Huot, Martin Léveillé, Anaïs Rinfret Pilon, Daniel Saint-Hilaire, and Louis-Marc Soyez took part in the surveys. Many thanks to all these people. Finally, special thanks go to Réhaume Courtois and Michel Lepage for their comments on the preliminary version.

Literature Cited

Beaulieu H. 1992. Liste des espèces de la faune vertébrée susceptibles d'être désignées menacées ou vulnérables. Québec: Ministère du Loisir, de la Chasse et de la Pêche.

Behler JL, King FW. 1979. The Audubon Society field guide to North American reptiles and amphibians. New York: Alfred A. Knopf.

Bider JR, Matte S. 1991. Atlas des amphibiens et des reptiles du Québec. Ste.-Anne-de-Bellvue, QE: Société d'histoire naturelle de la vallée du Saint-Laurent, and Québec: Ministère du Loisir, de la Chasse et de la Pêche.

Bishop CA. 1992. The effects of pesticides on amphibians and the implications for determining causes of declines in amphibian populations. In: Bishop CA, Pettit KE, editors. Declines in Canadian amphibian populations: designing a national monitoring strategy. Ottawa: Canadian Wildlife Service. Occasional Paper 76. p 6–7.

Blaustein AR, Wake DB. 1990. Declining amphibian populations: a global phenomenon? Trends in Ecology and Evololution 5:203–204.

Bleakney S. 1959. Postglacial dispersal of the western chorus frog in eastern Canada. Canadian Field-Naturalist 73:197–205.

Bogart JP. 1992. Monitoring genetic diversity. In: Bishop CA, Pettit KE, editors. Declines in Canadian amphibian populations: designing a national monitoring strategy. Ottawa: Canadian Wildlife Service. Occasional Paper 76. p 50–52.

Bonin J, Galois P. 1994. Rapport sur la situation de la rainette faux-grillon de l'ouest (*Pseudacris triseriata*) au Québec et au Canada. Québec: Ministère de l'Environnement et de la Faune.

Clark KL. 1992. Monitoring the effects of acidic deposition on amphibian populations in Canada. In: Bishop CA, Pettit KE, editors. Declines in Canadian amphibian populations: designing a national monitoring strategy. Ottawa: Canadian Wildlife Service. Occasional Paper 76. p 63–66.

Conant R, Collins JT. 1991. A field guide to reptiles and amphibians of eastern and central North America. Boston: Houghton Mifflin.

Cook FR. 1984. Introduction aux amphibiens et reptiles du Canada. Ottawa: Musés Nationaux du Canada.

Corn PS, Fogleman JC. 1984. Extinction of montane populations of the northern leopard frogs (*Rana pipiens*) in Colorado. Journal of Herpetology 18:147–152.

Crawshaw GJ. 1992. The role of disease in amphibian decline. In: Bishop CA, Pettit KE, editors. Declines in Canadian amphibian populations: designing a national monitoring strategy. Ottawa: Canadian Wildlife Service. Occasional Paper 76. p 60–62.

Froom B. 1982. Amphibians of Canada. Toronto: McClelland and Stewart.

Johnson B. 1992. Habitat loss and declining amphibian populations. In: Bishop CA, Pettit KE, editors. Declines in Canadian amphibian populations: designing a national monitoring strategy. Ottawa: Canadian Wildlife Service. Occasional Paper 76. p 71–75.

Labrecque J. 1987. Superficie drainée souterrainement par région agricole. Avril 1986 à mars 1987. Québec: Ministère de l'Agriculture, des Pêcheries et de l'Alimentation du Québec.

Ministère du Loisir, de la Chasse et de la Pêche. 1986. Les Parcs québécois - 7. Les régions naturelles, 1re édition. Québec: Ministère du Loisir, de la Chasse et de la Pêche.

Pechmann JHK, Scott DE, Semlitsch RD, Caldwell JP, Vitt LJ, Gibbons JW. 1991. Declining amphibian populations: the problem of separating human impact from natural fluctuations. Science 253:892–895.

Wilson CV. 1971. The climate of Québec. Part one. Climatic atlas. Ottawa: Canadian Meteorological Service.

©1997 by the Society for the Study of Amphibians and Reptiles
Amphibians in decline: Canadian studies of a global problem. David M. Green, editor.
Herpetological Conservation 1:78–86.

Chapter 9

FLOW CYTOMETRIC ANALYSIS OF AMPHIBIAN POPULATION COMPOSITION

TIMOTHY F. SHARBEL[1] AND LESLIE A. LOWCOCK[2]

Redpath Museum, McGill University, 859 Sherbrooke Street West, Montréal, Québec H3A 2K6, Canada

ROBERT W. MURPHY

Centre for Biodiversity and Conservation Biology, Royal Ontario Museum, 100 Queen's Park, Toronto, Ontario M5S 2C6, Canada

ABSTRACT.—The distribution and variability of genotypes which compose an amphibian population provide a framework upon which temporal and spacial fluctuations, and possible long-term decline, may be understood. Flow cytometry is a precise and cost-effective technique to analyze a suite of macrogenomic parameters. Species composition and introgression in sympatric, cryptic species can be estimated through analyses of C-value (pg DNA per haploid nucleus). Similarly, flow cytometric analysis can reveal abnormal DNA profiles indicative of polyploidy and aneuploidy. These phenomena may be naturally occurring, as with unisexual hybrid complexes composed of polyploid individuals, or induced, as with aneuploid mosaicism resulting from exposure to environmental contaminants. In amphibians, analyses of such genomic traits can be made from their nucleated erythrocytes, which may be obtained non-destructively from a considerable number of individuals with minimal time and expense. Large sample sizes enable the identification of rare genetic events. Repeated sampling of individual animals is also possible, enabling perception of genomic change over time.

RÉSUMÉ. — La répartition et la variabilité des génotypes qui composent les populations d'amphibiens fournit le cadre qui permet de comprendre les fluctuations spatiales et temporelles, voire le déclin potentiel à long terme des populations concernées. La cytométrie de flux est une technique précise et rentable qui permet d'analyser une série de paramètres macrogénomiques. La composition des espèces et l'introgression des espèces sympatriques et cryptiques peut être évaluée grâce à des analyses de la valeur-C (pourcentage d'ADN par noyau haploïde). Parallèlement, l'analyse par cytométrie de flux peut révéler des profils d'ADN anormaux qui indiquent l'existence de polyploïdie ou d'aneuploïdie. Ces phénomènes peuvent survenir naturellement, comme dans le cas des complexes hybrides unisexuels composés d'individus polyploïdes ou être induits, comme dans le cas du mosaïcisme aneuploïde résultant de l'exposition à des contaminants environnementaux. Chez les amphibiens, l'analyse de ces caractères génomiques peut se faire à partir de leurs érythrocytes nucléés qu'il est possible d'obtenir auprès d'un nombre considérable de spécimens sans le sacrificier, rapidement et à peu de frais. Plus l'échantillon est grand, plus il est facile d'identifier des manifestations génétiques rares. L'échantillonnage répété de spécimens est également possible et permet de percevoir les changements génomiques qui se produisent au fil du temps.

1 Present Address: *Arbeitsgruppe Michiels, Max-Planck-Institut für Verhaltensphysiologie, D-82319 Seewiesen (Post Starnberg), Germany.*

2 Present Address: *Centre for Biodiversity and Conservation Biology, Royal Ontario Museum, 100 Queen's Park, Toronto, Ontario M5S 2C6, Canada.*

Visible and acoustic traits are, for the most part, sufficient for the identification of amphibian species and their distribution. But in cases where species boundaries are clouded by morphological similarity and variability, introgression, or geographic isolation, additional characters are required in order to identify the taxonomic unit under study. For example, *Desmognathus fuscus* and *D. ochrophaeus*, are 2 closely-related species often found in sympatry whose morphological variability makes for difficult field identification (Sharbel and Bonin 1992). Yet, the species occupy distinct microhabitats (Conant and Collins 1991), which subject them to differing environmental regimes and potentially different responses to habitat disruption. Inferences regarding spatial and temporal fluctuations in these species can only be made after detailed investigations of their relative numbers at a particular site, and how they partition the habitat in sympatry. Their morphological resemblance precludes accurate identification and thus genetic characters have been used to resolve questions of species composition (Sharbel et al. 1995). This type of phenomenon is prevalent within the Amphibia, so a cost-effective and standardized method of genetic analysis would facilitate questions of species composition by providing additional traits useful in their diagnosis.

We have used measures of genome size and both quantitative and qualitative comparisons of macrogenomic characteristics to shed light upon questions of species composition, hybridization, and, recently, the genotoxic effects of environmental contaminants on amphibian populations. These characters can be precisely and efficiently measured using flow cytometry. Low cost of sample collection and preparation, large intra- and interindividual sample sizes, the identification of rare genetic events, non-destructive sampling, and inter-study standardization are strengths inherent to this technique.

Amphibians lend themselves in particular to flow cytometric analysis, because their nucleated erythrocytes may be obtained through non-destructive means and require minimal preparation for storage and analysis. Furthermore, blood samples are easily collected during non-related sampling procedures from toe-clips or dissections to obtain tissue samples, thus maximizing the information gathered when sampling a wild or laboratory population.

Flow Cytometry Methods

Measurements on somatic cells are most-easily performed with nucleated erythrocytes (Tank et al. 1987; Licht and Lowcock 1991; Lowcock et al. 1991, 1992). Heart punctures yield the highest quality samples with relatively low contamination from other bodily fluids. Alternatively, tail and toe-clips, if taken carefully, can yield adequate samples of blood for analysis (Licht and Lowcock 1991; Lowcock and Murphy 1991; Lowcock et al. 1992; Sharbel et al. 1995).

From a heart puncture or arterial nick, approximately 20 to 25 μl of blood should be collected in a heparinized microhematocrit capillary tube (approximately ¼ of the length of a 100-μl tube), and suspended in 200 μl of a sodium-citrate/sucrose freezing solution containing the anticoagulant dimethylsulfoxide (DMSO) in a 1.5-ml cryotube (Tank et al. 1987). Samples can be collected from toe and tail clips by bleeding the appendage into 200 μl of freezing solution set in the well of a depression plate. These can then be pipetted into cryotubes. A sample with the proper number of erythrocytes should be a dark pink colour; lighter samples mean inadequate numbers of erythrocytes and dark red samples will contain too many of these cells. Cryotubes should be well-agitated before freezing and storage. Samples in sodium-citrate solution must be immediately flash-frozen in liquid nitrogen (–196° C), where they can be left almost indefinitely or transferred to an ultracold freezer (–100° to –80°) for up to 5 yr until use. Frozen samples may also be shipped or transported in supercooled containers, with ultracold-frozen icepacks, or on dry ice. Care should be taken not to allow rapid temperature changes in samples until the time of analysis. Sample preparations not requiring flash freezing are more practical for field collections, and methods using alcohol-preservation at ambient temperatures (Vindeløv et al. 1983; Holtfreter and Cohen 1990) show great promise.

Staining for genome size, ploidy, CV variation and cell-cycle parameters uses an intercalating DNA dye, e.g., propidium iodide, adjusted to the hypotonicity of amphibian blood (Vindeløv et al. 1983; Tank et al. 1987) and the increased nucleic acid content of amphibian cells (Licht and Lowcock

1991). We generated DNA profiles of propidium iodide-stained erythrocyte nuclei using an EPICS Profile flow cytometer (Coulter Electronics, Hialeah, Florida), and recorded the relative index of red fluorescence under laser illumination for 10,000 to 40,000 individual nuclei, which were then plotted in the form of a frequency distribution (Fig. 1a) for each individual.

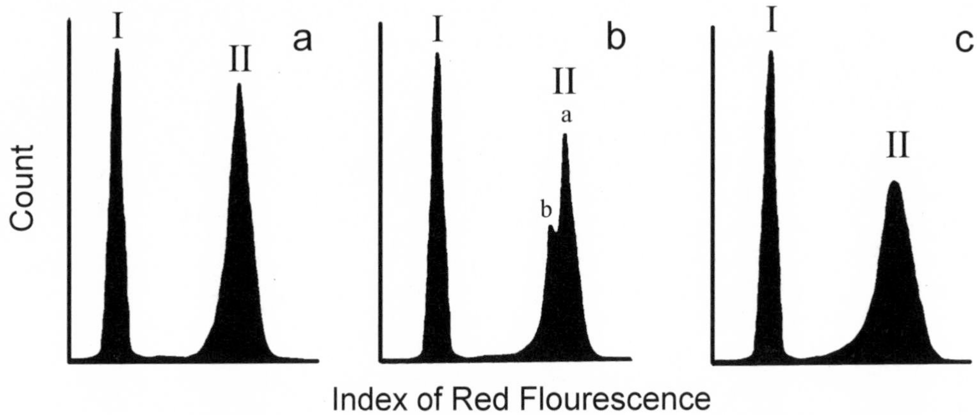

Figure 1. *DNA profiles generated on an EPICS Profile flow cytometer. In all samples, peak I is from a control individual (*Xenopus laevis*, 2C = 6.2 pg), and peak II is from the test sample. a) normal profiles in both control and test samples; b) anueploid mosaicism in the test sample; c) increased coefficient of variation in the test sample.*

Studying Population Composition Using Flow Cytometry

Despite the potentially confounding effects of selection, genome size in amphibians appears to be relatively stable intraspecifically, even among taxa which occupy extensive ranges (Olmo and Morescalchi 1978; Olmo et al. 1989; King 1990; Licht and Lowcock 1991). In contrast, genome size can vary significantly between different species in both anurans and caudates (King 1990; Licht and Lowcock 1991). Because of this attribute, DNA content data can provide relatively quick and cost effective insights into taxonomic conundrums. DNA content within populations tends to be highly conserved and, assuming little intraspecific variability, systematic applications to species identification are therefore possible.

Genome size can be used to assess proportions of differing genotypes in sympatric arrays of cryptic species. Genome size analysis in conjunction with isozyme electrophoresis have been used to differentiate between syntopic *Desmognathus fuscus* and *D. ochrophaeus* in the northern Adirondacks, where low levels of introgression and lack of distinguishing morphological traits confounded estimates of pure-species and hybrid proportions (Sharbel and Bonin 1992; Sharbel et al. 1995). Using tail clips, 187 individuals were sampled to generate an accurate estimate of genotype composition in 2 streams, with minimal effects on the populations in question (Sharbel et al. 1995). Genome sizes corresponded closely with previously-published values for the 2 species—18 pg DNA for *D. fuscus* and 16 pg for *D. ochrophaeus* (Licht and Lowcock 1991)—and coincided with allelic identifications at 4 diagnostic loci (Sharbel et al. 1995). Additionally, putative hybrids were characterized by allelic arrays at diagnostic loci and genome size values which reflected considerable backcrossing with the dominant genotype in the streams: *D. ochrophaeus*.

The *Rana pretiosa* complex of western North America, previously recognized as a single species, has been resolved using fixed allele differences between post-Pleistocene isolated populations (Green et al. 1996). Significant differences in genome size data (t = 7.32, P = 0.0026) were found between 2 geographically distinct genotypes characterized by fixed differences at 6 enzyme loci. Mean C-values were 8.45 pg and 9.08 pg for a Washington and a Nevada population, respectively, with no overlap in distribution. Genetic differentiation in both isozyme and DNA content has therefore occurred in the near absence of morphological divergence (Green et al. 1996).

*Columbia spotted frog (*Rana luteiventris*) a member of the* R. pretiosa *complex. Photo by Stephen Corn.*

Blood samples from a sizable collection of anuran specimens from Vietnam revealed several putative cryptic species differing significantly in DNA content, both within and among populations (Murphy et al., 1997). A closer examination of morphological attributes has revealed, in some cases, morphological characteristics useful in segregating the genotypes, even though these characteristics initially seemed rather trivial. Similar results have been reported by MacCulloch et al. (1996) on collections of both amphibians and reptiles from Guyana.

Although these data alone may not suffice to form the basis of species descriptions, they do point to directions for closer examinations of specimens, using both morphological and molecular evaluations. Therefore such studies of cryptic species should be designed with more than one method of analysis in mind, including flow cytometry.

In the absence of genomic instability or DNA adjustment in hybrid nuclei (Price et al. 1983, 1985), genome size values of F1 hybrid individuals are expected to exhibit intermediacy between those of the parental species (Fig. 2a). Backcrossing with members of either parental species would show progeny with genome size tending towards the characteristic value of the pure parent (Fig. 2b). Preliminary data are suggestive of these phenomena. Sympatric *Bufo americanus* and *B. fowleri* have been shown to undergo low levels of hybridization, as shown by morphological, acoustic, and genetic criteria (Green 1982, 1984; Jones 1973). Juvenile toads can be difficult to identify. Fifteen juveniles, collected from Long Point, Ontario during the 1993 breeding season by D.M. Green, were tentatively identified as *B. americanus* (n = 5), *B. fowleri* (n = 5), or F1 hybrids (n = 5) based upon varying degrees of belly-mottling, a putative diagnostic morphological character, and overall morphology (Green 1984). But results from the 3 diagnostic enzyme loci PGDH, IDH-2, and SOD (Green 1984) demonstrated 4 pure *B. americanus* and 9 pure *B. fowleri* genotypes among them, and these determinations were supported by genome-size values from the same individuals (4.29 pg DNA for *B. americanus* and 4.69 pg for *B. fowleri*). Two of the 15 individuals were characterized by allelic arrays representative of *B. americanus* but with a single *B. fowleri* allele at the SOD locus and genome size values diagnostic of *B. americanus* (4.25 and 4.29 pg). Only 1 of the 2 genetically determined hybrids had originally been diagnosed as a hybrid based on morphology, with the sec-

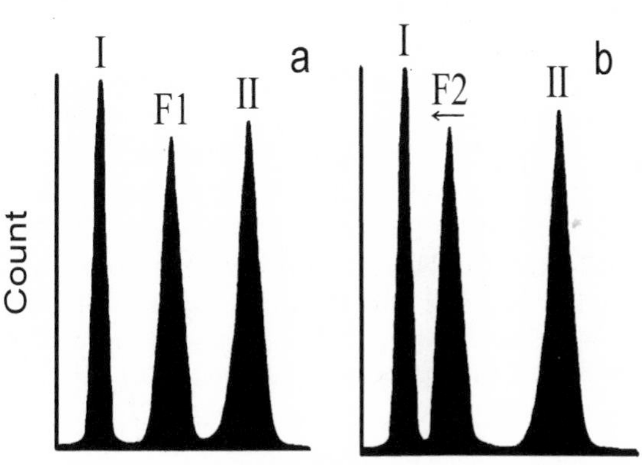

Figure 2. *Predicted shifts in genome size in a) an F1 hybrid resulting from a cross between hypothetical species I and II, and b) a hybrid individual backcrossed with hypothetical parental genotype I.*

ond having been identified as *B. americanus*. Both genetic characters suggest introgression in the direction of *B. americanus*, the most abundant species at the collection site.

Interspecific genome segregation is possible in hybridogenic complexes, and flow cytometry can be used to compare somatic and germ-line ploidy within the same individual. Two germ cell lineages have been identified through flow cytometry in individuals from certain populations of the hybridogenic frog *Rana esculenta* (Vinogradov et al. 1990, 1991). Flow cytometric evaluations of erythrocytes and sperm from certain males revealed hybrid amphispermy, a phenomenon in which parental genotypes remain segregated in the germ line (Vinogradov et al. 1991). Males existed as mosaics of 3 cell types—a single *R. esculenta* (i.e. *R. ridibunda* + *R. lessonae*) somatic cell lineage and 2 germ cell lineages of either pure *R. ridibunda* or *R. lessonae* genotypes (Vinogradov et al. 1991). These data exemplify the importance of utilizing a suite of independent characters when assessing introgression in amphibian populations.

It is often difficult to associate larvae with adults but this can be aided through an analysis of DNA content. Tadpoles can generally be identified at least to species group, thus reducing the chance of mis-identifying larval forms. In questions about species identification or kin association, appropriate molecular analyses using high molecular weight DNA extracted from the same blood samples (unpubl. obs.) can be used in their elucidation.

Unisexuality is frequently accompanied by elevations in ploidy level, and hence DNA content. At the population level, flow cytometry has proven invaluable for identifying polyploids, aneuploids, somatic mosaics, and in assessing parentage of unisexual fishes of the genus *Phoxinus* (Dawley and Goddard 1988). The extensive studies on populations of bisexual-unisexual complexes of ambystomatid salamanders (Lowcock et al. 1991, 1992; Lowcock 1994) best demonstrate the utility of flow cytometry in characterizing entire populations in regard to sex-ratio, ploidy-ratio, and hybridity. Applications of such data include generating baseline information about geographic variation of populations within a particular complex for correlation with genomic, environmental, ecological, or ecotoxicological factors.

Spontaneous heteroploidy, though generally rare in animals, is not uncommon in amphibians (Bogart 1980; Lowcock and Licht 1990). Flow cytometry is a simple way to quantify the occurrence of heteroploidy in natural or laboratory populations of normal, bisexual species (Lowcock and Licht 1990). This phenomenon may have either environmental (Bogart et al. 1989; Lowcock 1994) or ecotoxicological causations (Bickham et al. 1988; McBee and Bickham 1988; Lamb et al. 1991; Lowcock et al. 1997), and thus may be indicative of extrinsic factors having indirect effects at the genomic level.

Genome size data does not present phylogenetic information, however. DNA content tends to be highly species-specific and frequently, even sister species do not have the same quantity of DNA in the nuclei of their cells (Licht and Lowcock 1991). Consequently, if DNA content is used as the character in phylogenetic analyses, the character states become the amount of DNA itself, and virtually every taxon will have a unique character state, rendering the data uninformative. The only alternative is to order states based on the amount of DNA under the assumption of additivity, but this is theoretically very unappealing. Because there are only 2 directions for DNA content to evolve, either up or down, the likelihood of homoplasy is very high. Two taxa evolving independently and without selection will have a 50% probability of evolving in the same direction. With 3 taxa homoplasy is inevitable. Thus, there does not appear to be much promise for these data as characters for phylogenetic analyses. Rather, these data should be evaluated on strongly supported phylogenies to gain an understanding of the evolution of DNA content (Licht and Lowcock 1991).

Ecotoxicology Studies Using Flow Cytometry

Flow cytometry is a powerful tool for identifying and measuring aneugenic and clastogenic effects of environmental contaminants on the vertebrate genome. The majority of genotoxicity evaluations using flow cytometry has been performed under controlled laboratory conditions (Otto et al. 1981, 1984; Deaven 1982; Nüsse and Kramer 1984; Melcion et al. 1988; Dwarakanath and Jain 1990; Nüsse et al. 1992; Shen et al. 1992; Schreiber et al. 1992; Miller and Nüsse 1993; Tometsko et al. 1993). But a few studies have examined genomic damage in populations in nature (Bickham et al. 1988; McBee and Bickham 1988; Lamb et al. 1991). Flow cytometry-mediated field studies may not specifically pinpoint causal environmental mutagens or synergistic effects, but comparative experimental designs using control populations under different contamination regimes may enable the identification of sublethal genotoxic effects relative to the genomic norm (Bickham et al. 1988; McBee and Bickham 1988; Lamb et al. 1991). Therefore, the rapidity, accuracy and comparative inexpense of flow cytometry compared to other cytological methods makes it well-suited for screening large numbers of animals in preliminary ecotoxicological studies. Inter- and intrapopulational comparisons of coefficients of variation (CV) of genome size distributions (Otto et al. 1981, 1984; Bickham et al. 1988; McBee and Bickham 1988; Lamb et al. 1991), and micronucleus frequency (Risley and Pohorenec 1991; Nüsse et al. 1992; Rudek and Rozek 1992; Schreiber et al. 1992; Fernandez et al. 1993; Krauter 1993; Schultz et al. 1993; Weller et al. 1993) provide evidence of genomic disruption resulting from environmental contamination. Both types of analysis are possible using flow cytometry, although micronucleus counts are significantly more difficult.

Our preliminary work has shown flow cytometry to be informative in identifying genomic damage in anuran populations subjected to agricultural pesticides (Lowcock et al. 1997). Frogs from farming regions in Quebec suffer a suit of physical and physiological problems associated with the use of agricultural pesticides (Bonin et al., this volume). Flow cytometry was used to compare incidence of abnormal DNA profiles, CV, and variability in genome size between *Rana clamitans* from corn and potato fields, and control sites dissociated from agricultural practices, to infer possible genomic changes due to pesticide use (Lowcock et al. 1997). Statistical analyses showed a significant increase in abnormal DNA profile (i.e. aneuploid mosaicism, Fig. 1b) in individuals collected from agricultural plots relative to the control sites. Adult CV's (Fig. 1c) were higher for samples taken from both potato and corn fields relative to the control samples and juveniles showing physical deformity had significantly higher CV's than normal individuals. The presence of different classes of DNA profile abnormality, as identified by flow cytometry, allows inference into the general mechanisms of pesticide-induced DNA damage at the study sites. Aneuploid mosaicism implies an acute effect early in development resulting in 2 viable cell lineages of differing DNA content. The observed elevated frequencies of these irregularities are the likely result of aneugenic and clastogenic effects brought about through mitotic interference by environmental contaminants (Galloway 1994; Krepinsky and Heddle 1983). Elevated CVs result from cell populations of varying DNA content within an individual and are most often a result of chromosomal aberrations caused by chronic exposure to clastogenic agents (Galloway 1994). Bickham et al. (1988) also encountered aneuploid mosaicism and other abnormal DNA profiles in turtles exposed to radiation.

Among frogs subjected to pesticide exposure, measurable genomic effects were most apparent in individuals showing physiological and physical stress (Bonin et al., this volume). More striking though was the presence of a significant number of individuals showing no visible evidence of stress but which were characterized by genomic damage as measured by flow cytometry (Lowcock et al. 1997). Thus flow cytometry assessments of amphibian populations can provide complimentary data in cases where physical and physiological stress are apparent, or they may reveal genomic damage in individuals showing no outward signs of abnormality (Bonin et al., this volume).

Conclusions

Hybridization, cryptic speciation, and varying degrees of ploidy are fundamental questions regarding the composition and geographical range of an amphibian population. Rigorous methods must therefore be employed to identify units to be monitored, and specific criteria invoked to establish the uniqueness of these units if conservation is to be effective.

Flow cytometry is a cost-effective technique. Both qualitative and quantitative interpretations of flow cytometry-generated data can be used to resolve questions regarding population composition and extent, hybridization, and biodiversity. Additionally, flow cytometry can be used to screen populations for abnormal genomic profiles which may be correlated with the presence of environmental contaminants. The facility with which flow cytometry samples may be obtained, prepared, and stored warrant their collection whenever possible.

Acknowledgments

We thank C. Smith, F. Halwani, and J. L'Haridon for their assistance in flow cytometry techniques and interpretation of flow cytometry data, and D.M. Green for the use of his laboratory for sample preparation and storage. We also thank D.M. Green and Mr. Mosquito for their help in the preparation of this manuscript. This research was supported, in part, by a Canadian Wildlife Service (Quebec Region) grant to TFS and LAL, a Natural Sciences and Engineering Research Council grant (A3148) to RWM, and a Natural Sciences and Engineering Research Council post-doctoral fellowships to LAL. This is Contribution Number 6 from the Centre for Biodiversity and Conservation Biology of the Royal Ontario Museum.

Literature Cited

Bickham JW, Hanks BG, Stolen MJ, Lamb T, Gibbons JW. 1988. Flow cytometric analysis of low-level radiation exposure on natural populations of slider turtles (*Pseudemys scripta*). Archives of Environmental Contamination and Toxicology 17:837–841.

Bogart JP. 1980. Evolutionary implication of polyploidy in amphibians and reptiles. In: Lewis WH, editor. Polyploidy. New York: Plenum Press. p 341–378.

Bogart JP, Elinson RP, Licht LE. 1989. Temperature and sperm incorporation in polyploid salamanders. Science 246:1032–34.

Conant R, Collins JT. 1991. A field guide to reptiles and amphibians of eastern and central North America, 3rd ed. Boston: Houghton Mifflin.

Dawley RM, Goddard KA. 1988. Diploid-triploid mosaics among unisexual hybrids of the minnows *Phoxinus eos* and *Phoxinus neogaeus*. Evolution 42:649–659.

Deaven LL. 1982. Application of flow cytometry to cytogenetic testing of environmental mutagens. In: Hsu TC, editor. Cytogenetic Assays of Environmental Mutagens. Montclair, NJ: Allanheld. p 325–351.

Dwarakanath BS, Jain V. 1990. In vitro radiation responses of human intracranial meningiomas and their modifications by 2-deoxy-D-glucose. Indian Journal of Medical Research 92:183–188.

Fernandez M, L'Haridon J, Gauthier L, Zoll-Moreuxnothing C. 1993. Amphibian micronucleus test(s):a simple and reliable method for evaluating in vivo genotoxic effects of freshwater pollutants and radiations. Initial assessment. Mutation Research 292:83–89.

Galloway SM. 1994. Chromosome aberrations induced *in vitro*: mechanisms, delayed expression, and intriguing questions. Environmental and Molecular Mutagenesis 23, suppl. 24:44–53.

Green DM. 1982. Mating call characteristics of hybrid toads (*Bufo americanus* × *B. fowleri*) at Long Point, Ontario. Canadian Journal of Zoology 60:3293–3297.

Green DM. 1984. Sympatric hybridization and allozyme variation in the toads *Bufo americanus* and *B. fowleri* in southern Ontario. Copeia 1984:18–26.

Green DM, Sharbel TF, Kearsley J, Kaisernothing H. 1996. Postglacial range fluctuation, genetic subdivision and speciation in the western North American spotted frog complex, *Rana pretiosa*. Evolution. 50:374–390.

Holtfreter HB, Cohen N. 1990. Fixation-associated quantitative variations of DNA fluorescence observed in flow cytometric analysis of hemopoietic cells from adult diploid frogs. Cytometry 11:676–685.

Jones JM. 1973. Effects of thirty years hybridization on the toads *Bufo americanus* and *Bufo woodhousei fowleri* at Bloomington, Indiana. Evolution 27:435–448.

King M. 1990. Animal Cytogenetics 4, Chordata 2, Amphibia. Berlin: Gerbrüder Borntraeger.

Krauter PW. 1993. Micronucleus incidence and hematological effects in bullfrog tadpoles (*Rana catesbeiana*) exposed to 2-acetylaminofluorene. Archives of Environmental Contamination and Toxicology 24:487–493.

Krepinsky AB, Heddle JA. 1983. Micronuclei as a rapid and inexpensive measure of radiation-induced chromosomal aberrations. In: Ishihara T, Sasaki MS, editors. Radiation-Induced Chromosome Damage in Man. New York: Alan R. Liss. p 93–109.

Lamb T, Bickham JW, Gibbons JW, Stolen MJ, McDowell S. 1991. Genetic damage in a population of slider turtles (*Trachemys scripta*) in a radioactive reservoir. Archives of Environmental Contamination and Toxicology 20:138–142.

Licht LE, Lowcock LA. 1991. Genome size and metabolic rate in salamanders. Comparative Biochemistry and Physiology B 100:83–92.

Lowcock LA. 1994. Biotype, genomotype, and genotype: variable effects of polyploidy and hybridity on ecological partitioning in a bisexual-unisexual community of salamanders. Canadian Journal of Zoology 72:104–117.

Lowcock LA, Griffith H, Murphy RW. 1991. The *Ambystoma laterale-jeffersonianum* complex in central Ontario: ploidy structure, sex ratio and breeding dynamics in bisexual-unisexual communities. Copeia 1991:87–105.

Lowcock LA, Griffith H, Murphy RW. 1992. Size in relation to sex, hybridity, ploidy, and breeding dynamics in central Ontario populations of the *Ambystoma laterale-jeffersonianum* complex. Journal of Herpetology 26:46–53.

Lowcock LA, Licht LE. 1990. Natural autotriploidy in salamanders. Genome 33:674–678.

Lowcock LA, Murphy RW. 1991. Pentaploidy in hybrid salamanders demonstrates enhanced tolerance of multiple chromosome sets. Experientia 47:490–493.

Lowcock LA, Sharbel TF, Bonin J, Ouellet M, Rodrigue J, DesGranges J-L. 1997. Flow cytometric assay for *in vivo* genotoxic effects in green frogs (*Rana clamitans*). Aquatic Toxicology (in press).

MacCulloch RD, Upton DE, Murphy RW. 1996. Trends in nuclear DNA content among amphibians and reptiles. Comparative Biochemistry and Physiology B 113:601–606..

McBee K, Bickham JW. 1988. Petrochemical-related DNA damage in wild rodents detected by flow cytometry. Bulletin of Environmental Contamination and Toxicology 40:343–349.

Melcion C, Maratrat M, Grünwald D, Frelat G, Cordier A. 1988. Two-parameter flow cytometry to detect clastogenic agents. Mutation Research 202:76.

Miller BM, Nüsse M. 1993. Analysis of micronuclei induced by 2-chlorobenzylidene malonitrile (CS) using fluorescence in situ hybridization with telomeric and centromeric DNA probes, and flow cytometry. Mutagenesis 8:35–41.

Murphy RW, Lowcock LA, Smith C, Darevsky IS, Orlov N, MacCulloch RD, Upton DE. 1997. Flow cytometry in biodiversity surveys: methods, utility, and constraints. Amphibia-Reptilia 18:1–13.

Nüsse M, Kramer J. 1984. Flow cytometric analysis of micronuclei found in cells after irradiation. Cytometry 5:20–25.

Nüsse M, Kramer J, Miller BM. 1992. Factors influencing the DNA content of radiation-induced micronuclei. International Journal of Radiation Biology 62:587–602.

Olmo E, Morescalchi A. 1978. Genome and cell sizes in frogs: a comparison with salamanders. Experientia 34:44–46.

Olmo E, Capriglione T, Odierna G. 1989. Genome size evolution in vertebrates: trends and constraints. Comparative Biochemistry and Physiology 92:447–453.

Otto FJ, Oldiges H, Jain VK. 1984. Flow cytometric measurement of cellular DNA content dispersion induced by mutagenic treatment. In: Eisert WG, Mendelsohn ML, editors. Biological Dosimetry. Cytometric Approaches to Mammalian Systems. New York: Springer-Verlag. p 38–48.

Otto FJ, Oldiges H, Göhde W, Jain VK.nothing 1981. Flow cytometric measurement of nuclear DNA content variations as a potential *in vivo* mutagenicity test. Cytometry 2:189–191.

Price HJ, Chambers KL, Bachmann K, Riggs J. 1983. Inheritance of nuclear 2C DNA content variation in intraspecific and interspecific hybrids of *Microseris* (Asteraceae). American Journal of Botany 70:1133–1138.

Price HJ, Chambers KL, Bachmann K, Riggs J. 1985. Inheritance of nuclear 2C DNA content in a cross between *Microseris douglasii* and *M. bigelovii* (Asteraceae). Biologisches Zentralblatt 104:269–276.

Risley MS, Pohorenec GM. nothing 1991. Micronuclei and chromosome aberrations in *Xenopus laevis* spermatocytes and spermatids exposed to adriamycin and colcemid. Mutation Research 247:29–38.

Rudek Z, Roek M. 1992. Induction of micronuclei in tadpoles of *Rana temporaria* and *Xenopus laevis* by the pyrethroid Fastac 10 EC. Mutation Research 298:25–29.

Schreiber GA, Beisker W, Bauchinger M, Nüsse M. 1992. Multiparametric flow cytometric analysis of radiation-induced micronuclei in mammalian cell cultures. Cytometry 13:90–102.

Schultz N, Norrgren L, Grawé J, JohannissonA, Medhage O. 1993. Micronuclei frequency in circulating erythrocytes from rainbow trout (*Oncorhynchus mykiss*) subjected to radiation, an image analysis and flow cytometric study. Comparative Biochemistry and Physiology 105:207–211.

Sharbel TF, Bonin J. 1992. Northernmost record for *Desmognathus ochrophaeus*: Biochemical identification in the Chateaugay river drainage basin, Quebec Journal of Herpetology 26:505–508.

Sharbel TF, Bonin J, Lowcock LA, Green DM. 1995. Partial genetic compatibility and unidirectional hybridization in syntopic populations of the salamanders *Desmognathus fuscus* and *D. ochrophaeus*. Copeia.1995:466–469.

Shen RN, Crabtree WN, Wu B, Young P, Sandison GA, Hornback NB, Shidnia H. 1992. A reliable method for quantitating chromatin fragments by flow cytometry to predict the effect of total body irradiation and hyperthermia on mice. International Journal of Radiation Oncology, Biology, and Physics 24:139–143.

Tank PW, Charleton RK, Burns ER. 1987. Flow cytometric analysis of ploidy in the axolotl, *Ambystoma mexicanum*. Journal of Experimental Zoology 243:423–433.

Tometsko AM, Torous DK, Dertinger SD. 1993. Analysis of micronucleated cells by flow cytometry. 1. Achieving high resolution with a malaria model. Mutation Research 292:129–135.

Vindeløv LL, Christensen IJ, Jensen G, Nissen NI. 1983. Limits of detection of nuclear DNA abnormalities by flow cytometric DNA analysis. Results obtained by a set of methods for sample-storage, staining and internal standardization. Cytometry 3:332–339.

Vinogradov AE, Borkin LJ, Günther R, Rosanov JM. 1991. Two germ cell lineages with genomes of different species in one and the same animal. Hereditas 114:245–251.

Vinogradov AE, Borkin LJ, Gunther R, Rosanov. 1990. Genome elimination in diploid and triploid *Rana esculenta* males: cytological evidence from DNA flow cytometry. Genome 33:619–627.

Weller EM, Dietrich I, Viaggi S, Beisker W, Nüsse M. 1993. Flow cytometric analysis of bromodeoxyuridine-induced micronuclei. Mutagenesis 8:437–444.

©1997 by the Society for the Study of Amphibians and Reptiles
Amphibians in decline: Canadian studies of a global problem. David M. Green, editor.
Herpetological Conservation 1:87–92.

Chapter 10

THE VALUE OF MONITORING GENETIC DIVERSITY: DISTRIBUTION OF AMBYSTOMATID SALAMANDER LINEAGES IN ONTARIO

Leslie A. Rye, Wanda J. Cook, and James P. Bogart

Department of Zoology, University of Guelph, Guelph, Ontario N1G 2W1, Canada

Abstract.—Genetic analyses are essential to document the distribution of some amphibians, such as members of the *Ambystoma laterale–jeffersonianum* complex. Use of genetic analyses has extended the known ranges of *Ambystoma* [2] *laterale–jeffersonianum* (LLJ) and *Ambystoma laterale*–[2] *jeffersonianum* (LJJ) by approximately 80 km and 100 km, respectively, further north than indicated on current range maps. Failure to incorporate genetic monitoring can lead to inaccurate knowledge of distributions and, therefore, an inability to monitor declines. Very little information exists on the genetic variation in Canadian amphibian populations in general and opportunities to collect basic genetic information are disappearing quickly. A genetic analysis component should be incorporated into any long-term, intensive monitoring program.

Résumé.— Les analyses génétiques sont essentielles à l'étude de la répartition de certains amphibiens comme les membres du complexe *Ambystoma laterale–jeffersonianum*.*Ambystoma* [2] *laterale–jeffersonianum* (LLJ) et *Ambystoma laterale*–[2] *jeffersonianum* (LJJ) ont été retrouvés respectivement à environ 80 km au nord d'un secteur décrit antérieurement et à environ 100 km plus au nord du point indiqué sur les cartes actuellement disponibles. L'absence de surveillance génétique peut fausser les données sur la répartition et par conséquent empêcher la surveillance des déclins démographiques. Nous avons peu de données sur les variations génétiques des populations canadiennes d'amphibiens en général et les possibilités d'obtenir des données génétiques fondamentales se font rapidement de plus en plus rares. L'analyse génétique devrait être intégrée dans tout programme de surveillance intensif à long terme.

Genetic diversity is an important but often neglected component of amphibian monitoring programs (Bogart 1992). *Ambystoma laterale* and *A. jeffersonianum* are 2 of the 4 species of mole salamanders (Ambystomatidae) found in Ontario, where they are often associated with triploid (3n = 42 chromosomes), usually female, hybrids. Triploids associated with *A. laterale* almost invariably have 2 copies of the *A. laterale* genome and 1 copy of the *A. jeffersonianum* genome and are thus hybrid *A.* [2] *laterale–jeffersonianum*, hereafter abbreviated to LLJ following the nomenclature of Lowcock et al. (1987). These triploids have been identified from the southern part of northeastern Ontario using chromosomes and genetic analyses (Fig. 1). Similarly, those triploids found with *A. jeffersonianum* have 2 copies of the *A. jeffersonianum* genome and 1 copy of the *A. laterale* genome and are thus *A. laterale*–[2] *jeffersonianum*, abbreviated to LJJ. In Ontario, *A. jeffersonianum* and LJJ triploids have been documented as far north as Wellington County (Weller et al. 1979; Fig. 1). Both types of triploids lay mostly unreduced (3n) eggs in the spring that develop without incorporating the male's sperm. Tetraploid, and even pentaploid, hybrids have also been found (Lowcock and Murphy 1991), resulting from sperm incorporation into an unreduced egg. The frequency of ploidy elevation is influenced by temperature (Bogart et al. 1989).

Members of *A. laterale–jeffersonianum* complex are very difficult to distinguish morphologically, especially during their larval and juvenile stages. This makes distribution maps for individual members of this complex unreliable if based solely on morphological data and distribution maps which do not distinguish the various diploid and polyploid members of the complex are not as informative as they might be. Detecting declines in the various members of this complex would, therefore, be difficult if not impossile if identification relied solely on morphological information.

Juvenile Ambystoma jeffersonianum *from Hockley Valley, Ontario (JPB 21672). Photo by James P. Bogart.*

The hybrids are also unusual in that, regardless of the combination of nuclear genomes present, they share a common mitochondrial genome which is estimated to be at least 4,000,000 yr old (Hedges et al. 1992). This molecule, therefore, cannot be used to distinguish among the various hybrids. The nuclear genomes of *A. laterale* and *A. jeffersonianum* can be determined using the molecular technique of allozyme electrophoresis (Bogart 1982) but while allozymes can distinguish many members of the complex, they can not distinguish a diploid hybrid (LJ) from a symmetrical

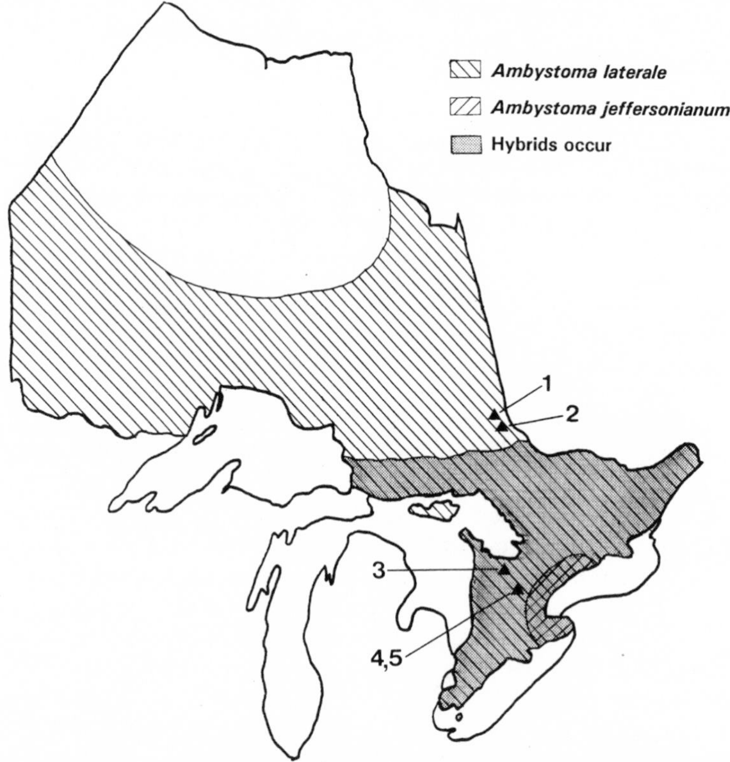

Figure 1. *Current known range of* Ambystoma laterale, A. jeffersonianum, *and their associated hybrids in Ontario (modified from Conant and Collins 1991; Uzell 1967a, 1967b, 1967c). Collection localities (triangles) represent range extensions of hybrid* A. [2] laterale–jeffersonianum *at Englehart (1) and New Liskeard (2), and of* A. Laterale–[2] jeffersonianum *at Kolapore Uplands Provincial Park (3), Mono Cliffs (4), and Hockley Valley Provincial Park (5), based on genetic analysis.*

tetraploid (LLJJ) or a triploid (LLJ or LJJ) from an asymmetrical tetraploid (LLLJ or LJJJ). However, in combination with karyotypic information, which ascertains the ploidy level of the individual salamander, allozymes can be used to reliably distinguish all members of the *A. laterale–jeffersonianum* complex (Bogart et al. 1987). These techniques can be applied to more accurately documenting the ranges of members of this complex.

Methods

Salamanders were collected in northeastern Ontario from Englehart (UTM 31 M/13 848973) during 1991 and from New Liskeard (31 M/12 925693 and 31 M/12 929685) in 1993 and 1994. Salamanders were also collected along the Niagara Escarpment in southern Ontario between 1990 and 1992 at 2 sites within Kolapore Uplands Provincial Park (41 A/8 456191 and 41 A/8 447172), at Mono Cliffs (41 A/1 728776), and from Hockley Valley Provincial Park (40 P/16 755607). At the time of collection, these sampling localities were thought to represent areas beyond the ranges of LLJ or LJJ or of *Ambystoma jeffersonianum* (Fig. 1).

Salamanders were collected as eggs, larvae or adult and transported to the University of Guelph for genetic identification. Eggs and larvae were raised to transformation in the laboratory. After metamorphosis, the salamanders were euthanized and liver, heart, spleen, and some skeletal muscle were removed and frozen. Any larva which died prior to transformation was also frozen. These tissues and whole larvae were then used for allozyme electrophoresis (Table 1). *Ambystoma laterale* and *A. jeffersonianum* show fixed differences for alternate allozymes for several enzyme systems.

Ploidy was determined by examining preparations of either tail tip epithelial cells from larvae or cloacal epithelial cells from adults. A small portion of the larval tail tip was removed and placed in a 7:4 solution of dilute Hank's Balanced Salt Solution (diluted 4:1 with deionized water): deionized water to which one part 1mg/ml colchicine solution had been added. Tail tips were left in this solution for 15.5 hr and then they were transferred to deionized water for about 3 hr. The tail tip was then

Table 1. *Diagnostic enzymes examined, buffers, and tissues used in identifying individuals of the* Ambystoma laterale–jeffersonianum *complex (modified from Bogart 1989).*

Enzyme name	Locus	Enzyme Commision number	Buffer[a]	Tissue[b]
Acid Phosphatase	AcPh-2	(EC 3.1.3.2)	Tris	Liver
Aconitase	Acon-1	(EC 4.2.1.3)	Tris	Liver
Creatine kinase	Ck-1	(EC 2.7.3.2)	C&T	HMS
	Ck-2		C&T	HMS
Glutamate-oxaloacetate transaminase	Aat-1	(EC 2.6.1.1)	C&T or Tris	Liver, e
(=Aspartate aminotransferase)	Aat-2		C&T	Liver
Lactate dehydrogenase	Ldh-1	(EC 1.1.1.27)	C&T	HMS, e
	Ldh-2		C&T	HMS, e
Malate dehydrogenase	Mdh-1	(EC 1.1.1.37)	C&T	HMS, e
	Mdh-2		C&T	HMS, e
	Mdh-3		C&T	HMS, e
Phosphoglucomutase	Pgm-1	(EC 5.4.2.5)	Tris	Liver
	Pgm-2		Tris	Liver
Superoxide dismutase	Sod-1	(EC 1.15.1.1)	C&T	Liver

[a]Tris= Tris-citrate, pH 6.7 (Selander et al. 1971): C&T = Amine-citrate, pH 6.5 (Clayton and Tretiak 1972)

[b]HMS = heart, muscle, and spleen combined; e = eggs or whole larvae

placed in a drop of 70% acetic acid on a siliconized coverslip and squashed. The slide was sealed with cover glass cement (Pfaltz & Bauer, Inc. Cat. No. C25974), examined under a microscope and the chromosomes counted. When using cloacal epithelial cells (necessary on metamorphosed individuals), the salamander was injected with a 1 mg/ml colchicine solution 48 hr prior to being killed in an overdose of MS222. The cloaca was removed and placed in deionized water for about 10 min. The tissue was then stored in a 3:1 ethanol:acetic acid fixative. Portions of the cloacal epithelium were squashed and examined as described above for tail tip squashes.

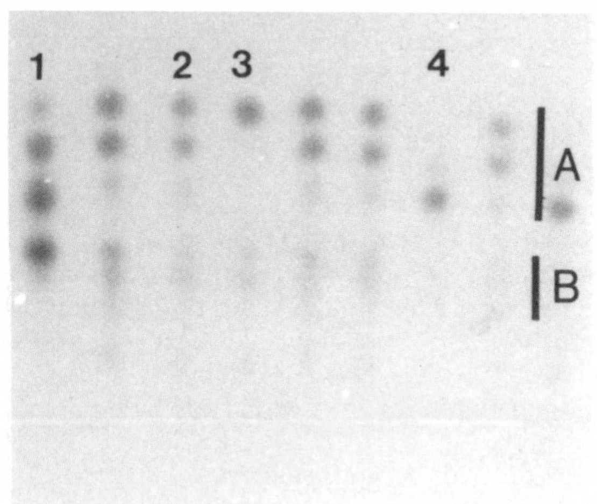

Figure 2. *Electrophoretic gel showing fixed differences between* A. laterale *and* A. jeffersonianum *(3,4) and reciprocal triploid hybrid genotypes (1,2), for malate dehydrogenase. A. Mdh-1. B. Mdh-2.*

Results

Genetic identities (Fig. 2) and chromosome numbers (Fig. 3) for all salamanders could be readily determined. The results of the genetic analyses (Table 2) revealed that the previously documented ranges for members of this complex must be amended. *Ambystoma jeffersonianum* and/or its associated hybrids (LJJ and LJJJ) were found approximately 80 km N of the previously described range for these salamanders and LLJ hybrids were found approximately 100 km further N than indicated on current range maps (Fig. 1).

Discussion

The current ranges for members of this salamander complex are not known well, yet a clear understanding of the distribution is necessary in order to assess any changes in distribution which may be associated with a decline. The use of genetic analyses to identify members of this salamander complex prompts a re-evaluation of their purported ranges. It is expected, for example, that the diploid, *A. jeffersonianum*, will be found

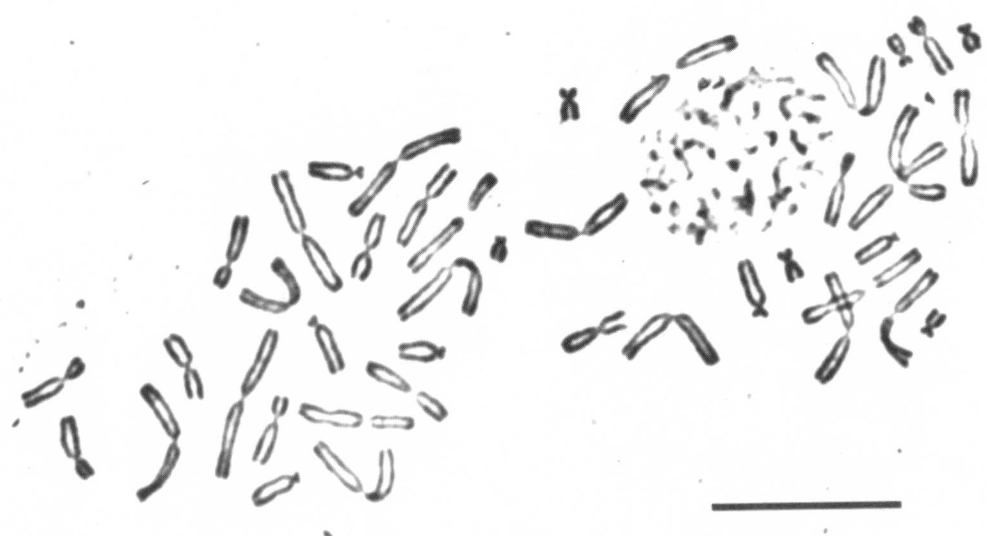

Figure 3. *Karyotypic preparation from* Ambystoma *tail tip epithelium showing the triploid condition of 42 chromosomes. Scale bar is 30 μm.*

Table 2. *Genotype and ploidy determinations of ambystomatid salamanders. LLJ = triploid* A. [2] laterale–jeffersonianum, *LJJ = triploid* A. laterale–[2] jeffersonianum, *LJJJ = tetraploid* A. laterale–[3] jeffersonianum.

Locality	Specimens	Genotype
Englehart	1 adult	LLJ
New Liskeard	10 adults	3 *A. laterale*, 7 LLJ
Kolapore Uplands Provincial Park	62 larvae raised from 9 egg masses	1) 2 LJJ; 2) 6 LJJ, 1 LJJJ; 3) 14 LJJ; 4) 18 LJJ, 6 LJJJ; 5)4 LJJ; 6) 4 LJJ; 7) 1 LJJ; 8) 3 LJJ; 9) 3 LJJ
Mono Cliffs	3 larvae	3 *A. jeffersonianum*
	1 adult	LJJ
	34 larvae raised from 5 egg masses	1) 13 LJJ, 5 LJJJ; 2) 3 LJJ; 3) 4 *A. jeffersonianum*; 4) 7 *A. jeffersonianum*; 5) 7 *A. jeffersonianum*
Hockley Valley Provincial Park	7 larvae	2 *A. jeffersonianum*, 5 LJJ

in Kolapore Uplands Provincial Park, because 2 of the 9 egg masses sampled contained both triploid LJJ and tetraploid LJJJ individuals. This ploidy elevation is indicative of sperm incorporation from *Ambystoma jeffersonianum*. These methods, because they can be used on eggs and/or larvae, also allow for sampling a population with minimum impact on the breeding population. These types of genetic analyses can also provide information on the presence of genetically differentiated populations throughout Ontario (Lowcock and Bogart 1989).

The identification of genetic biodiversity in Canadian populations of amphibians is meagre. In addition to the *Ambystoma* complex, geographical genetic population substructuring has been shown for only a few species in Canada. Zeyl (1993) documented considerable allozyme variation in *Rana sylvatica*, Green (1983; 1984) examined genetic variation in *Bufo americanus*, *B. fowleri*, and *B. hemiophrys*, Green et al. (1996) resolved species-level differentiation within *Rana pretiosa*, and Sharbel and Bonin (1992) could genetically distinguish between *Desmognathus fuscus*, *D. ochrophaeus*, and their hybrids. But beyond that, there is clearly little baseline data at present against which to document any decline in genetic diversity in Canadian amphibian populations.

Although we recognize that it is not practical to incorporate an analysis of genetic biodiversity into each, or even most, of the studies currently underway to monitor amphibian populations, we recommend that genetic analysis be incorporated into those involving long term, intensive monitoring. Individuals of the same species are not genetically identical, and 10 yr from now, we may find ourselves with no genetic baseline data—just as we now find a need for better abundance and distribution baseline data. It will be too late to obtain such information when the populations are small and in danger of extinction.

Acknowledgments

Graham Gables, James Johnston, Bill Lamond, Ian Mason, Jim McLister and Alison Taylor helped to collect specimens. Kathleen Gardiner and Jim McLister assisted with the raising of larvae and Alison Taylor helped with the chromosome analysis. We would like to thank the Niagara Escarpment Fund of the Ontario Heritage Foundation and the Natural Sciences and Engineering Research Council (NSERC operating grant to JPB) for financial assistance.

Literature Cited

Bogart JP. 1982. Ploidy and genetic diversity in Ontario salamanders of the *Ambystoma jeffersonianum* complex revealed through an electrophoretic examination of larvae. Canadian Journal of Zoology 60:848–855.

Bogart JP. 1989. A mechanism for interspecific gene exchange via all-female salamander hybrids. In: Dawley RM, Bogart JP, editors. Evolution and Ecology of Unisexual Vertebrates. Albany, NY: New York State Museum. Bulletin 466. p 170–179.

Bogart JP. 1992. Monitoring genetic diversity. In: Bishop CA, Pettit KE, editors. Declines in Canadian amphibian populations: designing a national monitoring strategy. Ottawa: Canadian Wildlife Service. Occasional Paper 76. p 50–52.

Bogart JP, Elinson RP, Licht LE. 1989. Temperature and sperm incorporation in polyploid salamanders. Science 246:1032–1034.

Bogart JP, Lowcock LA, Zeyl CW, Mable BK. 1987. Genome constitution and reproductive biology of hybrid salamanders, genus *Ambystoma*, on Kelley's Island in Lake Erie. Canadian Journal of Zoology 65:2188–2201.

Clayton JW, Tretiak DN. 1972. Amine-citrate buffers for pH control in starch gel electrophoresis. Journal of the Fisheries Research Board of Canada 29:1169–1172.

Conant R, Collins JT. 1991. A field guide to the reptiles and amphibians of eastern and central North America. 3rd. ed. Boston: Houghton Mifflin.

Green DM. 1983. Allozyme variation through a clinal hybrid zone between the toads *Bufo americanus* and *B. hemiophrys* in southeastern Manitoba. Herpetologica 39:28–40.

Green DM. 1984. Sympatric hybridization and allozyme variation in the toads *Bufo americanus* and *B. fowleri* in southern Ontario. Copeia 1984:18–26.

Green DM, Sharbel TF, Kearsley J, Kaiser H. 1996. Postglacial range fluctuation, genetic subdivision and speciation in the western North American Spotted frog complex, *Rana pretiosa*. Evolution 50:374–390.

Hedges SB, Bogart JP, Maxson LR. 1992. Ancestry of unisexual salamanders. Nature 356:708–710.

Lowcock LA, Bogart JP. 1989. Electrophoretic evidence for multiple origins of triploid forms in the *Ambystoma laterale - jeffersonianum* complex. Canadian Journal of Zoology 67:350–356.

Lowcock LA, Licht LE, Bogart JP. 1987. Nomenclature in hybrid complexes of *Ambystoma* (Urodela: Ambystomatidae): no case for the erection of hybrid "species". Systematic Zoology 6 :328–336.

Lowcock LA, Murphy RW. 1991. Pentaploidy in a hybrid salamander demonstrates enhanced tolerance of multiple chromosome sets. Experientia 47:490–493.

Selander RK, Smith MH, Yang SY, Johnson WE, Gentry JB. 1971. Biochemical polymorphism and systematics in the genus *Peromyscus*. I. Variation in the old-field mouse (*Peromyscus polionotus*). University of Texas Publication 7103:49–90.

Sharbel TF, Bonin J. 1992. Northernmost record of *Desmognathus ochrophaeus*: biochemical identification in the Chateauguay River drainage basin, Québec. Journal of Herpetology 26:505–508.

Uzzell TM. 1967a. *Ambystoma jeffersonianum*. Catalogue of American Amphibians and Reptiles 47:1–2.

Uzzell TM. 1967b. *Ambystoma laterale*. Catalogue of American Amphibians and Reptiles 48:1–2.

Uzzell TM. 1967c. *Ambystoma tremblayi*. Catalogue of American Amphibians and Reptiles 50:1–2.

Weller WF, Campbell CA, Lovisek J, Mackenzie B, Servage D, Tobias TN. 1979. Additional records of salamanders of the *Ambystoma jeffersonianum* complex from Ontario, Canada. Herpetological Review 10:61–62.

Zeyl CW. 1993. Allozyme variation and divergence among populations of *Rana sylvatica*. Journal of Herpetology 27:233–236.

Chapter 11

AMPHIBIANS OF NORTHERN AND NORTHEASTERN CANADA

Amphibians in decline: Canadian studies of a global problem. David M. Green, editor.
Herpetological Conservation 1:93–99.

AMPHIBIANS OF NEWFOUNDLAND AND LABRADOR: STATUS CHANGES SINCE 1983

JOHN E. MAUNDER

Natural History Section, Newfoundland Museum, 285 Duckworth Street, St. John's, Newfoundland A1C 1G9, Canada

ABSTRACT.—The amphibians of Newfoundland and Labrador remain poorly-known. At present, little can be said about temporal changes in their status. For Labrador, a complete species list may not yet exist, and range maps for both Labrador and Newfoundland remain fragmentary. Six species, all native, are known for Labrador: *Bufo americanus, Rana septentrionalis, R. sylvatica, R. pipiens, Ambystoma laterale*, and *Eurycea bislineata*. *Bufo americanus, R. clamitans*, and *R. sylvatica*, all introduced, are known to still exist in Newfoundland. Two additional introduced species *R. pipiens* and *Pseudacris triseriata* are now apparently extirpated there.

RÉSUMÉ.—Les amphibiens de Terre-Neuve et du Labrador restent méconnus. À l'heure actuelle, rien ne peut être véritablement affirmé sur les changements temporels intervenus dans leurs populations. La liste complète des espèces n'existe peut-être même pas pour le Labrador et les cartes du Labrador et de Terre-Neuve restent fragmentaires. Six espèces, toutes autochtones, sont connues au Labrador: *Bufo americanus, Rana septentrionalis, R. sylvatica, R. pipiens, Ambystoma laterale*, et *Eurycea bislineata*. *Bufo americanus, R. clamitans*, et *R. sylvatica* qui ont été introduites, existent encore à Terre-Neuve. Deux espèces introduites, *R. pipiens* et *Pseudacris triseriata*, semblent aujourd'hui avoir disparu de ces régions.

The only comprehensive survey of the amphibians of Newfoundland and Labrador published to date, by Maunder (1983; but see also Buckle 1971), recorded 5 species present on the Island of Newfoundland: American toad (*Bufo americanus*), green frog (*Rana clamitans*), wood frog (*Rana sylvatica*), northern leopard frog (*Rana pipiens*) and chorus frog (*Pseudacris triseriata*). All are introduced. Maunder (1983) also listed 6 native species for the mainland of Labrador: *B. americanus*, mink frog (*Rana septentrionalis*), *R. sylvatica*, *R. pipiens*, blue-spotted salamander (*Ambystoma laterale*) and northern two-lined salamander (*Eurycea bislineata*). Since 1983, the amphibians of the region have received very little additional attention. There has been some minor collecting, mainly by the author, and a few field workers have kept incidental notes. But a tremendous amount of work remains to be done. Labrador, in particular, is still *terra incognita*. Satisfactory range maps and probably even a complete species list remain elusive.

The present paper documents new collections and observations made by the author and others in Labrador and attempts to update changes in the status of the five introduced species on the Island of

Newfoundland, including the apparent demise of both *P. triseriata* and *R. pipiens*. I also describe the apparent sharp decline and recent slow recovery of the introduced *R. clamitans* in many parts of eastern Newfoundland.

Observations

Labrador

In July 1988, I made incidental collections of amphibians, including *B. americanus, R. septentrionalis, R. sylvatica, R. pipiens, A. laterale*, and *E. bislineata* (Newfoundland Mus. Nos. HE-105, HE-107 to HE-111, HE-117 to HE-132), in central Labrador (Appendix). These collections were made at eighteen localities along Rte. 520 between Happy Valley-Goose Bay and Northwest River and along the east-west Trans-Labrador Road (Rte. 500) between Happy Valley-Goose Bay and Esker (Fig. 1). In October 1984, M. Gillett (pers. comm.) collected one *E. bislineata* near Northwest River, Labrador (Royal Ontario Mus. No. H13270). During the summer of 1992, a credible sight record of "several, adult" *R. septentrionalis* was made along the western shore of Pocketknife Lake, Labrador (54°10′ N, 61°20′ W) by K. McAleese and R. Button (pers. comm.). The localities are new Labrador records (Appendix, Fig. 2).

A major gap in the distributions of at least 3 species of amphibian extends from where the Trans-Labrador Road rises away from the Churchill River toward the Labrador Plateau about 60 km W of Happy Valley-Goose Bay (about 61° W), to about 20 km W of Churchill Falls (about 64° W). This apparent gap is best illustrated by the distributions of *B. americanus, R. sylvatica*, and *A. laterale* (Fig. 2). The restricted eastern Labrador distribution of *R. pipiens*, and even its very occurrence in Labrador, appears to be primarily the result of locally favourable climatic conditions in the immediate vicinities of Goose Bay (Bleakney 1958; Maunder 1983) and Paradise River (pers. obs.). The gap is also strikingly impoverished of freshwater molluscs, particularly gastropods (J. Maunder, unpublished), despite the presence of innumerable water bodies of every size and description. Only *E. bislineata* appears to generally defy the pattern. However, a single *R. septentrionalis* collected from the banks of the Churchill River (Maunder 1983), within the gap area, may indicate that this species is also regularly present there, at least in the immediate vicinity of the river, whose waters flow at high volume from the enormous Smallwood Reservoir sprawling far to the north and west. Further study is required before anything more can be said on this matter.

Newfoundland

On the Island of Newfoundland, the introduced *B. americanus* has continued to spread. A credible sight record at Little Grand Lake (48°35′ N, 57°50′ W), about 25 km E of the Trans-Canada Hwy, Rte. 1, and the town of Gallants (J. Brazil, pers. comm.), represents a 25 km eastward advance since 1970. Toads have reached Badger (48°59′ N, 56°02′ W), a range extension of about 95 km E of Deer Lake since 1978 (M. Pitcher, pers. comm.). *Rana sylvatica* is still dispersing slowly throughout the greater Corner Brook/Steady Brook area in western Newfoundland. *Rana clamitans* were reported at Hawkes Bay (50°36′ N, 57°10′ W) on Newfoundland's Great Northern Peninsula (J. Buckle, pers. comm.), an apparent range extension of 115 km to the north-northwest from Gros Morne National Park since 1983. While amphibians have frequently been moved from place to place in Newfoundland by humans, it seems possible that at least the record of *B. americanus* at Little Grand Lake, which has no road connection, may reflect natural dispersal through upland areas. The Hawkes Bay *R. clamitans* population may or may not have originated on its own from an older population which once inhabited nearby Port Saunders (Cameron and Tomlinson 1962; Maunder 1983).

During the period 1978 to 1981, *B. americanus, R. pipiens, R. sylvatica*, and *P. triseriata* from existing western Newfoundland populations were intentionally transplanted to several locations within Gros Morne National Park by park officials, apparently to enrich the local fauna. Park surveys carried out between 1989 and 1991 indicated that only the toad introductions were successful (R. Prosper, pers. comm.). In the early 1980s, repeated attempts were also made by Newfoundland and Labrador Wildlife Division officials to introduce *B. americanus* and *R. sylvatica* from the Corner Brook area to vari-

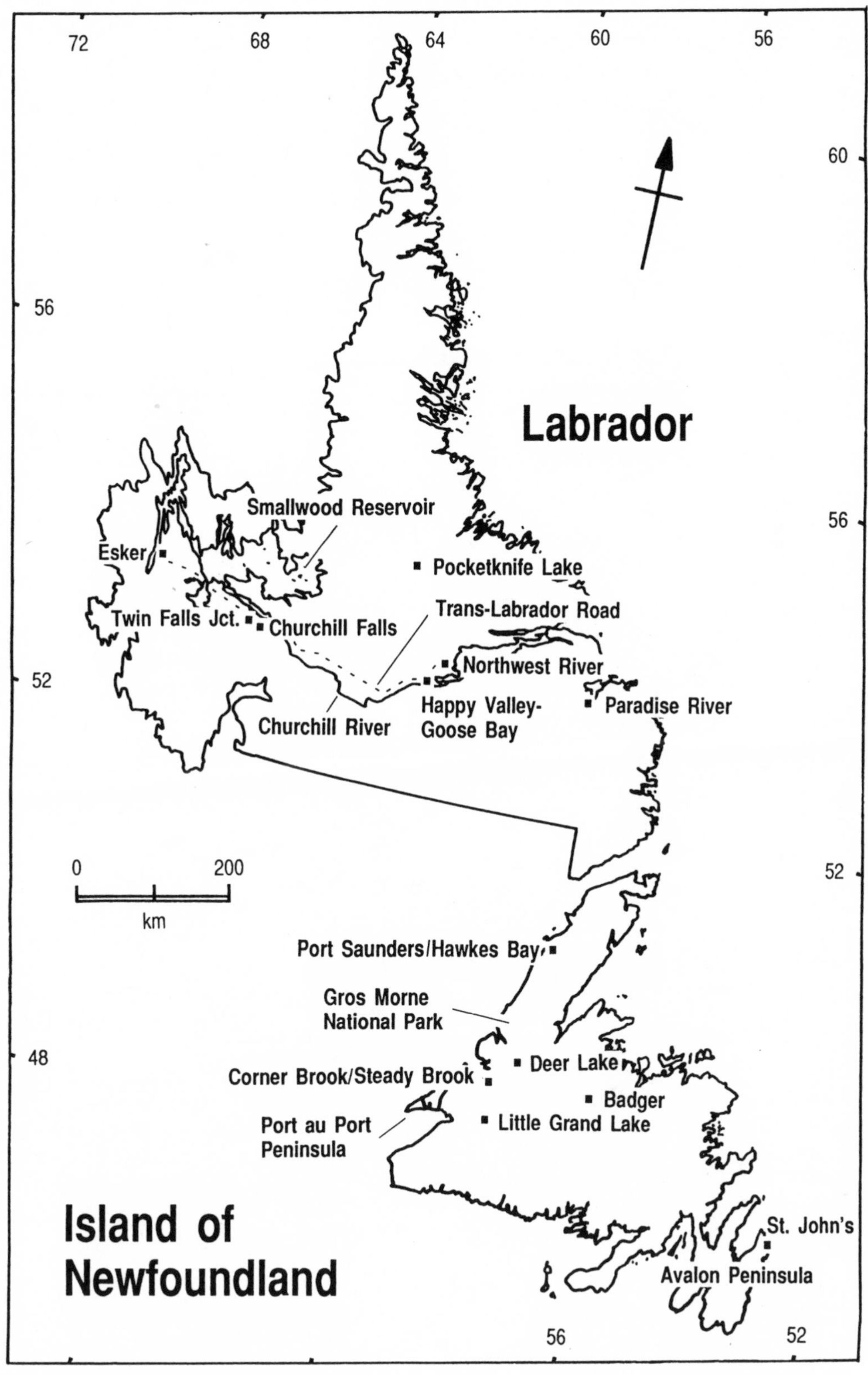

Figure 1. *Newfoundland and Labrador, including locations identified in text.*

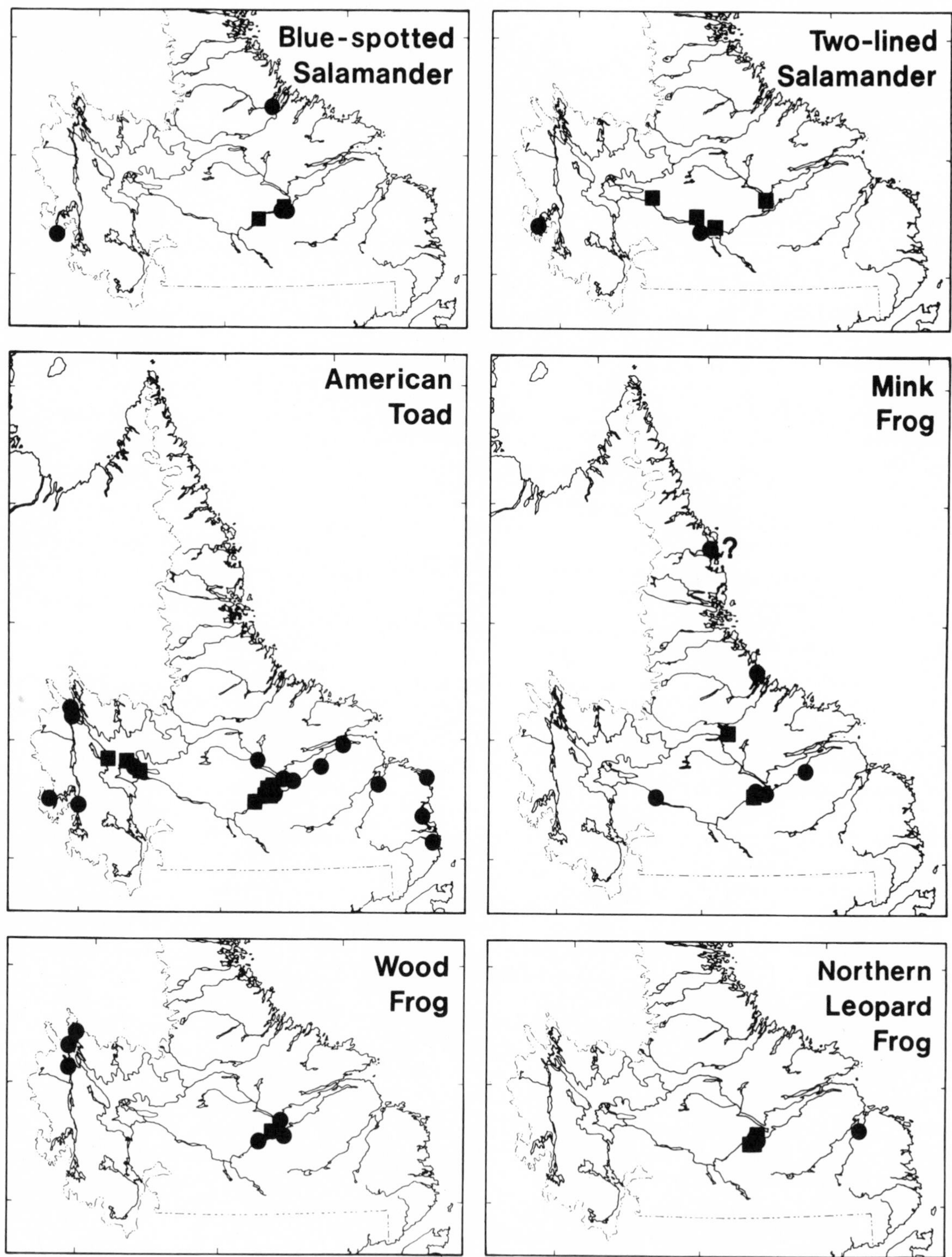

Figure 2. *Distribution of amphibian species known to occur in Labrador. Circles represent records previously documented in Maunder (1983). Squares represent new records documented in this paper.*

ous locations on the Avalon Peninsula, in particular to the centrally located Salmonier Nature Park and its vicinity. The intent of these endeavours was also, apparently, to enrich the local fauna. None of these attempts, which involved significant numbers of eggs (> 1000) and larvae (> 1300), as well as a few adults (< 15), appears to have been successful (M. Pitcher, pers. comm.).

Wood frogs, Rana sylvatica, *in amplexus. Photo by Martin Ouellet.*

The movement of amphibians by private individuals in Newfoundland continues to be a phenomenon. All 4 recently-introduced species have been transplanted a number of times from near their original introduction sites in the Corner Brook area to locations the length and breadth of the Island (J. Buckle, pers. comm.). However, outside a triangle anchored by Gros Morne National Park in the north, the Port au Port Peninsula in the southwest, and Deer Lake in the east (Fig. 1), none of these transplants appear to have been successful.

Pseudacris triseriata apparently no longer exist in Newfoundland. James Buckle (pers. comm.) has monitored introduction sites in western Newfoundland on a regular basis since 1963 but has reported no chorus frog sightings since June 1982. *Rana pipiens* may also no longer exist in Newfoundland. Originally introduced to clear water ponds west of Corner Brook (Buckle 1971), the populations never dispersed. *Rana pipiens* (2 yearlings) was last reported in Newfoundland during May and June 1989 (J. Buckle, pers. comm.).

Rana clamitans is well-established in Newfoundland and has had a long history of successful transplantation within the island (Cameron and Tomlinson 1962; Maunder 1983), but nothing is known of any movements since 1983. However, the species appears to have experienced a dramatic decline in at least the eastern regions of Newfoundland over the last decade or so. That trend may now be slowly reversing. During the last 2 yr, *R. clamitans* have been reported from traditional localities where they have apparently been absent, or at least very rare, for some time. Yet there is no quantitative baseline or recent amphibian population data of any sort that has ever been collected in Newfoundland. The few recent observations of *R. clamitans* that do exist are entirely qualitative, being based solely upon largely undirected incidental observations a number of reliable observers.

Discussion

Left unresolved by Maunder (1983) was the question of just how *Rana clamitans* was introduced to the Island of Newfoundland. It now appears that the original introduction(s) may have been intentional, rather than possibly accidental as was suggested by Johansen (1926) "... these frogs ... are supposed to have been introduced (with hay?) from Nova Scotia ...", apparently sometime after Jukes (1842) wrote that "Not a frog, nor a toad ... has ever been seen in the country" but before Maret (1867) found "frogs" in ponds and lakes in the St. John's area. Hardy (1869) reported, perhaps too pessimistically, and without elaboration, that "... more than once [on the Avalon Peninsula of eastern Newfoundland] has the experiment been tried of turning out some of the large green-headed frogs [ie. *R. clamitans*], to end in failure: in a few days they would all be found stiff on their backs." It seems reasonable to suggest that a few of these original animals may in fact have survived, since Johansen (1926) found the species to be common around St. John's.

It is very tempting to speculate further upon the apparent amphibian distribution gap in central Labrador, or the apparent demise of *R. pipiens* and *P. triseriata* in western Newfoundland, or the apparent recent fluctuation in *R. clamitans* numbers in eastern Newfoundland. Indeed, the latter topic may be of real interest to those studying the effects of acid precipitation, since general observations by myself and others suggest that *R. clamitans* populations on the limestone-rich west coast of the Island have remained relatively stable during the last decade. Climatic change may also be involved in the fluctuation of *R. clamitans* numbers, because in eastern Newfoundland, the species is apparently very near the limits of its ecological range (Bleakney 1958; Maunder 1983). But so little of a concrete nature is currently known about the amphibians of Newfoundland and Labrador, or their general ecology, that the resolution of these and other problems must be left to future workers. Opportunities for amphibian research in Newfoundland and Labrador are presently unlimited, since no specific studies are ongoing or anticipated.

Acknowledgments

I wish to thank Joël Bonin, Joe Brazil, James Buckle, Rex Button, Kevin McAleese, Mac Pitcher, Rob Prosper, Dave Snow, and Gerald Yetman, for their unpublished observations. I also wish to thank the Royal Ontario Museum for permission to reference the two-lined Salamander collection made by M. Gillett, and Ed Andrews for his continuing encouragement and assistance.

Literature Cited

Bleakney JS. 1958. A zoogeographical study of the amphibians and reptiles of eastern Canada. Bulletin of the National Museum of Canada 155:1–119.

Buckle J. 1971. A recent introduction of frogs to Newfoundland. Canadian Field-Naturalist 85:72–74.

Cameron AW, Tomlinson AJ. 1962. Dispersal of the introduced Green Frog in Newfoundland. Bulletin of the National Museum of Canada 183:104–110.

Hardy C Capt. 1869. Forest Life in Acadie. New York: D Appleton.

Johansen F. 1926. Occurrences of frogs on Anticosti Island and Newfoundland. Canadian Field-Naturalist 40:16.

Jukes JB. 1842. Excursions in and about Newfoundland during the years 1839 and 1840. Vol. 2. London: John Murray.

Maret E. 1867. Frogs on Newfoundland. Proceedings of the Nova Scotian Institute of Science 1:6.

Maunder JE. 1983. Amphibians of the Province of Newfoundland. Canadian Field-Naturalist 97:33–46.

Appendix. Labrador Amphibian Collections in 1988

Bufo americanus

1. 1.5 km W of Happy Valley-Goose Bay on Trans-Labrador Road, 53°17' N, 60°22' W, roadside pool, edge of trembling aspen (*Populus tremuloides*) woods, 12 July 1988 (tadpoles). 2. Gosling Lake, on Northwest River Road, 53°25' N, 60°23' W, edge of lake, grass, cow-parsnip (*Heracleum maximum*), horsetail (*Equisetum* sp.), 13 July 1988 (tadpoles). 3. 3.2 km W of Happy Valley-Goose Bay on Trans-Labrador Road, 53°17' N, 60°24' W, roadside pool, edge of mixed woods, 14 July 1988 (tadpoles). 4. 6.2 km W of Happy Valley-Goose Bay on Trans-Labrador Road, 53°17' N, 60°27' W, roadside pool, edge of mixed woods, 14 July 1988 (adults and tadpoles). 5. 21.8 km W of Happy Valley-Goose Bay on Trans-Labrador Road, slow place in brook, 53°16' N, 60°39' W, 14 July 1988 (tadpoles). 6. 29.3 km W of Happy Valley-Goose Bay on Trans-Labrador Road, 53°16' N, 60°45' W, slow place in brook, 14 July 1988 (tadpoles). 7. 32.5 km W of Happy Valley-Goose Bay on Trans-Labrador Road, 53°16' N, 60°45' W, temporary pool in gravel pit, 14 July 1988 (tadpoles). 8. 19.5 km W of Churchill Falls on road to Esker, 53°35' N, 64°15' W, water-filled depression, grass and sedges, 16 July 1988 (tadpoles). 9. 35.3 km W of Churchill Falls, 0.8 km W of Twin Falls Junction, on road to Esker, 53°37' N, 64°28' W, pebbly temporary pool, black spruce woods, 17 July 1988 (tadpoles). 10. 43.8 km W of Twin Falls Junction, on road to Esker, 53°43' N, 65°06' W, near pond, mixed black spruce-balsam fir-willow-alder woods, 17 July 1988 (adults and tadpoles). 11. 244 km E of Churchill Falls, 53°08' N, 61°03' W, under plywood, damp depression with sedges, moss and Labrador tea, black spruce-willow (*Salix* sp.)-white

birch woods, 18 July 1988 (tadpoles). 12. 2.3 km N, on Dump Road, Goose Bay, 53°22' N, 60°27' W, pond, 19 July 1988 (tadpole).

Rana pipiens

1. Gosling Lake, on Northwest River Road, 53°25' N, 60°23' W, edge of lake, grass, cow-parsnip (*Heracleum maximum*), horsetail (*Equisetum* sp.), 13 July 1988 (tadpoles). 2. 3.2 km W of Happy Valley-Goose Bay on Trans-Labrador Road, 53°17' N, 60°24' W, roadside pool, edge of mixed woods, 14 July 1988 (tadpoles and adults). 3. 6.2 km W of Happy Valley-Goose Bay on Trans-Labrador Road, 53°17' N, 60°27' W, roadside pool, edge of mixed woods, 14 July 1988 (tadpoles).

Rana sylvatica

Goose River, bridge on Northwest River Road, 53°23' N, 60°26' W, under leaf litter on sandy dry wash, alder (*Alnus* sp.)-white birch (*Betula papyrifera*) woods, 13 July 1988.

Rana septentrionalis

6.2 km W of Happy Valley-Goose Bay on Trans-Labrador Road, 53°17' N, 60°27' W, roadside pool, edge of mixed woods, 14 July 1988.

Ambystoma laterale

1. Birch Island, Happy Valley, 53°17' N, 60°18' W, under wood in disturbed grassy field, 14 July 1988. 2. 58.7 km W of Happy Valley-Goose Bay on Trans-Labrador Road, 53°09' N, 61°02' W, under plywood, edge of Labrador tea (*Ledum groenlandicum*) fen, black spruce (*Picea mariana*) woods, 15 July 1988.

Eurycea bislineata

1. 118.7 km W of Happy Valley-Goose Bay on Trans-Labrador Road, 53°06' N, 61°43' W, gravel seepage slope above brook, with sedges, black spruce forest, 15 July 1988. 2. 29 km E of Churchill Falls on Trans-Labrador Road, 53°29' N, 63°36' W, under rocks, seepage slope above river, 18 July 1988. 3. 152.3 km E of Churchill Falls on Trans-Labrador Road, 53°12' N, 62°15' W, seepage slope above brook, 18 July 1988.

©1997 by the Society for the Study of Amphibians and Reptiles
Amphibians in decline: Canadian studies of a global problem. David M. Green, editor.
Herpetological Conservation 1:100–106.

AMPHIBIANS IN THE NORTHWEST TERRITORIES

MICHAEL A. FOURNIER

Canadian Wildlife Service, P.O. Box 637, Yellowknife, Northwest Territories X1A 2N5, Canada

ABSTRACT. — Four species of amphibians are known to occur, and a 5th species may occur, within the Northwest Territories. *Pseudacris maculata* and *Rana sylvatica* are common and of relatively widespread distribution south of the treeline. *Bufo hemiophrys* and *Rana pipiens* are uncommon and of restricted distribution, occurring only in the vicinity of the lower reaches of the Slave River. *Bufo boreas* may occur in the extreme southwestern corner of the territory along the lower reaches of the Liard River. Basic research is required for all species in order to understand more fully distribution, abundance, population trends, and aspects of ecology which may limit population growth and/or range expansion.

RÉSUMÉ. — Quatre espèces d'amphibiens vivent dans les Territoires du Nord-Ouest et une cinquième espèce y est aussi possiblement présente. *Pseudacris maculata* et *Rana sylvatica* sont des espèces courantes et relativement bien distribuées au sud de la forêt. *Bufo hemiophrys* et *Rana pipiens* sont pas communs, leur répartition est restreinte et on ne les retrouve que dans le voisinage des passages de la rivière Slave. *Bufo boreas* vit tout à fait dans le sud-ouest des Territoires, le long des passages de la rivière Liard. Des recherches fondamentales s'imposent sur toutes les espèces pour mieux comprendre leur répartition, leur abondance, leurs tendances démographiques et les différents aspects écologiques qui limitent leur croissance démographique et(ou) leur expansion.

The Northwest Territories (NWT) undoubtedly contains some of the most pristine amphibian habitat remaining in North America. The quality of air, water, and soil is generally very high due to the presence of relatively few people and a low level of industrial activity. However, the distribution and abundance of amphibians in the NWT is severely restricted by the harsh climate.

Four species of amphibians are known to occur within the NWT including *Bufo hemiophrys* (Canadian toad), *Pseudacris maculata* (boreal chorus frog), *Rana sylvatica* (wood frog), and *Rana pipiens* (northern leopard frog). In addition, *Bufo boreas* (western toad) may occur in the extreme southwestern corner of the territory.

Pseudacris maculata and *R. sylvatica* are freeze-tolerant and hibernate terrestrially withing the frost zone (Schmid 1982; MacArthur and Dandy 1982; Storey 1984; Storey and Storey 1986). *Bufo hemiophrys* and *R. pipiens* are not freeze tolerant and must hibernate below the frost zone, on land and in water, respectively (Schmid 1982; Storey and Storey 1986). These differences in adaptation to cold weather have direct consequences in terms of distribution and abundance. Whereas the freeze-tolerant species are abundant and widespread, the remaining species are uncommon and of restricted distribution in the NWT.

METHODS

Historical and more recent literature were summarized in chronological order. Some published material was not included as earlier observations were often referred to by subsequent authors without the addition of new information. Published accounts were supplemented with data from the herpetological collections of the Canadian Museum of Nature (CMN), Ottawa; Royal Ontario Museum (ROM), Toronto; National Museum of Natural History (USNM), Washington D.C.; Alaska Fisheries Science Center, Auke Bay Laboratory (ABL), Juneau; and the Zoology Museum of the University of

Alberta (UAZM), Edmonton. This summary can be considered to include virtually all of the available information regarding amphibians in the NWT.

The islands in James Bay, which are legally part of NWT, were not included. Ecologically, these islands are best considered within the context of the herpetofauna of the adjacent provinces of Ontario and Quebec.

RESULTS

Bufo hemiophrys

Richardson (1851) reported toads at Great Bear Lake. Specimens supposedly collected there, by him, were subsequently listed by Gunther (1858), Boulenger (1882), and Logier and Toner (1955). However, reference to these specimens has disappeared from more recent literature (for example, Logier and Toner 1961; Cook 1984). The possibility of toads occurring at Great Bear Lake seems unlikely as none have been reported within 700 km of this location in the 145 year period since Richardson's visit. Richardson's specimens may have been improperly recorded and/or catalogued and were most likely confused with specimens he collected at Lake Winnipeg during the same expedition (F. Cook, pers. comm.). Gunther (1858) also recorded a specimen of *Hyla versicolor* supposedly collected by Richardson at Great Bear Lake. This is at least 1200 km beyond the currently accepted range of that species (Cook 1984). It also appears that Richardson was unable to distinguish between some species of amphibians. In his account of observations at Great Bear Lake he states: "a small frog (*Bufo americanus*) is common in every pond, ... A frog resembling it, but perhaps of a different species, abounds on the Saskatchewan, and its cry of love in early spring ... resembles the quack of a duck, ..." Despite naming this "frog" *B. americanus*, Richardson was undoubtedly referring to *R. sylvatica*, the species most often described as having a call resembling the quack of a duck. Preble (1908) provided the first verifiable evidence of the presence of toads in the NWT when he collected a specimen of *B. hemiophrys* at Fort Smith, 21 June 1901. The Canadian Museum of Nature possesses 2 specimens collected by Kuyt (1991) who reported a communal overwintering site used by several hundred toads, in Wood Buffalo National Park. There may have been as many as 1100 toads at this site in May, 1994 (K. Timoney, pers. comm.).

Bufo hemiophrys should be considered uncommon and of restricted distribution in the NWT (Fig. 1). It currently has no protected status under NWT legislation, but most of the territorial population may occur within Wood Buffalo National Park.

Bufo boreas

A specimen of *B. boreas* (USNM 48057) is recorded as having been collected by E.A. Preble at Fort Simpson, 13 August 1897. It appears this specimen has never been previously reported in literature on the distribution of amphibians in the NWT and Canada (for example, Logier and Toner 1955 1961; Hodge 1976; Cook 1984; Larsen and Gregory 1988). Identification of the specimen was recently verified by G. Zug (pers. comm.). However, the information for date, locality, and/or collector appears somewhat suspect, as Preble's (1908) own review of his travels does not provide evidence that he visited the Fort Simpson area prior to 1903.

The specimen may have been collected by someone other than Preble. One possible candidate appears to be A.J. Stone, who was sent by the American Museum of Natural History to collect mammal and bird specimens in northwestern North America in 1897 and travelled down the Liard River to Fort Simpson in the spring of 1898. However, a review of some literature relating the results of this expedition (Allen 1899; Stone 1900) has failed to produce any reference to amphibian specimens.

Similarly to Richardson's observations of toads at Great Bear Lake, there have been no subsequent observations of *B. boreas* to provide verification of its existence in the Fort Simpson region. However, unlike Richardson's supposed observation of toads hundreds of kilometres north of the currently accepted range, this specimen, if verified, would represent a much shorter extension of the recognized

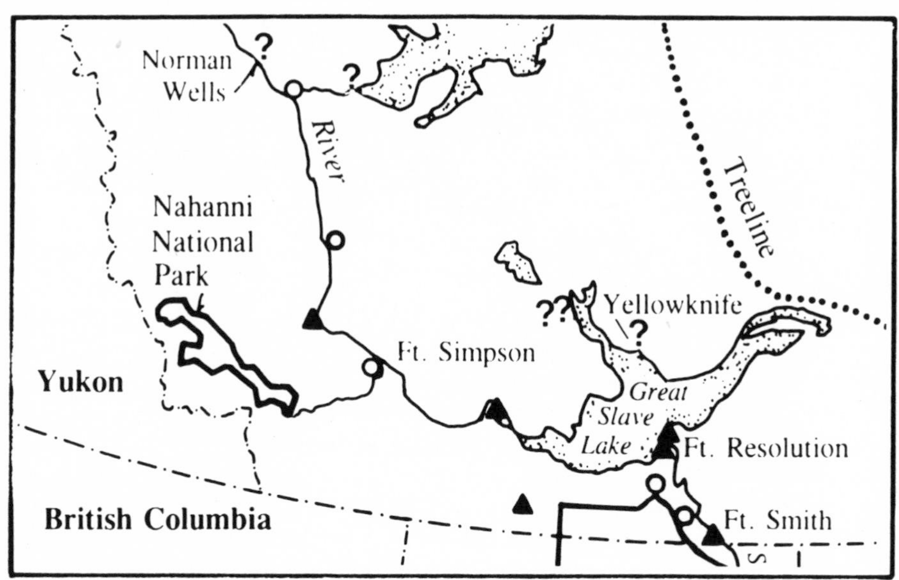

Boreal Chorus Frog

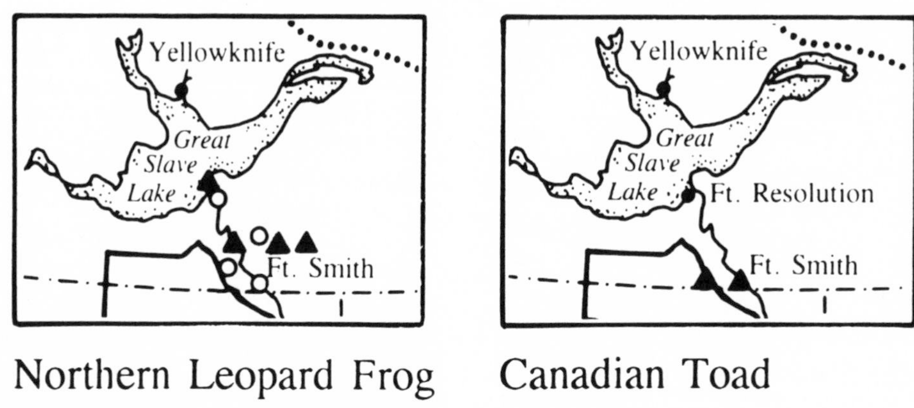

Figure 1. *Known distribution of* Pseudacris maculata, Rana pipiens, *and* Bufo hemiophrys *in the Northwest Territories. See Fig. 2 for the base map.*

range of *B. boreas*, which is known to occur in northern British Columbia and southern Yukon Territory (Cook 1984). Larsen and Gregory (1988) have suggested that additions to the herpetofauna of the NWT may yet be discovered and identified *B. boreas* as a candidate.

Pseudacris maculata

Richardson's specimens, reportedly collected at Great Bear Lake in 1849, included this species (Gunther 1858; Boulenger 1882) and the northern range thus established has been generally accepted (e.g., Preble 1908; Logier and Toner 1961; Cook 1984). Although there are no subsequent records of *P. maculata* from Great Bear Lake itself, and no adult specimens have been collected north of the mouth of the South Nahanni River, Richardson's observations are plausible given the observations of Preble (1908) and Williams (1933).

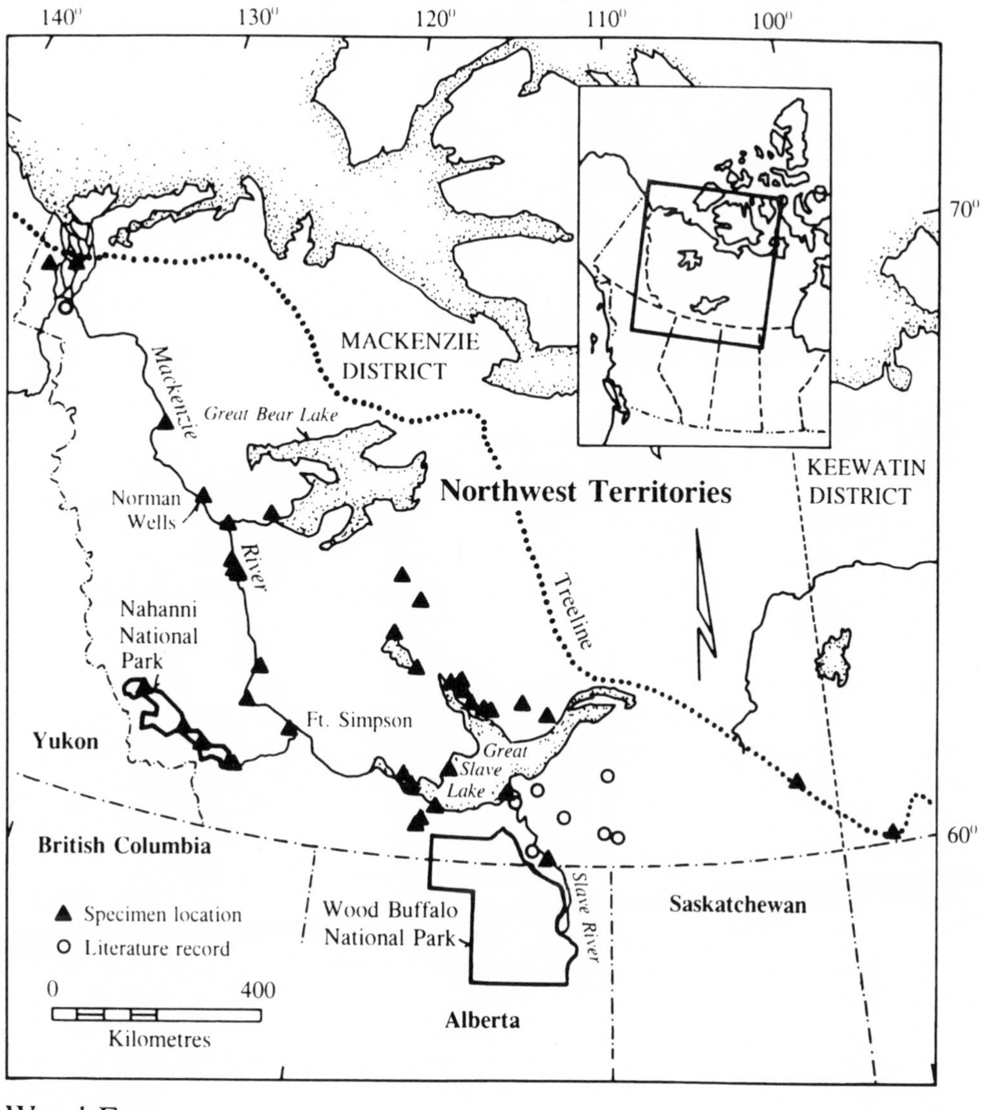

Figure 2. *Known distribution of* Rana sylvatica *in the Northwest Territories.*

Pseudacris maculata was collected by R. Kennicott at Fort Resolution in the 1860's (USNM). Preble (1908) collected specimens at Fort Smith and Fort Resolution in 1901, and heard its call at Fort Simpson on 3 May 1904 and "above Fort Norman" in early June, 1904. Harper (1931) collected specimens at Fort Resolution on 25 August 1914 and heard it calling at Slave River Delta, 26 August 1914. Williams (1933) reported "Pipers" in the Franklin Mountains east of Wrigley, 6 July 1922. A specimen was taken at Slave River Delta (Resdelta), 1 August 1944 (ROM). Stewart (1966) heard this species throughout the Grand Detour Portage area and along the Little Buffalo River, 23 to 30 May 1955. Tadpoles originally identified as *P. maculata* were collected 7 to 13 June 1965 at a number of locations, including Yellowknife, 27 km west of Rae, 48 km west of Rae, and 32 km south of Enterprise, and 27 June 1975, 2 km south of Norman Wells (CMN). However, none of these specimens could be verified as *P. maculata* upon subsequent examination (F. Cook, pers. comm.). Adult specimens were collected at the mouth of the South Nahanni River, 8 August 1953, 6 km south of Fort Smith on 7 September 1966, and at Mills Lake, 11 September 1988 (CMN).

Pseudacris maculata should be considered fairly common, and relatively widespread in the forested portions of the NWT (Fig. 1).

Rana sylvatica

Gunther (1858) and Boulenger (1882) reported specimens from Great Bear Lake, collected by Richardson in 1849. Preble (1908) "collected a large series, comprising specimens from the following localities: ...; Fort Smith; Fort Resolution; Fort Rae; Grandin River; Lake St. Croix; Fort Providence; Willow River; Fort Simpson; mouth of Nahanni River; Birch Island, 144 km below Fort Wrigley; Fort Norman; and the site of old Fort Good Hope." Harper (1931) observed small numbers at Hill Island Lake and near Kolethe Rapids on the Tazin River. He reported *R. sylvatica* as abundant along the Taltson River just above the junction with the Tazin, near Tsu Lake, at the mouth of Pierrot Creek, and at Fort Resolution where a specimen was taken, 22 to 26 August 1914.

Specimens were taken at Fort Resolution and Fort Simpson by R. Kennicott, and at Big Island by J. Reid, in the 1860's (USNM). The same institution holds specimens taken during the early to mid 1900's by various collectors at a number of locations including Hay River at Great Slave Lake in 1908, Mackenzie District in 1929, Windy River and Windy Lake in the District of Keewatin in 1947, and 16 to 64 km northwest of Yellowknife in 1962. Specimens were taken at Frank Channel in 1944, Yellowknife in 1944, Mackenzie Delta in 1956, Hay River in 1956, Lac la Martre in 1959, and Beaverlodge Lake in 1959 (ROM). Stewart (1966) reported this species to be common along the Little Buffalo and Sass rivers, 23 to 30 May 1955.

Martof and Humphries (1959) studied at least one NWT specimen from Husky River below the Peel River junction which does not appear to have been reported elsewhere. Weintraub (1967) collected specimens at Hinde Lake, 8 July 1966. Specimens were collected in 1970 at Stagg River, Louise Falls, and Fort Providence, and at Hay River in 1974 (ABL). Many specimens in the Canadian Museum of Nature were collected between 1896 and 1977 by a variety of collectors from a number of sites including Fort Resolution, Enterprise, Nahanni Butte, South Nahanni River, Yellowknife, Ross Lake, Desperation Lake, Fort Norman, Norman Wells, Fort Good Hope, Redstone River, 90 km from the mouth of Keele River, and the Mackenzie River Delta. A large number of specimens were collected between 1951 and 1975 from a number of sites, including Moraine Bay in Great Slave Lake, Hay River, Norman Wells, 48 km south of Inuvik, 8 km south of Fort Smith, Heart Lake, and Travaillant Lake (UAZM).

Rana sylvatica should be considered common and widely distributed throughout the forested regions of the NWT (Fig. 2).

Rana pipiens

Preble (1908) observed this species at Fort Smith in 1901. Harper (1931) observed several of these frogs and took a specimen at Natla Rapids on the Taltson River, 5 August 1914. He also observed several at a rapid on the Taltson River about 14 km below Tsu Lake, 10 August 1914, as well as on the Slave River below McConnell Island, 28 August 1914. A specimen was taken on the Slave River, downstream from Fort Smith (ROM, no date). Stewart (1966) provided a second-hand account of observations by R.P. Allen (National Audubon Society) of this species near the mouth of the Sass River, 26 May 1955. Another adult specimen was collected at the Slave River Delta, 20 July 1949 (CMN). One adult specimen collected 29 September 1986, on the Taltson River at Deskenatlata Lake (UAZM). The most recent observations of *R. pipiens* in the NWT occurred on 7 July and 15 September, 1994 (1 frog observed in each case) at Kuzo Lake near the Tazin River (D. Walker, pers. comm.).

Rana pipiens should be considered uncommon and of restricted distribution in the NWT (Fig. 1). It currently has no protected status under NWT legislation, but small numbers may occur within Wood Buffalo National Park (Parks Canada 1979).

Conclusions

The amphibian fauna of the NWT is undoubtedly among the least studied and most poorly understood in Canada. Amphibians are of little economic or gastronomic significance in the NWT. As a result, they are often considered of little importance by government wildlife management agencies and local people. But perhaps the most prominent factor limiting our understanding of amphibians is the nature of the land itself. The NWT consists of vast remote areas populated by few people and bisected by fewer km of road. As a result there are few resident scientists or naturalists and logistical and financial constraints to biological studies can be significant.

The relatively pristine nature of the NWT and the limited extent of our current knowledge of its herpetofauna present a myriad of research opportunities. Distribution of all species in the NWT requires clarification. For example, a relatively widespread distribution is generally accepted for *P. maculata* (Cook 1984) despite the existence of very few specimens. Although *R. sylvatica* has been recorded in tundra habitats in Alaska (Hodge 1976), data on distribution above treeline in the NWT are lacking. Uncertainties remain regarding the distribution of toads in the NWT.

Available data for all species of amphibians which occur in the NWT preclude assessment of population trends. However, populations of *R. pipiens* and *B. hemiophrys* in Alberta have experienced declines (Roberts 1981, 1992; Alberta Fish and Wildlife Division 1991) and this fact, combined with the apparent rarity and very restricted distribution of these species in the NWT provides reason for concern. Much basic research is required for both of these species. Important breeding areas and overwintering sites should be identified and protected. Local breeding populations, once identified, should be monitored to determine population trends and identify factors limiting their distribution and abundance.

Other potential research topics include a variety of aspects of the ecology of amphibians at the northern limit of their distribution daily and seasonal activity periods, rates of growth of juveniles and attainment of sexual maturity, and the mechanisms via which climate influences distribution and abundance (Hodge 1976, Larsen and Gregory 1988). Some amphibian populations in the NWT may be suitable as natural control groups for studies of anthropomorphic influences on amphibians elsewhere.

It is important that we improve the state of knowledge of amphibians in the NWT now, as the future is likely to bring considerable environmental change to the region. This is particularly true in the southern portions of the territory where the potential for forestry, agriculture, and industrial activities is greatest and, coincidentally, where most of the NWT's amphibians occur.

Acknowledgments

The author gratefully acknowledges the Canadian Museum of Nature, Royal Ontario Museum, National Museum of Natural History, Alaska Fisheries Science Center, and the Zoology Museum, University of Alberta for the provision of data on NWT specimens. R. P. Hodge provided valuable insight concerning the literature records and collections described in his 1976 book. Francis Cook provided encouragement and thought provoking discussion as well as information on several specimens. The staff of the Prince of Wales Northern Heritage Centre Archives in Yellowknife provided access to historical literature. Jim Hines, Chris Shank and Alison Welch provided valuable comments on the manuscript. It would be remiss not to pay tribute to the collector who has contributed most to our knowledge of the amphibians of the NWT, E. A. Preble. Preble appears to have been an incomparably thorough and knowledgeable naturalist. He collected extensively in the NWT, both in terms of geographic location and number of specimens.

Literature Cited

Alberta Fish and Wildlife Division. 1991. Alberta's threatened wildlife—northern leopard frog. Edmonton, AB: Alberta Fish and Wildlife Division Nongame Management Program.

Allen JA. 1899. On mammals from the Northwest Territory collected by Mr. A.J. Stone. Bulletin of the American Museum of Natural History 12:1–9.

Boulenger GA. 1882. Catalogue of the Batrachia Salientia s. Ecaudata in the collection of the British Museum, 2nd ed.

Cook FR. 1984. Introduction to Canadian amphibians and reptiles. Ottawa: National Museum of Canada.

Gunther A. 1858. Catalogue of the Batrachia Salientia in the collection of the British Museum. London: British Museum (Natural History).

Harper F. 1931. Amphibians and reptiles of the Athabasca and Great Slave Lake region. Canadian Field-Naturalist 45:68–70.

Hodge RP. 1976. Amphibians and reptiles in Alaska, the Yukon and Northwest Territories. Anchorage: Alaska Northwest Publishing.

Kuyt E. 1991. A communal overwintering site for the Canadian Toad, *Bufo americanus hemiophrys*, in the Northwest Territories. Canadian Field-Naturalist 105:119–121.

Larsen KW, Gregory PT. 1988. Amphibians and reptiles in the Northwest Territories. In: Kobelka C, Stephens C, editors. The natural history of Canada's North: current research. Yellowknife, NWT: Prince of Wales Northern Heritage Centre. Occasional Paper 3. p 31–51.

Logier EBS, Toner GC. 1955. Check list of the amphibians and reptiles of Canada and Alaska. Royal Ontario Museum Division of Zoology and Palaeontology Contributions 41:1–88.

Logier EBS, Toner GC. 1961. Check list of the amphibians and reptiles of Canada and Alaska. A revision of Contribution No. 41. Royal Ontario Museum Life Sciences Division Contributions 53:1–92.

MaCarthur DL, Dandy JWT. 1982. Physiological aspects of overwintering in the boreal chorus frog (*Pseudacris triseriata maculata*). Comparative Biochemistry and Physiology A 72:137–141.

Martof BS, Humphries RL. 1959. Geographical variation in the wood frog *Rana sylvatica*. American Midland Naturalist 61:350–389.

Parks Canada. 1979. List of amphibians and reptiles of Wood Buffalo National Park. Ottawa: Parks Canada.

Preble EA. 1908. A biological investigation of the Athabasca–Mackenzie Region. North American Fauna 27:1–574.

Richardson Sir J. 1851. Arctic searching expedition. London: Longman, Brown, Green, and Longmans.

Roberts W. 1981. What happened to the Leopard Frogs. Alberta Naturalist 11:1–4.

Roberts W. 1992. Declines in amphibian populations in Alberta. In: Bishop CA, Pettit KE, editors. Declines in Canadian amphibian populations: designing a national monitoring strategy. Ottawa: Canadian Wildlife Service. Occasional Paper 76. p 15–16.

Schmid WD. 1982. Survival of frogs in low temperature. Science 215:697–698.

Stewart RE. 1966. Notes on birds and other animals in the Slave River - Little Buffalo River area, NWT. Blue Jay 24:22–32.

Stone AJ. 1900. Some results of a natural history journey to northern British Columbia, Alaska, and the Northwest Territory, in the interest of the American Museum of Natural History. Bulletin of the American Museum Natural History 13:31–62.

Storey KB. 1984. Freeze tolerance in the frog, *Rana sylvatica*. Experientia 40:1261–1262.

Storey KB, Storey JM. 1986. Freeze tolerance and intolerance as strategies of winter survival in terrestrially- hibernating amphibians. Comparative Biochemistry and Physiology A 83:613–617.

Weintraub JD. 1967. Herpetological observations in Saskatchewan and the Mackenzie District. Canadian Field-Naturalist 81:106–109.

Williams MY. 1933. Biological notes, covering parts of the Peace, Liard, Mackenzie and Great Bear river basins. Canadian Field-Naturalist 47:23–31.

©1997 by the Society for the Study of Amphibians and Reptiles
Amphibians in decline: Canadian studies of a global problem. David M. Green, editor.
Herpetological Conservation 1:107–109.

AMPHIBIANS IN SOUTHWESTERN YUKON AND NORTHWESTERN BRITISH COLUMBIA

LEE MENNELL

Box 105, Carcross, Yukon Territory Y0B 1B0, Canada

ABSTRACT.—Twenty-three locations in southwestern Yukon and northwestern British Columbia were surveyed April to September, 1993 for amphibians. *Rana sylvatica* were observed breeding at all locations surveyed in late April to mid-May but were infrequently observed afterwards. *Rana luteiventris* were abundant within the Tutshi uplands and Chilkoot Trail National Historic Site in northwestern British Columbia. They were observed at 2 locations within the Yukon Territory. *Bufo boreas* were observed in Sloko Inlet at the south end of Atlin Lake, Fantail River in the region adjacent to its outfall into Tagish Lake, and at Lake Lindeman in the Chilkoot Trail National Historic Site.

RÉSUMÉ.—Vingt-trois sites dans le sud-ouest du Yukon et dans le nord-ouest de la Colombie-Britannique ont été étudiés entre avril et septembre 1993. La reproduction de *Rana sylvatica* a été observée dans plusieurs sites de la fin avril à la mi-mai puis moins fréquemment par la suite. *Rana luteiventris* est répandue dans les hautes-terres Tutshi et dans le Lieu historique national de la piste Chilkoot dans le nord-ouest de la Colombie-Britannique. Elle a été observée dans 2 sites dans les Territoires du Yukon. *Bufo boreas* a été observée à Sloko Inlet à l'extrémité sud du lac Atlin et à Fantail River dans la région qui jouxte la décharge du lac Tagish, ainsi qu'au lac Lindeman dans le Lieu historique national de la piste Chilkoot.

Little is know about northern amphibians, neither the species present, specifics of their ecology in the north, nor any quantitative data concerning the status of their populations. Wood frogs, *Rana sylvatica*, and western toads, *Bufo boreas*, have been documented north of 60° N latitude in the Yukon (Cook 1977) and the rough-skinned newt has been recorded along the Alaska coast as far as Prince William Sound (Hodge 1976). Other species whose ranges extend into far northwestern British Columbia include the Columbia spotted frog, *Rana luteiventris*, long-toed salamander, *Ambystoma macrodactylum*, northwestern salamander, *Ambystoma gracile*, and the tailed frog, *Ascaphus truei* (Green and Campbell 1984). Prompted by a recent report of *R. luteiventris* from subalpine meadows immediately south of the Yukon/British Columbia border (unpublished), I undertook a thorough search of this relatively unexplored region to determine which amphibian species are present and, in particular, to append the northern range limit of *R. luteiventris*.

METHODS

Ponds were surveyed in southwestern Yukon and northwestern British Columbia (Fig. 1) between Whitehorse, Carcross, and Tagish from 30 April to 15 May 1993 while ponds in the Tutshi upland were surveyed between 15 May and 30 May 1993. Surveys were then conducted through to September in the Tutshi upland, Chilkoot Trail National Historic Site, the west arm of Bennett Lake, Mush and Bates Lakes in Kluane National Park, the Fantail River at the southern end of Tagish Lake, and Sloko Inlet in Atlin Lake Provincial Park.

Ponds and wetlands to be surveyed were identified using topographic maps (National Topographic series, 1:50,000 scale). Visual and aural searches were made of each appropriate habitat and all ponds were searched for the presence of calling male frogs, amplexing pairs, egg masses, and/or tadpoles. Modified minnow traps were employed in ponds around Bare Loon Lake in the Chilkoot Trail National Historic Site to check for the presence of adult or larval salamanders or newts.

Results

Only 3 species of amphibians were found. *Rana sylvatica* were found in all areas surveyed, though not at every location. *Bufo boreas* were sighted at Sloko Inlet, the Fantail River area, and Chilkoot Trail National Historic Site, specifically at the mouth of the Mountain River. *Rana luteiventris* were found at Sloko Inlet, the Tutshi upland, and Chilkoot Trail National Historic Site, most notably at 2 isolated ponds within the Yukon Territory. These ponds (UTM MB895535 and MB999606). One was a small beaver pond on a tiny stream 100 m above the Partridge River valley, isolated from other appropriate habitat by 15 km of rugged, mountainous terrain. These are the most northerly records of *R. luteiventris* and the 1st known within the Yukon Territory (Fig. 1).

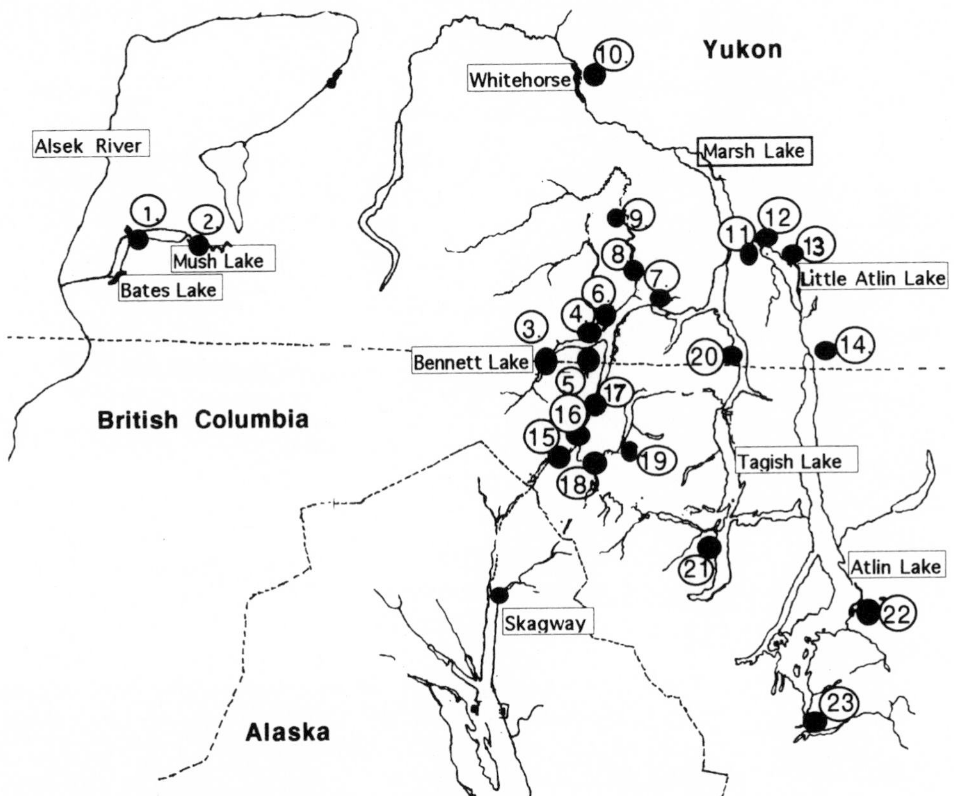

Figure 1. *Survey locations in southwestern Yukon and northwestern British Columbia. The localities, their habitats, and the amphibian species present are as follows: 1. Mush Lake outlet marshes. 2. Mush Lake/Alder Creek wetland complex and fen (*Rana sylvatica*). 3. West arm Bennett Lake/Partridge Lake river mouth, backwater sloughs and beaver pond (*Rana luteiventris*). 4. West arm Bennett Lake/Birch Pond shallow calcareous lake and marshy margins (*R. luteiventris*). 5a. Munro Lake outlet muskeg ponds. 5b. Munro Lake inlet meadering river course, backwater sloughs. 6. Wheaton wetland complex, muskeg bogs and fens (*R. sylvatica*). 7. Nares Lake pond (*R. sylvatica*). 8. Spirit Lake outlet beaver ponds and meandering creek (*R. sylvatica*). 9. Rat Lake shallow esker pothole with sedge margin (*R. sylvatica*). 10. Whitehorse esker pothole and small pond (*R. sylvatica*). 11. Chilkoot Lake swampy creek with beaver ponds (*R. sylvatica*). 12. ponds at Little Atlin Creek dwarf birch meadow and lowland (*R. sylvatica*). 13. Little Atlin Lake shallow sedge-filled bay (*R. sylvatica*). 14. Unidentified esker pothole pond with marsh fringe. 15. Deep Lake and adjacent ponds at treeline. 16. Lindeman Lake/Mountain River shallow silty bay, mouth of river, and backwater slough (*R. sylvatica, Bufo boreas, R. luteiventris*). 16. Bare Loon Lake small ponds, creek, and sedge meadows (*R. sylvatica, B. boreas, R. luteiventris*). 17. Bennett pond at mouth of Lindeman River (*R. luteiventris*). 18a. Tutshi uplands small lakes, ponds, and sedge meadows (*R. sylvatica, R. luteiventris*). 18b. Tutshi headwaters ponds (*R. luteiventris*). 18c. Tutshi backwater pond (*R. sylvatica, R. luteiventris*). 19. Tutshi River mouth spruce bog seepage (*B. boreas, R. luteiventris*). 20. west shore of Tagish Lake warm springs. 21. Fantail River backwater slough, river mouth, meadow, and pine forest (*R. sylvatica, B. boreas*). 22. Atlin warm springs pools and meandering creek (*B. boreas*). 23. Sloko Inlet shore of Atlin Lake and small ponds (*R. sylvatica, B. boreas, R. luteiventris*).*

Discussion

Rana luteiventris were noted in 1992 in the Tutshi River uplands (unpublished). Although this area is part of the Yukon River drainage, it lies within the Coast Mountain range. As a result of its proximity to the coast, there are higher snowfall, higher rainfall, and milder temperatures than in the interior plateau area directly to the east. Sites chosen to survey were within this area of higher precipitation and/or more moderated climate. The ecosystem features a mix of coastal alpine amd interior flora, with extensive sedge meadow and pond habitats of a type not found in the interior mountains. But little actual amphibian habitat is available in this area. It is mountainous with deeply cleft valleys. The occasional wetlands or ponds that do exist are widely separated from each other, often by mountain ranges. While *Rana sylvatica* were found in all areas surveyed throughout the interior plateau and Coast mountains, both *Bufo boreas* and *R. luteiventris* were found in locations within the Coast mountains which typically have higher precipitation.

Acknowledgments

I thank the Northern Research Institute for support of the field work. Research permits were granted by Parks Canada, British Columbia Parks, and the Yukon Heritage Branch. Personal thanks to Bob McLure and John Ward for help with boat access to remote sites and Pat Milligan for encouragement and advice.

Literature Cited

Cook FR. 1977. Records of the boreal toad from the Yukon and northern British Columbia. Canadian Field-Naturalist 96:185–186.

Green DM, Campbell RW. 1984. The amphibians of British Columbia. Victoria, BC: British Columbia Provincial Museum. Handbook 45.

Hodge RP. 1976. Amphibians and reptiles in Alaska, the Yukon and Northwest Territories. Anchorage, AK: Alaska Northwest Publishing Company.

Columbia spotted frog, Rana luteiventris. Photo by David M. Green.

©1997 by the Society for the Study of Amphibians and Reptiles
Amphibians in decline: Canadian studies of a global problem. David M. Green, editor.
Herpetological Conservation 1:110–116.

Chapter 12

STATUS OF AMPHIBIANS IN SASKATCHEWAN

ANDREW DIDIUK

Saskatchewan Amphibian Monitoring Project, P.O. Box 1574, Saskatoon, Saskatchewan S7K 3R3, Canada

ABSTRACT.—Very limited information is available regarding the distribution of the 7 species of amphibians in Saskatchewan and no data are available to describe population trends. Assessments of status for each species must be tentative at this time. However, populations of *Bufo cognatus* and *Rana pipiens* are possibly vulnerable to local extirpation because of their restricted and possibly highly localized distributions and perceived declines in population size. Populations of other species of amphibians exhibit no evident trends in population size or distribution changes. Increased knowledge of the distribution of amphibian species, development of monitoring programs, initiation of biological studies, identification and protection of critical habitats, inclusion of amphibian habitat requirements in land management plans, and enhanced education programs are all necessary for responsible protection and management of amphibian populations in Saskatchewan.

RÉSUMÉ.—Les données sur la répartition des 7 espèces d'amphibiens en Saskatchewan sont rares et aucune donnée ne permet de décrire leurs tendances démographiques. L'évaluation de la situation de chaque espèce est donc à l'heure actuelle tout à fait provisoire. Il semble néanmoins que les populations de *Bufo cognatus* et de *Rana pipiens* soient vulnérables à l'extinction locale car leur répartition est restreinte et probablement très localisée et parce que la taille de leur population respective semble avoir diminué. Les populations d'autres espèces d'amphibiens n'affichent aucune tendance évidente au titre de la taille ou des changements de répartition. Il faut multiplier les connaissances sur la répartition des espèces d'amphibiens, mettre au point des programmes de surveillance, entreprendre des études biologiques, identifier et protéger les habitats critiques, tenir compte de l'habitat des amphibiens dans les plans d'aménagement des sols et améliorer les programmes pédagogiques pour pouvoir protéger et gérer de façon responsable les populations d'amphibiens en Saskatchewan.

Seven species of amphibians have been recorded in Saskatchewan (Cook 1966). Compared to Canada's other prairie provinces, Saskatchewan has neither the western montane species of Alberta (Russell and Bauer 1993) nor the eastern species which are at or near the northeastern edges of their ranges in Manitoba (Preston 1982). *Pseudacris maculata, Rana sylvatica, Rana pipiens, Bufo hemiophrys*, and *Ambystoma tigrinum* are widespread and typical of northern grasslands and southern boreal forests and occur in all 3 prairie provinces. The remaining 2 Sasakatchewan species, *Bufo cognatus* and *Spea bombifrons*, have more restricted distributions at the extreme northern edges of their ranges.

Saskatchewan remains a region where there is very limited knowledge of the herpetofauna. One of the earliest summaries of historical records was provided by Logier and Toner (1961) but the most significant contributions towards our understanding of the presence and distribution of amphibians have been provided by Cook (1959, 1960, 1965) and limited additional knowledge of amphibian distributions in Saskatchewan has been derived from occasional reports of new range extensions or natural history observations (Nero 1959; Nero and Cook 1964; Driver 1971; Heard 1985). The more comprehensive works on herpetofauna of the province by Cook (1966) and by Secoy and Vincent (1976) are the most important summary documents addressing Saskatchewan species of amphibians. Only 1 regional summary of Saskatchewan herpetology has been prepared (Hooper 1992) and no detailed natural history studies or focused research projects have addressed any species of amphibian in Saskatchewan. The status and conservation of Saskatchewan species of amphibians have been addressed as part of national perspectives (Cook 1970, 1977; Stewart 1974; Secoy 1987; Seburn 1992). Some have addressed the unique opportunities for herpetological studies in Saskatchewan and all have noted that little substantial information is available regarding the distribution and population status of amphibian species in Saskatchewan. Because several species are at the northern edges of their ranges, there is considerable potential for research to address their adaptations to environmental conditions which may be marginal for the species.

It has been suggested that amphibian populations may be declining globally (Phillips 1990; Bishop and Pettit 1992; Blaustein et al. 1994; Pounds and Crump 1994) yet population size in some species of amphibians may normally fluctuate greatly over the short term in response to varying environmental conditions (Pechmann, et al. 1991). Many species utilize habitats which vary greatly among years in both quantity and quality. These fluctuations may occur in response to climatic trends but little information is available to demonstrate any periodicity. This lack of monitoring data is in part the result of the difficulty in detecting and estimating abundance of most amphibian species in either their aquatic or terrestrial stages (Heyer et al. 1994). Most Saskatchewan species are secretive most of the year and when they do congregate during breeding they remain difficult subjects to survey. Nevertheless, in this report I attempt to assess the population status for each species of amphibian in Saskatchewan and address past, current and potential threats to these populations.

Tiger salamander, Ambystoma tigrinum. *Photo by Lawrence Powell.*

Methods

I assessed the status of, and threats to, the 7 species of amphibians in Saskatchewan by considering all available sources. These include historical records from the National Museum of Canada and the Royal Museum of Saskatchewan, published literature as mentioned above and unpublished reports, perceptions and limited reporting from a few observers, recent reports from the Saskatchewan Amphibian Monitoring Project, anecdotal comments regarding abundance and persistence within known ranges, habitat trends, and perceptions of risk. The methods were not quantitative in any way due to the nearly complete lack of available hard data. Nevertheless, I could assign the terms "stable" or "vulnerable" to each species. I used "stable" to describe populations which may fluctuate in response to short-term climatic changes or other disturbances but do not exhibit any distinct trend in abundance. "Vulnerable" was used to describe populations which are small in size or very localized such that prolonged climate changes or local habitat disturbances could result in extirpation of these populations. But given the limited information regarding distribution and abundance, these assessments must be regarded as tentative.

Assessments of Status

There is no trend information available to quantitatively assess the population status of any species of amphibian in Saskatchewan.

Ambystoma tigrinum occurs throughout the grasslands and parklands of southern Saskatchewan with occasional reports along the extreme southern fringe of the mixed wood forest zone (Fig. 1). The subspecies *A. t. melanosticum* occurs in the more arid southwestern portions of southern Saskatchewan, primarily within and to the southwest of the Missouri Coteau region while *A. t. diaboli* occurs mainly to the north and east of the Missouri Coteau region. More detailed delineation of the ranges of these 2 subspecies, and the occur-

Figure 1. Ecoregions of Saskatchewan.

rence of hybrid forms, requires further study. *Ambystoma tigrinum* continues to be reported throughout its wide range on a regular basis. Populations overall may be stable although local populations in areas where available breeding sites are more temporary in nature may fluctuate greatly in response to longer periods of drought.

Pseudacris maculata is found throughout all but the northeastern portion of Saskatchewan. Throughout this wide range it is common, with populations particularly large and evident in southern Saskatchewan in years with significant spring and summer precipitation. It is able to exploit temporary wetland basins when they are filled by spring runoff. Populations of this species may be stable.

Rana sylvatica is widely distributed throughout the parklands and all but the most extreme northeast portion of northern Saskatchewan. It is absent throughout the more arid areas of the Missouri Coteau and southwestern Saskatchewan. Populations of this species may be stable.

Rana pipiens has been reported from localities within the boreal forest as far north as Lake Athabasca, but its primary range is the grassland, parkland and southern mixed forest regions. Populations have greatly declined in number since the late 1970s based upon extremely few reports of its occurrence since this time. Small local populations have continued to exist in isolated drainages and waterbodies. Recent increases in observations may reflect increased populations and/or increased observation effort. Populations are small and localized, and therefore vulnerable, but may be increasing.

Bufo hemiophrys is locally distributed throughout all but the northeastern portion of Saskatchewan. Although this species has been reported be declining at the western edge of its range in Alberta (Roberts 1992), there are no data to evaluate population trends in Saskatchewan. Populations of this species may be stable.

The distribution of *Bufo cognatus* is not well-defined, with only a few records restricted to extreme southern Saskatchewan. The degree of range overlap and occurrence of hybrids with *B. hemiophrys* have been little studied (Cook 1983). Populations may be rare, very localized, and therefore vulnerable.

Spea bombifrons appears to be restricted to the Missouri Coteau region and the extreme southwestern and southeastern corners of the province. It is a difficult species to detect due to its explosive breeding in response to irregular rainfall and few records are available. Populations are likely localized throughout its range and may be stable.

Canadian toad, Bufo hemiophrys. *Photo by David M.Green.*

Assessments of Threats

Habitat Loss And Degradation

The conversion of large areas of grassland to various agricultural crops has resulted in a loss of historic populations and habitat for amphibians, particularly foraging and overwintering habitats in close proximity to breeding wetlands. Turner et al. (1987) reported that 59% of all wetland basins and 78% of all wetland margins in southern Saskatchewan have been affected by agricultural practices. Native grassland margins of these wetlands have been greatly reduced in width, burned in the spring and autumn, and subjected to grazing which reduces plant biomass, soil surface moisture and near-surface humidity. Direct loss of wetlands in Saskatchewan greatly accelerated after World War II, with the development of road networks and their drainage ditches. Drainage of wetlands still occurs but is limited (Turner et al. 1987), because most remaining wetlands are large or distant from drainage corridors and more recent water control legislation restricts drainage opportunities for landowners. However, many smaller and more ephemeral basins are frequently tilled in dry years, affecting their ability to hold surface water, and there is a continuing impact upon wetland basins due to clearing of wetland margins and the filling of basins with the cleared material. These activities greatly accelerate the natural rate of filling of shallow prairie basins. Alteration of wetland habitat in forested areas of Saskatchewan during logging operations, the effects of possible water quality changes upon amphibian populations, and the effects of altered riparian habitat have not been studied. The potential for creation of new habitat, or improvement of existing habitat, arising from logging activities has not been addressed.

Population Isolation

The prairie landscape is now highly fragmented. Wetlands and native upland habitat are small islands within a sea of cropland in many areas of southern Saskatchewan. The implications of this habitat fragmentation upon survival of isolated populations is not clear. The ability of some species of amphibians to traverse croplands and recolonize areas where local populations have disappeared is unknown. This may be of particular concern for *B. cognatus, R. pipiens* and, possibly, *S. bombifrons* which appear to have very localized distributions. The potential of reintroductions as a means of restoring lost populations has not been explored (Roberts 1992).

Toxic Substances

The extensive use of pesticides and herbicides in the agricultural industry may pose risks to amphibians at all developmental stages through direct mortality, impaired reproduction or effects upon food and habitat quality (Bishop 1992). Although laboratory assessments of the toxicity of various chemicals have been conducted (Harfenist et al. 1989) there have been no field experiments directly addressing amphibians in prairie Canada. The effects of these chemicals upon other aquatic organisms important for food, as predators, or as competitors has been addressed (Pimental 1971). At least 5% of prairie wetlands contain concentrations of pesticides exceeding those considered deleterious to aquatic life (D. Donald, pers. comm.). *Rana pipiens* overwintering in more permanent waterbodies are in close contact with the bottom sediments and with any increases in agricultural chemicals which may occur during spring runoff. This may make this species more vulnerable to adverse effects of these chemicals. *Rana pipiens* also forages at considerable distances from wetlands during the summer (Seburn et al., this volume) and may be subjected to contact with a variety of agricultural chemicals in both native grasslands and cultivated crops. Neotenic adults of *A. tigrinum* and any overwintering larvae may also be subject to the effects these chemicals may impose. Only limited information is available regarding the effects of chemicals used in the forestry industry or in municipal mosquito control programs upon amphibians.

Predators

Amphibians which breed and/or overwinter in waterbodies which support fish populations, such as *R. pipiens* and neotenic *A. tigrinum*, may be affected by predation from those fish (Sprules 1974). Although there is an ongoing fish enhancement program in Saskatchewan, with the introduction of fish where natural reproduction may or may not be possible, there are no data to indicate if any amphibian populations have been, or are, at risk. Fisheries management activities to remove coarse fish through poisoning can produce extensive or complete mortality of amphibians (Orchard 1988). Introduction of bullfrogs, *Rana catesbeiana*, has been considered to be a factor in local declines of other species of amphibians (Hayes and Jennings 1986) but there is no evidence for this type of problem in Saskatchewan.

Other Factors

The harvesting of amphibians for subsistence or commercial use does not appear to be a problem in Saskatchewan nor have there been any significant harvests in the past. Although mortality arising from proximity of roads and highways to breeding and foraging areas for amphibians occurs widely (Langton 1989), no data are available to suggest particular species, or local populations, are at risk in Saskatchewan. The effects of ultraviolet radiation on amphibians has not been studied in the prairie regions.

Discussion

There have been no systematic surveys or monitoring programs in the past in Saskatchewan. Thus any attempt to assess either the current status or past trends in population size, and therefore to anticipate future trends, is tentative at best. The lack of information makes development of species conservation plans or evaluation of potential risks from proposed developments very difficult. Improving the provincial data-base to determine the distribution of each species of amphibian ideally can be achieved by a variety of means, including the development of a regional observer network among naturalists, solicitation of reports from government, conservation agency, and academic scientists engaged in amphibian research, extraction of records from extensive and intensive monitoring programs, and development of a distribution atlas of amphibians. In 1993, the Saskatchewan Amphibian Monitoring Project was initiated and forms the basis for a Saskatchewan Herpetology Atlas Project. These projects will greatly accelerate the development of a provincial data base and observer network.

The Saskatchewan Amphibian Monitoring Project is an aural survey, similar in design to several provincial and state projects to assess relative abundance and persistence of amphibian populations (Gartshore et al. 1992; Mossman et al. 1992; Bishop et al., this volume). A network of survey routes throughout the province may allow detection of the disappearance of species or long-term trends in presence or relative abundance of selected species of amphibians. But more intensive projects are also necessary to closely monitor environmental and anthropogenic changes and to obtain more detailed demographic data on particular amphibian populations. The Saskatchewan Herpetology Atlas Project will assist monitoring efforts by allowing detection of range contractions or disappearance of given species from portions of their range over time.

Biological studies are urgently needed to provide a basic understanding of the life history and habitat requirements of amphibians in Saskatchewan. Applied research addressing toxicology issues, identification of critical habitat requirements, habitat fragmentation, and dispersal capability are necessary to support conservation and management actions. Conservation actions required for the near future include the development of individual species evaluations, provisions to protect critical habitats, enhancement of the environmental impact assessment process to allow adequate evaluation of amphibian populations, and inclusion of amphibian populations in the development of stewardship of public lands. Cooperative efforts of various government agencies, conservation organizations, and individuals will be more effective in attaining these objectives. Finally, education will be an important component for amphibian conservation and must include the public in general, school children and youth groups, and land managers. Development of displays, preparation of resource materials, workshops for land managers, and effective use of various media can all assist in providing information and promoting interest in the amphibians of Saskatchewan.

LITERATURE CITED

Bishop CA. 1992. The effects of pesticides on amphibians and the implications for determining causes of declines in amphibian populations. In: Bishop CA, Pettit K, editors. Declines in Canadian amphibian populations: designing a national monitoring strategy. Ottawa: Canadian Wildlife Service. Occasional Paper 76. p 19–20.

Bishop CA, Pettit K, editors. 1992. Declines in Canadian amphibian populations: designing a national monitoring strategy. Ottawa: Canadian Wildlife Service, Occasional Paper 76.

Blaustein AR, Wake DB, Sousa WP. 1994. Amphibian declines: judging stability, persistence, and susceptibility to local and global extinctions. Conservation Biology 8:60–71.

Cook FR. 1959. Amphibians seen at Moose Mountain Park, Saskatchewan. Blue Jay 17:126.

Cook FR. 1960. New localities for the plains spadefoot toad, tiger salamander and the great plains toad in the Canadian prairies. Copeia 1960:363–364.

Cook FR. 1965. Additions to the known range of some amphibians reptiles in Saskatchewan. Canadian Field-Naturalist 79:112–120.

Cook FR. 1966. A guide to the amphibians and reptiles of Saskatchewan. Saskatchewan Museum of Natural History Popular Series 13:1–40.

Cook FR. 1970. Rare or endangered Canadian amphibians and reptiles. Canadian Field-Naturalist 84:9–16.

Cook FR. 1977. Review of the Canadian herpetological scene. In: Mosquin T, Suchal C, editors. Canada's Threatened Species and Habitats. Ottawa: Canadian Nature Federation. p 117–121.

Cook FR. 1983. An analysis of toads of the *Bufo americanus* group in a contact zone in central northern North America. National Museum of Canada Natural Science Publication 3:1–89.

Driver EA. 1971. Die-off of *Ambystoma tigrinum* in a prairie pond. Blue Jay 29:214–215.

Gartshore ME, Oldham MJ, Van der Ham R, Schueler FW. 1992. Amphibian road call counts: participants manual. Peterborough, ON: Ontario Field Herpetologists.

Harfenist A, Power T, Clark K, Peakall D. 1989. A review and evaluation of the amphibian toxicological literature. Canadian Wildlife Service Technical Report Series 61:1–222.

Hayes M, Jennings M. 1986. Decline of ranid frog species in western North America: are bullfrogs (*Rana catesbeiana*) responsible? Journal of Herpetology 20:490–509.

Heard S. 1985. Leopard frog at Bompas Lake, Saskatchewan. Blue Jay 43:17.

Heyer WR, Donnelly MA, McDiarmid RW, Hayek LC, Foster MS, editors. 1994. Measuring and monitoring biological diversity: standard methods for amphibians. Washington: Smithsonian Institute Press.

Hooper D. 1992. Turtles, snakes and salamanders of east-central Saskatchewan. Blue Jay 50:72–75.

Langton T. 1989. Amphibians and roads. Proceedings of the Toad Tunnel Conference 7-8 January 1989, Rendburg; Federal Republic of Germany. Shefford, UK: ACO Polymer Products Ltd.

Logier EBS, Toner GC. 1961. Check list of the amphibians and reptiles of Canada and Alaska. A revision of Contribution No. 41. Royal Ontario Museum Life Sciences Division Contributions 53:1-—92.

Mossman MJ, Huff JJ, Hine RM. 1992. Monitoring long-term trends in Wisconsin frog and toad populations. Madison, WI: Wisconsin Department of Natural Resources.

Nero RW. 1959. The spadefoot toad in Saskatchewan. Blue Jay 17:41–42.

Nero RW, Cook FR. 1964. A range extension for the Wood frog in northeastern Saskatchewan. Canadian Field-Naturalist 78:268–269.

Orchard S. 1988. Lake poisoning: its effects upon amphibians and reptiles. Bioline 7:12–13.

Pechmann JHK, Scott DE, Semlitsch RD, Caldwell JP, Vitt LJ, Gibbons JW. 1991. Declining amphibian populations: The problem of separating human impacts from natural fluctuations. Science 253:892–895.

Phillips K. 1990. Where have all the frogs and toads gone? Bioscience 40:422–424.

Pimental D. 1971. Ecological effects of pesticides on non-target organisms. Washington: Office of Science and Technology.

Pounds JA, Crump ML. 1994. Amphibian declines and climate disturbance: the case of the golden toad and the harlequin frog. Conservation Biology 8:72–85.

Preston W. 1982. The amphibians and reptiles of Manitoba. Winnipeg MB: Manitoba Museum of Man and Nature.

Roberts W. 1992. Declines in amphibian populations in Alberta. In: Bishop CA, Pettit K, editors. Declines in Canadian amphibian populations: designing a national monitoring strategy. Ottawa: Canadian Wildlife Services Occasional Paper 76. p 17–18.

Russell A, Bauer A. 1993. The amphibians and reptiles of Alberta. Edmonton, AB: University of Calgary Press; Calgary, AB: University of Alberta Press.

Seburn CNL. 1992. The status of amphibian populations in Saskatchewan. In: Bishop CA, Pettit K, editors. Declines in Canadian amphibian populations: designing a national monitoring strategy. Ottawa: Canadian Wildlife Services Occasional Paper 76. p 17–18.

Secoy DM. 1987. Status report on the reptiles and amphibians of Saskatchewan. In: Holroyd G et al., editors. Endangered species in the prairie provinces. Edmonton, AB: Provincial Museum of Alberta. Natural History Paper 9. p 139–141.

Secoy DM, Vincent TK. 1976. Distribution and population status of Saskatchewan's amphibians and reptiles. Regina, SK: Saskatchewan Department of the Environment.

Sprules W. 1974. Environmental factors and the incidence of neoteny in *Ambystoma gracile*. Canadian Journal of Zoology 52:1545–1552.

Stewart D. 1974. Canadian endangered species. Toronto: Gage Publishing Ltd.

Turner B, Hochbaum G, Caswell D, Nieman D. 1987. Agricultural impacts on wetland habitats on the Canadian prairies, 1981-1985. Transactions of the North American Wildlife and Natural Resources Conferennce 52:206–215.

©1997 by the Society for the Study of Amphibians and Reptiles
Amphibians in decline: Canadian studies of a global problem. David M. Green, editor.
Herpetological Conservation 1:117–127.

Chapter 13

HISTORICAL EVIDENCE DOES NOT SUGGEST NEW BRUNSWICK AMPHIBIANS HAVE DECLINED

DONALD F. MCALPINE

Natural Sciences Division, New Brunswick Museum, 277 Douglas Avenue, Saint John, New Brunswick E2K 1E5, Canada

ABSTRACT.—The accounts of early naturalists, combined with field notes and museum specimen records, do not suggest that there have been any general declines in amphibian abundance in New Brunswick over the past century. Species reported as common during the late 19th and early 20th Centuries remain so. Most of the 16 amphibian species present in the province are not habitat specialists. Some, such as *Rana sylvatica, Pseudacris crucifer, Bufo americanus*, and *Ambystoma* spp., have been able to exploit habitats created through extensive forest clearing and may have increased in abundance as a result. There has been little loss of freshwater wetland in New Brunswick during the past century, and many ponds and marshy areas have been created through road construction and by recovering beaver populations. *Hyla versicolor* may have expanded its range in New Brunswick in response to these changes. Historically, *R. pipiens* has been common in the province and shows no evidence of decline. Although *Plethodon cinereus* remains common, extensive clear cutting and shorter cutting cycles may have depleted populations. Single-species conifer plantations in New Brunswick support a lower diversity and abundance of amphibians than do naturally regenerated forest stands of comparable age.

RÉSUMÉ.—Les rapports des premiers naturalistes, les notes prises sur le terrain et les dossiers sur les spécimens de musée ne permettent pas d'affirmer qu'il y a eu un déclin généralisé des populations d'amphibiens au Nouveau-Brunswick durant le dernier siècle. Les espèces courantes à la fin du XIXe siècle et au début du XXe siècle le sont restées. La plupart des 16 espèces d'amphibiens qui vivent dans la province choisissent leur habitat sans trop de discrimination. Certaines, comme *Rana sylvatica, Pseudacris crucifer, Bufo americanus*, et *Ambystoma* sp., sont capables de coloniser des habitats où l'homme a pratiqué des coupes forestières importantes et leur nombre a parfois augmenté. Le Nouveau-Brunswick a conservé l'essentiel de ses terres humides et de nombreux étangs et marécages ont été créés à l'issue de la construction de routes et de la multiplication des populations de castors. *Hyla versicolor* semble même avoir colonisé le Nouveau-Brunswick à l'issue de ces changements. Historiquement, *R. pipiens* a toujours été répandue dans la province et n'affiche aucun signe de déclin. Même si *Plethodon cinereus* reste répandue, les coupes à blanc et les cycles de coupes plus courts ont probablement contribué au déclin de sa population. Les plantations d'une seule espèce de conifères au Nouveau-Brunswick freinent néanmoins la diversité et l'abondance des amphibiens, par rapport aux forêts régénérées d'âge comparable.

> "The great expanse of swamps and stagnant waters (in New Brunswick) presents admirable retreats for various species. In the still summer evenings, when the firefly is about, and the crickets, grasshoppers, and the like have just ceased their noisy utterances, then comes forth from the dank places such a medley of voices of different notes, that we can hardly believe that all are produced by frogs."
>
> - A. Leith Adams (1873)

Sixteen species of amphibians are native to New Brunswick (Table 1). Like virtually all of Canada, detailed historical records describing the abundance of New Brunswick amphibians are lacking. A historical record does exist, though much of it is not readily accessible, being scattered through the accounts of late 19th and early 20th Century naturalists and buried in unpublished manuscripts and field notes. This information, compared with specimen records generated during the past 3 decades in New Brunswick, provides the basis of the assessment of changes in amphibian status in the province presented here.

Table 1. *New Brunswick amphibians, their current status, and source of 1st New Brunsick report. Species reported as common are widespread and sometimes locally abundant. Ucommon refers to species which are not frequently encountered by herpetologists but may be widespread. Species described as local may be common and even abundant at specific sites but are very restricted in distribution within the province (Fig. 2).*

Species	Status	1st New Brunswick report
Ambystoma laterale	Uncommon	Adams (1873)
A. maculatum	Common	Adams (1873)
Notopthalmus viridescens	Common	Adams (1873)
Desmognathus fuscus	Local	Adams (1873)
Plethodon cinereus	Common	Adams (1873)
Hemidactylium scutatum	Single specimen	Woodley and Rosen (1988)
Eurycea bislineata	Common	Adams (1873)
Bufo americanus	Common	Gesner (1847)
Pseudacris crucifer	Common	Adams (1873)
Hyla versicolor	Local	Adams (1873)
Rana catesbeiana	Common	Adams (1873)
R. clamitans	Common	Gesner (1847)
R. septentrionalis	Common	Rowe (1899)
R. sylvatica	Common	Adams (1873)
R. pipiens	Common	Adams (1873)
R. palustris	Common	Adams (1873)

The limitations of this approach must be appreciated. The information available permits only broad, qualitative generalizations of long-term historical changes in species status. In few cases can one make any comments on regional changes in amphibian abundance within the province or determine whether populations are currently increasing, declining or stable (McAlpine 1992). On a local level, amphibian populations may undergo dramatic annual changes in abundance naturally (Pechmann et al. 1991; Green, this volume), making it extremely difficult to determine the importance of anthropogenic factors in amphibian declines. Distribution atlas programmes (for example, Oldham and Weller 1992), repeated at intervals over a period of decades, may be the only reliable means of determining large-scale amphibian declines. At present, the most prudent approach to conserving amphibians is protection of amphibian habitat generally. Fortunately, it may not be too late to follow this course in New Brunswick.

The Historical Record Reviewed

The 1st report on the amphibians of New Brunswick is a species list provided by Dr. Abraham Gesner (Fig. 1A), a naturalist best known as a geologist and as the inventor of the process for refining kerosene (Barkhouse 1980). Gesner (1847) notes the presence of 7 species of amphibians, including "toads, two varieties, salamanders, three varieties ... *R. flavi-viridis* - spring frog" [= *Rana clamitans*]. Although he mentions "*Rana pipiens* - bullfrog", it remains uncertain if he refers to the northern leopard frog or to the bullfrog, *R. catesbeiana*.

In 1899, after a long life as a naturalist resident in southwestern New Brunswick, George Boardman (Fig. 1B) reported 12 species of amphibians present in the province (Boardman 1903). Although the nomenclature he used has changed, it is possible to assign all his names to 11 species that we know to be present in New Brunswick today. Unfortunately, neither Gesner (1847) nor Boardman (1903) provided any comments on the relative abundance of any species.

It was A. Leith Adams (Fig. 1C), army surgeon, naturalist, and eventual authority on fossil elephants, who provided the 1st descriptive account of the abundance of New Brunswick amphibians. Adams spent the years 1866 to 1868 in central New Brunswick and appears to have been a remarkably keen observer. Adams (1873) reports 19 species of amphibians (8 frogs, 11 salamanders) as being present in New Brunswick. However, the nomenclature for some of the salamanders is confusing and, in some cases, it is not clear to which species he was referring. Adams reported that "*Bufo americanus* abound in all suitable localities." He pronounced *Rana clamitans, R. catesbeiana, R. pipiens, R. palustris*, and *Pseudacris crucifer* (current nomenclature) as "common". *Hyla versicolor* he considered "exceedingly rare". *Plethodon cinereus, Eurycea bislineata*, and *Notophthalmus viridescens* he also considered common, while *Desmognathus fuscus* was rare.

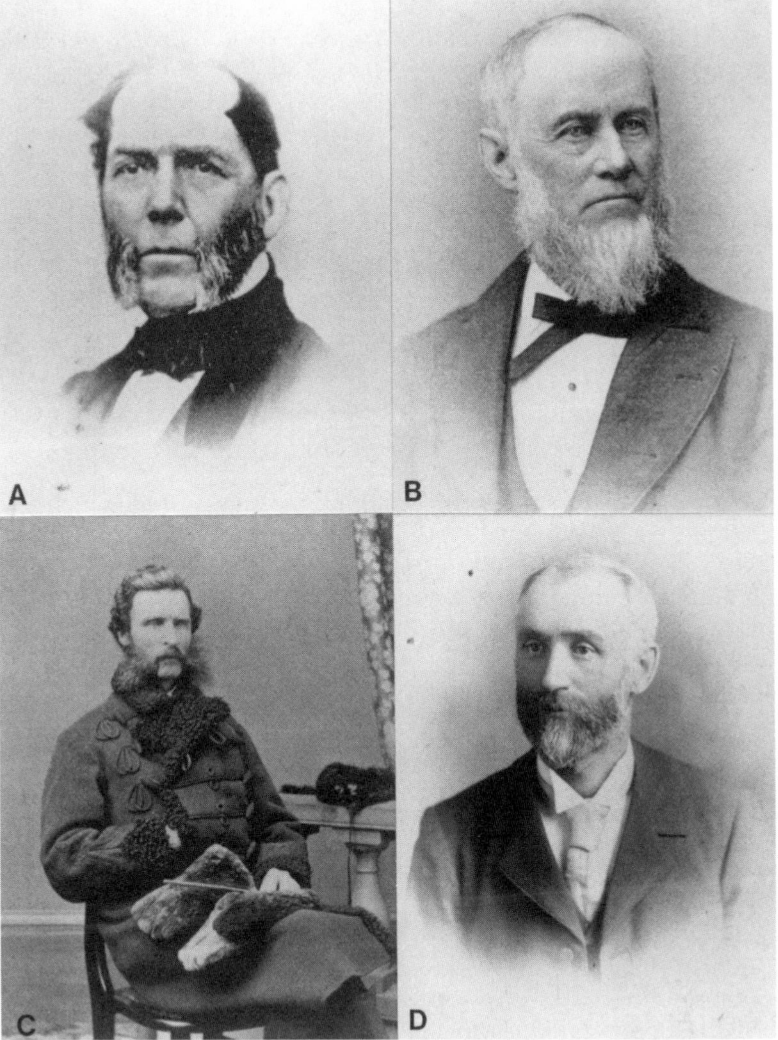

Figure 1. *The principal 19th and early 20th Century herpetologists of New Brunswick. A) Abraham Gesner (1797–1864). B) George A. Boardman (1818–1901). C) A. Leith Adams (1827–1882). D) Philip Cox (1847–1939).*

Between 1898 and 1907, Dr. Philip Cox (Fig. 1D), inspector of schools in Northumberland and Restigouche counties and eventually Professor of Natural Sciences at the University of New Brunswick, published 5 papers on the amphibians of the Atlantic region (Cox 1898, 1899a,b,c, 1907). These are the 1st truly scientific papers on the amphibians of the region. Cox (1898) recognized 22 species and subspecies of New Brunswick amphibians. Although he seems to have deferred to "splitters" on questions of nomenclature, recognizing 2 named New Brunswick forms for *P. cinereus, E. bislineata, N. viridescens*, and *R. pipiens* and 3 forms for *R. clamitans*, he appears to have done so reluctantly (Cox 1899b, 1907). Cox appeared to have confused brown colour forms of *R. pipiens* on Prince Edward Island with *R. palustris* (Cook 1967), but there is no indication he made this error with New Brunswick specimens.

The lists of Cox (1898, 1899a,b,c, 1907) are virtually complete with respect to the amphibians of New Brunswick. Only *Hemidactylium scutatum* is missing, but this species was not known in the province until a single specimen was found in 1983 (Woodley and Rosen 1988). Cox (1898, 1907) included *Desmognathus "ochrophaea"* on the authority of E.D. Cope, who identified two specimens that Cox collected on Oromocto Island near Fredericton. However, it is clear that Cox (1898), although considering himself "ever...the humblest plodder", doubted Cope's identification, because he included the material under *D. fuscus* when he added the dusky salamander to the provincial list for the 1st time (Cox 1899a). Later, Dunn (1926) examined these specimens and confirmed them to be *D. fuscus*. Nonetheless, there are some inconsistencies in species status among Cox's (1898, 1899a,b,c, 1907) lists. The list of Cox (1899a), identified as preliminary, reports that *R. fontinalis* (= *R. clamitans*) was "rather uncommon" and that *R. virescens* (= *R. pipiens*) and *R. sylvatica* were "rare" and "rather rare", respectively, whereas elsewhere Cox lists these species as "abundant", "not abundant" or "generally distributed". Because Cox was unlikely to consider a species as common or abundant when it was actually rare, and because there is consistency among all his lists except the one he considered preliminary (Cox 1899a), I have generally followed his later lists.

A hiatus in the study of New Brunswick amphibians followed Cox's research. Sherman Bleakney (1958), in a classic study, reported on the distribution of species in New Brunswick based on field work carried out in 1953 and 1955. Although Bleakney (1958) provided the 1st distribution maps for New Brunswick amphibians, there were few comments on the status of specific species either in Bleakney(1958) or in his field notes housed at the Canadian Museum of Nature. He did consider *D. fuscus, E. bislineata*, and *H. versicolor*, "rare to locally common", "common", and "rare", respectively, and his maps provide a useful comparison for those species earlier writers considered common or generally distributed.

The most detailed published account of the status of New Brunswick amphibians from the period 1950 to 1970 is Gorham (1964), in which he reported his observations made in the lower St. John River valley from 1948 to 1953. Like Bleakney (1958), his observations were consistent with earlier 19th Century naturalists. Extending his observations provincially, and based on the specimen collections he assembled at the New Brunswick Museum, Gorham (1970) produced a booklet on the amphibians and reptiles of the province. Gorham makes no mention of amphibian declines; the status of all species was again unchanged since the reports of Adams (1873) and Cox (1898, 1899b,c, 1907). The status of 2 less conspicuous species, *D. fuscus* and *E. bislineata*, had been expanded with further field work (Cook and Bleakney 1960).

Although I scanned without success the herpetological field notes for New Brunswick of F.R. Cook for 1955, C. Bruce Powell for 1962, and F.W. Schueler for 1976 for comments on abundance, their collections do provide a crude index of species status. Those species most frequently reported as present, common, or abundant by earlier authors continued to be the most readily and widely collected. Recent amphibian surveys of Fundy (Rosen and Woodley 1984) and Kouchibouguac (Bérubé and Tremblay 1993) National Parks provide a useful benchmark for species status on the Fundy coast and southeastern New Brunswick, respectively. Gorham and Bleakney (1983) similarly covered the southwest part of the province. Species identified as abundant through the previous 110 years remain so and there are no suggestions of any amphibian declines in these reports.

Current distribution maps for the 16 species of New Brunswick amphibians (Fig. 2) are based on specimen records assembled from 1965 to the present and housed in the New Brunswick Museum and the Royal Ontario Museum. Records housed at the Canadian Museum of Nature are currently computerized in an abbreviated form that made their use for this project impractical. Although these maps under-represent the known distribution of most of the species, they nonetheless conform generally with the observations made by the early naturalists to describe the abundance of these species.

Spring peeper, Pseudacris crucifer. *Photo by Donald F. McAlpine.*

Largely unexplained declines of *Rana pipiens* have been documented elsewhere in Canada (Koonz 1992, Roberts 1992, Seburn 1992) and, historically, *R. pipiens* has been

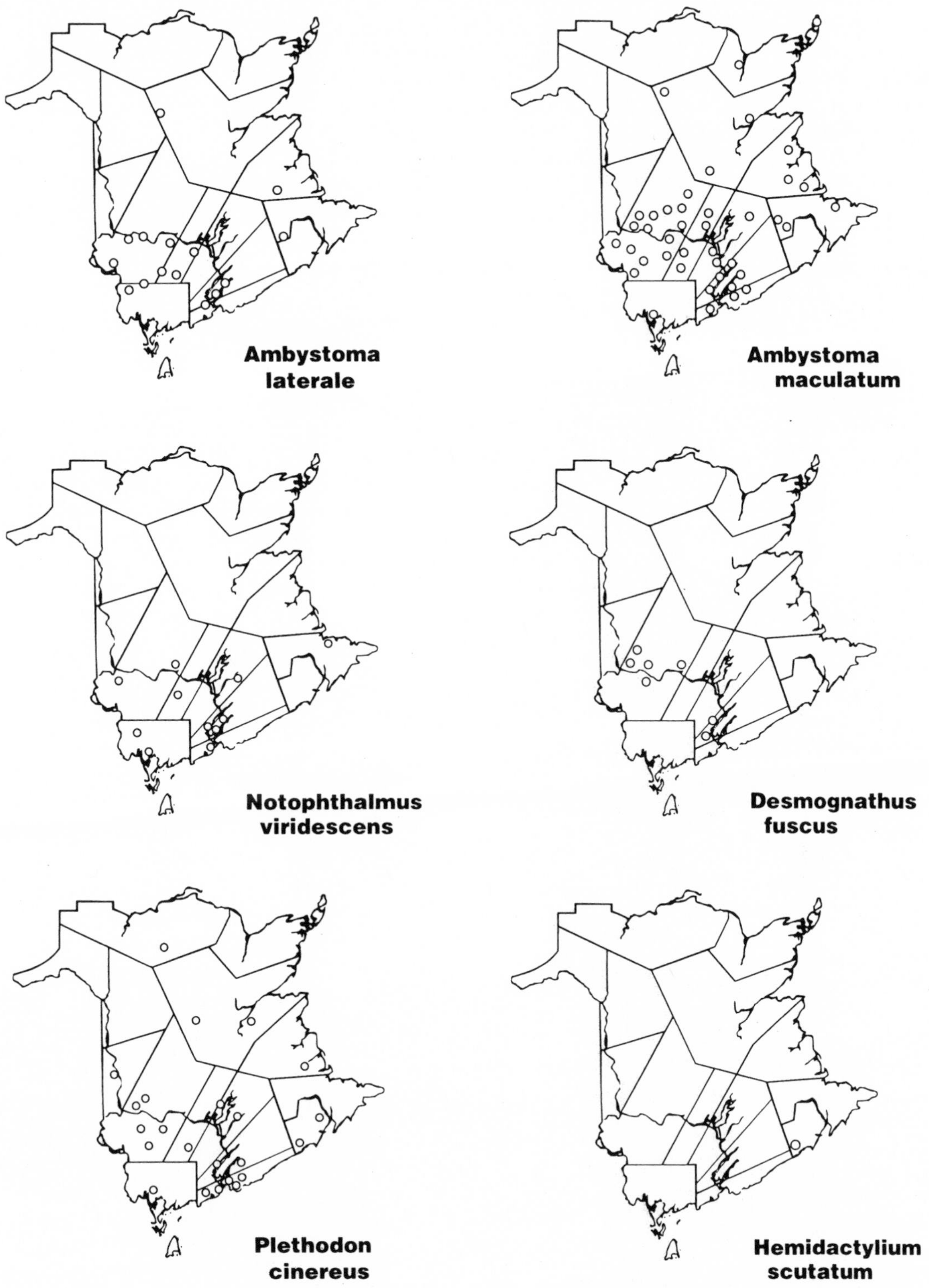

Figure 2. *Distributions of New Brunswick amphibians based on specimen records collected 1965 to the present and now in the New Brunswick Museum and Royal Ontario Museum. Records under represent the known species distributions in most cases, particularly in the north of the province where herpetological field work has been scanty. Nonetheless, these distributions conform generally to the status reported for New Brunswick amphibians by 19th Century herpetologists and the descriptions of status provided in Table 1. Lines mark county borders.*

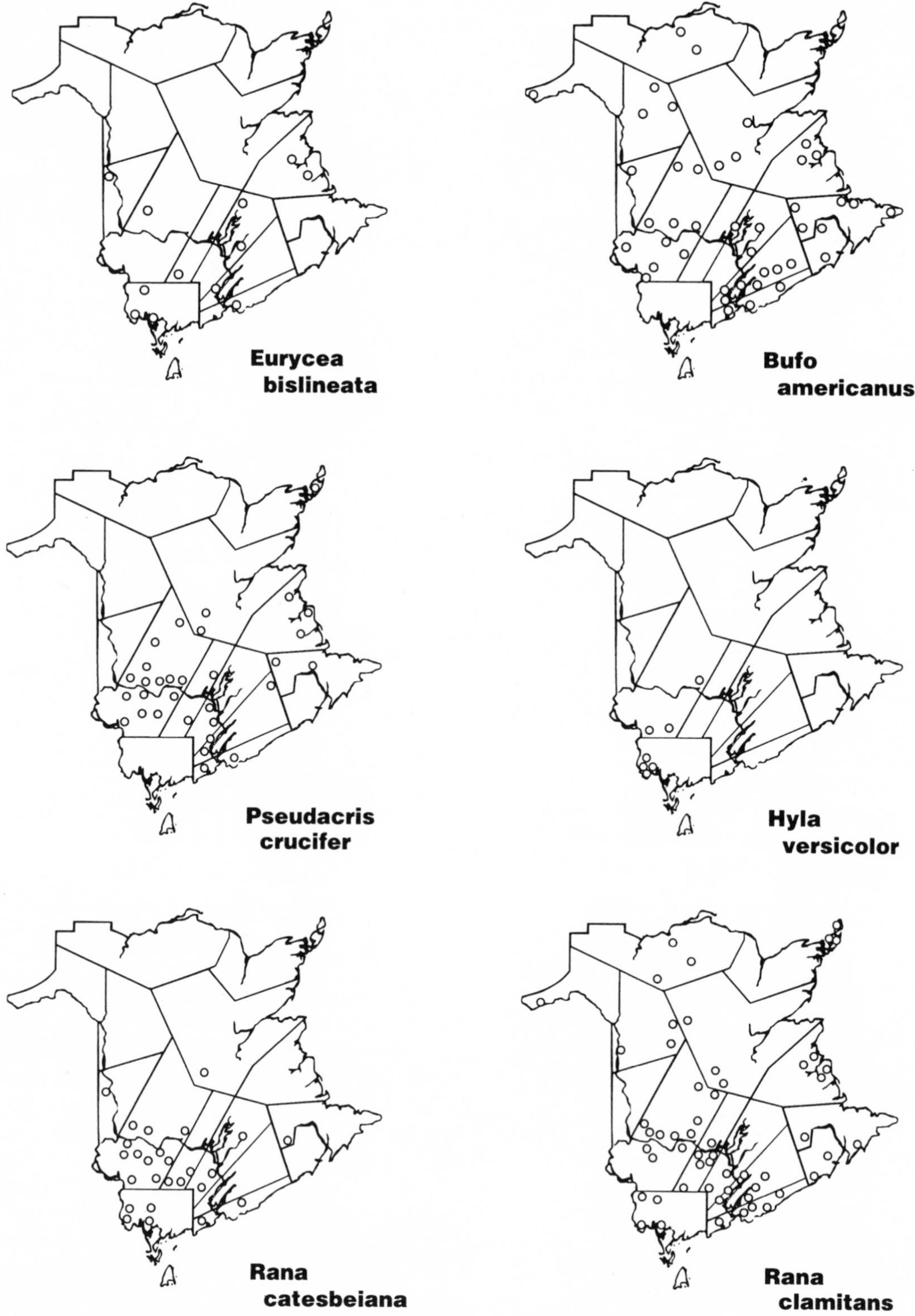

Figure 2. *Continued.*

reported as common in New Brunswick, although often not as common as the closely related *R. palustris*. *Rana pipiens* is a species of damp grassy areas and clearings, and it is scarce or absent in heavily forested areas where such habitats are not present. Thus, forest clearing for agriculture and

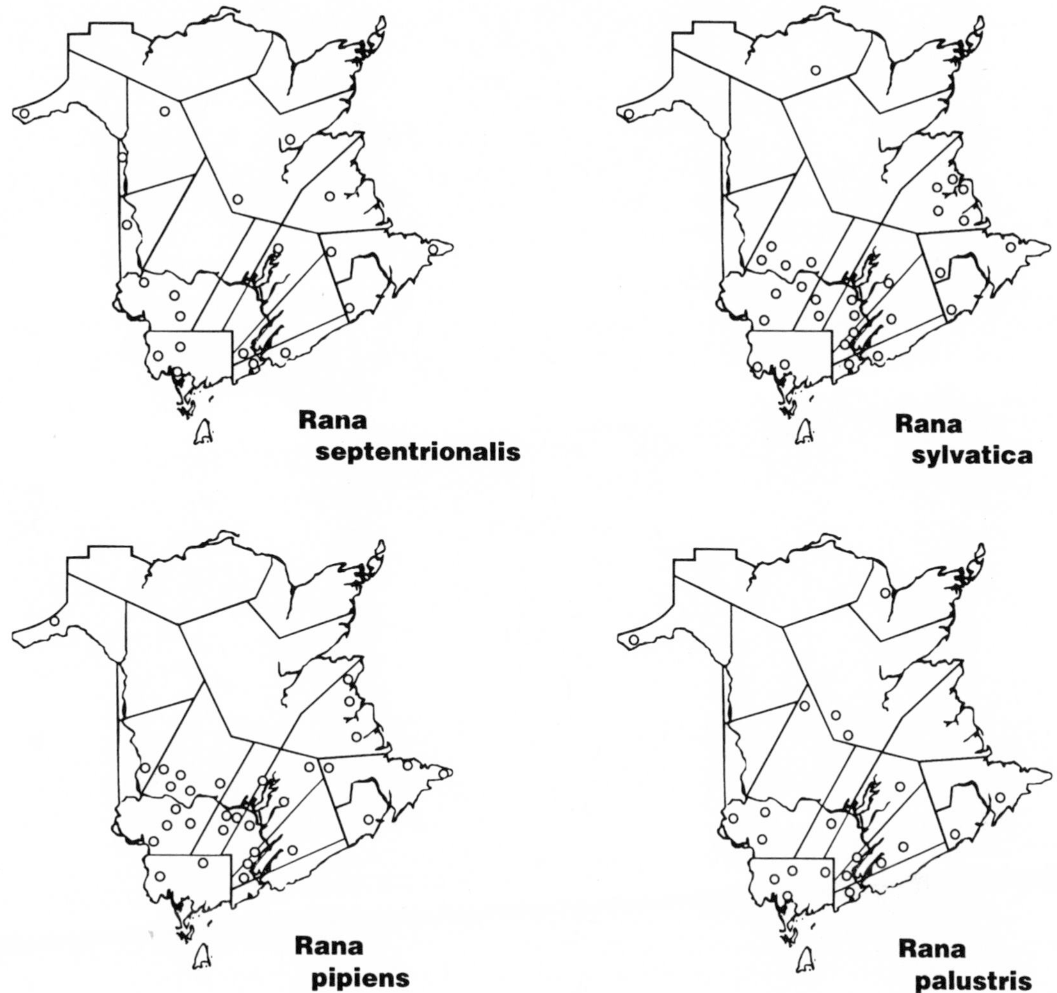

Figure 2. *Continued.*

settlement in New Brunswick over the last century may have favoured *R. pipiens* populations. Forest has steadily reclaimed New Brunswick agricultural lands since the 1880s (Kienholz 1984). Nonetheless 1989 and 1990 observations at southern and central New Brunswick sites where *R. pipiens* had been collected since 1965 have shown that the species remains common, and even abundant, in at least these areas of the province.

Evidence for Local Declines

Is there any suggestion that local declines of amphibians might have taken place in New Brunswick? Although the answer is yes, the long-term effect of these declines is unclear. Cox (1899b) reported that the bullfrog was "formerly well represented in the lakes and streams tributary to the lower St. John; but has grown quite scarce since the introduction of the eastern pickerel [ie. *Esox niger*] devours it and the spring frog [= *R. clamitans*] to such an extent as to render them even rare where they were a few years ago abundant." The exact time and place of the introduction of *E. niger* to New Brunswick has not been fully documented (Scott and Crossman 1959) but the species was apparently quite common on the lower St. John River by the mid-1890s (Cox 1896). *Esox niger* remains widespread through southern New Brunswick (Gorham 1970), as are *R. catesbeiana* and *R. clamitans*.

Heavy doses of DDT sprayed on 120 ha of northern New Brunswick forest in 1952 led to considerable mortality of larval amphibians (Pearce and Price 1975), and there is some evidence that long-term spraying of the insecticide fenitrothion on new Brunswick woodlands may have indirectly lowered *R. septentrionalis* numbers at some lakes in northern New Brunswick (McAlpine 1992).

Several workers have noted the negative impact of logging practices on amphibians (Bury and Corn 1988, Corn and Bury 1989, Petranka et al. 1993; Dupuis, this volume; Waldick, this volume). There are no historical data on the impacts of such activity on amphibians in New Brunswick, but these impacts have likely been mixed. Although breeding sites for those amphibians that rely on small, often ephemeral pools may be created during forest clearing, populations of woodland, stream, and cold-spring salamanders such as *P. cinereus, E. bislineata*, and *D. fuscus* are more likely to be reduced or eliminated as a result of clear-cutting or the associated siltation of streams (Waldick, this volume). The timber harvest of the 19th Century relied heavily on streams and rivers for moving logs, and nearly 600 mills in the province deposited vast quantities of sawdust into waterways, leaving many heavily clogged. This activity virtually destroyed the fishery in some areas (Allerdyce 1972) and undoubtedly had a negative impact on stream and stream-side amphibians as well. Nonetheless, *E. bislineata* remains widely distributed in New Brunswick.

Ash and Bruce (1994) have emphasised how little we know about the long term effects of clear-cutting on woodland salamander populations. Data presented by Waldick (1994, this volume) suggest that the diversity and abundance of amphibians are reduced where naturally regenerated New Brunswick forest has been replaced with conifer plantations. Although relatively little of New Brunswick forest land has been replaced with plantations, this acreage is likely to increase in the future.

Evidence for Population Increases

There is little evidence to suggest that amphibian populations have increased in New Brunswick. The data that are available, much like that suggesting some local declines have occurred, are largely anecdotal and circumstantial. All early workers reported *H. versicolor* to be rare, but it is now locally common in southwestern New Brunswick (McAlpine et al. 1991). The status of the species in the province is now considered secure by the New Brunswick Endangered Species Committee of the New Brunswick Department Natural Resources and Energy, although as recently as the early 1980s, it was a candidate for protection under provincial endangered species legislation (McAlpine 1980). It is not clear whether the presence of *H. versicolor* in southwestern New Brunswick was overlooked in the past or if the species has undergone a recent range expansion in the area. Nonetheless, nearly all ponds in New Brunswick where *H. versicolor* is known to occur have been created incidental to road construction and gravel excavation (Fig. 3). The range and abundance of *H. versicolor* in New Brunswick has been enhanced by human activities. It is likely that other amphibians in New Brunswick dependent on small or ephemeral ponds, such as *Ambystoma* spp., *R. sylvatica, P. crucifer*, and *B. americanus*, have also benefitted.

Several *H. versicolor* sites have been further altered by beaver, *Castor canadensis*, an animal with an important influence on wetland ecosystems (Naiman et al. 1988). By the early 19th Century, overtrapping had nearly extirpated beaver in New Brunswick (Squires 1946). However, through complete closure of the trapping season, 1st applied in 1897, beaver numbers increased greatly. The season was re-opened in 1944, and abundance of beaver continues to increase to this day (G. Redmond, pers. comm). Increasing New Brunswick beaver populations has benefitted wetland wildlife generally, including *N. viridescens, Ambystoma* spp., *H. versicolor, P. crucifer, B. americanus, R. clamitans, R. sylvatica, R. pipiens*, and *R. septentrionalis*.

Discussion

There is now substantial evidence for declines in amphibian populations in many parts of the world (Vial and Saylor 1993). However, it is overstating this record to propose that amphibian declines are a general phenomena affecting all species in the class or amphibians at all localities. In several studies where long-term data are available, declines have not been detected, although annual fluctuations have been dramatic (Pechmann et al. 1991, Hairston and Wiley 1993; Green, this volume). Although it is important to document amphibian declines, it is equally important to carefully document situations were populations have not declined or have increased.

Figure 3. The ponds in this overgrown gravel pit near McAdam, York County, are typical breeding ponds for Hyla versicolor in New Brunswick. Human activities appear to have had a positive influence on the distribution and abundance of this species in New Brunswick where it is at the northern limit of its range.

Is it likely that long-term province-wide declines in amphibians have gone undetected in New Brunswick? In spite of the subjective nature of the historical record, I do not believe so. Although there is evidence for some local declines, the pattern that emerges does not suggest that there has been any large scale general decline of amphibians in New Brunswick to date. On the contrary, observers have consistently reported most of the now common New Brunswick species to be common, and even abundant, over the past century and more. Most of the 16 amphibian species present in the province are not habitat specialists. Some have been able to exploit the numerous ephemeral ponds and 2nd-growth woodlands created through the extensive forest clearing that has dominated the landscape over the past 4 decades. *Rana sylvatica, P. crucifer, B. americanus*, and *Ambystoma* spp. may have increased in abundance as a result. Additionally, there has been little loss of freshwater wetland in New Brunswick during the past century and amphibians in New Brunswick have undoubtedly benefitted from the activities of recovering beaver populations.

The historical record for New Brunswick amphibians is too fragmentary to be a useful means of detecting anything less than a massive decline, which does not appear to have occurred. Nonetheless, the decline of amphibians in New Brunswick through the incremental loss of local populations should be of concern. Only long-term monitoring at a number of sites and province-wide atlasing of amphibians will provide the baseline data that will make it possible to reliably detect these more subtle trends in amphibian populations in New Brunswick in the future.

Acknowledgments

I am grateful to Dr. Francis Cook, formerly Curator of Herpetology at the Canadian Museum of Nature, for permitting me to examine field notes housed at the Museum. Michael Rankin, Assistant in Herpetology, Canadian Museum of Nature, expedited my access to the material and ensured I was able to make maximum use of my limited time in Ottawa. Ross McCulloch very quickly met my request for copies of New Brunswick amphibian records housed in the Royal Ontario Museum. New Brunswick Museum archivist Jane Smith and Librarian Janet Smith have patiently assisted with my various requests for access to historical volumes dealing with the natural history of New Brunswick. Bruce Bagnell and Alan Heward prepared the maps from my drafts. My early interest in the amphibians of New Brunswick was encouraged by the late Dr. Stanley W. Gorham, and I had the privilege of accompanying him on many trips to study New Brunswick amphibians in the field, an opportunity for which I will be ever grateful.

Literature Cited

Adams AL. 1873. Field and forest rambles. London: King and Company.

Allerdyce G. 1972. The vexed question of sawdust: river pollution in nineteenth century New Brunswick. Dalhousie Review 52:177–190.

Ash AN, Bruce RC. 1994. Impacts of timber harvesting on salamanders. Conservation Biology 8:300–301.

Barkhouse J. 1980. Abraham Gesner. Don Mills, ON: Fitzhenry and Whiteside.

Bérubé D, Tremblay E. 1993. Ecological notes and compilation of data on the amphibians of Kouchibouguac National Park, New Brunswick. Ottawa: Canadian Parks Service.

Boardman SL. 1903. The Naturalist of the Saint Croix. Bangor, ME: Privately printed, Charles H. Glass and Company.

Bleakney JS. 1958. A zoogeographical study of the amphibians and reptiles of eastern Canada. National Museum of Canada Bulletin 155:1–119.

Bury RB, Corn PS. 1988. Responses of aquatic and streamside amphibians to timber harvest: a review. In: Raedeke KJ, editor. Streamside management: riparian wildlife and forestry interactions. Seattle: Institute of Forest Resources, University of Washington. Contribution 59. p 165–181.

Cook FR. 1967. An analysis of the herpetofauna of Prince Edward Island. National Museum of Canada Bulletin 212:1–60.

Cook FR, Bleakney JS. 1960. Additional records of stream salamanders from New Brunswick. Copeia 1960:362–363.

Corn PS, Bury RB. 1989. Logging in western Oregon: responses of headwater habitats and stream amphibians. Forest Ecology and Management 29:39–57.

Cox P. 1896. Catalogue of the marine and freshwater fishes of New Brunswick. Bulletin of Natural History Society of New Brunswick 13:62–75.

Cox P. 1898. Batrachia of New Brunswick. Bulletin of Natural Historical Society of New Brunswick 41:64–66.

Cox P. 1899a. Preliminary list of the batrachia of the Gaspé peninsula and the Maritime provinces. Ottawa Field Naturalist 13:194–195.

Cox P. 1899b. The anoura of New Brunswick. Proceedings of the Mirimichi Natural History Association 1:9–19.

Cox P. 1899c. Freshwater fishes and batrachia of the peninsula of Gaspé, P.Q. and their distribution in the Maritime provinces of Canada. Transactions of the Royal Society of Canada 5:141–154.

Cox P. 1907. Lizards and salamanders of Canada. Proceedings of the Miramichi Natural History Association 5:46–55.

Dunn ER. 1926. The salamanders of the family Plethodontidae. Northampton, MA: Smith College. Smith College 50th Anniversary Publications 7.

Gesner A. 1847. New Brunswick; with notes for emigrants. London: Simonds and Ward.

Gorham SW. 1964. Notes on the amphibians of the Browns Flat area, New Brunswick. Canadian Field--Naturalist 78:154–160.

Gorham SW. 1970. Distributional checklist of the fishes of New Brunswick. Saint John, NB: New Brunswick Museum.

Gorham SW. 1970. The amphibians and reptiles of New Brunswick. New Brunswick Museum Monograph Series 6:1–30.

Gorham SW, Bleakney JS. 1983. Amphibians and reptiles. In: Thomas MLH, editor. Marine and coastal systems of the Quoddy Region, New Brunswick. Ottawa: Department of Fisheries and Oceans. Canadian Special Publication of Fisheries and Aquatic Science 64. p 230–244.

Hairston NGSr, Wiley RH. 1993. No decline in salamander (Amphibia: Caudata) populations: a twenty year study in the southern Appalachians. Brimleyana 18:59–64.

Kienholz E. 1984. Land use in the Atlantic region: agriculture. Ottawa: Lands Integration Progam Directorate, Atlantic Region, Environment Canada.

Koonz W. 1992. Amphibians in Manitoba. In: Bishop CA, Pettit KE, editors. Declines in Canadian amphibian populations: designing a national monitoring strategy. Ottawa: Canadian Wildlife Service. Occasional Paper 76. p 19–20.

McAlpine DF. 1992. The status of New Brunswick amphibian populations. In: Bishop CA, Pettit KE, editors. Declines in Canadian amphibian populations: designing a national monitoring strategy. Ottawa: Canadian Wildlife Service. Occasional Paper 76. p 26–29.

McAlpine DF, Fletcher TJ, Gorham SW, Gorham IT. 1991. Distribution and habitat of the tetraploid gray treefrog, *Hyla versicolor*, in New Brunswick and Eastern Maine. Canadian Field--Naturalist 105:526–529.

McAlpine DF, Gorham SW, Heward ADB. 1980. Distributional status and aspects of the biology of the gray treefrog, *Hyla versicolor*, in New Brunswick. Journal of New Brunswick Museum 1980:92–102.

Naiman RJ, Johnston CA, Kelly JC. 1988. Alteration of North American streams by beaver. BioScience 38:753–762.

Oldham MJ, Weller WF. 1992. Ontario herpetofaunal summary: compiling information on the distribution and life history of amphibians and reptiles in Ontario. In: Bishop CA, Pettit KE, editors. Declines in Canadian amphibian populations: designing a national monitoring strategy. Ottawa: Canadian Wildlife Service. Occasional Paper 76. p 21–22.

Pearce PA, Price IM. 1975. Effects on amphibians. In: Prebble ML, editor. Aerial control of forest insects in Canada. Ottawa: Environmental Canada. p 301–305.

Pechmann JHK, Scott DE, Semilitsch RD, Caldwell JP, Vitt LJ, Gibbons JW. 1991. Declining amphibian populations: the problem of separating human impacts from natural fluctuations. Science 253:892–895.

Petranka JW, Eldridge ME, Haley KE. 1993. Efects of timber harvesting on southern Appalachian salamanders. Conservation Biology 7:363–370.

Roberts W. 1992. Declines in amphibian populations in Alberta. In: Bishop CA, Pettit KE, editors. Declines in Canadian amphibian populations: designing a national monitoring strategy. Ottawa: Canadian Wildlife Service. Occasional Paper 76. p 14–16.

Rosen M, Woodley S. 1984. A survey of the amphibians and reptiles of Fundy National Park. Ottawa: Parks Canada.

Rowe CFB. 1899. First record of *Rana septentrionalis* for New Brunswick. Bulletin of the Natural History Society of New Brunswick 16:169–170.

Scott WB, Crossman EJ. 1959. The freshwater fishes of New Brunswick: a checklist with distributional notes. Royal Ontario Museum Division of Zoology and Palaeontology Contributions 51:1– 45.

Seburn CNL. 1992. The status of amphibian populations in Saskatchewan. In: Bishop CA, Pettit KE, editors. Declines in Canadian amphibian populations: designing a national monitoring strategy. Ottawa: Canadian Wildlife Service. Occasional Paper 76. p 17–18.

Squires WA. 1946. Changes in mammal populations in New Brunswick. Acadian Naturalist 2:26–44.

Vial JL, Saylor L. 1993. The status of amphibian populations: a compilation and analysis. Corvallis, OR: IUCN–The World Conservation Union Species Survival Commission, Declining Amphibians Populations Task Force. Working Document 1.

Waldick RC. 1994. Implications of forestry-associated habitat conversion on amphibians in the vicinity of Fundy National Park, New Brunswick [thesis]. Halifax, NS: Dalhousie University.

Woodley SJ, Rosen M. 1988. First record of the four-toed salamander, *Hemidactylium scutatum*, in New Brunswick. Canadian Field-Naturalist 102:712.

©1997 by the Society for the Study of Amphibians and Reptiles
Amphibians in decline: Canadian studies of a global problem. David M. Green, editor.
Herpetological Conservation 1:128–140.

Chapter 14

SURVEYING CALLING ANURANS IN QUÉBEC USING VOLUNTEERS

MICHEL LEPAGE, RÉHAUME COURTOIS, AND CLAUDE DAIGLE

Ministère de l'Environnement et de la Faune, Direction de la Faune et des Habitats, 150 boulevard René-Lévesque est, Québec, Québec G1R 4Y1, Canada

SYLVIE MATTE

St. Lawrence Valley Natural History Society, 21 111 Bord-du-Lac, Ste-Anne-de-Bellevue, Québec H9X 1C0, Canada

ABSTRACT.—In 1993, we began a pilot project whose main objective was to verify the reliability of annual monitoring from roads to detect long-term changes in anuran abundance. The pilot project permitted the evaluation of volunteer participation and the standardization of survey methods. Listening surveys involved 34 volunteers. Survey effort varied regionally and temporally but data were collected in 23 ponds and 31 routes, each route comprising 10 stations spaced at 0.8-km intervals. About 850 observations were collected at an approximative cost of $11.75 Cdn each. Volunteers correctly followed survey methodology. Ten anuran species were recorded, and peak calling periods were estimated for 8 species using pond survey data. Surveys conducted in 1993 showed that precise results could be obtained for most species by surveying about 30 routes (300 stations) per management unit, 3 times a year between mid-April and the end of July. We suggest continuing this programme and completing it by adding annual habitat descriptions at each listening station. This addition could eventually explain long-term abundance and diversity changes.

RÉSUMÉ.—En 1993, nous avons démarré un projet visant à vérifier la fiabilité d'un inventaire annuel des anoures le long des routes pour détecter leurs changements d'abondance à long terme. Le projet pilote a permis d'évaluer la participation de bénévoles et de standardiser les méthodes d'inventaire. Trente-quatre personnes ont collaboré au projet. L'effort d'échantillonnage a varié régionalement et temporellement mais des données ont été recueillies dans 23 étangs et le long de 31 routes, chacune comportant 10 stations d'écoute espacées de 0,8 km. Environ 850 observations ont été faites à un coût unitaire de 11.75 $ can. Les bénévoles ont suivi adéquatement le protocole expérimental. Les données de la première année d'opération montrent que des résultats précis peuvent être obtenus pour la plupart des espèces en inventoriant 30 routes (300 stations d'écoute) par unité de gestion 3 fois par année, entre la mi-avril et la fin de juillet. Nous suggérons de poursuivre ce programme et de le compléter en notant annuellement l'habitat de chaque station d'écoute pour évaluer les modifications du milieu qui pourraient expliquer d'éventuels changements d'abondance ou de diversité des espèces.

The possibility of a general and global decline among amphibian populations (Phillips 1990) is now a major concern to most naturalists and wildlife management agencies. However, before 1993, no long-term monitoring of amphibian populations had been attempted in Québec. Previous surveys had been primarily aimed at determining species distribution, the status of potentially vulnerable species, or the impact of contaminants and acidity on anuran populations (Vladikov 1941; Bleakney 1958; Denman and Lapper 1964; Bider 1976; Bracher and Bider 1982; Morency and Lafleur 1984; Bider and Matte 1994; Daigle 1992, 1994).

Recent works conducted in other parts of North America (Francis 1978; Anonymous 1991a,b; Huff 1991; M.E. Gartshore, M.J. Oldham, R. van der Ham, and F.W. Schueler, unpublished) suggested that annual monitoring of anuran breeding calls on roads representative of regional habitats could be useful in testing the long-term anuran change hypothesis. In 1993, we began a pilot project whose main objective was to verify the reliability of such a monitoring programme. We used volunteers for field sampling to limit survey costs and make the public aware of the anuran situation. Consequently, the pilot project was necessary to evaluate the volunteer participation level and standardize our survey methods. Listening surveys were conducted in ponds in order to determine the best dates and the range of acceptable atmospheric conditions for conducting such surveys. Finally, an anuran abundance and diversity census was conducted along representative routes in southern Québec, and abundance data were used to estimate the sample sizes required to obtain precise results suitable for detecting differences among years and regions.

Materials and Methods

We divided Québec into 3 large ecozones for this study: deciduous forest, mixed forest and coniferous forest (Fig. 1; Thibault and Hotte 1985). Listening points were established in various breeding sites, mostly at ponds and marshes, in order to identify the calling periods of various species and to verify the influence of certain climatic factors on listening quality and anuran mating call abundance. Volunteers visited these listening points 2 to 3 times per wk between spring thaw and August. Peak breeding was assumed to occur during the period when a given species was heard in the highest proportion of the ponds. Volunteers also were asked to survey 1 or 2 lightly urbanized routes at least 8-km long, selected in a non-random manner. The routes were identified on 1:20,000 topographic maps and 10 listening sites were chosen along each route every 0.8 km. Each volunteer observer received detailed instructions indicating dates, acceptable atmospheric conditions, and the proposed methodology to survey anurans. The methodology included the telephone number and address of a resource person and an audio cassette tape explaining how to identify Québec's anuran species and their approximate regional distribution (Elliot and Mack 1991).

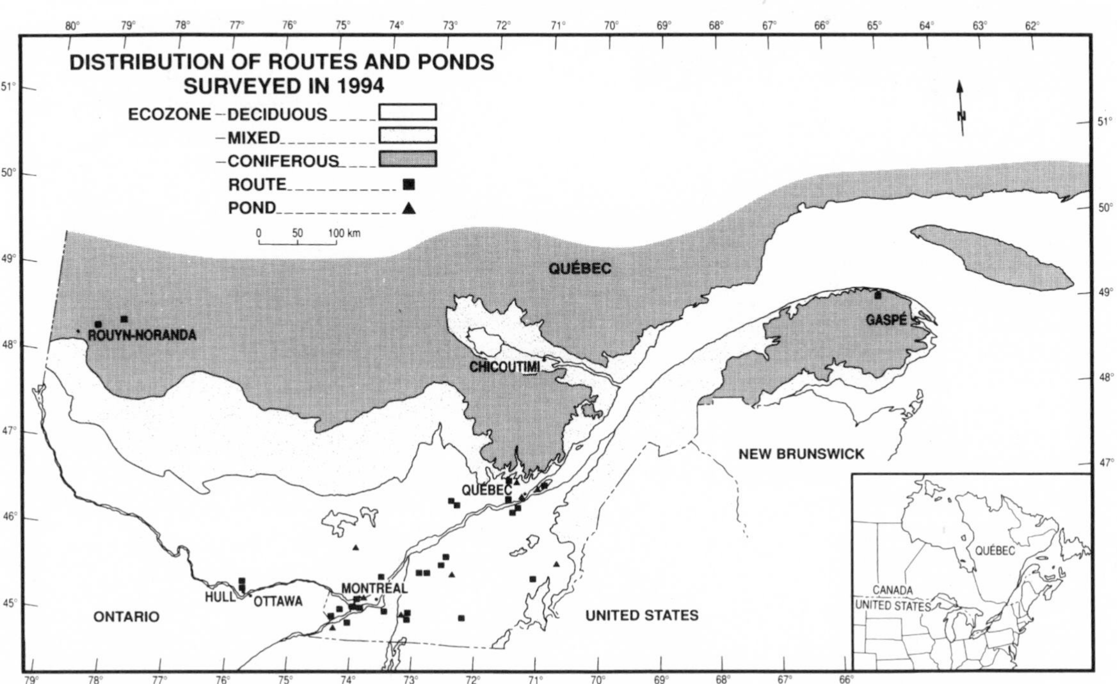

Figure 1. *Ecozones of southern Québec and locations of ponds and routes where calling anurans were surveyed in 1993.*

When surveying, observers stopped at each site along the road, noting all anurans heard during 3 min. Observers were asked to evaluate anuran abundance using the following classes: 0 = no individual heard, 1 = calls of individuals can be counted separately, 2 = calls of individuals are distinguishable with some calls overlapping, 3 = full chorus with calls being continuous, overlapping, and too numerous to be counted. Observers conducted 3 surveys. The 1st was for early calling frogs such as *Rana sylvatica* (wood frog), *Pseudacris triseriata* (chorus frog), and *Pseudacris crucifer* (spring peeper), the 2nd for mid-season species such as *Rana pipiens* (northern leopard frog), *Bufo americanus* (American toad), and *Rana palustris* (pickerel frog), and the last was for late species such as *Hyla versicolor* (gray treefrog), *Rana clamitans* (green frog), *Rana catesbeiana* (bullfrog), and *Rana septentrionalis* (mink frog). Timing of the 3 surveys varied according to zone in order to take probable, latitude-influenced breeding delays into account. Target dates were 15 to 30 April, 1 to 15 May, and 15 to 30 June in the deciduous forest zone, 1 to 15 May, 15 to 31 May, and 1 to 15 July in the mixed forest zone, and 15 to 30 May, 1 to 15 June, and 1 to 15 July in the coniferous forest zone. Some records were collected more than 1 wk later than these dates but were nonetheless retained to increase sample sizes.

Spring peeper, Pseudacris crucifer. *Photo by Jacques Brisson.*

Data were collected from half an hour after sunset till midnight. Observations included air and water temperatures noted at the beginning and the end of the survey routes, cloud cover percentage, and wind direction and intensity measured by the Beaufort scale. Volunteers were asked not to conduct their surveys during cold weather, during periods when wind intensity exceeded 3 on the Beaufort scale (≤ 20 km/hr), or when rainfall was sufficiently heavy to interfere with the hearing of anuran calls. Optimal air temperatures were considered to be 10°C at sunset for the 1st survey but during the 2nd and 3rd surveys, we targeted minimum daily temperatures of 10° and 15°, respectively.

In the deciduous forest zone, the project originally called for 6 ponds and 12 routes to be surveyed. In the other 2 zones, surveys of 2 ponds and 4 routes were planned. Less effort was foreseen for the mixed and coniferous forest zones because of their distance from urban centres and the anticipated difficulty in recruiting volunteers.

Data were input with dBASE software and analyzed using SAS statistical software (SAS Institute 1987). Four indices were used to evaluate anuran species abundance for each survey period and zone: 1) abundance class frequency (%), 2) presence index (percent of listening stations with at least 1 mating call), 3) mean abundance per observation, and 4) mean abundance per sample route. In the latter 2 cases, the abundance classes were converted to absolute values as suggested by Daigle (1992): class 0 = 0 individual, class 1 = 1 individual, class 2 = 5 individuals and class 3 = 15 individuals. Though arbitrary, especially for class 3, this allowed data to be better integrated and more easily interpreted. The 1st 2 indices, which are discrete variables, were presented in the form of frequency tables. Observed and expected values were tested with a Chi-square test (Zar 1974). The abundance indices were analyzed statistically as continuous variables but a non-parametric procedure (Kruskal-Wallis test) was used to compare abundance between regions. Spearman's rank correlation coefficients were calculated both among the abiotic data (ie. Julian date, cloud cover, time, air, and water temperatures) to illustrate their relationships, and between them and the abundance index per observation to determine their influence on anuran mating call activity. Spearman's rank correlation was used because it is more suitable for detecting nonlinear relationships (Legendre and Legendre 1978), and because we wanted to bypass the limitation of parametric tests, which assume data normality. Air and water temperatures at the beginning and at the end of the surveys were compared with the help of a t-test for paired samples to estimate changes during the survey and to determine which measure is the most suitable for explaining anuran calling activity.

Sample sizes necessary to reach a relative error of ± 5% and ± 20% at a significance level of $\alpha = 0.05$ and $\alpha = 0.10$ were estimated using methods described by Snedecor and Cochran (1971). For continuous variables, $n_{req} = t^2 \bullet S^2 / L^2$ where n_{req}= required number of listening stations, t = Student's t, S^2 =estimated variance, L = relative error accepted. For binomial proportions $n_{req} = t^2 \bullet P \bullet Q / L^2$, where P = estimated proportion and Q = 1 − P.

RESULTS

Survey Effort Per Zone And Costs

Response to our request for volunteer participation in the programme surpassed our expectations for deciduous forest zone ponds and marshes. Although only 6 habitats were planned in that ecozone, volunteers committed to conducting surveys in 21 habitats. However, partial data were returned for 13 habitats, and complete data were only returned for 6 habitats. It was difficult to find volunteers for mixed forest and coniferous forest zone breeding sites. The objective of 2 ponds per zone was not reached (Table 1).

Interest was also greater than anticipated in the survey routes of the deciduous forest zone. The programme was expanded from the original 12 routes to 31. Complete data were received for 14 routes and partial data for 12 others. It was more difficult to recruit volunteers for survey routes in the mixed and coniferous zones. Partial data were received for only 2 of the routes in each zone and complete data for only 1 route in the coniferous forest zone.

Table 1. Number of breeding sites (ponds and marches) and number of survey routes planned at the beginning of the programme and actually surveyed by volunteers, number of observers, and number of observations (listening stations).

Ecozone	Sites planned	Committed to by volunteers	Partial data	Complete data	Number of observers	Total observations
Breeding sites						
Deciduous	6	21	13	6	9	195
Mixed	2	5	3	1	2	26
Coniferous	2	1	0	0	1	1
Total	10	27	16	7	12	223
Survey routes						
Deciduous	12	31	12	14	16	560
Mixed	4	4	2	0	2	30
Coniferous	4	2	2	1	2	40
Total	20	37	16	15	20	630

Considerable effort was nonetheless spent and a substantial amount of data were collected. In the deciduous forest zone, 755 observations were made in ponds and along routes, compared to 56 in the mixed forest zone and 41 in the coniferous forest zone, for a total of 853 listening stations. This was the 1st time that such an extensive anuran survey has been conducted in Québec.

Observer participation in the project made it possible to collect reliable data for about $10,000 Cdn. This sum covered the costs of finding volunteers, buying maps, identifying routes and potential breeding sites, distributing maps, forms, and directions, and returning results. The project cost $11.75 per observation.

Methodology

Volunteers included many wildlife professionnals and, in general, survey instructions were correctly followed. Time was respected in 95.3% of the cases for ponds and marshes and in 93.3% of the cases for observations along survey routes. A few observations were made briefly after midnight, contrary to instructions. Approximately 40% of observations were made when cloud cover was less than 50%, and 23% of the observations were made when cloud cover was complete. However, this variable had not been standardized in the survey design. Most (84%) of the observations were made in dry weather. Light precipitation was recorded 16% of the time, with only 0.3% of the data being collected during heavy rainfall. Most (67%) of the observations were made when wind speed was 5 km/hr, 15% when wind speed was 6 to 11 km/hr, and 17% when winds were 12 to 19 km/hr. Thus, 98% of the observations were made when wind speed registered less than 3 on the Beaufort scale, as requested.

In the 3 ecozones, there was a significant temperature drop ($4.07 \leq t \leq 17.90$, $P = 0.001$, $20 \leq n \leq 540$) between the beginning and the end of each survey, though the differences were rather weak (air: 1.5 to 4.3°, water: 0.9 to 3.2°). The smallest temperature differences ($\bar{x} = 1.5°$) were observed in the deciduous zone. Although significant ($t = 3.457$, $P = 0.0017$, $n = 30$), the difference between air and water temperatures was not very great ($\bar{x} = 2.08°$, SE = 0.60). The 2 types of measures were strongly correlated ($0.64 \leq r \leq 0.92$, $P = 0.001$), indicating simiThe most frequent species in the decidu

Breeding Periods As Determined By Pond Surveys

Listening points in deciduous zone ponds permitted a more precise identification of the breeding periods of the majority of the species (Figs. 2 and 3). For *P. triseriata* and *R. sylvatica*, the peak mating call period was between 15 April and 1 May 1993 (Table 2). *Pseudacris crucifer*, *B. americanus*, *R. pipiens*, and *H. versicolor* had mating seasons that lasted for approximately 1 mo, with peak activity occurring 1 May to 15 June. *Rana clamitans* and *R. catesbeiana* also called for approximately 1 mo, with their peak breeding activity occurring between 20 June and 20 July. Data were insufficient or absent to determine peak mating call periods for *R. septentrionalis* and *R. palustris*.

Table 2. *Dates on which anuran species were heard in the deciduous forest zone and presumed calling peak in 1993, based on pond surveys.*

Species	Beginning	End	Peak
Pseudacris triseriata	17 Apr	14 May	17 to 25 Apr
Rana sylvatica	16 Apr	22 May	25 Apr to 9 May
P. crucifer	16 Apr	28 June	2 May to 13 June
Bufo americanus	16 Apr	6 July	2 May to 13 June
R. pipiens	2 May	5 July	2 to 23 May
Hyla versicolor	4 May	6 June	16 May to 27 June
R. clamitans	7 May	30 July	13 June to 4 July
R. catesbeiana	13 June	30 July	27 June to 18 July

Abundance indices showed that most species called in chorus during peak breeding because abundance classes of 2 and 3 and/or calling frequencies (percent of the listening stations) > 30% were observed for most species. However, *Rana pipiens* and *R. catesbeiana* did not seem to be highly concentrated at survey sites we visited, because only class 1 was noted. Their calling frequency, however, reached 30% during the breeding peak suggesting they were quite widespread but probably not in large numbers.

Regional Distribution

Sampling effort along the routes varied greatly depending on the ecozone. There were 560 observations in the deciduous zone, but only 30 in the mixed zone and 40 in the coniferous zone. Regional comparisons were thus limited to the calling frequencies of species at listening sites. Frequencies were calculated using all the available data, so as not to reduce the number of observations in the mixed and coniferous zones where monitoring was less extensive. Consequently, abundances were somewhat underestimated compared with the observations made during the breeding peak.

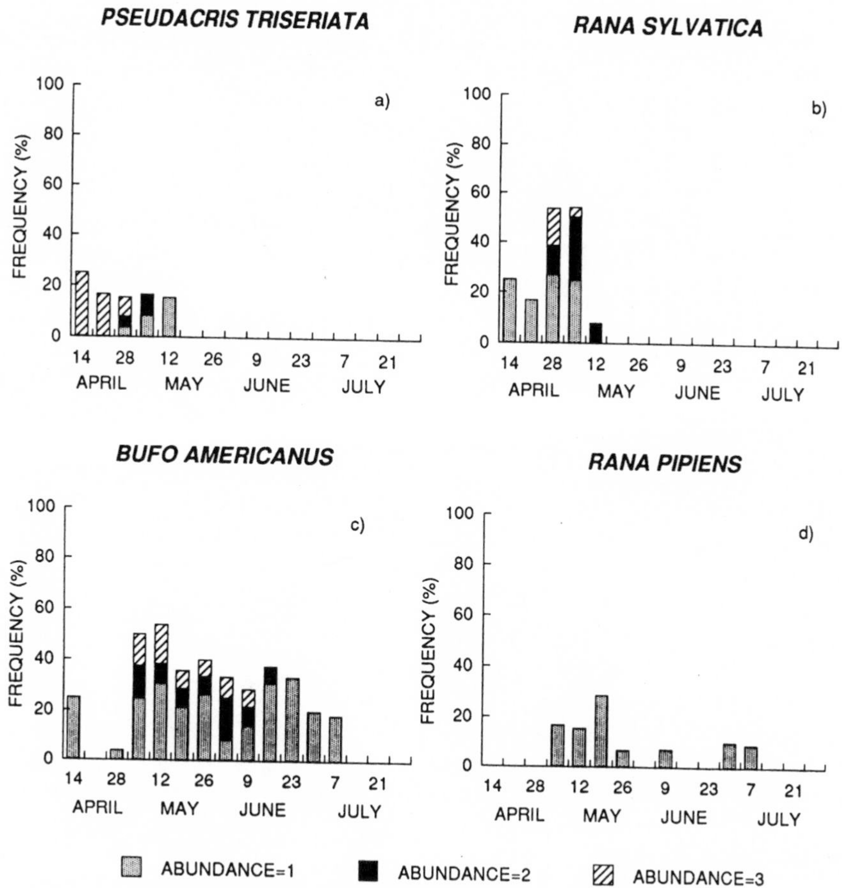

Figure 2. *Seasonal changes of mating call frequency (percent of the stations where the species were heard) in 1993.* Pseudacris triseriata *and* R. sylvatica *were classified as "very early" and* B. americanus *and* R. pipiens *as "early" calling species based on volunteer observations in survey ponds.*

Table 3. *Percentage of observations where different anuran species were heard by ecozone in 1993. Ponds = breeding site surveys. Routes = route surveys.*

Species	Deciduous zone (n)		Mixed zone (n)		Coniferous zone (n)	
	Ponds (196)	Routes (560)	Ponds (26)	Routes (30)	Ponds (1)	Routes (40)
Bufo americanus	27.6	25.7	34.6	16.7		
Hyla versicolor	12.2	13.8				
Pseudacris crucifer	53.6	67.0	53.8	60.0		40.0
P. triseriata	5.6	4.1				
Rana sylvatica	15.3	22.5	26.9	20.0		
R. palustris		0.3				
R. clamitans	30.1	7.7				
R. pipiens	7.1	3.8				10.0
R. septentrionalis	2.0	0.4				
R. catesbeiana	7.7	2.9				

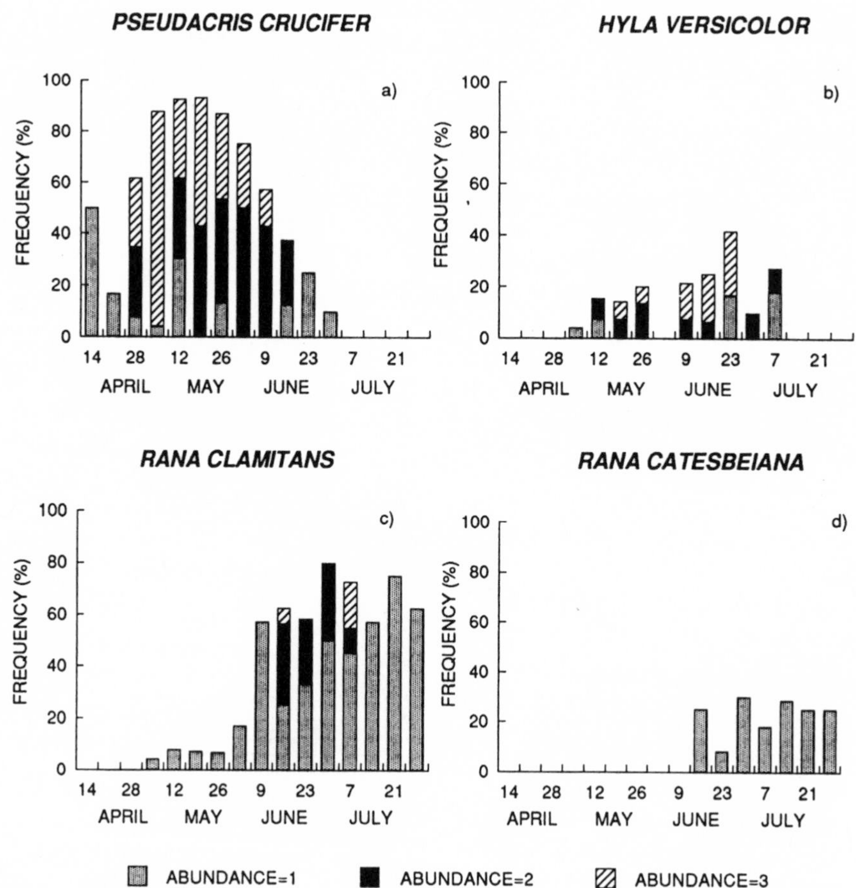

Figure 3. *Seasonal changes of mating call frequency (percent of the stations where the species were heard) in 1993.* Pseudacris crucifer *and* Hyla versicolor *were classified as "early" and* Rana clamitans *and* R. catesbeiana *as "late" calling species based on volunteer observations in survey ponds.*

Table 4. *Four abundance indices during peak breeding activity determined from deciduous zone survey routes (160 to 210 listening stations on 16 to 21 routes). Composite abundance index based on class 0 = 0 individuals, class 1 = 1 individual, class 2 = 5 individuals, and class 3 = 15 individuals.*

		Abundance class frequency (%)					Composite abundance	
Species	Period	0	1	2	3	Presence (%)	Per station ($\bar{x}$ ± SE)	Per route ($\bar{x}$ ± SE)
Bufo americanus	2	56.2	26.2	9.5	8.1	43.8	1.95 (0.12)	6.95 (1.46)
Hyla versicolor	3	68.1	15.6	8.8	7.5	31.9	1.72 (0.32)	5.56 (1.95)
Pseudacris crucifer	2	21.0	19.1	19.1	41.0	79.0	7.29 (0.46)	18.0 (1.72)
P. triseriata	1	91.6	3.7	1.1	3.7	8.4	0.64 (0.21)	1.68 (0.81)
Rana sylvatica	1	47.4	11.6	16.3	24.7	52.6	4.64 (0.45)	11.84 (1.72)
R. palustris	2	99.0	1.0			1.0	0.01 (0.01)	0.10 (0.10)
R. clamitans	3	78.1	15.6	5.0	1.3	21.9	0.50 (0.13)	3.00 (0.82)
R. pipiens	2	93.3	3.3	1.0	2.4	6.7	0.44 (0.16)	1.24 (0.66)
R. septentrionalis	3	98.8		1.3		1.3	0.06 (0.04)	0.25 (0.25)
R. catesbeiana	3	93.8	6.3			6.3	0.06 (0.02)	0.63 (0.24)

Ten species were surveyed in the deciduous zone, 4 in the mixed zone and only 2, *P. crucifer* and *R. pipiens*, in the coniferous zone (Table 3). Frequency indices estimated from habitat type (ponds vs. routes) were fairly similar for a given ecozone. The highest frequency was noted for *P. crucifer*, encountered during 40 to 60% of the observations. Next were *B. americanus* and *R. sylvatica* (20 to 35%), though they were found only in the deciduous or mixed zones. The only other fairly frequent species (12 to 30%) were *H. versicolor* and *R. clamitans*. The 4 species surveyed in the mixed zone had similar frequency indices to those found in the deciduous zone.

Species Abundances During Their Peak Breeding Periods

The most frequent species in the deciduous zone were *P. crucifer*, *R. sylvatica*, and *B. americanus* (Table 4). *Hyla versicolor* and *R. clamitans* were heard in 22 to 32% of the stations while *R. pipiens*, *P. triseriata*, and *R. catesbeiana* were present in 6 to 8% of the listening stations during peak breeding. The most abundant species (5 to 7 individuals/station) were *P. crucifer* and *R. sylvatica*. Four species (*P. triseriata*, *R. pipiens*, *R. clamitans*, and *R. catesbeiana*) were not very abundant (0.5 to 1 individuals/station), but their presence was regular because we found them in 6 to 22% of the listening stations. Other species were rare (presence ≈ 1%). The results per survey route gave an identical picture to those obtained through listening stations.

Our data suggest that *P. crucifer* was the only frequent (50 to 80%) and abundant (3 to 8 individuals/station) species in the mixed and coniferous zones. In the mixed zone, *R. sylvatica* seemed to be frequent (60%) but not too abundant (1 ± 0.47 individuals/station), and *B. americanus* and *R. clamitans* were heard occasionally (15%) in small numbers (0.15 ± 0.08/station). *Pseudacris crucifer* (presence = 75%, abundance = 6.8 ± 1.8/station) and *R. pipiens* (presence = 40% and abundance = 1.8 ± 1.5/station in the first sampling period) were the only two species sampled in the coniferous forest zone in 1993.

In spite of a weak sampling effort in the mixed and coniferous forest zones, we compared regional abundance to determine if statistical differences could be obtained with our sampling strategy. Regional differences (Table 5) were tested when available data met statistical test assumptions (Chi-square tests: all expected values > 1.0 and < 20% of expected values < 5.0; ANOVA: normality and homogeneity of variances; Zar 1994). Abundance classes could only be used to compare the regional differences of *P. crucifer*; differences were only slightly significant ($P = 0.046$). *Bufo americanus* was significantly ($P = 0.014$) more frequent in the deciduous (presence = 44%) than in the mixed zone (15%). *Pseudacris crucifer* was as frequent in the coniferous zone (75%) as in the deciduous zone (79%), but less frequent in the mixed zone (50%). The composite abundance index showed that *R. sylvatica* (4.64 ± 0.45 vs. 1.00 ± 0.47) and *R. clamitans* (0.50 ± 0.13 vs. 0.15 ± 0.08) were more abundant in the deciduous zone than in the mixed zone.

Table 5. *Comparison of anuran abundance among the 3 ecozones based on 3 abundance indices. Only comparisons respecting statistical assumptions were conducted.*

		Abundance class frequency[a]		Presence[a]		Composite abundance[b]	
Species	Period	χ^2	P	χ^2	P	χ^2	P
Bufo americanus	2			6.25	0.014	1.84	0.175
Pseudacris crucifer	2	12.89	0.046	8.57	0.015	2.41	0.300
Rana sylvatica	1			0.21	0.674	9.26	0.002
R. clamitans	2 and 3			0.44[c]	0.511	3.93[d]	0.048

[a]Chi-square test

[b]Kruskal-Wallis test

[c]tested for period 3

[d]tested for period 2

Influence Of Abiotic Variables On Anuran Mating Calls

When data from all periods and all ecozones were used, significant correlations between Julian date and the abundance index showed that *H. versicolor, R. clamitans*, and *R. catesbeiana* called later in the season ($0.31 \leq r \leq 0.55$, $P \leq 0.05$) while *P. crucifer, P. triseriata*, and *R. sylvatica* ($-0.57 \leq r \leq -0.78$, $P \leq 0.05$) were more active at the beginning of the survey period. Anurans did not seem to be greatly influenced by cloud cover, except that *P. triseriata* and *R. sylvatica* called more often when it was cloudy. *Rana pipiens,* on the other hand, called more often under clear skies. However, even for these 3 species the correlations were weak ($r < 0.3$, $P \leq 0.05$). Survey time was not correlated with mating calls ($P > 0.05$) within the limits of the experimental survey design. Air and water temperatures were positively correlated with abundance index in *H. versicolor, R. clamitans,* and *R. catesbeiana* ($0.38 \leq r \leq 0.60$, $P < 0.05$), whereas *P. crucifer, P. triseriata*, and *R. sylvatica* seemed to call less during warm weather ($-0.30 \leq r \leq -0.60$, $P < 0.05$). These relationships were the opposite of those observed for Julian dates and simply represented seasonal changes.

Correlations were conducted using only the data gathered during the peak activity of each species so as to remove the influence of the survey period. This analysis was possible only for the deciduous forest zone since too few data were available in the other ecozones. *Pseudacris crucifer, P. triseriata,* and *R. pipiens* called more intensively at the beginning of their respective peak calling periods since their abundance was inversely correlated with the Julian date (Table 6). We detected no influence of cloud cover on calling activity. *Bufo americanus* seemed to call later at night, its abundance being positively correlated with time whereas *P. crucifer* and *R. palustris* called more often at dusk. Air and water temperature influenced only *P. triseriata*, which called less with higher temperatures, and *R. clamitans,* which did the reverse.

Discussion

Observers followed their directions well. The only discrepancies were a small number of surveys carried out a few days later than planned (probably due to poor weather) and some poorly noted cloud cover data. The experimental survey design seemed to function well, but minor modifications should be considered. The differences between air and water temperatures were significant but negligible.

Table 6. *Spearman's rank correlation between composite abundance of anurans during peak calling and abiotic variables after route surveys of the deciduous forest zone (n = 14 to 21).*

				Temperature (°C)	
Species	Julian date	Cloud cover (%)	Time	Air	Water
Bufo americanus	0.21	0.23	0.37[a]	0.18	0.37
Hyla versicolor	–0.40	0.27	0.33	0.16	–0.16
Pseudacris crucifer	–0.57[b]	–0.22	–0.42[a]	0.00	–0.21
P. triseriata	–0.60[b]	0.11	0.14	–0.42[a]	–0.45[a]
Rana sylvatica	0.36	0.22	0.24	–0.36	0.04
R. pipiens	–0.52[b]	–0.36	0.29	–0.15	–0.31
R. palustris	–0.14	0.02	–0.38[a]	0.26	–0.20
R. clamitans	0.17	0.26	–0.05	0.23	0.61[a]
R. septentrionalis	0.31	–0.04	–0.36	0.19	–0.09
R. catesbeiana	0.00	0.24	0.06	0.14	–0.52[a]

[a]$0.10 \leq P \leq 0.05$

[b]$0.05 \leq P \leq 0.01$

Only 1 variable was thus necessary for analysis. In future, air temperature should be used because it is easier to measure than water temperature. Absence of correlation between cloud cover and each species abundance index, and weak correlations between these indices and water or air temperatures, suggest that atmospheric conditions had little influence on results within the limits proposed in the methodology. However, correlations with Julian date suggest that calling activity is more intensive at the beginning of the target periods at least for *P. crucifer, P. triseriata*, and *R. pipiens*. In the future, observers should conduct their surveys as soon as these species can be heard in certain areas (i.e. at a pond near their home). Some correlations with time also suggest that surveys not be conducted too late at night. The midnight target proposed in the methodology should be respected.

The suggested route survey dates were adequate. However, *P. triseriata* could have been surveyed earlier, and whereas we had expected *P. crucifer* to be more active during the 1st survey and *H. versicolor* to be more active during the last survey, both were more frequent during the 2nd survey. Observers correctly identified species with the help of available tools (Cook 1984; Conant and Collins 1991; Elliot and Mack 1991). The only questionable records were 2 reports of *R. septentrionalis* taken in May, which could have been *R. sylvatica*.

Our survey suggests that most of the species were not very abundant. This was true even in the deciduous forest zone, which was the most intensively surveyed (Table 4). Only *P. crucifer* and *R. sylvatica* had composite abundance indices > 4/listening station. *Bufo americanus* did not seem to be as abundant as expected with only about 2 individuals/station, but its occurrence index, which was > 44%, showed that it was widespread. Likewise, *H. versicolor, R. clamitans* and *P. triseriata* seemed to be moderately abundant (≥ 0.5/station) and relatively frequent (about 8 to 30%). The other species were considerably more rare. We reported only 4 species in the mixed zone and 2 species in the coniferous zone. This apparent lack in diversity in the last 2 zones is due to insufficient sampling effort. In their distribution atlas of amphibians in Québec, Bider and Matte (1994) reported the presence of 10 anuran species in the deciduous forest zone and 8 in each of the 2 other ecozones. However, the real abundance of each species in the mixed and coniferous forest zones is not known due to a lack of information.

Huff (1991) described anuran population tendencies between 1984 and 1991 in Wisconsin using similar surveys to our own. Most of the species were stable (23 cases) or in decline (7 cases). Only 1 case of a population increase was noted. Among species heard during our surveys, Huff (1991) detected a decrease in numbers for *P. crucifer, R. pipiens*, and *H. versicolor*.

To determine whether temporal tendencies of anuran populations over years are real, it seems necessary to estimate the sample size required to obtain reliable results. To solve this problem, a suitable level of precision must first be established in order to determine what changes are biologically significant and thus sufficient enough to warrant intervention. These choices must take into account the normal annual fluctuations of wild animal populations, their heterogeneous distributions, the financial and human costs of surveys, and the inevitable imprecision that comes with atmospheric conditions, technical errors, and the inherent difficul-

American toad, Bufo americanus. *Photo by John Mitchell.*

ties of all field work. Take, for example, the target thresholds of aerial moose surveys, where well-polished methods have been in use for decades. In these surveys, wildlife managers generally set relative error at ± 20%, with a probability of $\alpha = 0.10$ (Gasaway and Dubois 1987; Courtois 1991).

By using the estimated variance for the anuran survey in the deciduous zone, one can draw several important methodological conclusions (Table 7). The number of observations to be made depends on the index chosen to monitor populations. The presence index is much less demanding as a variable than is the composite abundance index since it does not take into account the number of anurans heard at each station. Significant variations in species presence would be detectable (± 5%, α = 0.10) for all species using 300 listening stations (30 routes). With the composite abundance index, the same sample size would only detect changes of > 20% (α = 0.10) and only for 3 species: *B. americanus, P. crucifer*, and *R. sylvatica*. A sample size of 300 stations would be required for each sampling zone, either by ecozone (for example, the 3 forest zones of the present study) or for the entire province of Québec depending on the management units defined by those in charge of monitoring.

Table 7. *Number of observations (listening stations) required to estimate presence at ± 5% and composite abundance at ± 20% at significance thresholds of α = 0.05 and 0.10, based on deciduous forest zone surveys in 1993.*

	Presence		Composite abundance	
Species	α = 0.05	α = 0.10	α = 0.05	α = 0.10
Bufo americanus	378	275	80	54
Hyla versicolor	334	242	1716	1159
Pseudacris crucifer	255	185	84	57
P. triseriata	118	86	2039	1376
Rana sylvatica	383	278	179	121
R. pipiens	96	70	2756	1860
R. palustris	15	11	1960	13,230
R. clamitans	263	191	1076	726
R. septentrionalis	20	14	9344	6308
R. catesbeiana	91	66	1736	1172

The volunteer participation level and an estimation of the required sample size show that this type of survey is technically possible. Even with a minimum sample size in the mixed forest and coniferous forest zones, it was possible to detect significant regional differences. Three indices appear to be useful in long-term monitoring: frequencies per abundance class, presence index, and composite abundance index per station. Abundance variations can be estimated using the 1st and last indices, whereas the 2nd can detect distribution changes. The abundance index appears to be very sensitive. This might become a handicap if it is used alone since it may detect biologically insignificant differences. Abundance per route was not useful since it gave the same portrait as abundance per station while considerably reducing sample sizes.

Habitat quality must influence anuran abundance and species diversity. Consequently, surveyed routes must cover a habitat range large enough to produce a representative picture of each management unit. Moreover, it is necessary to choose different habitat types to limit the possibility of abundance changes being due to changes within a certain type, such as young ponds. However, this would appear unlikely given a large number of listening stations appropriately distributed throughout the study area. In the future, it will be extremely important to note habitat type at each listening station at least roughly (for example, pristine, disturbed, highly disturbed). This will help to explain temporal changes that could eventually occur in anuran abundance. However, many species of anurans have been shown to have some cycles in overall breeding activity. Some cycles may be more than 10-yr long. Breeding may occur annually in a species but the intensity and success rate rise and fall for unknown reasons (R. E. Ashton, pers. comm.). Consequently, it is necessary to work with the same routes over a long period of time, in the same way, to insure that all the variables that cause changes in species and frequency are leveled out. Finally, barometric pressure is another important factor for anuran calling activity. Greater activity and increased diversity have been noted when barometric pressure is 759 mm mercury and falling (R. E. Ashton, pers. comm.). This variable should also be noted for each survey.

We have concluded that it is possible to adequately monitor most of our species through route surveys that use at least the presence index. The programme worked well since we collected a respectable amount of data for the 1st yr of operation and volunteers adequately respected the survey methodology. Only *R. palustris* and *R. septentrionalis* did not seem to be adequately detected through our route surveys. Alternative sources of information must be considered for these species. Nonetheless, it would be beneficial to have more volunteers in the mixed and coniferous ecozones.

Acknowledgments

We thank the 34 volunteers who participated in this first survey of anuran populations in Québec. Though they are unfortunately too numerous to be named individually, their generous contribution was invaluable and essential to the project's success. We would also like to thank Roger Bider and Joël Bonin for their suggestions on the survey's methodology, as well as the members of the Ontario Task Force on Declining Amphibian Populations whose work served as a basis for the development of our participant's manual. We would particularly wish to emphasize the contribution of M. E. Gartshore, M. J. Oldman, R. Van der Ham, and F. W. Schueler.

Literature Cited

Anonymous. 1991a. Illinois frog and toad survey. Participant's manual. Springfield, IL: Illinois Department of Conservation Natural Heritage Division.

Anonymous. 1991b. Wisconsin frog and toad survey. Participant's manual. Madison, WI: Wisconsin Department of Natural Resources Bureau of Endangered Resources.

Bider JR. 1976. The distribution and abundance of terrestrial vertebrates of the James and Hudson Bay regions of Quebec. Cahiers Géog. Québec 20:393–407.

Bider JR, Matte S. 1994. Atlas des amphibiens et reptiles du Québec 1988-89-90. Québec: Société d'histoire naturelle de la vallée du Saint-Laurent et Ministère du Loisir, de la Chasse et de la Pêche du Québec.

Bleakney JS. 1958. A zoogeographical study of the amphibians and reptiles of eastern Canada. National Museum of Canada Bulletin 155:1–119.

Bracher GA, Bider JR. 1982. Changes in terrestrial animal activity of a forest community after an application of aminocarb (Matacil). Canadian Journal of Zoology 60:1981–1997.

Conant R, Collins J. 1991. A field guide to reptiles and amphibians of eastern and central North America. (3rd ed.) Boston: Houghton Mifflin.

Cook FR. 1984. Introduction aux amphibiens du Canada. Ottawa: Musées nationaux du Canada.

Courtois R. 1991. Résultats du premier plan quinquennal d'inventaires aériens de l'Orignal au Québec, 1987-1991. Ministère du loisir, de la chasse et de la pêche du québec, Direction de la gestion des espèces et des habitats, Québec. SP 1921:1–36.

Daigle C. 1992. Inventaire de la Rainette faux-grillon du l'ouest dans le sud-est du Québec. Ministère du loisir, de la chasse et de la pêche du Québec, Direction de la gestion des espèces et des habitats, Québec. SP 2131:1–26.

Daigle C. 1994. Inventaire de la Rainette faux-grillon de l'ouest dans les régions de Montréal et de l'Outaouais. Ministère de l'environnement et de la faune du Québec, Direction de la faune et des habitats, Québec. SP 2426:1–33.

Denman NS, Lapper IS. 1964. The herpetology of Mont St.Hilaire, Rouville County, Quebec, Canada. Herpetologica 20:25–30.

Elliot L, Mack T. 1991. Les sons de nos forêts. Montréal: Le centre de conservation de la faune ailée de Montréal.

Francis G R. 1978. Roads transects to record the occurence of frogs and toads in Wilmot Township, Waterloo region, Southern Ontario. Ontario Field Biologist 32:78.

Gasaway WC, Dubois SD. 1987. Estimating moose population parameters. Swedish Wildlife Research Supplement 1:603–617.

Huff J. 1991. Frog and Toad Survey 1991. Madison: Wisconsin Department of Natural Resources. Technical Report. p 114–123.

Legendre L, Legendre P. 1978. Écologie numérique. Vol. 2. La structure des données écologiques. Paris: Masson; Montréal: Les presses de l'Université du Québec.

Morency R, Lafleur Y. 1984. Inventaire de l'herpétofaune, années 1978-81. Shawinigan: Parcs Canada, Service de la conservation des resources naturelles,.

Phillips K. 1990. Where have all the frogs and toads gone? BioScience 40:422–424.

Snedecor GW, Cochran WG. 1971. Méthodes statistiques. Paris: Association de coordination de technique agricole.

SAS Institute. 1987. SAS/STAT guide for personnal computers version 6. Cary, NC: SAS Institute Inc.

Thibault M, Hotte D. 1985. Les régions écologiques du Québec méridional-deuxième approximation. Québec: Ministère de l'énergie et des resources du Québec.

Vladikov VD. 1941. Preliminary list of amphibia of the Laurentides Park in Province of Quebec. Canadian Field-Naturalist 55:83–84.

Zar JH. 1974. Biostatistical analysis. Englewood Cliffs, NJ: Prentice-Hall Inc.

©1997 by the Society for the Study of Amphibians and Reptiles
Amphibians in decline: Canadian studies of a global problem. David M. Green, editor.
Herpetological Conservation 1:141–149.

Chapter 15

ANURAN SPECIES RICHNESS IN AGRICULTURAL LANDSCAPES OF QUÉBEC: FORESEEING LONG-TERM RESULTS OF ROAD CALL SURVEYS

JOËL BONIN[1], JEAN-LUC DESGRANGES, JEAN RODRIGUE AND MARTIN OUELLET[1]

Canadian Wildlife Service, Québec Region, P.O.Box 10100, Ste-Foy, Québec G1V 4H5, Canada

ABSTRACT.—Breeding call surveys were conducted at 157 sites in southern Québec to assess the relationship between agricultural landscape features and the occurrence of 9 species of anurans. Using canonical discriminant analyses, the presence of 5 species—*Pseudacris crucifer, Rana sylvatica, Rana pipiens, Hyla versicolor*, and *Rana clamitans*—was related to ≥1 terrestrial or aquatic landscape features. The toad, *Bufo americanus*, was ubiquitous while 3 other species were too scarce to draw any conclusions about their preferences. In general, species richness was inversely related to the occurence of monocultures where pesticides were usually applied. The effect of increased monoculture, however, could not be dissociated from the loss of essential habitats such as permanent waterbodies, forests, and old fields. These landscape features have diminished during the past 25 yr and it is highly probable that anuran populations have changed correspondingly. In our results, > 65% of the variability in anuran occurrences remained unexplained. This might be partly due to the crude nature of the data obtained from the road call surveys and the landscape features evaluations. To investigate factors involved in a global decline of amphibians, we will need more detailed data on habitat and breeding locations, together with sampling strategies oriented toward testing hypotheses.

RÉSUMÉ.—Des inventaires de chants de reproduction ont été réalisés dans le sud du Québec pour déterminer la relation entre les éléments du paysage agricole et la présence de 9 espèces d'anoures. À partir d'analyses canoniques discriminantes, la présence de 5 espèces,*Pseudacris crucifer, Rana sylvatica, Rana pipiens, Hyla versicolor*, et *Rana clamitans*, fut associée à un ou plusieurs éléments terrestres ou aquatiques. Le crapaud, *Bufo americanus*, était ubiquiste tandis que 3 autres espèces n'étaient pas assez fréquentes pour tirer des conclusions sur leurs préférences. En général, la richesse spécifique décroissait avec l'augmentation de la superficie en monoculture où des pesticides sont généralement appliqués. Toutefois, l'effet d'une augmentation de la monoculture ne pouvait être dissocié de la perte d'habitats essentiels (plan d'eau permanent, forêt et friche). Des changements dans ces éléments du paysage sont survenus au cours des 25 dernières années et il est fort probable que les populations d'anoures aient également changé. Dans nos résultats, une grande part de la variabilité dans la présence des anoures est demeurée inexpliquée (> 65%). Cela pourrait être dû en partie à l'imprécision des données provenant de l'inventaire routier des chants de reproduction et de l'évaluation des éléments du paysage. Afin de rechercher les facteurs impliqués dans un déclin global des amphibiens, nous aurons besoin de données plus détaillées sur les habitats et les sites de reproduction, ainsi que des stratégies d'échantillonnage permettant la vérification d'hypothèses.

In the past 50 yr, the intensification of agriculture in southern Québec has modified the landscape and adversely affected wildlife (Freemark and Boutin 1994; Boutin et al. 1994; DesGranges et al.

1 Present Address: *Redpath Museum, McGill University, 859 Sherbrooke St., W., Montréal, Québec, Canada*

1995). In the most productive lands of the Saint Lawrence Lowlands, wetlands and woodlands have been replaced by large scale agricultural monocultures. Irrigation has resulted in the replacement of natural wetland habitats by temporary ditches. Corn production, which was favoured by subsidies for underground drainage, increased in Québec from 43,000 ha in 1973 to 228,000 ha in 1987 (Jobin et al. 1994). On the other hand, cultivation of marginal lands located in the foothills of the Appalachians and the Laurentians has been partly abandoned, leaving old fields and forests in their place (Boutin et al. 1994; Jobin et al. 1994).

Ten species of anurans live in southern Québec (Bider and Matte 1994). They require permanent or temporary waterbodies for breeding (Collins and Wilbur 1979), and marshes, fields or forests for summer foraging. Anurans may be unable to find suitable habitats in the present agricultural landscape. This has been the case in the Netherlands where a decline in amphibians was noticed after a decrease in the number of pools, mainly drinking ponds for cattle, between 1960 and 1980. Now, out of 12 native species, 2 are extinct and 5 are endangered (Laan and Verboom 1990). Berger (1989) noted a similar decline of amphibians in the agricultural landscape of Poland, and found that chemical pollution of aquatic habitats caused the death of larvae.

The recent awareness of a possible amphibian declines in Canada raised the need for population monitoring (Bishop and Pettit 1992). Programs based on breeding call surveys were recently set up in various part of the country. Though methodological and analytic aspects have been addressed (Bishop et al., this volume; Lepage et al., this volume), little is known on the usefulness of this approach, especially to identify causes of decline.

Bullfrog, Rana catesbeiana. *Photo by John Mitchell*

The objectives of this study were to determine anuran species richness in agricultural lands, and to infer the conditions required for their survival. For those purposes, we applied the methodology of breeding call surveys over an area where conditions varied from rustic and heterogeneous landscape to large scale monocultures. We felt that this range of habitat conditions would reflect the extent of landscape changes that occurred in southern Québec since agricultural use intensified. Therefore, we expected to find indications of a possible decline of anuran in the past, as well as information concerning the usefulness of road call surveys to evaluate causes of decline.

Material and Methods

We conducted the survey along 10 Breeding Bird Survey (BBS) routes used to study the response of birds to changes in agricultural landscapes over the past 25 yr (Jobin et al. 1994). Most routes were located in the productive St. Lawrence lowlands, but some routes reached marginal lands located in the foothills of the Appalachians and the Laurentians (Fig. 1). Most routes were within the range of 10 species of anurans (Bider and Matte 1994). From each BBS route, consisting of 50 stops set at every 800 m, we selected some stops according to the landscape features present, as described in Jobin et al. (1994).

Following the methodology described in Lepage et al. (this volume), we conducted anuran call counts after sunset, during warm nights. We listened for 3 min at each stop, noting the abundance of each species that was calling from any direction. We ranked abundance by using a calling index: 0 = no calls, 1 = few males calling, 2 = calls overlapping but some individual calls still distinguished from the chorus, 3 = a full chorus. We repeated surveys during the year to correspond with the breed-

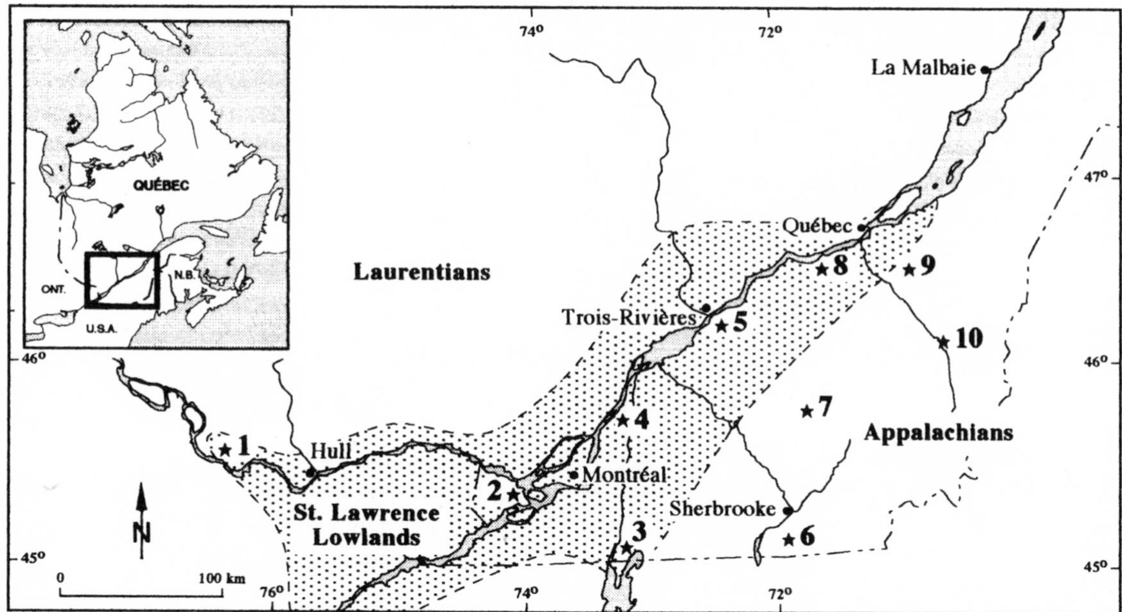

Figure 1. *Location of routes surveyed in 1992 and 1993 in southern Québec.*

ing period of each species (Lepage et al., this volume). If there was too much wind, or if the temperature dropped suddenly and the frogs stopped calling, we re-surveyed the route the next suitable night. We neglected stops where other noises from cars or barn fans impaired our ability to hear. In 1992, 157 stops were surveyed (Table 1), and we re-surveyed most routes at least once in 1993 to detect either the early breeding species like *Rana sylvatica* or other species that had been missed the 1st time. We surveyed routes 1, 3, and 4 more frequently in 1993 to measure variation in the presence and abundance of species from one year to another.

The presence of each species was related to landscape features by the mean of canonical discriminant analyses. Variables included the presence or absence of permanent waterbodies or slow flowing watercourses (slope < 10 m over 500 m), intermittent slow flowing watercourses, old fields, and forests identified using recent aerial photographs (1:15,000) and topographical maps (1:20,000). These features were counted if within 400 m from the listening stop. Forests located up to 100 m away from the 400 m radius were also taken into account on the assumption that frogs could return to these forests after breeding. The routes which were located out of the distribution range of a species or which were not surveyed during the species breeding period, were removed from the analysis for that species.

Species occurrence and richness (total number of species) were also compared to the percentage of the area in monoculture and the size of wood lots. We identified crops in the field during 1992 (corn, wheat, oat, barley, soybean, or vegetables) and we evaluated the percentage of

Table 1. *Number of stops and dates of surveys for each route in 1992 and 1993.*

Route	Stops	Survey Dates 1992	Survey Dates 1993
1. Eardley	13	8 May; 2 June	29 April; 8,14,30 May; 6 July
2. Hudson	6	7 May, 14 June; 10 July	
3. Lacolle	16	29 April; 10, 30 May; 3 July	25 April; 6 May; 2,7,13 July
4. Verchères	25	28 April; 9 May; 1,30 June	28 April; 10 May; 11 June; 7 July
5. St-Célestin	22	13 May; 30 May	30 April
6. Coaticook	13	20 May; 17 June	2 May
7. Ham Sud	10	20 May; 17 June	2 May
8. St-Antoine	29	12 May; 8 June	26 May; 29 June
9. Ste-Marguerite	18	11 May; 18 June	
10. Morisset	5	11 May; 18 June	

the area of each stop represented by monocultures in which pesticides were usually applied. We distinguished stops with small wood lots (< 5 ha, n = 25) from those with woodlands of greater area (n = 82). Fifty stops had no forest at all. We used contingency table analyses (Fisher's exact test) to evaluate the relationship between a landscape variable and the species presence. Rank tests were used when the variables were not binary (percent of area in monocultures, and species richness).

We calculated rank-correlations (Kendall's τ corrected for ties [Kendall and Gibbons 1990]) among habitat variables in order to determine how these variables were linked together and if multicolinearity (Pedhazur 1982) could have impaired their simultaneous analysis by the canonical discriminant method. The test of significance corresponded to the Mann-Whitney *U* test when one of the correlated variables was dichotomous, while it was equivalent to Fisher's exact test when both variables were dichotomous (Burr 1960).

Results

Nine species were heard during the 2 yr of survey. *Pseudacris crucifer* was found at 135 stops (86% of all stops), *Bufo americanus* at 133 (85%), *R. sylvatica* at 80 (51%), *Rana clamitans* at 56 (36%), *Rana pipiens* at 41 (26%), *Hyla versicolor* at 28 (18%, or 34% of the locations within its distribution range), *Rana catesbeiana* at 8 (5%), *Rana septentrionalis* at 4 (3%), and *Pseudacris triseriata* at 2 (1%, or 3 to 6% of the locations within its distribution range).

Calling index values varied greatly between years. Different values were obtained for 49 to 60% of the stops surveyed both years for *P. crucifer* (51%, n = 41 stops where the species was present ≥ 1 yr), *B. americanus* (49%, n = 30), and *R. pipiens* (60%, n = 55). The estimated abundance at 1 stop varied from one yr to another but no general trends were noticed. On the other hand, presence/absence data differed in 11 to 50% of the stops: 22% for *P. crucifer*, 11% for *B. americanus*, and 50% for *R. pipiens*. This suggests that presence/absence would be a more stable index than that of abundance and we therefore used only presence/absence data in the subsequent analysis.

In intensive agriculture, wildlife habitats (forest, old field, wetland) were rarer while intermittent ditches were more frequent (Fig. 2). The correlation of the percent of land in monoculture with the presence of all other variables precluded any conclusion about its direct effect on frog occurrence. We had to remove the variable "percent of monoculture" from the canonical discriminant analysis in order to limit the problem of multicolinearity (e.g. Pedhazur 1982). Therefore, each variable used in the analysis represented a complex of variables that subsumed the direct effects of monoculture on anurans.

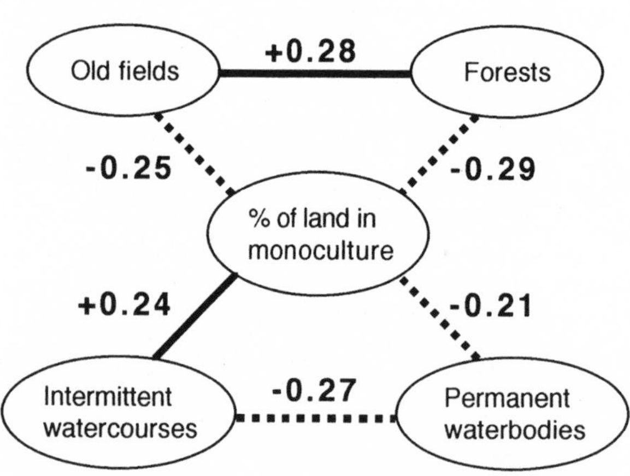

Figure 2. *Significant correlations (Kendall τ, P < 0.001) among landscape features of the 157 stops surveyed in agricultural lands of southern Québec.*

Contingency table analyses and canonical discriminant analyses (CDA) gave concordant results about the relationship between species and landscapes (Table 2). *Bufo americanus*, was ubiquitous and its presence was unrelated to any landscape feature (CDA and Fisher's exact tests, $P <$ 0.05). The occurrence of 5 frog species was significantly related to ≥ 1 landscape features (CDA, $P < 0.001$), though a large portion of the variability in species presence or absence remained unexplained ($R^2 < 0.35$ in all cases). The presence of *R. sylvatica* was related solely to the presence of forest, while *H. versicolor* presence was related to forest and, to a lesser extent, old field (Table

Table 2. *Relative importance of landscape features (standardized coefficients) in the canonical variates discriminating for the presence or absence of the 6 species considered. R is the canonical correlation coefficient.*

		Landscape Features					
Species	Stations	Intermittent watercourse	Permanent waterbody	Old field	Forest	R	*P*
Pseudacris crucifer	157	−0.22	−0.17	0.61*	0.55*	0.43	< 0.001
Rana sylvatica	151	0.19	0.24	−0.08	0.98*	0.42	< 0.001
Rana pipiens	134	0.58*	0.62*	0.40	0.41	0.42	< 0.001
Hyla versicolor	82	−0.22	0.02	0.38*	0.81*	0.58	< 0.001
Rana clamitans	157	0.34	0.60*	0.43*	0.57*	0.41	< 0.001
Bufo americanus	157	−0.82	0.55	0.49	0.22	0.07	> 0.05

*Significant relationship ($P < 0.05$) between the presence of a species and the occurrence of a landscape feature

2). *Pseudacris crucifer* presence was related to old field and, secondarily, to forest. The presence of *R. clamitans* was related to the presence of permanent waterbodies and, as well, to the presence of forest and old field. *Rana pipiens* was associated mainly with permanent and intermittent waterbodies. No statistical analysis could be performed for the 3 less common species because of small sample sizes. *Rana catesbeiana* and *R. septentrionalis* were encountered in a few permanent waterbodies while *P. triseriata* was found in a pasture and a hay field located in marginal lands.

Pseudacris crucifer, *R. sylvatica*, and *R. clamitans* were usually found where a low percentage of the area was in monoculture (Mann-Whitney U, $P < 0.05$ in all tests). Species richness at each stop was also negatively correlated to the percentage of the area in monoculture (Kendall τ, $P < 0.001$). The number of species tended to be higher where there were permanent waterbodies, old fields, and forests (Mann-Whitney U, $P < 0.01$ in all tests). None of the species was found less frequently at stops adjoining only small woodlots < 5 ha than at stops with larger forests (Fisher's exact test, $P < 0.05$ in all tests).

Discussion

The strong relationships between landscape features and the presence of many anuran species indicate that forest and aquatic habitats are prerequisites for survival. These predictable relationships represent known habitat preferences of each species (e.g. Collins and Wilbur 1979). Beebee (1985) also found similar facts using discriminant analysis to determine habitat preference of amphibians in south-east England.

Because correlations among landscape features (Fig. 2) agreed with the history of agricultural changes, the geographical gradient surveyed reflected the temporal trends in agricultural practices during the past 25 yr, as described by Jobin et al. (1994). We therefore suspect that frog occurrences also varied over this period. In the lowlands, intensification of agriculture has probably affected frog populations by the loss of habitat. Although fewer species were usually found at stops with large scale monocultures, the direct effect of monocultures and associated pesticides could not be evaluated because of the strong correlations with all other landscape features measured. Nevertheless, the association between the semi-riparian species, *R. clamitans* (Martof 1953) and the presence of forest and old field might reflect that waterbodies in rustic landscapes were more suitable than those in areas of intensive agriculture. The wide use of herbicides to maintain agricultural monocultures may degrade habitat quality for anurans by reducing the diversity of plants, soil organisms, and other invertebrates (Sotherton et al. 1988). In addition, chemical pollution of aquatic habitat may affect survival of anuran larvae (Berger 1989). Genetical damages and diseases were also found in *R. clamitans* populations from cultivated lands of southern Québec (Bonin et al., this volume).

The level of forest fragmentation (decreased to wood lots ≤ 5 ha), did not seem to affect the occurrence of woodland species of frogs. We selected 5 ha to have a sufficient number of stops with small wood lots, but this might not be a critical threshold. Both forest history and their distribution in agricultural landscapes can influence amphibian presence (Bonin 1991; Waldick, this volume) and dispersal (Laan and Verboom 1990) and should be considered in studying the impact of forest fragmentation on amphibians. Over the last 25 yr in Québec, forest loss and fragmentation took place in the most productive agricultural lands (Jobin et al. 1994). In the marginal lands, the increase of forested areas may benefit woodland frogs, but could not be verified in this study since forests were already present 25 yr ago in most of the marginal lands surveyed.

Usefulness Of Road Call Surveys For Detecting Causes Of Decline

Although the road call surveys permit large areas to be surveyed and geographical variations in species distribution and abundance to be noted (Lepage et al., this volume), there are many limitations in using this method to assess causes of decline. With the road call count technique used, it was rarely possible to locate the chorus from the listening point and, therefore, breeding habitats remained unknown. Habitat characteristics had to be restricted to broad landscape features discernible on aerial photographs or on maps. Therefore, temporary woodland ponds and streams had to be neglected. Intermittent watercourses were distinguished from permanent ones following indications on maps. In some instances, calls could have been heard from a breeding site located out of the 400-m radius and toads could have been heard calling while migrating through the stop toward a breeding site located outside of the survey area. Inter-year variations in calling data were another source of error with this method. Variation in climate was probably the major factor affecting road call count efficiency to evaluate amphibian presence or abundance. For example, 1992 was a bad year to survey *H. versicolor* in a breeding population in southern Ontario (Bertram and Berrill, this volume). Full choruses were seldom heard because of cool temperatures during the species usual breeding period. In 1992, a few males called during several good nights spread over 8 wk, while in 1993, a larger chorus could be heard on many nights within a shorter breeding period (Bertram and Berrill, this volume). From our results, the use of presence/absence data rather than estimated abundance wih the calling index reduced the inter-year variation and, therefore, was considered more accurate.

The efficiency of the call counting technique varied also among locations and among species. Listening conditions vary from one site to another because of topography, buildings, wood lots, or simply the presence of a nearby chorus which masks the sound of frogs calling from further away. Changes in listening conditions at the stops will probably result in erroneous fluctuations in calling data. Species having loud calls, and breeding over a long period, have more chance to be detected, as is the case for the 2 most common species, *P. crucifer* and *B. americanus*. This bias may be increased by the distance at which calls are required to be heard (up to 400 m).

Our study using road call surveys pointed out the obvious influence of habitat loss on anuran survival. However, only a low percentage ($R^2 = 0.17$ to 0.34) of the variability associated with frog presence was explained by the landscape features used in the canonical discriminant analysis. This might be due to the crude nature of the data collected, the omission of important factors such as habitat characteristics, pollutants, zoogeography or interspecific relations, or to the ecological plasticity of each species. Hecnar (this volume), who built predictive models of amphibian diversity in farmland ponds of Southern Ontario, was also left with much unexplained variability (about 60%), even though he considered numerous habitat variables from breeding sites and surrounding areas. Similarly, Beebee (1985) found that landscape features were better predictors of amphibian presence than most measurements relating specifically to breeding ponds. Beebee (1985) and Hecnar (this volume) had canonical correlation levels comparable to ours even if they established amphibian presence from captures at breeding ponds rather than from road call surveys. This suggests the variability depends on other factors than low precision in habitat and population assessment inherent to road call surveys.

Ever since the recent reports of unexplained die-offs and decline in various amphibian populations throughout the world (Wake 1991), scientists and wildlife managers have been seeking to answer if there is a global trend not simply due to the loss of habitat. The road call surveys initiated in Canada (Bishop et al., this volume, Lepage et al., this volume) might not contribute to that question mainly because they will be conducted in areas where changes in habitats and landscapes are highly probable in the future. These habitat changes will bring fluctuations in anuran populations and hence might mask the effect of other global factors. To investigate the presumed roles of pesticides, atmospheric pollutants, UV-B radiation etc. in a global decline of amphibians, we need good data on these factors and on breeding habitat conditions. Furthermore, a sampling strategy would have to be oriented toward testing hypotheses in order to investigate matters such as the effect of pesticides on species occurrence. A long-term monitoring program based on the road call count technique without a sampling strategy would not provide any more valuable information on the causes of decline than the few predictable results presented here for agricultural lands.

Acknowledgments

This study was funded by the Canadian Wildlife Service, Québec region. We wish to thank Bernard Tardif for his help in statistical analysis and for his participation, along with Yves Bachand, Céline Boutin, Patrick Galois, Josée Malo, Paul Messier, and Simon Nadeau, in some of the route surveys. We are grateful to Suzanne Comer who edited the writing.

Literature Cited

Beebee TJC. 1985. Discriminant analysis of amphibian habitat determinants in south-east England. Amphibia-Reptilia 6:35–43.

Berger L. 1989. Disapearance of amphibian larvae in the agricultural landscape. Ecology International Bulletin 17:65–73.

Bider JR, Matte S. 1994. Atlas des amphibiens et des reptiles du Québec. Québec: Société d'histoire naturelle de la vallée du Saint-Laurent et Ministère de l'environnement et de la faune du Québec, direction de la faune et des habitats.

Bishop CA, Pettit KE, editors. 1992. Declines in Canadian amphibian populations: designing a national monitoring strategy. Ottawa: Canadian Wildlife Service. Occasional Paper 76.

Bonin J. 1991. Effect of forest age on woodland amphibians and the habitat and status of stream salamanders in southwestern Québec [thesis]. Montréal: McGill University.

Boutin C, Jobin B, DesGranges JL. 1994. Modifications of field margins and other habitats in agricultural areas of Québec, Canada, and effects on plants and birds. In: Boatman N, editor. Field margins: integrating agriculture and conservation. Farnham, U.K.: British Crop Protection Council. Monogragh 58. p 139–144.

Burr EJ. 1960. The distribution of Kendall's score S for a pair of tied rankings. Biometrika 47:151–171.

Collins JP, Wilbur HM. 1979. Breeding habits and habitats of the amphibians of the Edwin S. George reserve, Michigan, with notes on the local distribution of fishes. University of Michigan Museum of Zoology Occasional Paper 686:1–34.

DesGranges JL, Jobin B, PlanteN, Boutin C. 1995. Effets du changement du paysage rural québécois sur les oiseaux champêtres, In: Domonet G, Falardeau J, editors. Méthodes et réalisations de l'écologie du paysage pour l'aménagement du territoire. Québec: Actes du 4e Congrès national de la Société canadienne d'écologie et d'aménagement du paysage, juin 1994. p 177–180.

Freemark KE, Boutin C. 1995. Impacts of agricultural herbicide use on terrestrial wildlife in temperate landscapes: a review with special reference to North America. Agriculture Ecosystems and Environment 52:67–91.

Jobin B, DesGranges JL, Plante N, Boutin C. 1994. Relations entre la modification du paysage rural, les changements de pratique agricole et les fluctuations des populations d'oiseaux champêtres du sud du Québec (Plaine du Saint-Laurent). Québec: Service canadien de la faune, région du Québec. Série de rapports techniques 191.

Kendall M, Gibbons JD. 1990. Rank correlation methods. 5th ed. London: Edward Arnold.

Laan R, Verboom B. 1990. Effects of pool size and isolation on amphibian communities. Biological Conservation 54:251–262.

Martof B. 1953. Home range and movements of the green frog, *Rana clamitans*. Ecology 34:529–543.

Pedhazur EJ. 1982. Multiple regression in behavioral research: explanation and prediction. 2nd ed. New York: Holt, Rinehart and Winston.

Sotherton NW, Dover JW, Rands NRW. 1988. The effects of pesticide exclusion strips on faunal populations in Great Britain. Ecology Bulletin 39:97–199.

Wake D. 1991. Declining amphibian populations. Science 253:860.

American toad, Bufo americanus. *Photo by Donald F. McAlpine.*

Amphibians in decline: Canadian studies of a global problem. David M. Green, editor.
Herpetological Conservation 1:149–160.

Chapter 16

EXTENSIVE MONITORING OF ANURAN POPULATIONS USING CALL COUNTS AND ROAD TRANSECTS IN ONTARIO (1992 TO 1993)

CHRISTINE A. BISHOP AND KAREN E. PETTIT[1]

Environment Canada, Environmental Conservation Branch, Canadian Wildlife Service, 867 Lakeshore Road, Box 5050, Burlington, Ontario L7R 4A6, Canada

MARY E. GARTSHORE

R.R. # 1, Walsingham, Ontario N0B 1X0, Canada

DUNCAN A. MACLEOD

P.O. Box 3248, Station C, Ottawa, Ontario K1Y 4J5, Canada

ABSTRACT.—Amphibian populations in southern Ontario, Canada, were monitored in 1992 and 1993 using call counts of male anurans at points on roadside transects. Ten species were recorded calling on 62 route transects on rural roads. We asked volunteer observers to conduct 3 surveys from 1 April to 15 June during specific weather conditions and dates. They were to record presence or absence and the relative density of calling males heard. Observers did not consistently adhere to the survey protocols. We found occurrence of calling for several species was significantly related to weather conditions and time of survey chosen by the observer. *Rana clamitans* and *Bufo americanus* were reported more frequently when surveys were conducted later in the day. *Bufo americanus* occurrences were positively related to air temperature. When the data were adjusted for those variables, trends between years indicated by the adjusted data differed from the unadjusted data in some cases. Inter-observer variance was relatively low for identification of presence of calling, however, variance was high for ratings of the density of calling. Therefore, only presence or absence data could be used for analysis.

RÉSUMÉ.—Les populations d'amphibiens dans le sud de l'Ontario (Canada) ont été surveillées en 1992 et 1993 par décompte des coassements des anoures mâles à divers points de transects routiers. Dix espèces ont été répertoriées sur 62 transects sur des chemins ruraux. Nous avons demandé à des observateurs bénévoles d'effectuer 3 recensements entre le 1er avril et le 15 juin moyennant des conditions météorologiques et des dates spécifiques. Ils devaient consigner la présence ou l'absence de mâles qui coassaient ainsi que leur densité relative. Les observateurs n'ont pas adhéré uniformément aux protocoles de recherche. L'occurrence des coassements pour plusieurs espèces dépend de façon significative des conditions météorologiques et de l'heure choisie par l'observateur. *Rana clamitans* et *Bufo americanus* sont signalés plus fréquemment lorsque les recensements ont eu lieu à la fin de la journée. L'occurrence de *Bufo americanus* est positivement liée à la température de l'air. Lorsque les données ont été ajustées en fonction de ces variables, les tendances entre les différentes années révèlent que les données ajustées diffèrent des données non ajustées dans certains cas. Les variations inter-observateurs sont relativement faibles au titre de l'identification des coassements mais les variations sont importantes pour les évaluations de la densité des coassements. Seules les données sur la présence ou l'absence de mâles ont pu être utilisées pour l'analyse.

---

1 Present Address: *Department of Animal Science, University of British Columbia, Vancouver, BC V6T 1Z4, Canada*

Reports of declines in the occurrence and size of amphibian populations throughout the world have incited scientists and governments to review available data on recent and historical amphibian population trends (Vial and Saylor 1993; Blaustein et al. 1994). These efforts are impeded by either a lack of long-term population studies (Blaustein et al. 1994) or a lack of knowledge about the present and historical population sizes of amphibians. Museum records may not be a useful resource for determining trends in amphibian populations (Obbard and Scheuler 1992), because they have a variety of problems, including precision of location data, assumptions about breeding populations, and lack of information about natural extinction and recolonization (Corn 1994).

Extensive monitoring programs utilizing volunteer observers and professional biologists have been recognized as potentially useful tools for detecting fluctuations in the sizes and geographic distributions of amphibian populations in Canada (Freedman and Shackell 1992), but until recently this approach has not been employed. It is recommended as a method to assess gross changes in occurrence and abundance of amphibian populations on a broad geographic scale, and potentially to detect trends between the few intensively studied populations (Freedman and Shackell 1992; Bishop and Kay 1992).

The only extensive monitoring method using volunteers currently in use involves monitoring calls of male anurans at intervals along road transects. This has been used to indicate the presence of amphibian breeding sites and the relative density of the populations from 1984 to 1991 in Wisconsin (Huff 1991), from 1987 to 1989 in Illinois, and since 1992 in Iowa (M. Mossman, pers. comm.). The long-term goals of monitoring along road transects are to establish a database on the annual occurrence of anurans at locations throughout a large geographic area and to collect data in a standardized manner that would be used to assess broad-scale spatial and temporal fluctuations in amphibian populations. Trends in presence or absence or calling density in Wisconsin from 1984 to 1991 found that the spring peeper (*Pseudacris crucifer*), 1 of 12 species monitored, had significantly declined (Huff 1991). During 1992 and 1993, we initiated a monitoring program in Ontario using volunteers to count calling male anurans along linear road transects. Methods were similar to those employed in Wisconsin.

In this paper, we report on the adherence of the observers to the monitoring instructions and the inter-observer variability in identification of species' calls and densities. Although variation in occurrence of anurans among the 2 yr is probably not biologically meaningful, comparing the results between years is an example of statistical analysis of call count data that incorporates variations in results due to observers' lack of strict adherence to the methodological protocol.

Materials and Methods

Study Area, Transect Routes, And Survey Protocols

Sixty-two survey transect routes were established on rural roads in 9 counties in southwestern Ontario during 1992 and 1993 (Fig. 1). One hundred and sixteen surveys were conducted from 15 March to 25 July on 57 routes in 1992, and 104 surveys were conducted on 55 routes in 1993.

Survey routes were chosen along quiet secondary roads, avoiding sources of noise such as traffic, aircraft, or factories. Seasonal roads were not used because the route had to be accessible in early spring when these may not have been drivable. Routes were set up in a single direction, starting from an obvious landmark such as an intersection. If a Breeding Bird Survey route (Droege 1990) was known for the area, the amphibian call survey transect was set up along a portion of it. Routes were kept at least 3.2 km apart to avoid overlapping counts. The transects were established in a variety of habitats.

Each 7.2-km transect comprised 10 stops located 0.8 km apart. Stops were measured as accurately as possible using an automobile odometer calibrated by comparison to a known distance from a topographic map. That information was recorded on the Route Survey form (Fig. 2A), and a permanent marker, such as white paint on a tree or bicycle reflector tape, was placed at each survey point.

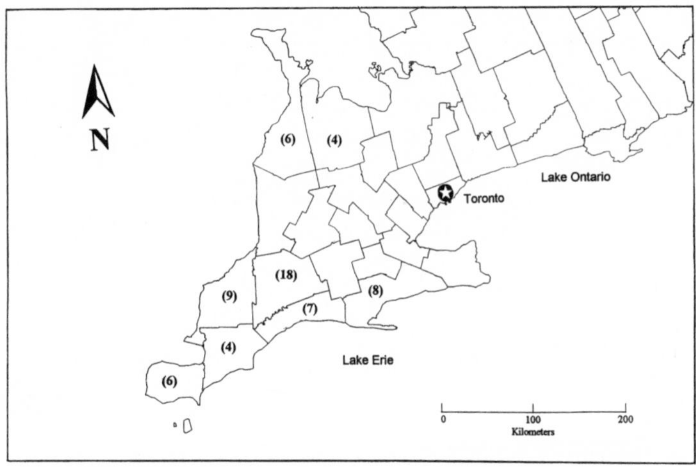

Figure 1. *Number of surveys routes (n) established in each county of southern Ontario in the Ontario Amphibian Call Monitoring Programme, 1992 to 1993.*

Before attempting a survey, all observers were to familiarize themselves with calls of all anurans that occur in Ontario, using audiotapes of anuran calls we provided. If observers were unable to identify calls in the field, we recommended that the calls be recorded and then compared to the audiotape.

The same road transect was to be surveyed 3 times during a specific range of dates for each survey (Table 1). Each survey was to be conducted between specified air temperatures, at wind speeds ≤ 3 on the Beaufort Wind Scale (World Meteorological Organization 1970), and between ½ hr after sunset and midnight, 2400 hr. Other pertinent weather and location data were recorded (Fig. 2B). These guidelines were developed to ensure that data collection was standardized and therefore comparable among monitoring points and transects. All 3 surveys of each road transect in the 1st and subsequent years were to be completed by the same person(s) to avoid observer bias. Although a partner was recommended for safety, only 1 person was to complete the actual call counts on each route.

Table 1. *Survey dates and conditions recommended to volunteers in the Ontario Amphibian Monitoring protocol, 1992 and 1993.*

	Recommended date			
Survey	Southern region	Central region	Northern region	Temperature range (°C)
1	1–15 April	15–30 April	1–15 May	8–12°
2	1–15 May	15–30 May	1–15 June	13–20°
3	1–15 June	15–30 June	15 June–1 July	21–28°

Observers were asked to stop at each point for 3 min and estimate the density of each amphibian species calling. Densities of individual calls were to be rated as one of 4 designations: 0 = no calls heard, 1 = individual calls not overlapping, 2 = individual calls overlapping but readily distinguishable, or 3 = a continuous chorus of calls in which individual calls could not be discerned. All calling heard at any distance from the monitoring point was to be rated.

Data Analysis

The annual peak calling period for each species was determined by plotting the percent occurrence of each species among all road transects for Ontario and the date of the survey. Within the analysis, the routes were all assumed to peak at the same time regardless of geographic or microclimatic differences between the various counties. The peak period for each species was defined as the range of dates containing 95% of presence data for that species.

In 1993, to determine the inter-observer variation in the ratings assigned to calling density, 2 or 3 observers simultaneously conducted surveys at the same points on 12 transect routes. Eleven surveys were conducted from 25 May to 10 June 1993 by 2 summer employees or experienced volunteers who recorded their observations separately but simultaneously (n = 110 point count comparisons). A 12th survey was conducted by the 2 summer employees and 1 experienced volunteer observer. That survey was treated as 3 separate surveys for the analysis, 1 for each pair of observers (n = 30 point count comparisons). Thus, paired observations were compared a total of 140 times. All surveys per-

A

ROUTE SURVEY FORM

Route Name: ______
Nearest Town: ______
Township: ______ Municipality: ______
Map Sheet #: ______ Direction: ______
Description of permanent marker system: ______

Point #	UTM	Description
1		
2		
3		
4		
5		
6		
7		
8		
9		
10		

Comments: ______

B

AMPHIBIAN ROAD CALL COUNT

Route Name: ______ Survey # (1-3): ______
Location: ______ Date: ______
Observer: ______ Assistant: ______
Cloud Cover (%): ______ Wind (0-3): ______ Precip. Descrip: ______
Air Temp. Start: ______ End: ______ Water Temp. 1): ______ 2): ______
Map #: ______ UTM Start: ______ UTM End: ______

0 or blank = none heard
1 = individuals can be counted; no overlapping calls
2 = calls distinguishable; some calls overlapping
3 = full chorus; calls contuinuous and overlapping

Points	1	2	3	4	5	6	7	8	9	10
Time										
Noise (X)										
Frog Noise (X)										
I.D. Probl. (Y/N)										
Species										
American Toad										
Chorus Frog										
Spring Peeper										
Gray Treefrog										
Wood Frog										
Leopard Frog										
Green Frog										
Mink Frog										
Bullfrog										

(Add Cricket Frog, Fowler's Toad, Pickerel Frog, Cope's Treefrog, as needed)
(Check box if you are familiar with calls of nocturnal: birds [ ]
mammals [ ]
Please list species and record actual numbers heard, if any.)

Remarks: ______

Figure 2. *Forms used in the Ontario Amphibian Call Monitoring Programme, 1992 to 1993. (A) Route Description Form, (B) Survey Data Collection Form.*

formed to determine inter-observer variation were conducted after the employee and volunteer observers had completed several surveys in 1993 and had become fully familiar with the data collection methods and species calls.

The number of times the rating of calling density agreed among observers was compiled. The number of times the observers concurred on presence of each species (rating = 1, 2, or 3) or its absence (rating = 0) was also tabulated. Transect results were considered unsuitable for calculating inter-observer variation if the percentage of points where an amphibian species occurred was less than 10% or greater than 90%. In those cases, the value for the 2 observers would be artificially close and would not reflect the variation normally expected under typical conditions. No adjustments were made for the effects of time or temperature in the analysis, because they would be expected to be the same for both observers on each survey.

We calculated the rate at which all observers adhered to the guidelines for date, time, air temperature, and wind speed during each survey. The relationships between those parameters and the log-transformed percentage of the occurrence of each species among all road transects were determined. If the relationship

Gray treefrog, Hyla versicolor. *Photo by Jacques Brisson.*

was significant, a linear transformation was used to adjust the occurrence of that species in each year for comparison to unadjusted field data (Cochran 1977).

For time of day and air temperature, we used standardized values within the regression equation to calculate the adjusted occurrence of each species. For time of day at which the survey was conducted, 2230 hr, midway between ½ hr after sunset and 2400 hr, was considered to be the average survey time for southern Ontario. Therefore, the regression equation was:

$$L_{adj} = L + m(2230\text{ hr} - T),$$

in which L_{adj} was the log-transformed adjusted occurrence of the species on the transect route, L equalled the log-transformed occurrence of the species on the transect route, m was the slope of the regression, and T was the actual time at which the survey was conducted. The average temperature for the survey route was calculated using the air temperature at the start and end of the survey, if available. The difference from seasonal temperature, t, as determined by 20-yr averages from local weather stations, was included in the equation:

$$L_{adj} = L + m(t - t_s),$$

in which t_s was the average air temperature during which the survey was conducted. Standardized values were not used for wind speed.

To compare the occurrence of amphibians between years, we analyzed the results from the 360 monitoring points from all road transects conducted in both years and, separately, the 36 transect routes surveyed in both 1992 and 1993. We calculated and compared the occurrence of each species in each year as reported and also calculated the occurrence of each species after adjusting for the effects of significant variables on the percentage occurrence of each species.

The number and percentage of monitoring points and transect routes where each species was present in each year were calculated. The number of observations where a species occurred in 1 yr but not the other was also determined. The Chi-square value (χ) and the Phi coefficient (ϕ, Fleiss 1981) were calculated for each species at each monitoring point to test for significant association of presence and absence values between years ($P < 0.05$). For transect routes, Chi-square analyses were used to test for significant differences between years, assuming that the data from each point on a route were independent of each other (SAS 1988). All data analyses were conducted using SAS for PC (version 6.04).

Table 2. *Peak calling periods of anuran species in Ontario in 1992.*

Species	Peak begins	Peak ends
Rana sylvatica	Start of survey (1 April)	17 May
Pseudacris crucifer	Start of survey (1 April)	29 May
R. pipiens	Start of survey (1 April)	31 May
Bufo americanus	25 April	20 May
P. triseriata	25 April	20 May
Hyla versicolor	9 May	8 July
R. catesbeiana	17 May	8 June
R. clamitans	20 May	End of survey (15 July)

RESULTS

Inter-observer Variation

Ten species were recorded on at least 1 survey (Fig. 3). Five species, the American toad (*Bufo americanus*), gray treefrog (*Hyla versicolor*), green frog (*Rana clamitans*), bullfrog (*Rana catesbeiana*), and *P. crucifer* were calling during the inter-observer call count comparisons. Although the surveys were conducted past the peak season of calling activity for *B. americanus* and *P. crucifer* (Table 2), the

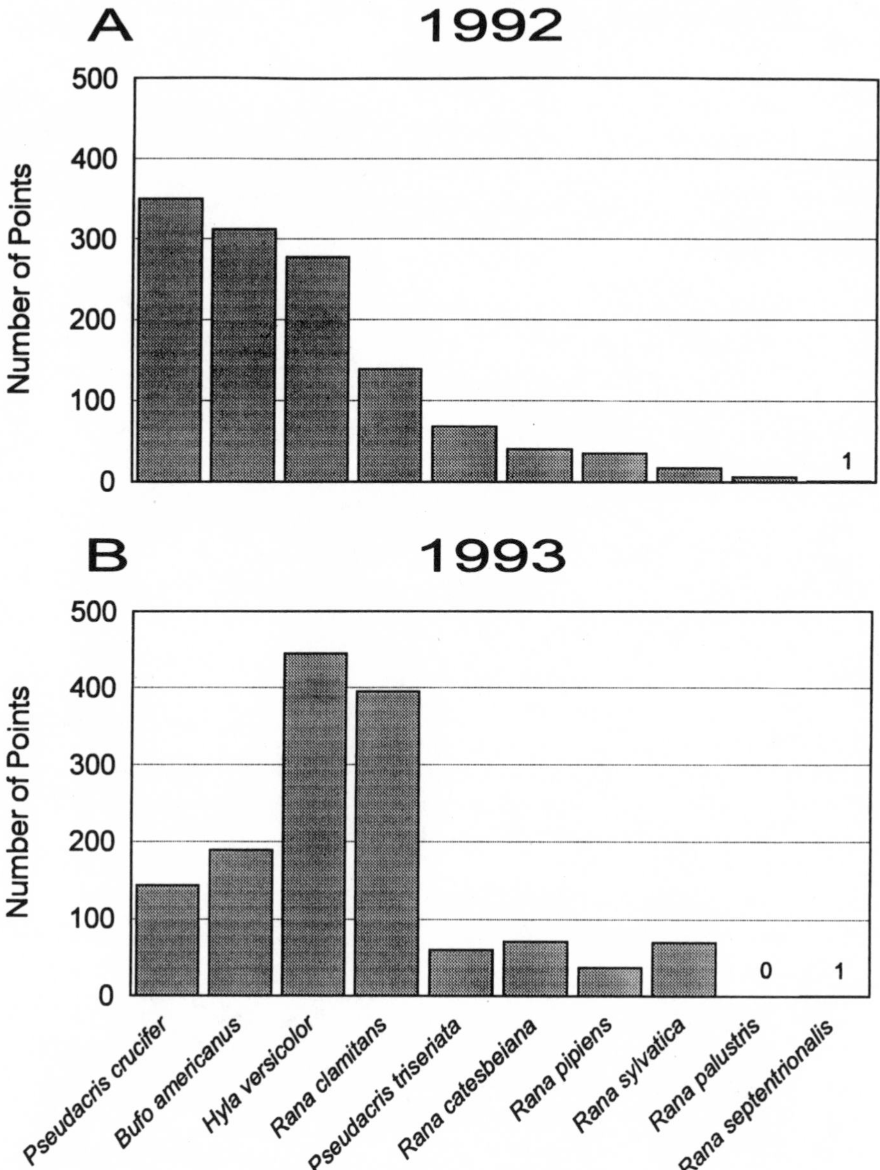

Figure 3. *Total number of points where amphibian species were heard calling on surveys conducted for the Ontario Amphibian Call Monitoring Programme in 1992 (n = 1145 points) and 1993 (n = 1036).*

off-peak effect on each survey was the same for both observers, and these species were recorded at sufficient numbers of points to allow analysis of data.

The rate of agreement between observers was 76 to 92% for identification of presence or absence of each species (Fig. 4) but only 56 to 83% for concurrence in call density ratings of 1, 2, or 3. At 2 of 140 points surveyed, one observer recorded a species as absent while the other recorded it in full chorus. Because ratings of presence or absence were the most highly consistent among observers, all subsequent data analyses were performed on the basis of these data only.

Variation In Data Collection Among Observers

Surveys were performed from 15 March to 25 July 1992 and 12 April to 20 July 1993, usually later than the recommended dates (Table 1). Surveys in 1992 were generally conducted earlier than in 1993. Some 1st and 2nd surveys occurred during the recommended times for the 2nd and 3rd surveys,

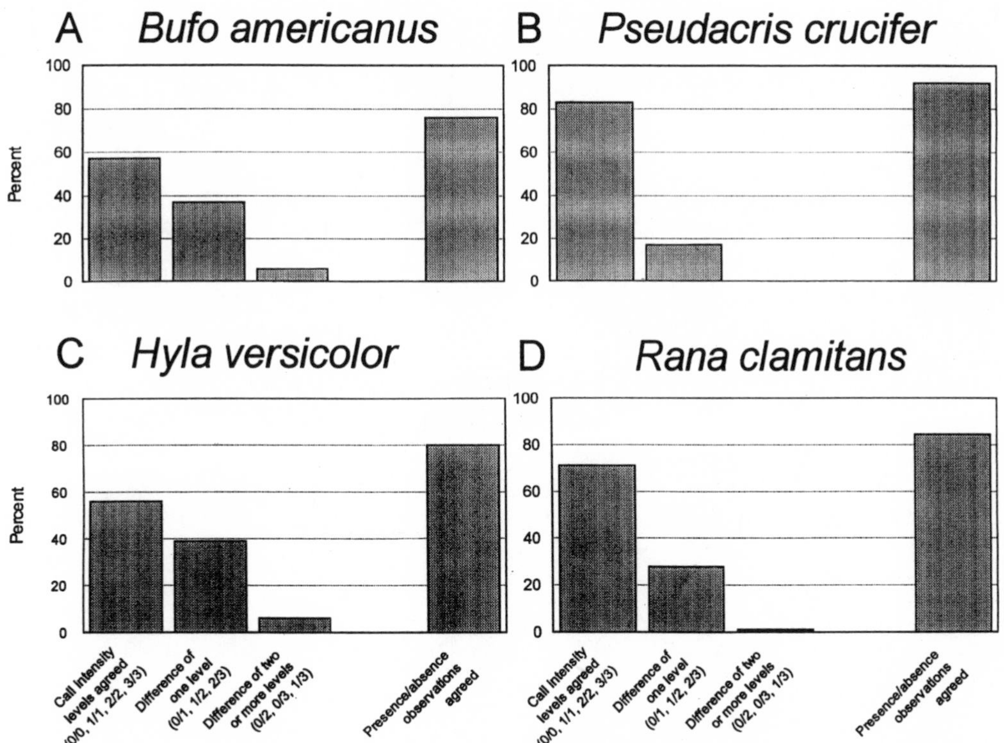

Figure 4. *Effects of observer bias on estimation of call density for anuran species in the Ontario Amphibian Call Monitoring Programme, 1992 to 1993. Surveys were omitted from the analysis if the species was present or absent at all 10 stations.*

respectively. Three to 55% of the surveys were conducted within the advised dates (Fig. 5). In 1993, only 1 route was surveyed 3 times. Seven surveys were run during the time period for the 1st surveys, and an 8th was run 1 day later.

In 1992, most surveys were conducted as specified after ½ hr after sunset and before 2400 hr, but in 1993, 47 of 117 surveys (41%) finished after 2400 hr. Start times were generally as requested, with 47 of 99 surveys (48%) in 1992 and 104 of 117 surveys (89%) in 1993 starting after 2200 hr. Many surveys in 1992 (88 of 115) were not conducted during the recommended ranges in air temperature (Fig. 5).

Spatial and Temporal Variation in Occurrence of Amphibians

For *R. clamitans* and *B. americanus*, a significant positive relationship was found between percent occurrence and time of day that the survey was performed (Fig. 6A, B). Records of occurrence of *B. americanus* and *H. versicolor* had a significant positive correlation with temperature (Fig. 6C).

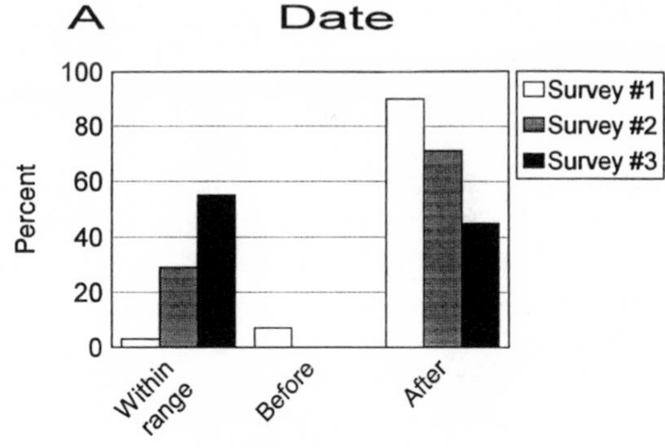

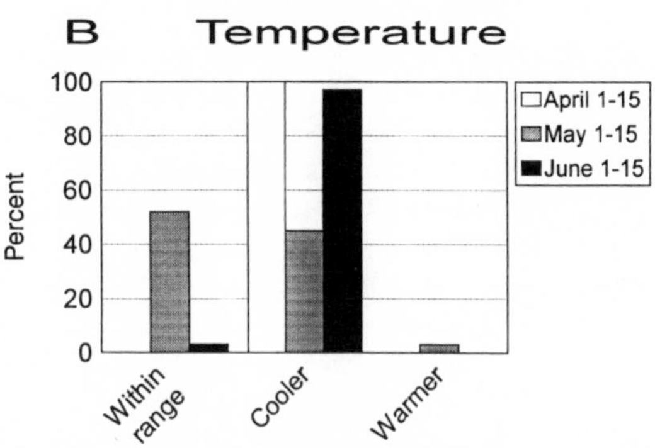

Figure 5. *Adherence by observers to survey guidelines in the Ontario Amphibian Call Monitoring Programme, 1992 to 1993.*

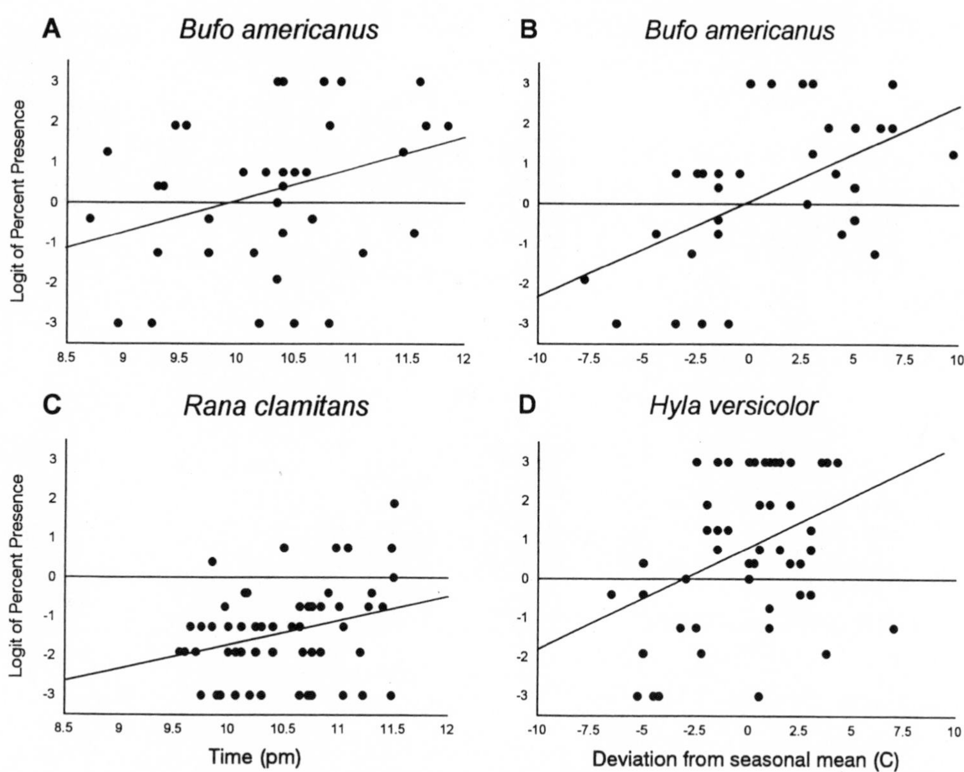

Figure 6. *Relationships between survey conditions and observed presence of anuran species in the Ontario Amphibian Monitoring Programme, 1992. (A)* Bufo americanus *presence vs. time of day* ($y = 0.784x - 7.821$, $P = 0.03$, $r^2 = 0.116$), *(B)* B. americanus *presence vs. deviation of average survey temperature from seasonal mean* ($y = 0.238x + 0.018$, $P < 0.001$, $r^2 = 0.278$), *(C)* Rana clamitans *presence vs. time of day* ($y = 0.617x - 7.911$, $P = 0.03$, $r^2 = 0.08$), *(D)* Hyla versicolor *presence vs. deviation of average survey temperature from seasonal mean* ($y = 0.256x - 0.792$, $P = 0.003$, $r^2 = 0.159$).

Because very few surveys were conducted during April and May of 1993, analysis of changes in amphibian presence between years was limited to those surveys conducted from 7 to 30 June in both years. These included 44 routes in 1992 and 54 routes in 1993, of which 36 routes were identical. But because time and temperature were not recorded on all routes, adjusted data reflected changes on only 27 routes for temperature using *H. versicolor* and 29 routes for time using *R. clamitans* (Fig. 7). Although all species were found on at least 1 of these routes in at least 1 yr (Fig. 7), only *H. versicolor* and *R. clamitans* were calling at their peak during this period in both years. Using only occurrence information, regressions applied only to average percent presence among all points within transects.

For *H. versicolor,* temperature was adjusted using $L_{adj} = L + 0.0450(\text{temperature} - 21)$ for 1992, and $L_{adj} = L + 0.261(\text{temperature} - 21)$ for 1993. *Hyla versicolor* showed no change between years at 240 points (66%), absence in 1993 at 28 points (8%) where it was present in 1992, and presence in 1993 at 92 points (26%) where it was absent in 1992 ($\chi^2 = 51.3$, $P < 0.01$; $\phi = 0.367$). Unadjusted data showed that 8 routes had significant increases in occurrence and only 1 route showed a significant decrease between years. Those trends were similar to the adjusted data accounting for temperature, in which 9 routes had a significant increase in occurrence and 1 route showed a significant decrease.

For *R. clamitans,* time of the survey was adjusted using $L_{adj} = L + 0.777(10.5 - \text{Time})$ for 1992, and $L_{adj} = L + 0.729(10.5 - \text{Time})$ for 1993. *Rana clamitans* showed no change for 259 points (72%), absence in 1993 from 18 points (5%) where it was present in 1992, and presence in 1993 at 83 points (23%) where it was absent in 1992 ($\chi^2 = 41.1$, $P < 0.01$; $\phi = 0.329$). The unadjusted data showed a significant change in occurrence between years on 4 routes, all increases, whereas data adjusted for time of the survey showed only 1 route with a significant increase.

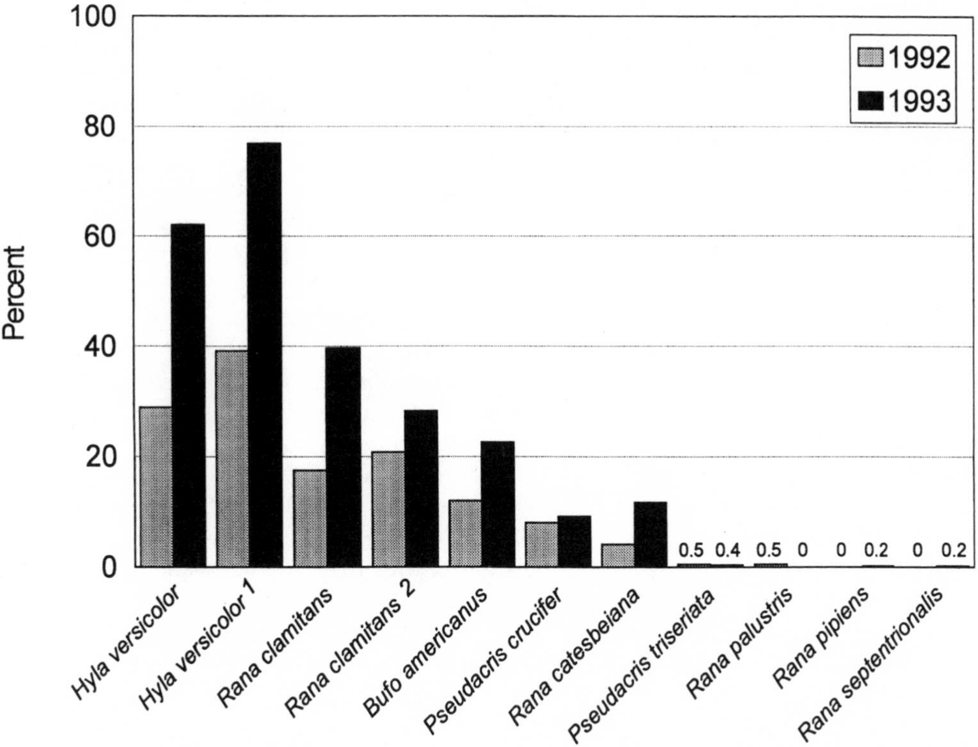

Figure 7. *Average percent of points with anurans species calling 7 to 30 June, 1992 and 1993 of the Ontario Amphibian Call Monitoring Programme (n = 44 routes in 1992, n = 54 routes in 1993), including data adjusted for average temperature and time of day of survey of survey for* Hyla versicolor *and* Rana clamitans. [1] *data adjusted for average temperature of survey (n = 34 routes in 1992, n = 54 routes in 1993);* [2] *data adjusted for time of survey (n = 36 routes in 1992, n = 54 routes in 1994).*

Discussion

Our 2 yr of extensive monitoring have recorded the occurrence and calling density of all 10 species of anurans commonly occurring in southwestern Ontario. Species significantly tended to be present or absent at the same monitoring points from 1 yr to the next (n = 360 points).

Although observers recorded all the species expected to commonly occur, we suspect that call counts of some species were biased. The monitoring results suggested that northern leopard frog (*Rana pipiens*), pickerel frog (*Rana palustris*), and *R. catesbeiana* were not as common as most other species, as has been suggested before (Weller and Oldham 1988). However, *R. pipiens* and *R. palustris* calls are much lower in volume compared to other species that call at the same time, and the distribution of *R. catesbeiana* in southern Ontario is known to be patchy (Weller and Oldham 1988). It is probably necessary to greatly increase the number of transect routes surveyed in order to acquire a sample size adequate to detect trends and occurrence of those species.

The only southern Ontario species not found on the transect routes were Fowler's toad (*Bufo fowleri*) and the cricket frog (*Acris crepitans*) despite surveys conducted within their geographic ranges. *Bufo fowleri* has a very limited geographic distribution in Canada along the Lake Erie shoreline (Green 1989). *Acris crepitans* is considered a rare species in Canada, restricted to Pelee Island, and was recently designated as endangered (Oldham and Campbell 1986). It has not been recorded since 1987 (Oldham 1992). For such rare or geographically restricted species, transects may need to be specifically established within very local and known areas of their occurrence. However, the transects should still otherwise conform to the standard guidelines.

Our sample sizes used to examine inter-observer variation were small, but the results emphasize the need to train observers to accurately and consistently identify calling density. Although different observers were only able to reliably assess presence and absence, it would be valuable to obtain information on the density of the chorus at a location over time. A population that produces a full chorus for many years and then changes to one which only produces a single or very few calling individuals over more than 1 generation represents an incremental change that may be quantifiable by call density records. This would provide a valuable measure of insight into population change beyond simple occurrence. However, variation among observers appears to be a problem with any auditory survey. For example, on Breeding Bird Survey routes, accuracy in identifying the number of individual birds calling decreases steadily with increasing density of species (Droege 1990), and species are occasionally recorded that are not calling (Bart and Schoultz 1984). Observer variation is estimated to cause errors of 25 to 33% and tend to underestimate the magnitude of a decline in a species (Bart and Schoultz 1984). Bart and Schoultz (1984) made no recommendations for adjustment of such data except to increase the sample size of routes surveyed and to determine the direction of the bias made by observers. Where possible, co-variates based on the ability of observers to identify birds are also included in the statistical analysis of Breeding Bird Survey data (Droege 1990; Geissler and Sauer 1990). For amphibians, it also should be possible to decrease inter-observer variation by better training of observers. The identification of amphibian species should be less problematical than for birds since only 14 species of anurans are found in Ontario, they do not all call simultaneously, and the ones that do so usually have dissimilar calls.

The lack of adherence by observers to the protocols for the surveys influenced our results. Although adjustment after the fact is possible, it is not desirable. That point must be emphasized to observers at the outset of their participation. Many surveys were conducted later than suggested, or not at all. Even if all observers appear to have conducted their surveys properly, screening the data is valuable to determine if such variables as changes in the weather might influence results.

Beyond observations of occurrence of amphibians on linear transects, in the future it would be worthwhile to record information on the habitat types where amphibians are calling relative to the monitoring point. Falardeau and DesGranges (1991) incorporated Breeding Bird Survey results with habitat surveys to characterize habitat preferences and recent population fluctuations (1974 to 1986) for birds in Quebec farmlands.

Our amphibian call count data provided information on occurrence of species. With refinement, the methods may also be able to provide temporally and spatially comparable information on density of species. But to enhance the value of that information, how the counts of calling males compare to the size of the breeding and non-breeding population of males and females must be determined. This can only be accomplished by comparing call count information to a population of anurans of known size.

Other types of volunteer amphibian population monitoring could provide data comparable to road call counts. The population trends found using a variety of volunteer bird counts such as Christmas Bird Counts, the Breeding Bird Surveys, migration counts and bird checklist projects are generally concordant with one another (Temple and Cary 1990). Although quantifiable population counts and/or density estimates are not conducted to confirm the results of volunteer counts for birds, general agreement in trends for individual bird species and for geographic locations using different monitoring methods suggests that such methods do indeed reveal actual changes. Our results on amphibian species occurrence were consistent with the number of anuran species commonly reported in southwestern Ontario in the Ontario Herpetofaunal Atlas (Weller and Oldham 1988). However, atlas data are compiled from non-random sightings of amphibians, and population trends cannot be derived from this information. In 1993, the Hamilton Region Conservation Authority initiated a 2nd type of anuran call count program known as "backyard surveys". Volunteers stand in their backyard for 3 min each night from 1 April to 1 August and record the occurrence of anuran calling. Backyard surveys can potentially provide more accurate information on the peak calling periods of each species.

Call counts ignore species that do not call and may overlook species that call quietly or are rare. Those species require other or more intensive monitoring methods, although volunteer observers could still be used. Further refinement of call count monitoring combined with other surveys for more cryptic species have the potential to greatly improve our current information on amphibian population fluctuations.

Acknowledgments

We would like to thank all the volunteers and summer employees that collected data. We thank Michael J. Oldham, Ruth van der Ham, and Fred W. Scheuler for participating in writing the participants manual for the road call counts. We thank Peter Carson, and the following Conservation Authorities for their participation in managing the program: Hamilton Region, Saugeen Valley, St. Clair River, Upper Thames River, and Essex Region. We also thank funding supporters of the collection of the data and production and distribution of the participant manuals: the Ontario Field Herpetologists, Ontario Ministry of Natural Resources/Environmental Youth Corps, IUCN Task Force on Declining Amphibian Populations, Great Lakes Action Plan, and Norfolk Field Naturalists.

Literature Cited

Bart J, Schoultz JD. 1984. Reliability of singing bird surveys: changes in observer efficiency with avian density. Auk 101:307–318.

Bishop CA, Kay DF. 1992. Intensive and extensive monitoring of anthropogenic stresses associated with amphibian survivorship. In: Bishop CA, Pettit KE, editors. Declines in Canadian amphibian populations: designing a national monitoring strategy. Ottawa: Canadian Wildlife Service. Occasional Paper 76. p 109–110.

Blaustein AR, Wake DB, Sousa WP. 1994. Amphibian declines: Judging stability, persistence and susceptibility of populations to local and global extinction. Conservation Biology 8:60–71.

Cochran WG. 1977. Sampling techniques. 3rd ed. New York: John Wiley & Sons.

Corn PS. 1994. What we know and don't know about amphibian declines. In: Covington WW, DeBano LF, technical coordinators. Sustainable ecological systems: implementing an ecological approach to land management. Fort Collins, CO: USDA Forest Service. General Technical Report RM-247. p 59–67.

Droege S. 1990. The North American breeding bird survey. In: Sauer JR, Droege S, editors. Survey designs and statistical methods for the estimation of avian population trends. Washington: USDI Fish and Wildlife Service. Biological Report 90(1). p 1–4.

Falardeau G, Desgranges JL. 1991. Sélection de l'habitat et fluctuations récentes des populations d'oiseaux des milieux agricoles du Québec. Canadian Field-Naturalist 105:469–482.

Fleiss JL. 1981. Statistical methods for rates and proportions. 2nd ed. New York: John Wiley & Sons.

Freeman B, Shackell NL. 1992. Amphibians in the context of a National Environmental Monitoring Program. In: Bishop CA, Pettit KE editors. Declines in Canadian amphibian populations: designing a national monitoring strategy. Ottawa: Canadian Wildlife Service. Occasional Paper 76. p 101–104.

Geissler PH, Sauer JR. 1990. Topics in route-regression analysis. In: Sauer JR, Droege S, editors. Survey designs and statistical methods for the estimation of avian population trends. Washington: USDI Fish and Wildlife Service. Biological Report 90(1). p 54–57.

Green DM. 1989. Fowler's toad, *Bufo woodhousii fowleri*, in Canada: biology and population status. Canadian Field-Naturalist 103:486–496.

Huff J. 1991. Frog and toad survey 1991. Madison, WI: Wisconsin Department of Natural Resources. Technical Report. p 114–123.

Obbard M, Scheuler F. 1992. A historical data base of amphibian data and data sources in Canada. In: Bishop CA, Pettit KE, editors. Declines in Canadian amphibian populations: designing a national monitoring strategy. Ottawa: Canadian Wildlife Service. Occasional Paper 76. p 105.

Oldham MJ. 1992. Declines in Blanchard's cricket frog in Ontario. In: Bishop CA, Pettit KE, editors. Declines in Canadian amphibian populations: designing a national monitoring strategy. Ottawa: Canadian Wildlife Service. Occasional Paper 76. p 30–31.

Oldham MJ, Campbell CA. 1986. Status report on Blanchard's cricket frog, *Acris crepitans blanchardi*, in Canada. Ottawa: Environment Canada, Committee on the Status of Endangered Wildlife in Canada.

SAS Institute. 1988. SAS/STAT user's guide, release 6.03. Cary, NC: SAS Institute.

Temple SA, Cary JR. 1990. Using checklist records to reveal trends in bird populations. In: Sauer JR, Droege S, editors. Survey designs and statistical methods for the estimation of avian population trends. Washington: USDI Fish and Wildlife Service. Biological Report 90(1). p 98–104.

Vial JL, Saylor L. 1993. The status of amphibian populations: a compilation and analysis. Corvallis, OR: IUCN/SSC Declining Amphibian Populations Task Force. Working Document.

Weller WF, Oldham MJ. 1988. Ontario herpetofaunal summary 1988. Peterborough, ON: Ontario Field Herpetologists.

World Meteorological Organization. 1970. The Beaufort Scale of wind force. Marine Science Affairs Report 3:1–22.

Bullfrog, Rana catesbeiana. *Photo by Jacques Brisson.*

©1997 by the Society for the Study of Amphibians and Reptiles
Amphibians in decline: Canadian studies of a global problem. David M. Green, editor.
Herpetological Conservation 1:161–174.

Chapter 17

NON-DISRUPTIVE MONITORING OF TERRESTRIAL SALAMANDERS WITH ARTIFICIAL COVER OBJECTS ON SOUTHERN VANCOUVER ISLAND, BRITISH COLUMBIA

THEODORE M. DAVIS

Department of Biology, University of Victoria, Victoria, British Columbia V8W 3N5, Canada

ABSTRACT.—Terrestrial salamander populations are often sampled using unit-effort searches of natural cover objects. Because such searches can disturb microhabitats, they may be unacceptable where multiple searches of the same site are required or where destructive sampling is prohibited. Where sites differ in structure, they may be undependable for estimating the relative abundance of salamanders across sites. The use of artificial cover objects can overcome some of these difficulties. For 2 field seasons, among 9 sites on Vancouver Island, 228 artificial cover objects containing complex multiple microhabitats were checked approximately every 2 wk. Spatial and temporal variation in abundances were observed for *Plethodon vehiculum*, *Aneides ferreus*, and *Taricha granulosa*. *Ensatina eschscholtzii* and *Ambystoma macrodactylum* were found in low numbers. Abundances of all species were reduced in clearcuts but there was no evidence that local disturbance has long-term population consequences. However, abundances among species varied across sites. This suggests that differences in site structure, climatic conditions, forest fragmentation, or stochastic events may influence metapopulation dynamics. Most species were usually found under the artificial cover objects on the soil but *A. ferreus* was most often found on wood. The use of artificial cover objects with complex multiple microhabitats is recommended where species differ in their microhabitat use, where repeated searches must be relatively non-destructive, where equal effort is required across sites that differ in structure, or where variation among collectors is a confounding variable.

RÉSUMÉ.—Les populations de salamandres terrestres sont souvent échantillonnées en recherchant systématiquement les couvertures naturelles. Dans la mesure où ces recherches peuvent perturber les microhabitats, elles sont souvent inacceptables surtout s'il faut procéder à plusieurs recherches sur un même site ou si les échantillonnages destructeurs sont interdits. Lorsque la structure des sites diffère, ces recherches peuvent ne pas être suffisamment fiables pour évaluer l'abondance relative de salamandres entre les différents sites. L'utilisation de couvertures artificielles permet de surmonter certaines de ces difficultés. Pendant 2 saisons et sur 9 sites de l'île de Vancouver, 228 couvertures artificielles abritant de multiples microhabitats complexes ont été vérifiées environ toutes les 2 semaines. Les variations spatiales et temporelles au titre de l'abondance ont été observées pour *Plethodon vehiculum*, *Aneides ferreus*, et *Taricha granulosa*. *Ensatina eschscholtzii* et *Ambystoma macrodactylum* ont été décelés en petits nombres. Il y avait moins d'espèces, quelles qu'elles soient, dans les endroits où des coupes à blanc avaient été pratiquées mais rien ne permet de prouver que la perturbation locale a des conséquences démographiques à long terme. Toutefois, l'abondance intra-espèces varie selon les sites. Cela donne à penser que les différences au titre de la structure du site, des conditions climatiques, de la fragmentation forestière ou des phénomènes stochastiques peuvent influencer la dynamique des métapopulations. La plupart des espèces vivent généralement sous des couvertures artificielles à même le sol mais *A. ferreus* se retrouvait le plus souvent à même le bois. L'utilisation de couvertures artificielles avec des microhabitats multiples complexes est recommandée lorsque les espèces exploitent différemment les microhabitats disponibles, lorsque des recherches répétées doivent être relativement non destructives et lorsque des efforts équivalents sont exigés entre plusieurs sites à structure différente ou encore lorsque les variations entre collecteurs constituent une variable confondante.

Estimation of temporal and spatial variation in abundance is of fundamental importance in ecology and conservation biology. Estimates of relative abundance of terrestrial amphibians in space and time are often made using unit-effort searches of natural cover objects and microhabitats (Corn and Bury 1990). Methods include time or area-constrained searches, surveys of coarse woody debris (CWD), and quadrat, transect and patch sampling (Corn and Bury 1990; Heyer et al. 1994). However, because of disturbance of the natural habitat, these methods may be unsuitable where repeated searches of the same area are needed or where disturbance of the natural habitat is unacceptable or prohibited. For example, the clouded salamander, *Aneides ferreus*, is typically found under bark on logs or within logs (Davis 1991) and one thorough search of this microhabitat can be very destructive. Such destructive sampling is unacceptable where > 1 sample is needed and may be prohibited in parks and reserves. Also, searches of natural cover among sites that differ in the amount and type of CWD may not be comparable because some types of cover may be difficult to search efficiently resulting in unequal search effort. Finally, search effort may vary among individuals (Heyer et al. 1994) or through time with the same individual (unpublished data).

To overcome some of these difficulties, unit-effort pitfall trapping may be used to estimate relative abundance. This has the advantage that the amount and type of CWD and individual effort are independent of the trapping effort. However, the capture rate varies widely among species (Bury and Corn 1987; Buhlmann et al. 1988; Corn and Bury 1990; Welsh and Lind 1988; Welsh 1990), and many species, probably because they are relatively site tenacious, are not readily trapped (Welsh and Lind 1988). Also, the traps must be checked frequently to prevent accidental mortality of the various vertebrates that may be caught.

In principle, artificial cover objects (ACO) can overcome many of these difficulties (Grant et al. 1992; Heyer et al. 1994). They are especially well suited to long-term monitoring, are relatively easy to sample once in place, result in little or no damage to the natural habitat, and can attract species that are difficult to trap in pitfall traps. Unlike pitfall traps, sampling can be opportunistic with no risk of mortality from failure to check frequently (Grant et al. 1992). Because ACO can be checked repeatedly over long periods, rare species can eventually be detected without damage to the natural habitat that would result from repeated searches of natural cover. In conjunction with mark-recapture methods, they can be used to monitor individual movements and to estimate life-history parameters and absolute abundance. Also, they can be used to investigate the relationship between habitat characteristics and abundance, and differential microhabitat use among species. Because ACO can be of a standard size and number and are independent of the amount and type of CWD or individual searching effort, they should give more dependable estimates of relative abundance across sites that differ in structure than do searches of natural cover.

On southern Vancouver Island, there are 6 species of salamanders that are either entirely terrestrial or have an adult terrestrial stage. Of these, the western red-backed salamander (*Plethodon vehiculum*), the roughskin newt (*Taricha granulosa*), *A. ferreus* and the ensatina (*Ensatina eschscholtzii*) can be locally abundant and easy to capture, but the terrestrial stages of the long-toed salamander (*Ambystoma macrodactylum*) and northwestern salamander (*Ambystoma gracile*) are difficult to locate (Ovaska 1987; Davis 1991, pers. obs.). I used ACO to detect variation in relative abundance among 9 sites, seasonal variation in surface abundance, and variation in patterns of microhabitat use among these species.

MATERIALS AND METHODS

Study Sites

Artificial cover objects were placed at nine sites on southern Vancouver Island. Four sites were in the Greater Victoria Watershed (GVW; 48°34'N, 123°39'W), 3 near Rosewall Creek Provincial Park (Rosewall Creek: 49°27'N, 124°46'W; McNaughton Creek: 49°27'N, 124°46'W; Cook Creek: 49°27'N,

124°45'W), and 1 each at Goldstream Provincial Park (48°28'N, 123°32'W), and the University of Victoria research property (Simpson property) on Marble Bay, Lake Cowichan (48°50'N, 124°10'W, Lot 29). The GVW sites are within the following polygon numbers from the Greater Victoria Water District Forest Cover Maps: GVW clear-cut: 753; GVW immature: 640; GVW mature:699/V; GVW old-growth: 648 and 650. Polygon numbers from Forest Cover Map 92F.47 (Ministry of Forests, British Columbia) for the Rosewall, McNaughton, and Cook Creek sites (RMC) are 205 and 207, 202, and 194, respectively.

The GVW clear-cut site was logged and burned in 1985 and planted with forest seedlings in 1986. The McNaughton Creek site was logged in 1977 and planted with forest seedlings in 1988. All the other sites were forested, but had been selectively logged or otherwise disturbed within the last 100 yr. Nevertheless, all the forested sites contained at least a few very old trees (100 yr) and a similar suite of dominant tree species and understory plants. The dominant trees were Douglas-fir (*Pseudotsuga menziesii*) and western hemlock (*Tsuga heterophylla*), but western red cedar (*Thuja plicata*), red alder (*Alnus rubra*), and broadleaf maple (*Acer macrophyllum*) were common as well. Little light penetrated the forest canopy, ground vegetation was generally sparse, and some areas were virtually devoid of undergrowth. At all the forested sites, sword fern (*Polystichum munitum*) was usually the dominant ground species, but in some areas salal (*Gaultheria shallon*) was dominant. Logs and woody debris, in various states of decomposition, were common. The GVW sites were selected by Forestry Canada to study the effect of forestry practices on carbon and nutrient dynamics and biodiversity in Douglas-fir and western hemlock coastal forests and approximately represent a chronoseries: GVW clear-cut 10 to 20 yr, GVW immature 40 to 60 yr, GVW mature 80 to 100 yr and GVW old-growth > 150 yr).

The climate on southern Vancouver Island is characterized by mild, wet winters with some snow and temperatures typically near or below freezing from December to February. Summers are generally dry, but under the cover of a dense forest canopy, spaces under bark and in decaying wood can remain relatively moist.

Artificial Cover Objects (ACO)

Individual ACO consisted of a base board and 2 cover boards. Because a large and massive cover object is needed to resist moisture loss and temperature changes, I judged that a 1.8-m long × 30.5-cm wide × 5-cm thick board to be about the largest base board that could be reasonably carried by 1 person into the study sites. The space beneath the base board was cleared of vegetation, and the board was placed flat on the soil surface. Two 1.8-m long × 15.3-cm wide × 2.5-cm thick cover boards were placed on top of the base board. Strips of cedar lathe separated the cover boards from the base board in such a way as to create a wedge-shaped space between the cover boards and the base board (Fig. 1). Rain water dripped through the crack between the two cover boards into this space. This created a complex microhabitat so that a salamander could be found on the soil under the base board or between the baseboard and the top boards. Commercial lumber is often treated with a fungicide that

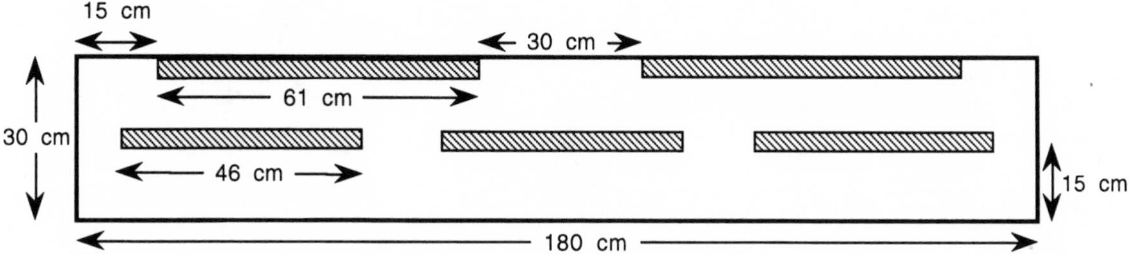

Figure 1. *Diagram of the baseboard portion (viewed from above) of an artificial cover object (after Heyer et al., 1994, p. 148). Strips of cedar lathe (6 × 38 mm) in lengths of 46 cm and 61 cm are attached with galvanized nails along the middle and one edge of the base board, respectively. The strips along the edge are doubled, so that the lathe raises above the baseboard about 12 mm. The baseboard is placed on the ground with the lathe strips facing up. Top boards (2.5 × 15 × 180 cm) are placed on top of the lathe strips, creating wedge-shaped spaces.*

may harm amphibians so untreated, rough cut, full dimensional lumber was used. Each ACO was individually labeled with both a metal tag and a marking pen.

ACO Plots

At every site except Goldstream, 6 ACO were arranged around natural logs and other debris within 10-m diameter plots. Each of the GVW sites and the Lake Cowichan site had 6 such plots (36 ACO/site). At the RMC sites, there were 2 plots per site (12 ACO/site). At the Goldstream site, 12 ACO were arranged within a single 13.8-m diameter plot. At all sites, the edge of any 1 ACO was at least 2 m from the edge of any other ACO. Logs and other large natural cover objects were not disturbed, and the ACO were placed among them as space allowed. This resulted in a roughly uniform, but haphazardly oriented, distribution of ACO within the 10-m plots.

Sampling Frequency

At the Goldstream, Lake Cowichan and Watershed sites, ACO were put in place in December 1992, and were checked during daylight hours, usually in the morning, about once every 2 to 3 weeks from the beginning of March 1992 until the end of September 1993, except when the temperature was below freezing. At the RMC sites, ACO were put in place in February 1993 and checked about once every 2 to 3 wk from the beginning of March 1993 until the end of October 1993.

Handling of Salamanders

Salamanders were weighed to the nearest 0.1 g with a Pesola® spring scale and measured from the tip of the snout to the anterior angle of the vent (snout-vent length, SVL) to the nearest 0.1 mm with vernier calipers. If it was raining or windy, salamanders were not weighed. If eggs were visible through the abdominal wall, I estimated the number present, although I did not attempt to verify these estimates by dissection. I sexed adult *P. vehiculum* by sliding a moistened finger anteriorly on the underside of the snout (Ovaska 1987). The protruding premaxillary teeth can be felt in males, but not in females. No attempt was made to sex the other species. For each salamander I recorded the ACO microhabitat (on wood or on soil) and the distance to the nearest 10 cm from the tagged end of the baseboard. To mark salamanders individually, I clipped a unique combination of 1 to 3 toes from each salamander, but never more than 1 toe from each foot. I released each salamander at its original location within 15 min of capture.

Clouded salamander, Aneides ferreus. *Photo by David M. Green*

Statistical Analysis

The data were non-normal (probability plots and the Lilliefors test; Wilkinson et al. 1992) and standard transformations (Zar 1984) were unsuccessful, so I used nonparametric tests. I used the Mann-Whitney U test for testing differences between the median number of salamanders per search between years. Pairwise comparisons for each site were done with multiple Mann-Whitney U tests and I reduced the α level to α/n where n was the number of such comparisons (Zar 1984).

I selected searches of ACO that were ≤ 5 days apart among sites to compare relative abundance among sites (Table 1). If there were equal numbers of ACO at each site, Friedman's test was used (Zar 1984). If the null hypothesis of equal medians was rejected, observations were considered independent and pairwise comparisons of medians were done with χ^2 goodness-of-fit (GoF), rather than with ordered rank sums, because the number of blocks was always < 15 (Gibbons 1993). If there were

Table 1. *Number of salamanders found on matched searches that were ≤ 5 days apart.*

Group	Sites[a]	Interval	Number of individual salamanders	Number of matched searches	Days between searches		
					$\bar{x}$	SD	Range
1	All sites	8 Apr – 12 Aug 93	458	8	3.0	1.07	2–5
2	All forested sites	8 Apr – 12 Aug 93	449	8	3.0	1.07	2–5
3	IM, MA, OG, LC, RC, MC, CO	8 Apr – 12 Aug 93	268	8	3.0	1.07	2–5
4	IM, MA, OG, RC, MC, CO	8 Apr – 12 Aug 93	194	8	3.0	1.07	2–5
5	LC, GS	29 Mar 92 – 18 Sep 93	640	34	1.3	1.55	0–5
6	CC, IM, MA, OG	4 Mar 92 – 26 Sep 93	289	30	0	0.18	0–1
7	RC, MC, CO	9 Mar – 19 Oct 93	84	12	0	0	0

[a] CC = GVW clearcut, IM = GVW immature, MA = GVW mature, OG = GVW old-growth, LC = Lake Cowichan, GS = Goldstream, RC = Rosewall Creek, MC = McNaughton Creek, CO = Cook Creek.

unequal numbers of ACO at each site, χ^2 GoF tests were used. Expected frequencies were based on the number of ACO inspected at each site. Recaptures were not included in these analyses. Contingency tables were used to analyze proportions of species among sites. All searches were used in these analyses, but recaptures were not included. I used contingency tables for microhabitat comparisons among species, and the the normal approximation for proportions (Zar 1984) for microhabitat comparisons for a single species across sites. I applied the Yates correction for continuity in χ^2 tests whenever df = 1 (Zar 1984), and the minimum level of significance was set at $\alpha = 0.05$.

RESULTS

Seasonal Variation in Surface Abundance

Salamanders virtually disappeared from the surface during the coldest part of the winter, and few salamanders were on the surface during the driest part of the summer in July and August (Fig. 2). To test the hypothesis that the number of salamanders found per search was greater in 1993 than in 1992, I restricted my analysis to all searches between 1 April and 1 October in both years. Based on 9 searches at each site in each year that scored the highest number of salamanders per search, there was an overall increase in the median number of salamanders found per search ($U_{54,54} = 1032$, $P = 0.009$). An even more extreme result was obtained when all searches in those intervals were used ($U_{57,78} = 1443.5$, $P = 0.0005$). The GVW immature and the Goldstream sites differed between years, the GVW mature and GVW old-growth sites were very close to significantly different and only the GVW clearcut and Lake Cowichan sites showed no differences between years (Table 2).

Variation in Abundance Among Sites

Abundances of salamanders (all species combined) varied greatly among sites (Table 1: Group 1, GoF $\chi^2 = 1085.5$, df = 8, $P < 0.001$; Fig. 3). There were many fewer salamanders captured in the clearcut sites compared to the forested sites (Table 1: Group 6; Friedman's test, B = 30, T = 4, $\chi^2 = 22.8$, df = 3, $P = 0.00004$; Table 1: Group 7, B = 12, T = 4, $\chi_r^2 = 8.167$, df = 2, $P = 0.017$). However, there were significant differences in the numbers of salamanders (all species combined) among forested sites as well (Table 1: Group 2, GoF $\chi^2 = 779.7$, df = 6, $P < 0.001$; Table 1: Group 5, $\chi^2 = 494.8$, df = 1, P 0.001). When the Goldstream site was excluded from the analysis, there were still significant differences in the relative abundances of salamanders among forested sites (Table 1: Group 3, GoF $\chi^2 = 26.4$, df = 5, $P < 0.001$). However, if both Goldstream and Lake Cowichan were excluded from the analysis, there were no significant differences among forested sites (Table 1: Group 4, GoF $\chi^2 = 5.6$, df = 4, $0.10 < P < 0.25$).

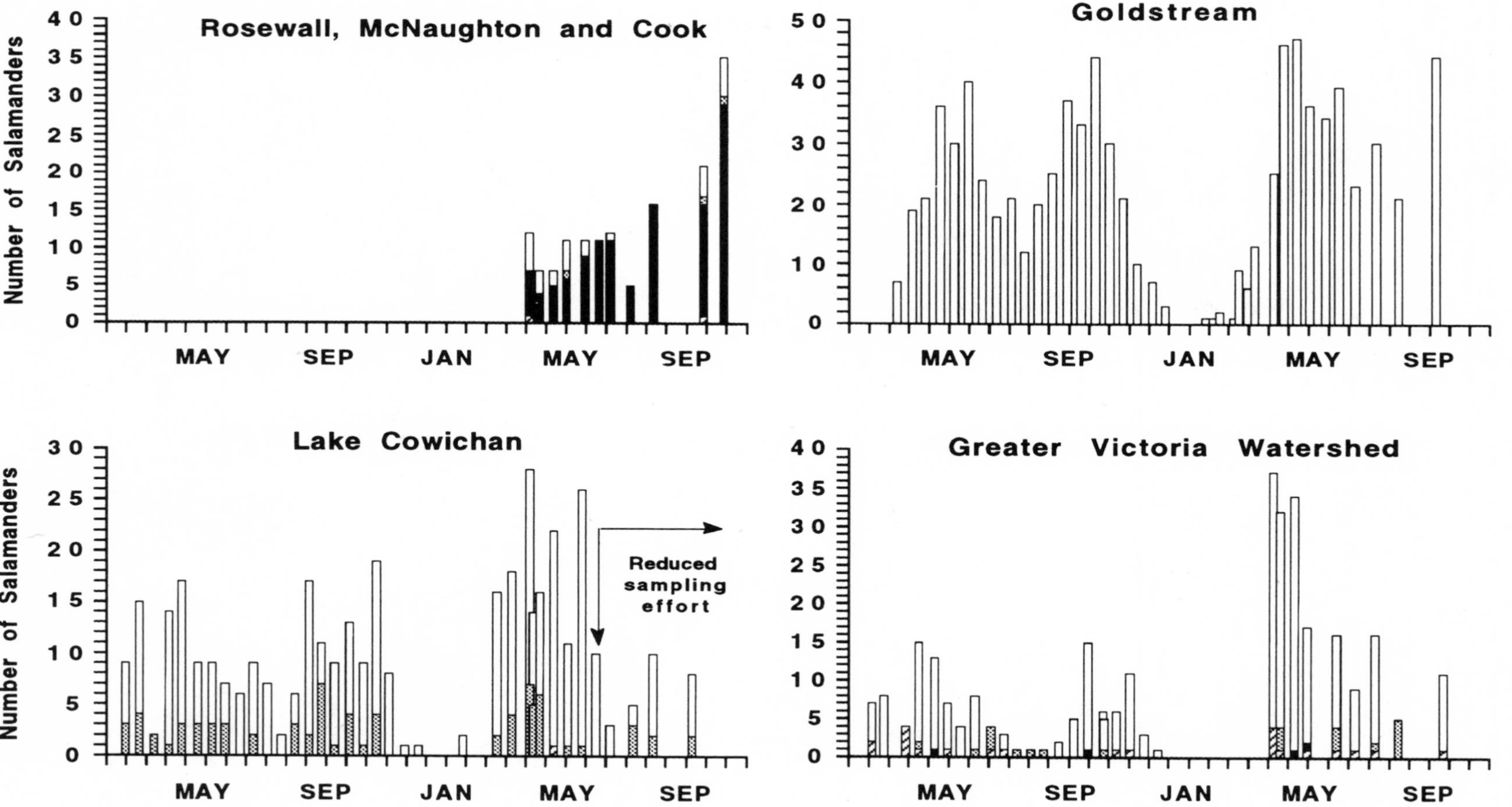

Figure 2. *Number of salamanders, including recaptures, found under artificial cover objects (ACO) at 9 sites on Vancouver Island, B.C. The 4 Greater Victoria Watershed (GVW clearcut, GVW immature, GVW mature, GVW old-growth) sites are combined, and the Rosewall, McNaughton and Cook Creek sites (RMC) are combined. Each column represents a search of all ACO at that particular site. The area in the Lake Cowichan plot labeled "Reduced sampling effort" indicates that only 18 of the 36 ACO were checked during this period because ½ the ACO were involved in experiments not reported here. The columns are subdivided by species: open =* Plethodon vehiculum, *stippling =* Taricha granulosa, *black =* Aneides ferreus, *and cross-hatched =* Ensatina eschscholtzii. *Ticks on the x-axis indicate 20-day intervals starting on 9 February 1992 and ending 1 December 1993. At least 1 salamander was found on every search, so every search is represented by a column.*

Table 2. *Pairwise comparisons of the median number of salamanders found per search by site from 1 April to 1 October in 1992 and 1993, between years, and among sites.*

Site	1992				1993				Mann-Whitney $U_{57,78}$[a]	P
	$\bar{x}$	SD	Median	n	$\bar{x}$	SD	Median	n		
Goldstream	18.9	8.0	17	13	34.4	9.7	35	10	14.5	0.002[b]
Lake Cowichan	9.5	4.4	9	13	11.6	8.0	11	11	69.0	0.884
GVW clearcut	0.2	0.4	0	13	0.7	0.9	0	9	39.5	0.104
GVW immature	2.2	2.4	1	13	9.2	4.1	9	9	8.0	0.001[b]
GVW mature	2.2	2.1	2	13	6.7	4.7	7	9	23.5	0.018
GVW old growth	2.3	1.7	2	13	6.3	4.5	5	9	23.0	0.016

[a] Overall difference between years: Mann-Whitney $U_{57,78} = 1443.5$, $P = 0.0005$. Pairwise comparisons were made with $\alpha = 0.05/6 = 0.008$ (Zar 1984).

[b] Significant difference.

Variation in Abundance Among Species

At most sites, *P. vehiculum* was the most abundant species, but relative abundances varied among species (Fig. 4). Only *P. vehiculum* was found under artificial cover objects (ACO) at Goldstream, but I have found *A. ferreus* elsewhere in the park, and an *E. eschscholtzii* was found in the stomach of a common garter snake, *Thamnophis sirtalis,* in this area (P. T. Gregory, pers. comm.). At Lake Cowichan, both *P. vehiculum* (69.7%, n = 168) and *T. granulosa* (30.3%, n = 73) were common, but no *A. ferreus, E. eschscholtzii,* or *A. macrodactylum* were found under ACO, although these species are known from other searches of this site. The proportions of *P. vehiculum* and *T. granulosa* at Lake Cowichan were significantly different from the forested GVW sites ($\chi^2 = 25.1$, df = 3, $P < 0.0001$) which were similar to each other in the proportions of *P. vehiculum* (72.6 to 81.3%, n = 217) and *T. granulosa* (7.7 to 13.1%, n = 30; Fig. 3) ($\chi^2 = 1.6$, df = 2, P = 0.44). They were also similar to each other in the proportions of *P. vehiculum* , *T. granulosa*, and *E. eschscholtzii* (4.9 to 11.9%, n = 22) ($\chi^2 = 4.7$, df = 4, $P = 0.32$). *Aneides ferreus* (0 to 2.4%, n = 5 and *A. macrodactylum* (0 to 2.4%, n = 3) were relatively rare at these sites. In the GVW clearcut, only *P. ve-*

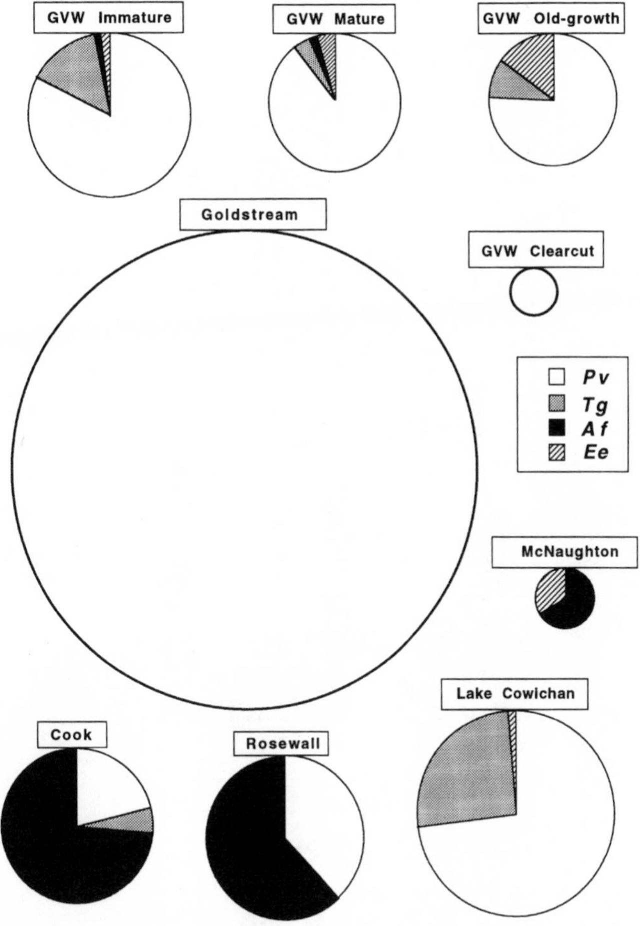

Figure 3. *Salamander abundance among 9 sites on Vancouver Island on 8 searches from 8 April 1993 to 12 August 1993 (Table 1: Group 1). Pv* = Plethodon vehiculum, *Tg* = Taricha granulosa, *Af* = Aneides ferreus, *Ee* = Ensatina eschscholtzii. *The area of each pie represents the number of salamanders per artificial cover object, not including recaptures within this period.*

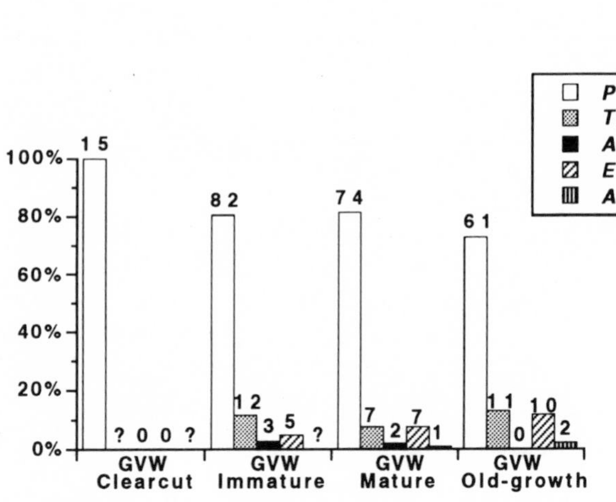

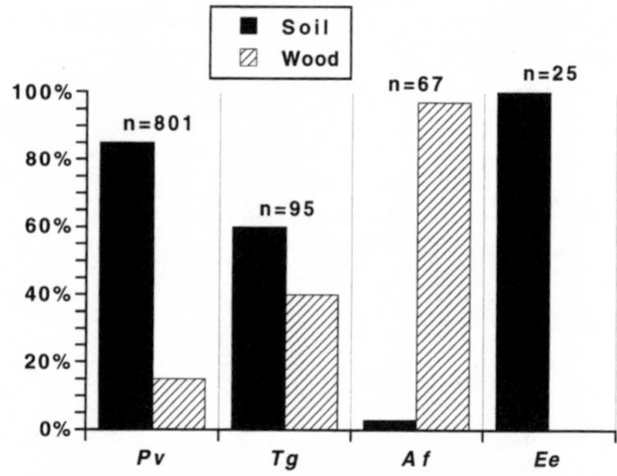

Figure 5. *Use of microhabitats within artificial cover objects (n = number of salamanders for each pair of columns). Each pair of columns sums to 100%. Species labels are as in Fig. 3.*

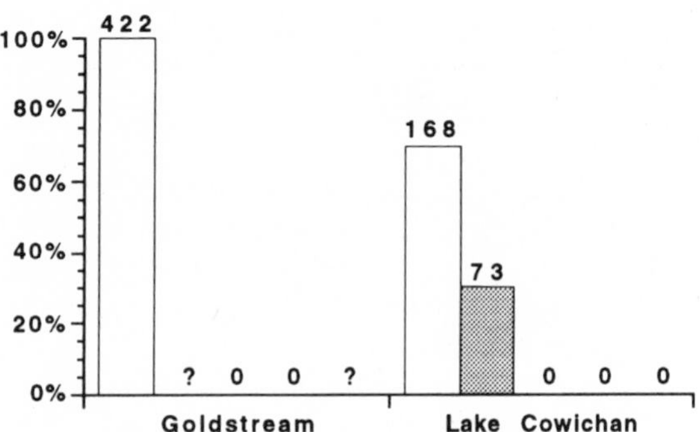

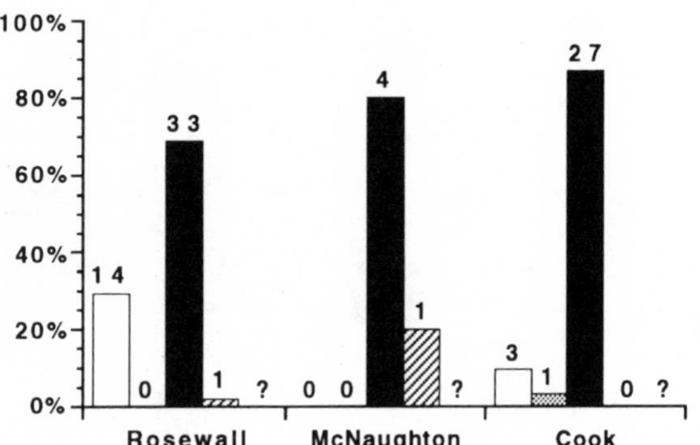

Figure 4. *Diversity of salamander species among sites. Columns indicate percentages of each species by site, not including recaptures. Species are labelled as in Fig 3. Numbers at the top of the columns indicate the number of individuals. "?" indicates that the species has not been found at that site.*

hiculum (n = 15) was found under ACO, although I found *A. ferreus* and *E. eschscholtzii* on searches of natural cover.

The proportions among species were very different at the RMC sites (Fig. 4). *Aneides ferreus* (68.9 to 87.1%, n = 64) was more abundant than *P. vehiculum* (0 to 29.17%, n = 17) (GoF $\chi^2 = 26.1$, df = 1, $P < 0.001$) and the 2 forested sites were similar in the proportions of these 2 species ($\chi^2 = 3.1$, df = 1, $P = 0.08$), although more individuals of *P. vehiculum* were found at Rosewall Creek than at Cook Creek. *Taricha granulosa* (0 to 3.2%, n = 1) and *E. eschscholtzii* (0 to 2.1%, n = 2) were rare. *Ambystoma macrodactylum* was not found at these sites.

Variation in Microhabitat Use Among Species

There were significant differences among species in their use of soil and wood microhabitats (Fig. 5, $\chi^2 = 262.7$, df = 3, $P = 0.0001$; *P. vehiculum* compared to *T. granulosa*: $\chi^2 = 34.9$, df = 1, $P = 0.0001$). *Plethodon vehiculum* (n = 801) was usually found on soil (85.0%), whereas *A. ferreus* (n = 67) was strongly associated with wood (97.0%). About 60% of the *T. granulosa* (n = 95) were found on soil. All the *E. eschscholtzii* (n = 25) were found on soil. However, there was significant variation in the proportion of each microhabitat used by each species across sites (sites with at least one capture in Fig. 6; *P. ve-*

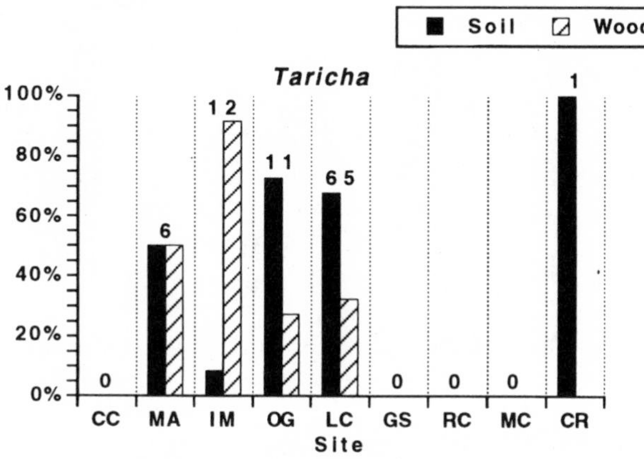

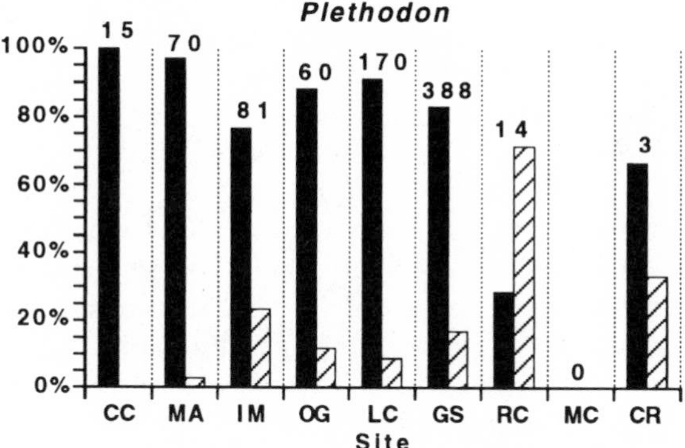

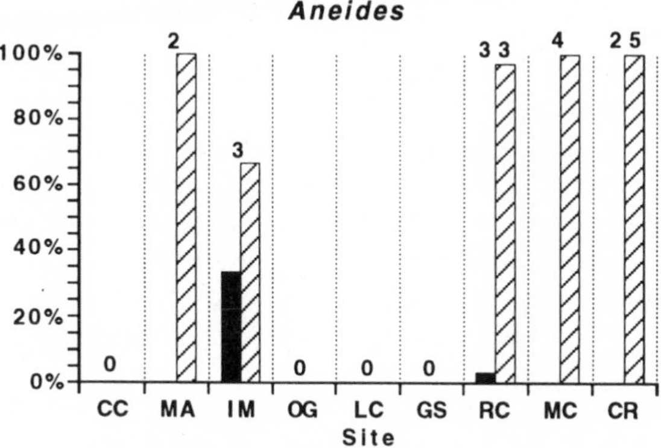

Figure 6. *Variation in use of microhabitats within artificial cover objects across sites. The number at the top of each pair of columns indicates the number of salamanders found at that site, not including recaptures. Each pair of columns sums to 100%. CC = GVW clearcut; MA = GVW mature, IM = GVW immature; OG = GVW old-growth; LC = Lake Cowichan; GS = Goldstream; RC = Rosewall Creek; MC = McNaughton Creek; CO = Cook Creek.*

hiculum; $\chi^2 = 57.9$, df = 7, $P < 0.001$; *T. granulosa*: $\chi^2 = 16.6$, df = 4, $0.001 < P < 0.005$; *A. ferreus*: $\chi^2 = 10.5$, df = 4, $0.025 < P < 0.05$).

Discussion

Seasonal Variation in Surface Abundance

Surface activity is important because it presumably reflects foraging behavior, and at least for some plethodontid salamanders, such as *Plethodon cinereus, P. hoffmani, Batrachoseps attenuatus*, and *B. major*, little or no feeding is thought to take place below the surface (Frazer 1976; Maiorana 1976). Thus, the length of time that a terrestrial salamander spends near the surface is probably directly related to its ability to grow and reproduce (Houck 1977; Jaeger 1980; Semlitsch and West 1983).

Patterns of surface activity in terrestrial salamanders are strongly influenced by climate. In the eastern United States, most rainfall occurs from late March through October, but during the winter (December through March), temperatures are near or below freezing. Thus, eastern terrestrial salamanders in upland areas tend to be active from April through October, but are inactive during the winter. In the western United States, rains usually begin in the autumn (September or October) and continue through the relatively mild winter until spring (April or May). Therefore, most western species (especially in California) are active from October through April, but are inactive through the relatively dry summer (Houck 1977; Bury and Corn 1987).

In Oregon, Peacock and Nussbaum (1973) found *P. vehiculum* on the surface in all months except August. On Vancouver Island, where winter conditions are somewhat more severe, Ovaska and Gregory (1989) found that surface activity in *P. vehiculum* was confined to the autumn (September to November) and spring (March to June) but occasional salamanders were seen on mild winter days, suggesting a direct role for temperature in influencing activity. Below-freezing temperatures in the winter and dry conditions in July and August probably forced the salamanders underground. However, Davis (1991) found that although *A. ferreus* also disappeared during the coldest part of the winter, individuals were fairly abundant even during the driest part of the summer.

These 2 patterns of surface activity may be related to microhabitat use. Davis (1991) found that the soil under most natural cover on the ground became dry during the summer, but wood under the bark on larger logs remained moist throughout the year. Because individuals of *P. vehiculum* were largely absent from the bark-on-log microhabitat and their usual microhabitat under coarse woody debris on the soil was relatively dry, they were difficult to find in July and August. In contrast, *A. ferreus* was common in the relatively moist bark-on-log microhabitat.

That local microhabitat conditions can influence seasonal activity patterns was demonstrated by Rosenthal (1957). In a natural area in California, he could not find individuals of *Aneides lugubris* near the surface during the dry season (June to October) but found that they remained near the surface of masonry walls in an urban area if the walls were artificially moistened.

Although the aquatic stage of *T. granulosa* has been well studied (Efford and Mathias 1969; Oliver and McCurdy 1974; Taylor 1984) and Pimentel (1960) investigated movements into and out of ponds, I know of no studies of the terrestrial ecology of this species. The common observation that large migrations of *T. granulosa* may be seen crossing roads in the spring and autumn is thought to represent movements from the terrestrial habitat to ponds for breeding (Leonard et al. 1993). I was unable to detect evidence of such migrations. Rather, surface abundance was strongly correlated with variations in moisture and temperature. However, there was considerably less variation in surface abundance of *T. granulosa* relative to *P. vehiculum* and *A. ferreus*. *Taricha granulosa* has a larger body size relative to the other species (reduced surface to volume ratio) and a thicker skin, so they are better able to resist moisture loss under desiccating conditions and may remain on the surface longer than the plethodontids. This suggests that a larger proportion of the *T. granulosa* population is on the surface at times when temperature and moisture have caused many *P. vehiculum* to disappear underground.

Variation in Abundance Among Sites

Populations of terrestrial salamanders vary hugely from one site to the next. Abundances (Fig. 2) cannot be compared directly across all plots. The Goldstream site yielded more salamanders at any particular time than any other site, although the sample was based on only 12 ACO (Fig. 3). The Greater Victoria Watershed (Fig. 2) represents 144 ACO divided among 4 sites, 1 of which was a clearcut, whereas the RMC sites had 36 ACO divided equally among 3 sites, 1 of which was a clearcut. The Lake Cowichan site had 36 ACO. Thus, the Goldstream site supports a much larger population of salamanders than any of the other sites (Fig. 3). That the GVW forested sites yielded about equal numbers of salamanders suggests that abundance may be similar across wide areas and that areas of relatively high abundance are local and patchy. The extent of the highly dense population of *P. vehiculum* at the Goldstream site has not been determined.

Variation in Abundance Among Species

A striking result of this study is the extreme difference in the ratio of *P. vehiculum* to *A. ferreus* between the forested RMC sites and the other forested sites. This may represent a difference in local site conditions such as amount and size of CWD, a geographical difference that may be related to climatic conditions, metapopulation phenomena related to landscape effects, chance historical events, or a combination of any or all of these factors. All age classes of *A. ferreus* were relatively more abundant in clearcuts near forested sites of high *A. ferreus* abundance than in any of the other forested sites in this study (T. Davis, unpublished). *Aneides ferreus* suspends eggs in cavities under bark on logs and in cavities within logs (Dunn 1942; Guppy 1946; Storm 1947; Davis 1991), but *P. vehiculum* lays eggs in burrows or cavities below the surface. This suggests that large logs of a particular decay class may be important in maintaining a source population (Corn and Bury 1991). A site that contains a source population in 1 decade may become a sink population in the next, as nesting logs pass beyond the optimum decay class. This is analogous to Gill's (1978) spatially shifting reproductive source ponds for the red-spotted newt, *Notophthalmus viridescens*. If such ephemeral source populations are not contiguous, then the overall population may eventually decline. Thus, where forest practices result in relatively evenly aged forests that contain uniform types, amounts, and decay classes of CWD, separated by clearcuts, populations of salamanders may decline. Populations of salamanders were reduced in clearcuts but there was no difference among the three forested GVW sites. This suggests that local disturbance has no long-term population consequences but leaves open the question of landscape and metapopulation effects. Also, because the GVW sites are small, contiguous with sites of various ages, and except for the GVW clearcut, contain stands of various ages, they represent a chronoseries only approximately. For example, the GVW immature and old-growth sites are part of the same contiguous forest and only represent local variation within that forest.

Clouded salamander, Aneides ferreus, *eggs. Photo by Stephen Corn.*

Such metapopulation dynamics do not explain the high densities of *A. ferreus* on small isolated islands adjacent to Vancouver Island nor why *P. vehiculum* is absent from these islands. Also, it appears that *P. vehiculum* is relatively scarce where *A. ferreus* is abundant. Species of similar-sized terrestrial salamanders are able to coexist if they each have slightly different life histories, feed on different size prey, and use different microhabitats (Burton 1976; Fraser 1976). For example, in Oregon, sympatric species of terrestrial salamanders (*A. ferreus, E. eschscholtzii, P. vehiculum*, and *Batrachoseps wrighti*) differed markedly in their use of different decay classes of CWD (Bury and Corn 1988; Corn and Bury 1991). Such microhabitat partitioning may result from each species being better able to exploit critical resources in slightly different microhabitats (Schoener 1974) or because one species, through interspecific competition, forces another into a different microhabitat (Hairston 1987; Schoener 1983). These factors may operate concurrently.

Several eastern North American species of salamanders are known to partition microhabitats through interspecific interference competition (Hairston 1987). For example, *P. cinereus* excluded *P. shenandoah* from areas of deep soil through interspecific aggression (Jaeger and Gergits 1979; Wrobel et al. 1980). As a result, *P. shenandoah* is confined to islands of talus unsuitable for *P. cinereus* (Jaeger 1971a,b). Similarly, *Desmognathus fuscus*, moved to microhabitats farther from a stream when in the presence of *D. monticola* (Keen 1982), but juvenile *D. monticola* shifted from rocks to wood in the presence of adult *D. fuscus* (Southerland 1986). The proximate mechanism is aggressive behavior as a result of competition for burrows. Among species of *Desmognathus*, this aggression can grade into predation if the sizes of individuals are sufficiently different (Hairston 1987; Southerland 1986). Interspecific predation is known to affect microhabitat use and abundance in this genus (Hairston 1986; Hairston 1987; Krzysik 1979). In North Carolina, a more balanced situation exists between *Plethodon glutinosus* and *P. jordani*, 2 terrestrial species that compete for space (Hairston 1987; Nishikawa 1985). There was an increase in the numbers of a species when either species was removed (Hairston 1980; Hairston 1987). A similar process may occur between *A. ferreus* and *P. vehiculum* on Vancouver Island.

Although interspecific interactions among terrestrial salamanders may profoundly affect local abundance and diversity, detailed studies of such interactions have not been carried out for any species in western North America. Maiorana (1978) found that *A. lugubris* and *B. attenuatus* may compete for food, but the availability of suitable burrows was probably more important in regulating their relative abundance. Lynch (1985) found that *A. lugubris, A. flavipunctatus, B. attenuatus*, and *E. eschscholtzii*, overlaped widely in the size of their prey. Competition among these species is plausible, but no experimental work has been done to test the hypothesis. Ovaska and Davis (1992) found that *Plethodon dunni* and *P. vehiculum* recognized and displayed toward each others' fecal pellets (which are used in chemical communication), and K. Ovaska and I (unpublished data) have observed attacks and agonistic display behavior by *P. dunni* toward *P. vehiculum*. However, much work needs to be done to clarify the extent of their interactions in nature.

Variation in Microhabitat Use

Davis (1991) found that *A. ferreus* could be found under bark on logs or in cracks in logs, but *P. vehiculum* was almost never found in those microhabitats. Instead, *P. vehiculum* was almost always found under bark or wood on the ground, a microhabitat rarely used by *A. ferreus* (Corn and Bury 1991). This observation is consistent with the use of ACO microhabitats and clearly demonstrates the importance of suitable cover objects when attempting to monitor a particular species. If the wood microhabitat had not been made available, very few *A. ferreus* would have been found, even in areas where they are relatively abundant. This suggests that the low numbers of *E. eschscholtzii* may be an artifact of sampling and a different artificial microhabitat would attract them. Aubry et al. (1988) found *E. eschscholtzii* most often in bark piles at the bases of moderately decayed snags, whereas *P. vehiculum* were most often found under logs. However, *E. eschscholtzii* can be easily located at some sites (for example, Cultus Lake on the British Columbia mainland, K. Ovaska, pers. comm.), but searches of the sites in this study failed to produce significant numbers of this species. Little is known of microhabitat use among the other species or whether the presence of one species affects microhabitat selection in others as it does in some eastern plethodontids (Krzysik 1979; Keen 1982; Hairston 1986; Wrobel et al. 1980; Southerland 1986).

This study demonstrated the utility of using ACO for repeated sampling of salamander populations. As discussed above, multiple microhabitats targeted for particular species proved important. However, it is clear that target species require ACO with species-specific microhabitats if the target organisms are to be successfully sampled. For this reason, ACO can only provide indices of population size among sites and not among species within sites. Also, through processes of decay and invasion by fungi and arthropods, the nature of the ACO may change through time, although this should be consistent across similar sites. It is not known if the increase in numbers of salamanders per search from 1992 to 1993 was due to differences in weather patterns between years, an improvement in the microhabitat conditions under the ACO, discovery by the resident population, or a combination of these factors. Finally, it is not known to what extent the availability of natural cover objects may influence capture rates.

Acknowledgments

I thank Sophie Boizard, Jeannine Caldbeck, Logan Caldbeck, Aziza Cooper, Lisa Crampton, Christian Engelstoft, Jennifer Harris, and Kristiina Ovaska for their assistance in the field, the authorities of the British Columbia Ministry of Parks who allowed me access to the site in Goldstream Park, the staff of the Greater Victoria Water District for access to sites under their control, and Douglas Pollard, Valin Marshall, and Tony Trofymow of Forestry Canada for their support and use of equipment. Patrick Gregory provided valuable advice and encouragement, and kindly reviewed the manuscript. Two anonymous reviewers provided useful comments on a previous draft. Funding for this study was provided through a grant to P.T. Gregory, University of Victoria, from the Forestry Practices Component of Forestry Canada's Green Plan, and through the British Columbia Ministry of Environment, Lands and Parks, Wildlife Branch. This manuscript is derived from work done in partial fulfillment of the requirements for a Ph.D. degree from the University of Victoria, Victoria, B.C., Canada.

Literature Cited

Aubry KB, Jones LLC, Hall PA. 1988. Use of woody debris by plethodontid salamanders in Douglas-fir forests in Washington. In: Szaro RC, Severson KE, Patton DR, technical coordinators. Management of amphibians, reptiles, and small mammals in North America. Fort Collins, CO: USDA Forest Service. General Technical Report RM-166. p 32–37.

Buhlmann KA, Pague CA, Mitchell JC, Glasgow RB. 1988. Forestry operations and terrestrial salamanders; techniques in a study of the Cow Knob salamander, *Plethodon punctatus*. In: Szaro RC, Severson KE, Patton DR, technical coordinators. Management of amphibians, reptiles, and small mammals in North America. Fort Collins, CO: USDA Forest Service. General Technical Report RM-166. p 38–44.

Burton TM. 1976. An analysis of the feeding ecology of the salamanders (Amphibia, Urodela) of the Hubbard Brook Experimental Forest, New Hampshire. Journal of Herpetology 10:187–204.

Bury RB, Corn PS. 1987. Evaluation of pitfall trapping in northwestern forests: trap arrays with drift fences. Journal of Wildlife Management 51:112–119.

Bury RB, Corn PS. 1988. Douglas-fir forests in the Oregon and Washington Cascades: relation of the herpetofauna to stand age and moisture. In: Szaro RC, Severson KE, Patton DR, technical coordinators. Management of amphibians, reptiles, and small mammals in North America. Fort Collins, CO: USDA Forest Service. General Technical Report RM-166. p 11–22.

Corn PS, Bury RB. 1990. Sampling methods for terrestrial amphibians and reptiles. Portland, OR: USDA Forest Service. General Technical Report PNW-GTR-256.

Corn PS, Bury RB. 1991. Terrestrial amphibian communities in the Oregon Coast Range. In: Ruggiero LF, Aubry KB, Carey AB, Huff MH, technical coordinators. Wildlife and vegetation of unmanaged Douglas-fir forests. Portland, OR: USDA Forest Service. General Technical Report PNW-GTR-285. p 305–317.

Davis TM. 1991. Natural history and behaviour of the clouded salamander, *Aneides ferreus* Cope [thesis]. Victoria, BC: University of Victoria.

Dunn ER. 1942. An egg cluster of *Aneides ferreus*. Copeia 1942:52.

Efford IE, Mathias JA. 1969. A comparison of two salamander populations in Marion Lake, British Columbia. Copeia 1969:723–736.

Fraser DF. 1976. Empirical evaluation of the hypothesis of food competition in salamander of the genus *Plethodon*. Ecology 57:459–471.

Gibbons JD. 1993. Nonparametric statistics: an introduction. London: Sage Publications.

Gill DE. 1978. The metapopulation ecology of the red-spotted newt, *Notophthalmus viridescens* (Rafinesque). Ecology Monograph 48:145–166.

Grant BW, Tucker AD, Lovich JE, Mills AM, Dixon PM, Gibbons JW. 1992. The use of coverboards in estimating patterns of reptile and amphibian biodiversity. In: McCullough DR, Barrett RH, editors. Wildlife 2001: populations. London: Elsevier Science Publishers Ltd. p 379–403.

Guppy R. 1946. Some salamander observations. Victoria Naturalist 3:45.

Hairston NG. 1980. The experimental test of an analysis of field distributions: competition in terrestrial salamanders. Ecology 61:817–826.

Hairston NG. 1986. Species packing in *Desmognathus* salamanders: experimental demonstration of predation and competition. American Naturalist 127:266–291.

Hairston NG. 1987. Community ecology and salamander guilds. Cambridge: Cambridge University Press.

Heyer WR, Donnelly MA, McDiarmid RW, Hayek LC, Foster MS, editors. 1994. Measuring and monitoring biological diversity: standard methods for amphibians. Washington: Smithsonian Inst. Press.

Houck LD. 1977. Life history patterns and reproductive biology of neotropical salamanders. In: Taylor DH, Guttman SI, editors. The Reproductive biology of amphibians. New York: Plenum Press. p 43–72.

Jaeger RG. 1971a. Competition exclusion as a factor influencing the distribution of two species of terrestrial salamanders. Ecology 52:632–637.

Jaeger RG. 1971b. Moisture as a factor influencing the distributions of two species of terrestrial salamanders. Oecologia 6:191–207.

Jaeger RG. 1980. Fluctuations in prey availability and food limitation for a terrestrial salamander. Oecologia 44:335–341.

Jaeger RG, Gergits WF. 1979. Intra- and interspecific communication in salamanders through chemical signals on the substrate. Animal Behaviour 27:150–156.

Keen WH. 1982. Habitat selection and interspecific competition in two species of plethodontid salamanders. Ecology 63:94–102.

Krzysik AJ. 1979. Resource allocation, coexistence, and the niche structure of a streambank salamander community. Ecological Monographs 49:173–194.

Leonard WP, Brown HA, Jones LLC, Storm RM. 1993. Amphibians of Washington and Oregon. Seattle: Seattle Audubon Society.

Lynch JF. 1985. The feeding ecology of *Aneides flavipunctatus* and sympatric plethodontid salamanders in northwestern California. Journal of Herpetology 19:328–352.

Maiorana VC. 1976. Size and environmental predictability for salamanders. Evolution 30:599–613.

Maiorana VC. 1978. Difference in diet as epiphenomenon: space regulates salamanders. Canadian Journal of Zoology 56:1017–1025.

Nishikawa KC. 1985. Competition and the evolution of aggressive behavior in two species of terrestrial salamanders. Evolution 39:1282–1294.

Oliver MG, McCurdy HM. 1974. Migration, overwintering, and reproductive patterns of *Taricha granulosa* on southern Vancouver Island. Canadian Journal of Zoology 52:541–545.

Ovaska K. 1987. Social behavior of the western red-backed salamander, *Plethodon vehiculum* [dissertation]. Victoria, BC: University of Victoria.

Ovaska K, Davis TM. 1992. Fecal pellets as burrow markers: intra- and interspecific odour recognition by four western plethodontid salamanders. Animal Behaviour 43:931–939.

Ovaska K, Gregory PT. 1989. Population structure, growth, and reproduction in a Vancouver Island population of the salamander *Plethodon vehiculum*. Herpetologica 45:133–143.

Peacock RL, Nussbaum RA. 1973. Reproductive biology and population structure of the western red-backed salamander, *Plethodon vehiculum* (Cooper). Journal of Herpetology 7:215–224.

Pimentel RA. 1960. Inter- and intrahabitat movements of the rough-skinned newt, *Taricha torosa granulosa* (Skilton). American Midland Naturalist 63:470–496.

Rosenthal GM. 1957. The role of moisture and temperature in the local distribution of the plethodontid salamander *Aneides lugubris*. University of California Publications in Zoology 54:371–420.

Schoener TW. 1974. Resource partitioning in ecological communities. Science 185:27–39.

Schoener TW. 1983. Field experiments on interspecific competition. American Naturalist 122:240–285.

Semlitsch RD, West CA. 1983. Aspects of the life history and ecology of Webster's salamander, *Plethodon websteri*. Copeia 1983:339–346.

Southerland MT. 1986. Behavioral interactions among four species of the salamander genus *Desmognathus*. Ecology 67:175–181.

Storm RM. 1947. Eggs and young of *Aneides ferreus*. Herpetologica 4:60–62.

Taylor J. 1984. Comparative evidence for competition between the salamanders *Ambystoma gracile* and *Taricha granulosa*. Copeia 1984:672–683.

Welsh HH, Jr. 1990. Relictual amphibians and old-growth forests. Conservation Biology 4:309–319.

Welsh HHJr, Lind AJ. 1988. Old growth forests and the distribution of the terrestrial herpetofauna. In: Szaro RC, Severson KE, Patton DR, technical coordinators. Management of amphibians, reptiles, and small mammals in North America. Fort Collins, CO: USDA Forest Service. General Technical Report RM-166. p 439–458.

Wilkinson L, Hill M, Vang E. 1992. SYSTAT for the Macintosh: Statistics, Version 5.2. Evanston, IL: SYSTAT, Inc.

Wrobel DJ, Gergits WF, Jaeger RG. 1980. An experimental study of interference competition among terrestrial salamanders. Ecology 61:1034–1039.

Zar JH. 1984. Biostatistical analysis. Englewood Cliffs, NJ: Prentice-Hall.

©1997 by the Society for the Study of Amphibians and Reptiles
Amphibians in decline: Canadian studies of a global problem. David M. Green, editor.
Herpetological Conservation 1:175–179.

Chapter 18

THE USE OF ARTIFICIAL COVERS TO SURVEY TERRESTRIAL SALAMANDERS IN QUÉBEC

JOËL BONIN[1] and YVES BACHAND

St. Lawrence Valley Natural History Society, 21 125 Chemin Ste-Marie, Ste-Anne-de-Bellevue, Québec H9X 3L2, Canada

ABSTRACT.—Four types of coverboard installations, either single boards on the ground or a pile of 2, 3, or 4 boards partially buried, were compared to survey terrestrial salamanders in forests of Québec. Wet leaf litter was added underneath boards to favour their rapid use by salamanders. *Plethodon cinereus* was the sole species encountered in numbers, reaching 23% of boards within 2 mo after installation. Every type of installation was used, but boards with wet leaf litter were preferred by salamanders. The installation of single boards on the ground is preferable to other buried installations because they require less material, are easier to install, and do not disrupt the ground surface. Despite the use of buried installations, the coverboard technique was inadequate to survey fossorial *Ambystoma* spp.

RÉSUMÉ.—Quatre types d'installation de planchettes ont été comparés pour l'échantillonnage de salamandres terrestres des forêts du Québec, soit une planchette au sol ou une pile de 2, 3, ou 4 planchettes partiellement enfouies. De la litière humide était ajoutée sous les planchettes pour favoriser leur utilisation rapide par les salamandres. *Plethodon cinereus* fut la seule espèce retrouvée en nombre; son taux d'occupation atteignait 23% des planchettes 2 mois après leur installation. Tous les types d'installation furent utilisés avec une préférence pour les planchettes dont la litière était humide. L'emploi de planchettes déposées sur le sol semble toutefois préférable aux installations enfouies parce que cela nécessite moins de matériel, s'avère plus facile à installer et ne perturbe pas la surface du sol. Malgré l'emploi d'installations enfouies, la technique des planchettes semble inadéquate pour échantillonner les salamandres fouisseuses *Ambystoma* spp.

A potentially useful technique for long-term monitoring of salamanders populations in North American forests is the use of artificial coverboards. Stewart and Bellis (1970), DeGraaf and Yamasaki (1992), Grant et al. (1992), and Davis (this volume) used boards to simulate logs in the forest understory and salamanders were found beneath these artificial shelters. This non-destructive technique is easy to apply and permits standardized survey with little between-observer variability (Fellers and Drost 1994).

The use of coverboards was not effective for surveying the most fossorial species such as *Ambystoma* spp. (DeGraaf and Yamasaki 1992; J. Bonin, pers. obs.), but we believed that this could be improved by providing shelters set deeper into the ground. The age of the coverboards could also affect the encounter rate (Grant et al. 1992) because amphibians preferentially use wetter boards (Grant et al. 1992) and old boards tend to retain more humidity. Furthermore, the soil fauna often takes time to reestablish itself under new cover. It has been usual procedure to leave the boards undisturbed for a year, allowing them to weather (DeGraaf and Yamasaki 1992). However, during a long-term monitoring program, data from young and old boards may not be comparable. Thus we tested the effec-

---

1 Present Address: *Redpath Museum, McGill University, 859 Sherbrooke St., W., Montréal, Québec, Canada*

tiveness of adding wet leaf litter under new coverboards to potentially increase their use by salamanders.

There is a risk of an artificial population increase because the addition of coverboards. Salamanders compete for natural cover objects to use as daytime retreats, foraging grounds, and even as nesting sites (Jaeger 1980; Smith and Pough 1994). This would be problematic for long-term monitoring objectives if the boards themselves attract salamanders from surrounding areas. We experimented with this technique in southern Québec.

Materials and Methods

Experiments were performed in Mont Orford Park, situated in the Appalachian Mountains of southern Québec, Canada (45°19'N, 72°11'W). This park is located within the northeast hardwood forest region characterised by *Acer saccharum, Fagus grandifolia*, and *Betula lutea*. The study area consisted of 4 ha of relatively homogeneous deciduous forest located near a stream and beaver pond.

Following Stewart and Bellis (1970), we used relatively small boards, because we wanted them to be light, easy to transport, and easy to install in various locations. Each board was made of a 30 × 30 × 1-cm chipboard sheet to which 2 sticks 1-cm thick were tacked on underneath in such a way as to provide space between overlying boards (Fig. 1). These boards were tried 1st during the summer and fall of 1992 to define sampling procedures and obtain gross information on encounter rates. We set 160 coverboard installations at regular intervals (about every 10 m). An installation consisted of a pile of 2 boards placed on the ground with leaf litter added between the 2 overlying boards. They were installed at the end of June, checked on 24 to 25 July, 7 to 10 August, 16 September, and 8 October 1992, and then removed.

We compared different types of installations in 1993. We set 351 coverboard installations in 4 types of installations. Type 1 consisted of a single board on the ground, and Types 2, 3, and 4 consisted of piles of 2, 3, or 4 boards, respectively, partially buried (Fig. 1). For greater efficiency, we added wet leaf litter beneath the boards instead of allowing them to weather, to test if this would provide suitable wet conditions, as well as encourage the kinds of soil fauna preyed on by salamanders. Leaf litter was collected in the forest understory, soaked in water, and put between each board. During each visit, we qualitatively

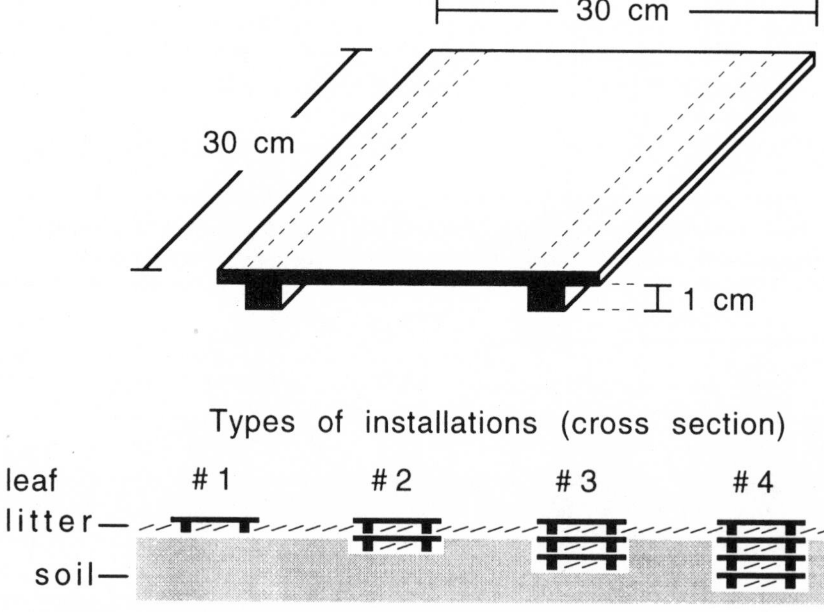

Figure 1. *Sketch of a coverboard and of the 4 types of installations used to survey terrestrial salamanders in Mont Orford Park, Québec, Canada, in 1993.*

noted if the litter was dry (loose and friable) or wet (sticky and supple), and wet it again. To avoid a change in the holding capacity of the habitat, we installed coverboards for the shortest possible period and only after the egg-laying season. Coverboards were installed during mid-August, visited 9, 14, and 16 October, and removed in November 1993. Two arrays of 30-m long drift fences with pitfall traps were set nearby to assess the salamander species present in the area.

Results

In 1992, we found 56 *Plethodon cinereus* (redback salamander), 15 *Eurycea bislineata* (two-lined salamander), 3 *Desmognathus fuscus* (northern dusky salamander), and 1 *Ambystoma maculatum* (spotted salamander). On 5 occasions, 2 *P. cinereus* were found together in the same installation. Greater numbers were found in September and October (Fig. 2). Encounter rate of *P. cinereus* reached 20% (30 individuals/153 installations) in September.

In October 1993, 113 *P. cinereus*, 12 *E. bislineata*, and 9 *Ambystoma laterale* (blue-spotted salamander) were encountered in board installations. No other species were found in the pitfall traps set in the study area. Only once was > 1 salamander found under the same board—a *P. cinereus* was found with an *A. laterale*. The encounter rate for *P. cinereus* decreased between 9 and 16 October: 23% (80 individuals/351 installations) on 9 October, 7% (24/351) on 14 October, and 5% (9/165) on 16 October.

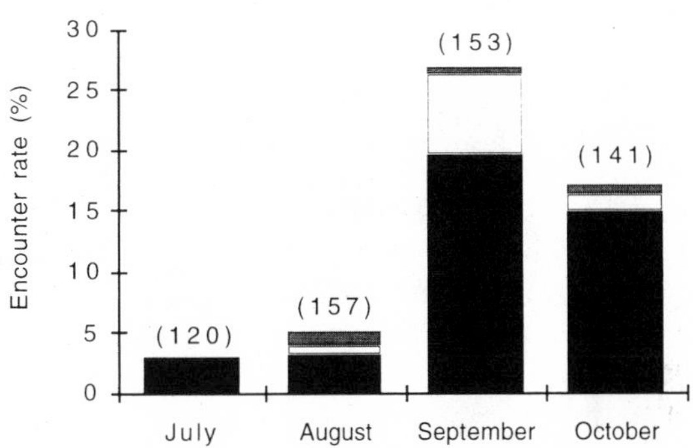

Figure 2. *Variation in encounter rates of* Plethodon cinereus *(black),* Eurycea bislineata *(white), and other species (stipple) during summer and fall 1992 in Mont Orford Park. The number of board installations checked is indicated in parentheses.*

Distribution of *P. cinereus* was not random within installations ($\chi^2 = 60.31$, df = 3, $P < 0.005$; Table 1). They were found mostly in wet spots but did not show a preference for being underneath compared to between boards ($\chi^2 = 2.18$, df = 1, $P > 0.05$; Table 1). More *P. cinereus* were found in installation Type 4 (n = 46 salamanders) than under single boards (Type 1, n = 29) or other installations (Table 2). On the other hand, the number of salamanders per board was higher with single board installation than with multiple board installations. Taking only available wet sites into consideration, *P. cinereus* encounter rates per board were 16% (29 salamanders/181 boards with wet condition underneath) for Type 1 and 11% (26/234), 2% (8/382), and 9% (42/460) for Types 2, 3, and 4, respectively. Captures of

Table 1. *Number of salamanders found in wet and dry sites and their position within the installation in Mont Orford Park in 1993. Encounter rates (% of boards) are in parentheses.*

	Wet (n)		Dry (n)	
Species	On ground (762)	Between boards (495)	On ground (105)	Between boards (807)
Plethodon cinereus	71 (9%)	34 (7%)	0	8 (1%)
Eurycea bislineata	11 (1%)	1 (< 1%)	0	0
Ambystoma laterale	4 (1%)	3 (1%)	0	2 (< 1%)

Table 2. *Distribution of salamanders among the 4 types of coverboard installations.*

	Coverboard type (n)			
Species	1 (88)	2 (87)	3 (88)	4 (88)
Plethodon cinereus	29	26	12	46
Eurycea bislineata	3	2	1	6
Ambystoma laterale	4	1	0	4

other species were insufficient to test for differences in position within installations or among types of installations.

Discussion

Despite the use of buried installations, the coverboard technique may be inefficient to survey species of *Ambystoma*. The small numbers we found are probably insufficient to enable year to year comparisons of densities. Although the actual size of the *Ambystoma* population in the study area is unknown, the area presents habitat conditions, including a nearby pond, sufficient to support a fairly large population of these salamanders. However, *Ambystoma* are usually less abundant than *P. cinereus* in northeastern forests (Burton and Likens 1975), and the encounter rates observed here may reflect the relative abundance of each species. A survey of *Ambystoma* might be constrained to examine breeding sites during the spring to sample greater numbers of individuals.

No newts (*Notophthalmus viridescens*) were found in our study. DeGraaf and Yamasaki (1992) stated that newts were seldom encountered under coverboards, which appear to be more efficient for surveying terrestrial plethodontids. Stewart and Bellis (1970) had success with the technique for surveying *Eurycea* and *Desmognathus* along streams in Pennsylvania, and Davis (this volume) encountered numerous *Plethodon* and *Aneides* under coverboards in forests of British Columbia. Our experiments conducted in southern Québec during 1992 and 1993 indicate that coverboards are useful for surveying *P. cinereus*. Highest encounter rates for that species were 20% (30 out of 153 installations) in September 1992 and 23% (80 out of 351 installations) in October 1993. These rates are relatively high compare to other surveys (DeGraaf and Yamasaki 1992; Grant et al. 1992).

Plethodon cinereus used new coverboards extensively within only 2 mo after installation. Humidity was the main factor determining their use of coverboards. Wet leaf litter seemed to provide suitably humid conditions, without the need to let the boards weather. However, from this short term study, we can not state that deploying new coverboards each year is advantageous to limit variation between new and old boards. For a long-term monitoring program, consistency of sampling sites from year to year is critical (C. Drost, pers. comm.). Deploying new coverboards would be more time and labor intensive, and would introduce an additional element of variability. To attempt to limit the humidity and aging variability, experiments should be conducted using non-decaying coverboards (made of plastic) and artificial sponges. Experiments should involve testing for an increase in the holding capacity of the habitat.

Redback salamander, Plethodon cinereus. *Photo courtesy of Canadian Museum of Nature.*

The use of buried stacked boards did not improve success in surveying fossorial *Ambystoma* species, and although more *P. cinereus* were found in installations with 4 stacked boards (Type 4), this installation required more boards per salamander sampled than single boards on the ground (Type 1). Furthermore, the ground surface around boards was often disrupted with the buried installations. Type 1 was the easiest one to install and was efficient to survey *P. cinereus*, and hence should be prefered over buried installations.

Salamanders are usually found under surface objects during wetter periods of the year, and encounter rates may vary with recent weather conditions (Fellers and Drost 1994). In Québec, the moister conditions in the fall may favour greater counts. The decrease in encounter rate during October 1993 probably corresponded to the reduction of ground surface activity of *P. cinereus* as winter approached. Disturbance caused by checking the boards may have affected encounter rate as well. Thus, many visits in the fall may be needed and only the highest counts must be retained to obtain the best population estimates.

Acknowledgments

This study was supported by the Ministère de loisir, de la chasse et de la pêche of the province of Québec. We thank J. Boulé, D. Fontaine, and I. Tartier for fieldwork assistance, R. Cook and the employees of Mont-Orford Park for technical assistance.

Literature Cited

Burton TM, Likens GE. 1975. Salamander populations and biomass in the Hubbard Brook experimental forest, New Hampshire. Copeia 1975:541–546.

DeGraaf RM, Yamasaki M. 1992. A nondestructive technique to monitor the relative abundance of terrestrial salamanders. Wildlife Society Bulletin 20:260–264.

Fellers GM, Drost CH. 1994. Sampling with artificial cover. In: Heyer WR, Donnelly MA, McDiarmid RW, Hayek LC, Foster MS, editors. Measuring and monitoring biological diversity: standard methods for amphibians. Washington: Smithsonian Institution Press. p 146–150.

Grant BW, Tucker AD, Lovich JE, Mills AM, Dixon PM, Gibbons JW. 1992. The use of coverboards in estimating patterns of reptile and amphibian biodiversity. In: McCullough DR, Barrett RH, editors. Wildlife 2001: populations. London: Elsevier Science Publishers Ltd. p 379–403.

Jaeger RG. 1980. Microhabitats of a terrestrial forest salamander. Copeia 1980:265–268.

Smith EM, Pough FH. 1994. Intergeneric aggression among salamanders. Journal of Herpetology 28:41–45.

Stewart GD, Bellis ED. 1970. Dispersion patterns of salamanders along a brook. Copeia 1970:86–89.

Amphibians in decline: Canadian studies of a global problem. David M. Green, editor.
Herpetological Conservation 1:180–184.

Chapter 19

A SIMPLE TRANSECT TECHNIQUE FOR ESTIMATING ABUNDANCE OF AQUATIC RANID FROGS

DONALD F. MCALPINE

Natural Sciences Division, New Brunswick Museum, 277 Douglas Avenue, Saint John, New Brunswick E2K 1E5, Canada

ABSTRACT.—A simple and inexpensive approach for running line transects along pond and lake margins provided quantitative estimates of mink frogs, *Rana septentrionalis*, during a study to gauge the impact of the insecticide fenitrothion on frog abundance at sites in northern New Brunswick. The technique was a modified area-constrained method for counts of aquatic frogs using a portable measuring device. The technique suggested that areas that had been most heavily sprayed over the previous 5 yr had a lower abundance of mink frogs.

RÉSUMÉ.—Une méthode simple et peu coûteuse qui permet d'effectuer des transects le long des étangs et des lacs a permis d'évaluer quantitativement les grenouilles du Nord, *Rana septentrionalis*, dans le cadre d'une étude visant à déterminer l'impact de l'insecticide fénitrothion sur l'abondance des grenouilles dans certaines régions du nord du Nouveau-Brunswick. Cette technique pour décompter les grenouilles aquatiques dans un endroit déterminé utilisait unappareil de mesure portatif. Grâce à cette technique, il semble que les secteurs où il y a eu la plus grande application d'insecticides au cours des 5 dernières années sont ceux où l'abondance de grenouilles du Nord est la moins grande.

Fenitrothion, an organophosphate insecticide, was first applied to New Brunswick forests in an experimental application in 1965. Since then it has been used to control spruce budworm, *Choristoneura fumiferana*, on approximately 10^6 ha of New Brunswick forests annually. Aquatic invertebrate responses in forest streams and ponds following fenitrothion spraying have been quite variable (Fairchild et al. 1989). However, invertebrates in bog ponds have been shown to be especially sensitive to fenitrothion. Impacts are of such magnitude as to substantially alter energy flow through these ecosystems (Fairchild 1990; Fairchild and Eidt 1993). This has raised concerns that the abundance of aquatic amphibians in forest ponds and lakes may be reduced indirectly through a reduction in prey species.

The abundance of aquatic frogs in areas of New Brunswick subject to long-term spraying with fenitrothion were compared to sites with moderate or minor fenitrothion applications. The transect method described here was developed to support field surveys done in 1991. Although several investigators have described techniques for estimating the abundance of amphibians in various habitats (Corn and Bury 1990; Bury and Corn 1991; Heyer et al. 1994) none of these approaches met the constraints of our survey situation.

MATERIALS AND METHODS

All sites selected for frog surveys were located in Northumberland and Restigouche counties of northern New Brunswick. The 9 sampled water bodies varied from about 1 to 21 ha, and were within 35' latitude of each other at 244 to 1550 m elevation (Table 1). Vegetation bordering ponds and lakes varied in composition and density among sites, but all were situated in the Acadian forest region as defined by Rowe (1972).

The remoteness of the study area, the potential for human disturbance (there was evidence that several of the sites were visited regularly by fishermen), the high probability of water level change over the June to August sampling period, and the 3 days available each month for sampling by only 2 field workers, placed constraints on our sampling approach. To meet these constraints, I developed a survey technique that is rapid, non-labour intensive, and does not involve the establishment of permanent or marked transects. The technique should be readily applicable to other species of anurans that inhabit shallow waters bordering ponds and lakes, particularly where vegetation bordering sites is generally uniform and not so dense as to obscure the vision of the surveyor (Fig. 1).

Figure 1. *Shoreline at Berry Brook Pond in northern New Brunswick. Habitats like this were readily surveyed for frogs using the technique described here.*

Survey Apparatus and Method

A meter stick bolted to 1 arm of a 33-mm diameter PVC right angle pipe was mounted on a 50-cm length of 19-mm diameter wooden dowel (Fig. 2). The base of the dowel was sharpened so it could be more easily forced into the substrate. The meter stick could be unbolted from the wooden rod for ease of transport to the field. The device was used to define a transect 2 × 50 m along the margin of each study pond or lake. The dowel was forced into the ground or pond bottom and the area scanned for frogs along the length of the meter stick. The dowel was then twisted left or right 90° and the procedure repeated. The forward perimeter of each quadrat was marked visually with reference to vegetation, the dowel lifted, and then moved ahead 1 m. In this way, the 50-m transect was surveyed in 1-m blocks.

We found young of the year more difficult to approach for counting than larger frogs. However, by moving slowly and scanning ahead along the transect so that young could be anticipated, it was possible to survey transects without causing frogs in quadrats ahead to disperse before counting. We attempted to carry out all transects in warm, sunny weather when wind speeds were low. Two to 6 transects were run at each site on each visit. Transects were run at regular, non-overlapping intervals along the margins of water bodies. We usually ran transects along the terrestrial side of pond or lake margins, but where heavy alder thickets were present, transects were run by wading through shallow water. Number of frogs present was recorded quadrat by quadrat along each 50-m tran-

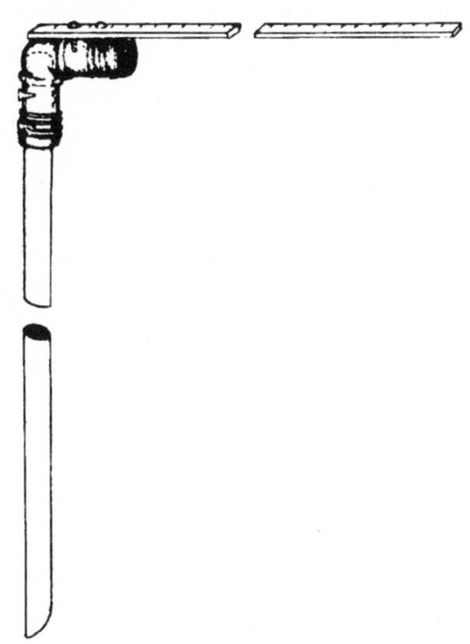

Figure 2. *Device used for surveying line transects for aquatic frogs.*

Table 1. *Estimated densities of mink frogs/100m^2 ± SE at New Brunswick study sites in 1991. Monthly means are based on 2 to 6 transects (n) per site. W is the test statistic for Kendall's coefficient of concordance where H_0 = no significant difference in frog densities among transects, α = 0.05.*

	June					July					August				
Site	Density	SE	n	W	*P*	Density	SE	n	W	*P*	Density	SE	n	W	*P*
Low spray zone															
Black Lake	6.5	1.2	4	0.008	0.743	12.8	2.8	4	0.027	0.257	5.8	2.5	4	0.046	0.075
South Lake	34.2	6.3	4	0.052	0.049	57.0	34.1	4	0.449	< 0.0001	9.8	3.5	4	0.089	0.004
Medium spray zone															
Berry Brook Pond	88.8	7.4	4	0.029	0.229	53.5	22.5	2	0.286	< 0.001	63.8	17.8	4	0.234	< 0.001
Juniper Lake	21.2	8.6	4	0.201	< 0.001	14.8	4.0	5	0.083	0.002	20.5	10.1	4	0.320	< 0.001
High spray zone															
Camel Back Lake						2.0	0.7	6	0.024	0.315	17.0	5.0	4	0.074	0.011
Forty Mile Pond						8.2	1.9	6	0.050	0.029	3.2	2.1	4	0.152	< 0.001
Forty Mile Lake						3.3	1.1	6	0.026	0.252	8.0	3.6	4	0.112	0.001
Indian Lake	0	–	2	–	–	10.0	2.5	6	0.009	0.832	20.0	11.3	4	0.130	< 0.001
McCormack Lake	0.7	0.3	6	0.016	0.549	0.7	0.3	6	0.020	0.416	1.5	0.3	4	0.005	0.861

sect on field sheets, and tallies were summed at the conclusion. Monthly means and standard errors were calculated for each pond or lake, and analysis of variance using Kendall's coefficient of concordance, **W**, with correction for ties, was used to test for consistency in density estimates across transects at individual sites on each survey date.

Results

Although the occasional toad, *Bufo americanus*, and several wood frogs, *Rana sylvatica*, were observed on transects, only *R. septentrionalis* occurred in numbers sufficient to estimate population density. Survey time for each 50 m transect ranged from 10 to 30 min, depending on the number of frogs present and the density of the vegetation. There was general consistency in frog densities recorded at individual sites throughout the study period (Table 1). Lakes and ponds showing relatively high densities in June continued to show high numbers in July and August while sites showing low numbers of frogs early in the study period usually did not show a marked increase in frog numbers later in the season. However, there were exceptions to this pattern. South Lake had very high densities in June and July but showed a dramatic drop in numbers in August. Camelback Lake, a site added and first surveyed in July, increased from 2.00 ± 0.7 frogs/100 m^2 to 17.00 ± 5.0 frogs/100 m^2. In 13 of the 23 survey cases, there was a significant difference in monthly population estimates among transects at a given site (Table 1).

Discussion

Of the techniques Heyer et al. (1994) describe for inventorying and monitoring amphibians, transect sampling is most appropriate for the circumstances described here, despite the recommendation (Jaeger 1994) that transects not be used in relatively homogeneous habitat but be reserved for studies of known habitat gradients. Neither the visual encounter nor quadrat method reported in Heyer et al. (1994) is well suited for comparing amphibian abundances along a series of pond and lake margins. Visual encounter surveys, because they are time-constrained, are limited in their utility in that abundances can be compared only among sites of the same habitat type. In cases where frogs are sparsely distributed along a lake, pond, or stream margin, it may be impractical to survey the hundreds of randomly selected individual quadrats needed to estimate abundances at a site. Although it may be possible to refine the quadrat method so that it can be applied in our circumstances, the transect method has been used with success for surveying amphibians along streams and riparian strips (Green and Tessier 1990; Bury and Corn 1991). Jaeger (1994) notes that randomly placed transects tend to run into each other, causing problems with replicated samples. This problem can be avoided when surveying lake or pond margins by dividing each margin into a series of sections roughly corresponding to the planned transect length and then randomly selecting transect locations from among these sections.

At Berry Brook Pond from June to August and South Lake from July to August, the differences among transects were the result of young-of-the-year congregating along certain sections of shoreline. At Juniper Lake, there was a section of near-shore habitat with heavy emergent vegetation where frog densities were higher than surrounding sphagnum-dominated shore line. At other sites, the reasons for the significant differences among transects was unclear. Shoreline habitat was generally similar among transects at these sites but there may have been microhabitat difference that were not apparent.

Mink frog, Rana septentrionalis. *Photo by Martin Ouellet.*

The approach I have described allowed us to survey sites on a first pass through a section of habitat, thereby minimizing disturbance. We could easily move transect lines as water levels changed during the summer and the risk of vandalism or disturbance to permanently marked transect lines was avoided. As a cheap, rapid method for estimating the abundance of aquatic frogs, our approach appeared to work well under the limited circumstances in which it was used. Our results suggested frog densities were significantly different between fenitrothion spray zones and between ponds within spray zones. Sites in the high spray zone had lower mean frog densities than ponds in either the medium or low spray zones (D.F. McAlpine, N.M. Burgess, and D.G. Busby, unpublished).

Undoubtedly, there are ways in which the method could be improved. Standard errors are highest where numbers of transects surveyed are lowest. Our results suggest that > 4 transects are preferable, particularly where young-of-the-year are abundant or where shoreline habitat varies. The use of aerial photos and a planimeter would permit the calculation of the number of transects required should one wish to standardize the percentage of total shoreline surveyed at each site. Our sampling effort at large lakes was less than that at smaller ponds, although there was no apparent correlation between frog density and the areas of the water bodies we surveyed. Where shoreline or near-shore habitats vary, transects can be laid out in a manner proportional to habitat types or, where habitat types are extensive enough, densities can be estimated for each type separately.

Acknowledgments

Neil Burgess, Tim Fletcher, and Cathy Wagg provided assistance in the field and useful discussions relating to the technique described here. Funding for this work was provided through contract KR203-1-0058 via the Canadian Wildlife Service, Environment Canada.

Literature Cited

Bury RB, Corn PS. 1991. Sampling methods for amphibians in streams in the Pacific Northwest. Portland: USDA Forest Service. General Technical Report PNW-GTR-275.

Corn PS, Bury RB. 1989. Logging in western Oregon: Responses of headwater habitats and stream amphibians. Forest Ecology and Management 29:39–57.

Corn PS, Bury RB. 1990. Sampling methods for terrestrial amphibians and reptiles. Portland: USDA Forest Service. General Technical Report PNW-GTR-256.

Fairchild WL. 1990. Perturbation of the aquatic invertebrate community of acidic bog ponds by the insecticide fenitrothion [thesis]. Fredericton, NB: University of New Brunswick.

Fairchild WL, Eidt DC. 1993. Perturbation of the aquatic invertebrate community of acidic bog ponds by the insecticide fenitrothion. Archives of Environmental Contamination and Toxicology 25:170–183.

Fairchild WL, Ernst WR, Mallet VN. 1989. Fenitrothion effect on aquatic organisms. In: Ernst WR, Pearce PA, Pollock TL, editors. Environmental effects of fenitrothion use in forestry. Dartmouth, NS: Environment Canada, Atlantic Region. p 109–166.

Green DM, Tessier C. 1990. Distribution and abundance of Hochstetter's frog, *Leiopelma hochstetteri*. Journal of the Royal Society of New Zealand 20:261–268.

Heyer WR, Donnelly MA, McDairmid RW, Hayek LC, Foster MS, editors. 1994. Measuring and monitoring biological diversity: standard methods for amphibians. Washington: Smithsonian Institution Press.

Jaeger RG. 1994. Transect sampling. In: Heyer WR, Donnelly MA, McDiarmid RW, Hayek LC, Foster MS, editors. Measuring and monitoring biological diversity: standard methods for amphibians. Washington: Smithsonian Institution Press. p 103–107.

Rowe JS. 1972. Forest regions of Canada. Ottawa: Canadian Forest Service. Publication Number 1300.

Amphibians in decline: Canadian studies of a global problem. David M. Green, editor.
Herpetological Conservation 1:185–190.

Chapter 20

EFFECTS OF LOGGING ON TERRESTRIAL AMPHIBIANS OF COASTAL BRITISH COLUMBIA

LINDA A. DUPUIS

Centre for Applied Conservation Biology, Faculty of Forestry, University of British Columbia, Vancouver, British Columbia V6K 1Z4, Canada

ABSTRACT.—The abundance of terrestrial-breeding amphibians was compared between old-growth forests and young and mature post-harvest stands of coastal British Columbia. Habitat features required by salamanders were contrasted between old growth and managed stands to indicate the specific effects of logging operations. Clearcut harvesting has a negative effect on terresrial amphibian populations. Salamander densities in managed stands were similar to those in old growth only within 10 m of streams. Managed stands lacked large downed wood, an important source of cover for salamanders, notably the western redback salamander, *Plethodon vehiculum*. The proximity of old growth to managed stands may also be crucial. Second-growth habitats isolated from old growth had lower densities of amphibians than those adjacent to old growth. To consider the needs of salamanders in timber harvest activities, moist habitat in the form of riparian buffers must be protected, large downed wood, snags and trees must be retained to ensure future supplies of moist cover, and 300-yr cutting cycles should be incorporated into standard rotations to allow for the continual production of large downed wood. Old growth reserves may facilitate the recolonization of post-harvest stands by amphibians.

RÉSUMÉ.—L'abondance des salamandres terrestres dans les forêts vierges de la côte ouest de la Colombie britannique a été comparée à celle que l'on retrouve dans les coupes à blanc récentes et plus âgées. Les micro-habitats utilisés par les salamandres ont eux aussi été comparés entre les parcelles vierges et exploitées afin de mieux comprendre les effets du déboisement. Les coupes à blanc affectent d'une façon négative les populations de salamandres terrestres. Seules les densités de salamandres à 10 m des ruisseaux dans les sections exploitées étaient équivalentes aux densités que l'on retrouvaient dans les sections vierges. Ce phénomène peut être relié à une réduction de la disponibilité des micro-habitats humides. D'ailleurs, le bois en stage intermédiaire de décomposition était relativement rare dans les zones exploitées. Il est cependant important comme source d'abri pour les salamandres, notamment la salamandre à dos rayé, *Plethodon vehiculum*. La proximité des sections vierges peut elle aussi être importante. L'abondance des salamandres était toujours moins élevée dans les sections exploitées et isolées des parcelles vierges. Dans la régie des activités forestières, il importe donc d'assurer la protection des zones riparliennes, de favoriser la rétention de bois et d'arbres vivants et morts afin d'assurer la présence permanente de micro-habitats humides, et de permettre l'incorporation de longs cycles (300 ans) dans les planifications d'exploitation afin d'assurer une source continuelle de bois adéquat. Les sections vierges protégées faciliteraient peut-être la recolonisation par les amphibiens des parcelles déboisées.

With the rapid rate of timber harvest in the Pacific Northwest of North America, there is a need to consider the requirements of resident species. An understanding of how species can persist in altered habitats and landscapes can help managers preserve biodiversity and maintain ecosystem integrity. Large amounts of time and money have been invested into the study of a few animals, such as spotted owls (*Strix occidentalis*) and grizzly bears (*Ursus horribilis*). However, these large-scale studies have yielded little insight on important forest characteristics for wildlife communities. Smaller species serve this purpose because they can be numerous and occupy small home ranges.

Terrestrial-breeding amphibians, primarily plethodontid salamanders, are particularly suitable for study because they are year-round residents of forest habitats and seldom range more than a few meters (Jaeger 1979; Ovaska 1988). Salamanders are indicators of habitat health, and they are sensitive to large-scale disturbances because of their physiological restrictions and their long life spans. Amphibians, as a whole, are also poorly understood in forest ecosystems, and their potentially worldwide declines make them important organisms in conservation (Johnson 1992). Recent studies have demonstrated that amphibian species are more abundant in old growth than second growth in Washington, Oregon, and northern California, and that some species are strongly associated with old growth (Bury and Corn 1988; Welsh and Lind 1988; Welsh 1990; Corn and Bury 1991). Similar results were obtained in eastern North America (Pough et al. 1987; Buhlmann et al. 1988; Petranka et al. 1991; Waldick, this volume) and New Mexico (Ratmonik and Scott 1988).

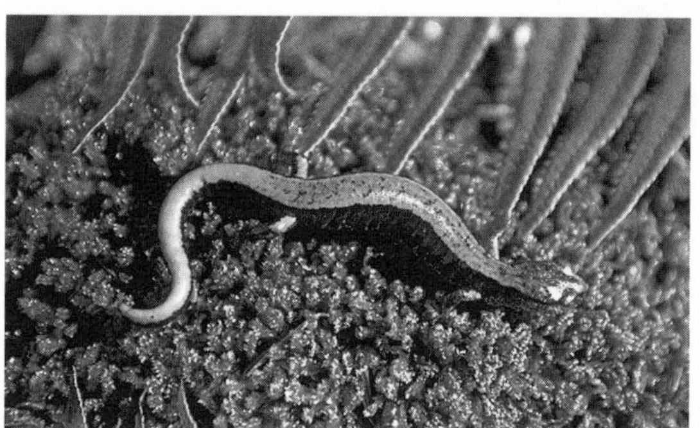

Western redback salamander, Plethodon vehiculum. *Photo by Martin Ouellet.*

This study examines the abundance of terrestrially breeding amphibians in old-growth forests and in stands with histories of clearcut logging on Vancouver Island, Canada, and investigates the microhabitat features used by amphibians in both old growth and second growth.

Materials and Methods

I used 3 replicates of each of 3 forest types: 1) moist, nutrient-rich Douglas-fir (*Pseudotsuga menziesii*)/western hemlock (*Tsuga heterophylla*) old growth; 2) mature, 54- to 75-yr old stands that were clearcut during the establishment of the railway; and 3) young, 17- to 18-yr old stands. Most replicates were on the western, windward side of Vancouver Island, British Columbia, but 1 site of each age class was situated on the drier, leeward side. The study sites were distributed over an area of roughly 900 km^2 near Port Alberni, central Vancouver Island. Some amphibian surveys were carried out in clearcuts to sample a range from harvest to stand re-establishment. Study sites were all in the Coastal Western Hemlock Biogeoclimatic Zone of British Columbia (Green et al. 1988).

I sampled terrestrial amphibians with area-constrained searches (ACS), which involved thorough searches of 1 × 2 m quadrats randomly placed on the forest floor of each site. Quadrats were placed at least 30 m from streams and 50 m from forest edges and roads. ACS were supplemented by downed wood surveys where randomly selected wood in 3 stages of decomposition (after Bury and Corn 1988) was thoroughly searched. These stages were: 1) fresh, hard logs with tight intact bark; 2) logs with a softening heartwood and sloughing bark; and 3) spongy or chunky, reddish logs with relatively no bark. ACS were also done within 10 m of streams (20 quadrats per site). Salamander densities in these moist, riparian habitats were compared to 20 randomly-selected quadrats from the upland (30 m) samples, within natural and managed stands.

Results

There were 3 to 6 times more amphibians in old growth than in managed stands (Fig. 1). These results were consistent between years and methods. There was very little variation, particularly in mature stands. The young stands had the lowest abundances, and clearcuts had virtually no amphibians. Almost all encounters with amphibians (95%) were with terrestrial species (Fig. 1), of which 92% were western redback salamanders, *Plethodon vehiculum*. Ensatinas, *Ensatina eschscholtzii*, and clouded salamanders, *Aneides ferreus* were also found.

Amphibian abundance fell in 1992 when precipitation decreased. May 1991 received 55.8 mm of rain compared with 6.4 mm in 1992. Amphibian abundance was higher in the spring of 1991 than in 1992; there was no apparent relationship between abundance and precipitation in 1991. Similarly, amphibian abundance decreased as soil moisture declined in late spring. Amphibians were also scarce in drier sites and in young stands, which had significantly drier soils even immediately after snow melt (Fig. 2). In managed stands, terrestrial-breeding amphibians were more concentrated along streams (Table 1). This phenomenon was not observed in old growth.

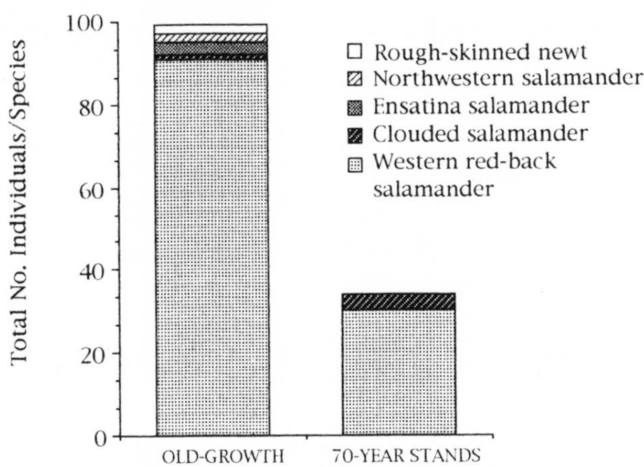

Figure 1. *The abundance and diversity of amphibians in old-growth forests and mature, 54- to 72-yr old stands. Data collected on Vancouver Island in the spring of 1992.*

Salamanders used a variety of cover objects, including vegetation, rocks, and wood. About 50% of the salamanders were associatated with wood or bark in both 1991 and 1992. Salamanders were most often associated with logs in the intermediate and advanced stages of decomposition (Fig. 3a). In old growth, nearly 85% of the individuals were encountered in the soil interface under logs and bark (Fig. 3b). In mature managed stands, 50% of the individuals were found under logs, and < 10% of the salamanders were under bark.

Table 1. *Density of salamanders (individuals/ha) near and away from streams in old-growth forest and managed stands in May 1992.*

	Near (< 10 m)		Away (> 30 m)	
Stand age (yr)	$\bar{x}$	SE	$\bar{x}$	SE
Old growth	170.0	50.0	150.0	50.0
54–72	162.5	37.5	37.5	0.0
17–18	62.5	62.5	12.5	12.5

Wood availability differed in undisturbed and logged forests. The volume of wood was greatest in old growth and lowest in the young managed stands (Fig. 4). Clearcuts had higher levels of wood than the young managed stands but < ½ of the wood volume found in old growth. The volume of intermediately decayed wood was particularly reduced in all harvested stands. This stage of decomposition is characterized by sloughed bark, which is also infrequent in managed stands. Large downed trees were most abundant in old growth and least abundant in young stands. Conversely, young stands had a larger number of small downed trees than old growth. Managed stands had about 55% less large wood and 10 to 20% more small wood than old growth.

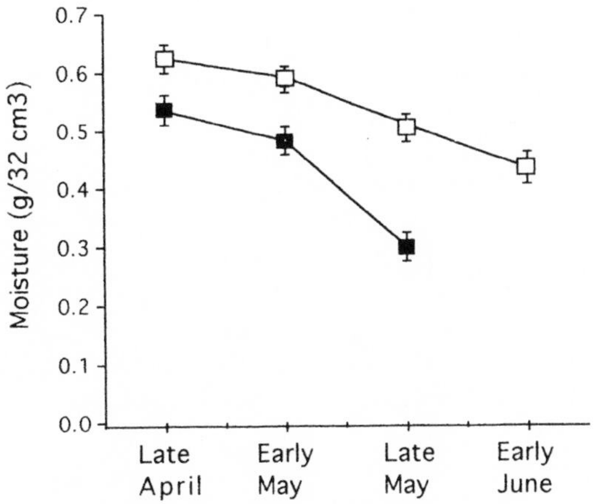

Figure 2. *Soil moisture in old growth (open squares) and young managed stands (black squares) in the spring of 1992.*

The mature managed stands furthest removed from old growth harboured the lowest densities of salamanders, in both 1991 and 1992. The same pattern was observed in young stands, which were only sampled in 1992.

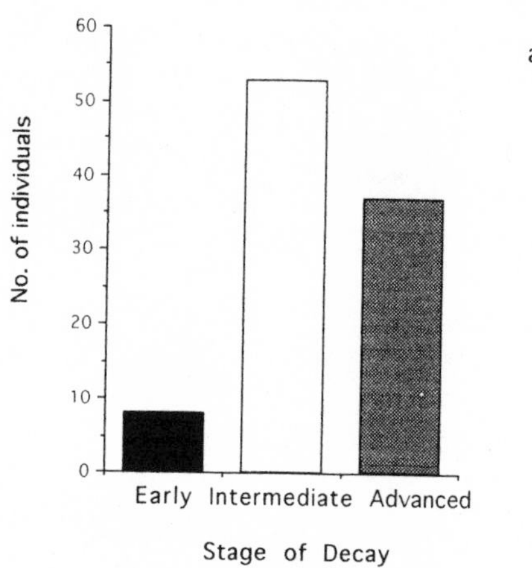

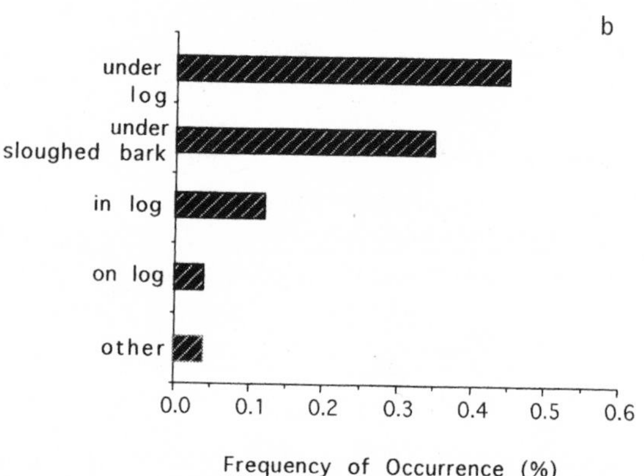

Figure 3. *a) The abundance of salamanders associated with hard, fresh logs (black bars), intermediately decayed logs (clear bars) and soft reddish logs (shaded bars), in the spring of 1991. b) the position of salamanders on logs in 1991 and 1992.*

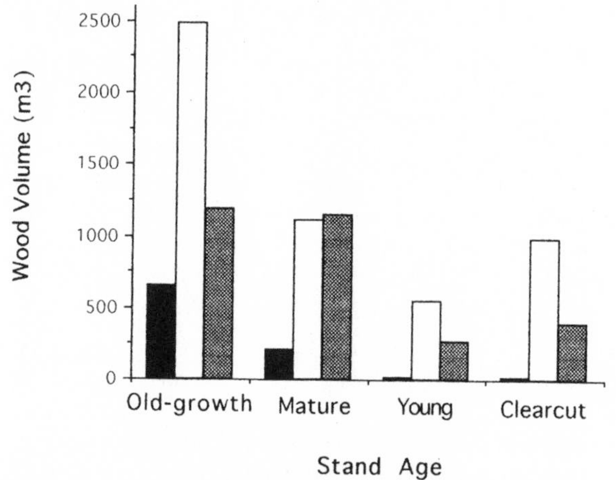

Figure 4. *The volume of wood in the 3 stages of decomposition, in old growth, managed stands, and clearcuts.*

Discussion

Lungless salamanders were negatively affected by clearcut logging on central Vancouver Island. There was a 3- to 6-fold difference in abundance of salamanders observed between managed stands and old growth (Fig. 1). These findings agree with other similar surveys (Corn and Bury 1991; Petranka et al. 1991; Welsh 1990).

Low amphibian densities indicate unfavourable conditions in managed stands, which may be related to moisture stress. The soils of young managed stands were drier than those of old-growth forests throughout the spring and summer (Fig. 2). Amphibians are susceptible to dehydration, which limits their foraging activities unless conditions are moist and wind-free (Jaeger 1971; Spotila 1972; Maiorana 1978). The important role that soil moisture plays in salamander activity is reflected in this study. Salamander numbers above ground were drastically reduced when precipitation approached zero in 1992, and above ground activity declined as soils dried up in late spring. Moreover, drier stands harboured the lowest salamander abundances. This association between terrestrial-breeding amphibian abundance and soil moisture was also suggested by streamside surveys in 1992. Salamander abundances were 4 to 6 times higher along the streambanks of managed stands and similar to those found in old growth.

The volume of wood increased with forest age (Fig. 4). Mature stands have more downed wood than present young stands will be able to generate. In the earlier part of the century only Douglas-fir and the highest grades of other species were selected for cutting, and large numbers of felled trees were left on site. This is not the case today. When today's young stands reach 60 yr old, they will have very low volumes of downed wood. The existing logs will have decayed and there will not have been enough time for new supplies of downed wood to accumulate (Spies et al. 1988; Maser and Trappe 1984). Downed wood in the intermediate stages of decay is particularly reduced in second growth (Fig. 4). Young, managed stands characteristically have new logs arising from self-thinning processes within forests and highly decayed remnants of the primeaval forest. Because logging rotations occur at 60- to 120-yr intervals, managed stands may never accumulate intermediate downed wood of large size. Salamanders are most often associated with intermediate and decayed logs (Fig. 3a), which are soft and moist because of increased microbial activity.

Although about 50% of salamanders are found under logs, the soil/surface interface provided by bark also represents critical habitat in old growth (Fig. 3b). Sloughed bark is a characteristic of intermediately decayed wood, which necessarily implies that it is largely absent from managed stands. The low frequency of encounters of salamanders under bark in mature managed stands reflects this loss of a microhabitat.

Size of downed wood is also important, because larger pieces can accomodate more individual salamanders and provide them with more protection. They are also more likely to provide soil interface habitat because of their weight. Large downed wood has a long residence time on the forest floor and offers some continuity to salamanders by linking old growth with newly established stands. Today, large downed wood is uncommon in managed stands and will become rarer if current harvesting practices and policies do not change.

Salamander abundance may decrease with increasing isolation from old growth. Managed stands may be sink habitats where plethodontid salamander populations can only be maintained through continual recolonization from neighbouring old growth. This needs further investigation, particularly in landscapes now dominated by managed stands.

My study demonstrates that clearcut logging has a negative effect on terrestrial-breeding amphibians due to reduced microclimatic stability and a decrease in the availability of large intermediately decayed downed wood with sloughed bark. The vicinity of old growth tracts may also influence salamander densities in managed stands. Several management recommendations can be made to protect amphibians in the Douglas fir forests of Vancouver Island.

Rivers and streams in forests need protective buffers, likely a minimum of 20 to 30 m. To date, small streams are only protected if they directly or indirectly affect commercial fish or their spawning grounds, and that protection consists of the retention of understory and large wood (British Columbia Ministry of Forests et al. 1992). This provides little or no protection against climate fluctuations.

Cable logging systems bring felled trees to roads where economically valuable logs are selected and loaded. This process produces large piles of undesirable wood near landings. Choker setters (who hook logs to the cables) could be trained to do some log selection within the cutblock, so that downed wood is left in the forest. Patches of trees and snags should be maintained along cutblock edges and within cable configuration centres as a future supply of downed wood for wildlife.

Old growth should be protected as a source of colonizers for marginal managed habitats. Connectivity in the landscape, between old growth patches, will help to prevent local extirpation of terrestrially breeding amphibians. British Columbia's Coastal Biodiversity Guidelines have introduced FENS, or Forest Ecosystem Networks (British Columbia Ministry of Forests and British Columbia Ministry of Environment 1995), made up of swamps, high elevation forests, unstable slopes, and areas of unmerchantable timber, to link up natural systems. FENS may be useful for salamander dispersal if they include moist forested areas.

Terrestrially breeding amphibians clearly are sensitive to logging disturbances. Large parts of Vancouver Island dominated by post-harvest second growth may represent sink habitats for salamanders. Future studies of amphibian forest associations should consider landscape level characteristics such as patch size and degree of isolation. By considering wood quality, streambanks, and landscape configurations, managers may better meet the needs of species and communities, thereby contributing to ecosystem integrity and avoiding widespread, irreversible damage to forest communities.

Acknowledgments

This project was funded by the Ministry of Forests, Victoria, British Columbia. Volunteers were provided by the Canadian Wildlife Service in Delta, British Columbia, for assistance in the field. The Sproat Lake Division of MacMillan Bloedel offered logistical support during a 2-yr period. Finally, critical reviews of early drafts of this manuscript were done by Dr. Jamie Fowler, Department of Zoology, and Dr. Fred Bunnell, Department of Forestry, at the University of British Columbia.

Literature Cited

British Columbia Ministry of Forests, British Columbia Ministry of Environment. 1995. Biodiversity guidebook. Forest practises code of British Columbia. Victoria: British Columbia Ministry of Forests.

British Columbia Ministry of Forests, British Columbia Ministry of Environment, Federal Department of Fisheries and Oceans, Council of Forest Industries. 1992. British Columbia coastal fisheries/forestry guidelines. 3rd ed. Victoria: British Columbia Ministry of Forests.

Bury RB, Corn PS. 1988. Douglas-fir forests in the Oregon and Washington Cascades: relation of the herpetofauna to stand age and moisture. In: Szaro RC, Severson KE, Patton DR, editors. Management of amphibians, reptiles and small mammals in North America. Fort Collins, CO: USDA Forest Service. General Technical Report RM-166. p 11–22.

Corn PS, Bury RB. 1991. Terrestrial amphibian communities in the Oregon Coast Range. In: Ruggierio LF, Carey AB, Huff MH, editors. Wildlife and vegetation of unmanaged Douglas-fir forests. Portland, OR: USDA Forest Service. General Technical Report PNW-GTR-285. p 305–318.

Green RN, Courtin PJ, Klinka K, Slaco RJ, Ray CA, editors. 1988. Site diagnosis, tree species selection, and slashburning guidelines for the Vancouver forest region. Vancouver, BC: British Columbia Ministry of Forests, British Columbia Forest Products Ltd., and University of British Columbia.

Jaeger RG. 1971. Moisture as a factor influencing the distribution of two species of terrestrial salamanders. Oecologia 6:191–207.

Jaeger RG. 1979. Seasonal spatial distributions of the terrestrial salamander *Plethodon cinereus*. Herpetologica 35:90–93.

Johnson B. 1992. Habitat loss and declining amphibian populations. In: Bishop CA, Pettit KE, editors. Declines in Canadian amphibian populations: designing a national monitoring strategy. Ottawa: Canadian Wildlife Service. Occasional Paper 76. p 71–75.

Maiorana VC. 1978. Differences in diet as an epiphenomenon: space regulates salamanders. Canadian Journal of Zoology 56:1017–1025.

Maser C, Trappe JM. 1984. The seen and unseen world of the fallen tree. Portland, OR: USDA Forest Service and USDI Bureau Land Management.

Ovaska K. 1988. Spacing and movements of the salamander *Plethodon vehiculum*. Herpetologica 44:277–386.

Petranka JW, Elridge ME, Haley KE. 1991. Effects of timber harvesting on southern Appalachian salamanders. Conservation Biology 7:363–368.

Pough HF, Smith EM, Rhodes DH, Collazo A. 1987. The abundance of salamanders in forest stands with different histories of disturbance. Forest Ecology and Management 20:1–9.

Ratmonik CA, Scott NJ Jr. 1988. Habitat requirements of New Mexico's endangered salamanders.In: Szaro RC, Severson KE, Patton DR, editors. Management of amphibians, reptiles and small mammals in North America. Fort Collins, CO: USDA Forest Service. General Technical Report RM-166. p 54–63.

Spies TA, Franklin JF, Thomas TB. 1988. Coarse woody debris in Douglas-fir forests of western Oregon and Washington. Ecology 69:1689–1702.

Spotila JR. 1972. Role of temperature and water in the ecology of lungless salamanders. Ecological Monographs 42:95–125.

Welsh HH Jr. 1990. Relictual amphibians and old-growth forests. Conservation Biology 4:309–319.

Welsh HH Jr, Lind A. 1988. Old growth forests and the distribution of the terrestrial herpetofauna.In: Szaro RC, Severson KE, Patton DR, editors. Management of amphibians, reptiles and small mammals in North America. Fort Collins, CO: USDA Forest Service. General Technical Report RM-166. p 439–455.

©1997 by the Society for the Study of Amphibians and Reptiles
Amphibians in decline: Canadian studies of a global problem. David M. Green, editor.
Herpetological Conservation 1:191–205.

Chapter 21

EFFECTS OF FORESTRY PRACTICES ON AMPHIBIAN POPULATIONS IN EASTERN NORTH AMERICA

RUTH WALDICK[1]

Biology Department, Dalhousie University, Halifax, Nova Scotia B3K 4H1, Canada

ABSTRACT.—Amphibian populations in eastern North America are continually subjected to the loss, fragmentation, and conversion of habitats as a result of forestry. Forestry operations ranging from low intensity selective harvests to high intensity clearcutting and site preparation cause landscape-level changes. By increasing exposure and reducing available refugia, these activities detrimentally affect amphibian species, especially plethodontid and ambystomatid salamanders. Increased evaporative losses lead to reduced pond hydroperiods, dry soil conditions, and lowered abundance and quality of ground surface litter. The distribution of plethodontid and ambystomatid salamanders is reduced by the progressive loss of mixed-species forests. Conditions can return to near pre-harvest levels if natural regeneration occurs but the establishment of conifer plantations permanently alters humidity levels and soil and ground surface conditions. Small, isolated populations such as those created over managed landscapes will be highly susceptible to local extinction events. The limited dispersal capabilities of amphibians may preclude dispersal across exposed clearcuts or plantation habitats > 1.5 km. Smaller cut-over areas that are proximal to source populations for recolonization can facilitate the recovery of resident amphibians. Connectivity of existing and potential habitable patches is required to sustain amphibian populations over their historic distributions. If key habitat features are maintained at levels that will support the most sensitive amphibian species, amphibian communities can persist in managed landscapes.

RÉSUMÉ.—Les populations d'amphibiens dans l'est de l'Amérique du Nord sont victimes de la disparition, du morcellement et de la conversion des habitats attribuables à l'exploitation forestière. L'exploitation forestière prend diverses formes : des récoltes sélectives de faible intensité aux coupes à blanc de grande intensité en passant par l'aménagement des sites, ce qui modifie radicalement le paysage. En augmentant l'exposition et en détruisant les refuges, ces activités nuisent aux espèces d'amphibiens et surtout aux salamandres pléthodontides et ambystomatides. L'accroissement de l'évaporation réduit les hydropéroides des étangs, assèche les sols et amoindrit la quantité et la qualité de la litière. La répartition des salamandres pléthodontides et ambystomatides diminue au fur et à mesure que disparaissent les forêts mixtes. Tout pourrait revenir à la normale si la régénération naturelle survenait mais la plantation de conifères modifie de façon permanente les niveaux d'humidité ainsi que les sols et surfaces. Les populations petites et isolées comme celles que l'on trouve dans des sites aménagés sont très vulnérables aux phénomènes locaux d'extinction. Les capacités réduites de dispersion des amphibiens peuvent prévenir leur dispersion dans des zones de coupes à blanc ou dans les habitats de plantation 1,5 km. De plus petites zones défrichées à proximité des populations sources pourraient faciliter la recolonisation par les amphibiens. Des liens entre les zones habitables potentielles et existantes s'imposent pour maintenir les populations d'amphibiens à leur niveau antérieur. Si les caractéristiques clé de l'habitat sont maintenues à des niveaux capables de satisfaire les espèces d'amphibiens les plus vulnérables, les communautés d'amphibiens pourront alors survivre dans des territoires aménagés.

---

1 Present address: *Department of Biology, McMaster University, Hamilton, Ontario L8S 4K1, Canada*

Conserving biodiversity and maintaining the integrity of natural ecosystems are key issues directing global biological research. Anthropogenic stressors are the most immediate threat to species of wildlife in natural systems and fall into several broad categories: pollution, over-exploitation, and habitat loss or alteration (Soulé 1980; Dodd and Seigel 1991; Andren 1994). Preserving species in the presence of these stresses requires conservation efforts based on understanding their short and long term effects upon wildlife, as well as understanding the basic requirements of individual taxa.

Amphibians not only have great intrinsic value, but they also play an important role in energy flow and the trophic dynamics or structure of most terrestrial ecosystems (Jaeger 1981; Brandon and Huheey 1975; Burton and Likens 1975; Ash 1988; Wake 1991). Their complex life-histories mean that amphibians play the role of intermediate level carnivore as well as provide an abundant food source for a variety of birds, mammals, other herpetofauna and fish.

Apparent widespread declines in amphibian populations have aroused concerns regarding their long-term persistence and have raised questions about the integrity of our remaining natural ecosystems (Wake 1991). Amphibians are especially vulnerable to certain environmental stresses and perturbations because of their limited physiological requirements and complex life-histories (Dunson and Connell 1982; Freda and Dunson 1984, 1985; Wake 1991; Feder and Burggren 1992; Sadinski and Dunson 1992; Blaustein et al. 1994). The potential cause(s) of the amphibian declines are numerous. They include habitat loss most importantly (Blymyer and McGinnis 1977; Ash 1988; Blaustein et al. 1994a,b), so it is imperative that we understand how and to what extent these changes pose threats to amphibians. Natural ecosystems are subject to conversion into less-suitable habitats for amphibians by forestry. Clearcutting has been shown to cause local extirpation of some amphibian species in North America (Bennett et al. 1980; Pough et al. 1987; Bury and Corn 1988b; Petranka et al. 1993) and prevailing forestry practices have caused, and continue to cause, landscape-level habitat conversion.

Forestry Practices

Because tree harvesting activities and management approaches vary, the severity of their impacts will also vary spatially and regionally. The variability occurs because of differences in Provincial and State forestry regulations, topographical differences between sites, and differences in forest-type. The intensity of tree removal can vary from clearcutting followed by intense site-preparation to selective tree harvesting with natural regeneration. The effect of clearcutting, which dramatically alters habitat structure, is even more severe when branches and foliage are also removed, leading to large-scale habitat fragmentation at the landscape-level (Freedman, et al. 1986). Thus, dramatic changes occur in the structure of the habitat and the microclimatic conditions immediately following a harvest (Table 1). Recently, less-disruptive methods such as selective harvesting, which involves removing individual trees or small groups of trees, have been used with increasing frequency in some regions (Kimmins 1992).

Redback salamander, Plethodon cinereus. *Photo by David M. Green.*

Tree harvesting is frequently followed by some form of site preparation, which may include slash--burning of coarse woody debris (CWD), physical crushing of slash using heavy equipment, or herbicide applications. Generally, sites are then planted with a single-species crop, which, in eastern North America, tends to be a fast-growing conifer species. Although other tree species may naturally become established over time in these uniform-age plantations, these sites tend to remain dominated by the planted species (T. Flemming, pers. comm.). When areas left to regenerate naturally are compared to planted sites, they tend to contain a greater mix of tree species and more structural complexity (B. Freedman, pers. comm).

Table 1. *Habitat conditions in some forest stands in eastern North America.*

Stand classification	Soil moisture (%)	Ground cover (%)	Litter depth (cm)	Study location
Clearcut	22	69	1.9	New York[a]
Conifer plantation	19	10	2.7	New York[a]
Firewood (selective harvest)	28	57	4.8	New York[a]
Secondary forest	20	24	4.2	New York[a]
Old growth	26–31	43–72	2.8–3.2	New York[a]
Conifer plantation (open canopy, < 10-yr old)		42–55		New Brunswick[b]
Conifer plantation (closed canopy, > 14-yr old)		81		New Brunswick[b]
Secondary forest		78		New Brunswick[b]

[a]Pough et al (1987)

[b]Waldick (1994)

Once trees in a given area reach maturity, they may be harvested again. A short duration between harvest rotations can reduce the abundance of CWD and impoverish the soil, reducing the suitability of the site as amphibian habitat or for the re-establishment of vegetation (Freedman et al. 1986; Graham et al. 1994). The impoverishment of a site due to over-harvesting may prevent natural forest assemblages from re-establishing in an area, further reducing the quality of the site. However, the length of time between successive harvests will depend on the characteristics of the site and the rate of development of the regenerating trees.

In eastern North America, large expanses of the landscape are composed of managed areas. As a result, 'old-growth' forests, such as those characteristic of the Pacific Northwest, are extremely rare. Little of the remaining forest is represented by mature, naturally regenerated stands, and most forested areas consist of plantations and fragmented forest patches resulting from decades of harvesting and replanting. In eastern Canada, at least 18% of all cut-over areas are re-planted with conifers and receive intensive management annually (Canadian Forestry Service 1992).

Habitat Features Influencing Amphibian Populations

Although the amount of resources present in a habitat can influence amphibian presence and abundance, it is the quality of these resources that ultimately determines the suitability of a site. In New Brunswick and other eastern regions, natural, mixed-species forests provide an abundant, heterogenous mixture of leaf litter that carpets most of the ground surface. The high structural complexity and moderate pH associated with this layer permits an assortment of invertebrate species to inhabit structurally complex habitats, thus increasing the quality of forage available in these habitats (Taub 1961; Hurlbert 1969; Maiorana 1978; Jaeger et al. 1981; Pough et al. 1987; Parmelee 1993). In contrast, conifer-dominated plantations produce structurally simple litter in lower quantities (Jaeger et al. 1981; Ash 1988; Waldick 1994). As a result, these stands are less able to moderate their soil surface conditions and provide a smaller invertebrate prey base (R. Waldick, unpublished).

To support a stable amphibian population, a habitat must provide the basic resource requirements for all life history stages (Soulé 1983; Gilpin 1987; Lande 1987), including actively reproducing adults. But even if a habitat appears large enough to support a minimum viable population (Simberloff 1988), it may be unable to sustain an adult population or support succesful juvenile recruitment if the necessary habitat resources are absent or are suboptimal in quality (Jaeger 1972, 1978, 1980; Pough et al. 1987; Pechmann et al. 1989). Canopy removal detrimentally alters the microclimate in clearcut sites by reducing shade, reducing vertical and horizontal structural complexity, increasing decomposition rates, reducing ground litter, and increasing daily temperature fluctuations, erosion and surface evaporation (Bury 1983; Enge and Marion 1986; Pough et al. 1987; Ash 1988; Raphael 1988).

Humidity Levels and Soil Moisture

The permeable nature of amphibian skin, key to its role as an organ for gas exchange, makes amphibians particularly susceptible to dehydration. Most amphibians require relatively humid habitats with refuges that allow them to avoid exposure during extreme conditions. The local distribution and activity patterns for many amphibians, including *Rana sylvatica*, *Plethodon cinereus*, and *Ambystoma maculatum* are influenced by moisture levels (Bellis 1962; Heatwole 1962; Spotila 1972; Petranka 1993; Parmelee 1993), particularly as dry surface conditions impede surface foraging (Heatwole 1962; Jaeger 1980; Sinsch 1990). Jaeger (1980) observed in *P. cinereus* that a greater proportion of salamanders collected from drier habitats had empty stomachs compared with those obtained from more humid areas. Stomach content was found to decline quickly following a rainfall, with most salamanders having empty stomachs after only 3 days. Heatwole (1962) found that the moisture levels at the soil-surface interface affected the density of *R. sylvatica* and observed a positive association between desiccation-tolerance and body size. This increased desiccation risk corresponds with the increased surface area-to-volume ratios, meaning that juvenile and smaller amphibians are most susceptible to dehydration (Bellis 1962; Spotila 1972; Shoop 1974; Jaeger 1978, 1980; Parmelee 1993). Amphibian population densities are also influenced by ambient humidity and soil moisture levels (Table 2).

Coarse Woody Debris, Leaf Litter Depth, and Cover Objects

Leaf litter and cover objects are important to amphibians as refuges to avoid adverse environmental conditions and potential predators (Heatwole 1962; Herrington 1988; Jones 1988a,b; Corn and Bury 1989a,b; Fauth et al. 1989; Allmon 1991). Leaf litter, CWD, and cover objects are also important habitat for estivation, hibernation, or breeding activities in some species (Cagle 1942; Ash 1988; Parmelee 1993; Waldick 1994). The importance of the quality of CWD is of particular importance for plethodontid salamanders which rely on it for breeding habitat, nursery grounds, and refuges from predation and climatic stress (Fauth et al. 1989; Allmon 1991). The high exposure in clearcut and young plantation habitats causes downed wood to become dehydrated, making it unsuitable for amphibians as a refuge (Aubry et al. 1988; Bury and Corn 1988a; Fauth et al. 1989; Allmon 1991; Parmelee 1993). During CWD searches in clearcuts, I have repeatedly found this wood to be highly

Table 2. *Density of* Plethodon *(number/200 m^2) in uncut mixed forest and various forest treatments.*

Species	Location	Mixed forest	Conifer forest	Old growth	Selective harvest	Clearcut (< 10-yr old)	Conifer plantation (< 10-yr old)	Conifer plantation (≥ 10-yr old)
P. cinereus[a]	New Brunswick	6				0	0	0
P. cinereus[b]	New York	7.7		1.5–4	1.6	0.3		0.7
P. cinereus[c]	Virginia	51[g]			0[g]			
P. jordani[d]	North Carolina	13–52[h]			0[h]			
P. jordani[e]	North Carolina	3.2–6			0–2			
P. glutinosus[d]	North Carolina	2.4–4.6[h]			0[h]			
P. glutinosus[e]	North Carolina	0.4–0.8			0–0.8			
P. glutinosus[f]	South Carolina	51	7–9					

[a]Waldick (1994); [b]Pough et al. (1987); [c]Blymer and McGinnis (1977); [d]Ash (1988); [e]Bennett et al. (1980); [f]Petranka et al. (1993)

[g]Relative abundance

[h]Based on Lincoln Index (capture-recapture) estimate

desiccated internally and incapable of supporting terrestrial plethodontid salamanders, particularly as breeding habitat. As a result, clearcut habitats preclude reproduction by plethodontids (Bishop 1941; Pough et al. 1987; Fauth et al. 1989; Allmon 1991).

Even those species capable of withstanding greater environmental extremes may require access to these resources seasonally or at particular life history stages (Taub 1961; Husting 1965; Pough and Wilson 1977; Spotila 1972; Shoop 1974; Jaeger et al. 1981; Pough et al. 1987). Heatwole (1962) noted that although leaf litter was not preferred habitat for *R. sylvatica* it was nevertheless, necessary when ponds or other available habitats became dry and serves as overwintering habitat (Pinder et al. 1992).

Redback salamander, Plethodon cinereus, *eggs. Photo by John Mitchell.*

Shading

The amount of shading afforded by vegetation can be used as an indirect or relative measure of canopy closure in habitats. Closed canopies act as wind breaks and sources of shade, which efficiently moderates ground surface conditions (Bury 1983). When canopy is removed, soil exposure increases and the environment becomes susceptible to greater temperature extremes and higher evaporative rates (Blymyer and McGinnis 1977; Bennett et al. 1980; Bury 1983; Pough et al. 1987; Ash 1988; Beiswenger 1988). Canopy removal also indirectly affects the abundance and quality of resources (Blymyer and McGinnis 1977; Bormann and Likens 1979; Pough et al. 1987; Harmon et al. 1991; Heinen 1993; Waldick 1994). Thermal and dehydration stress associated with an open canopy in cutovers reduces the overall suitability of these sites for most amphibians. Although temperature stress may be avoided by amphibians if suitable refuges are available, critical maximum temperatures (Taub 1961; Heatwole 1962; Pough and Wilson, 1970; Parmelee 1993) may be exceeded if refuges are low quality. Aquatic breeding habitat is also affected by the increased exposure and evaporative losses associated with canopy removal, because the prevailing conditions shorten pond hydroperiods (Holomuzki 1986; Semlitsch 1987; Loman 1988; Pechmann et al. 1989; Waldick 1994).

Soil pH

The pH of the layer of soil that amphibians inhabit is largely influenced by inputs of litter from the surrounding vegetation (Frisbie and Wyman 1992; Wyman and Jancola 1992). Forests with greater densities of conifers have lower soil pH than those containing a greater proportion of hardwood trees. Furthermore, soil pH is reduced by dry soil conditions, such as those prevalent in clearcut areas. The intolerance of *P. cinereus* to acidic conditions excludes them from soils with pH levels below 3.7 (Wyman and Jancola 1992), due to their inability to maintain their internal sodium balance under acidic conditions (Frisbie and Wyman 1992). Whereas *P. cinereus* is limited or excluded from conifer-dominated stands (Frisbie and Wyman 1992; Wyman and Jancola 1992), the more alkaline litter from hardwood trees or herbaceous vegetation buffers the acidic conifer litter, accounting for the higher densities of *P. cinereus* observed in mixed forests (Wyman and Jancola 1992).

Aquatic Habitat

Most amphibian species in eastern North America require access to either permanent or semi-permanent standing water for reproduction. Breeding ponds must provide an adequate habitat for developing eggs and larvae, including hydroperiods of sufficient length to ensure that larvae can reach metamorphosis (Enge and Marion 1986; Pechmann et al. 1989; Waldick 1994). The tradeoff between extended hydroperiods with high predation pressure and low food abundance versus shorter hydroperiods with fewer predators and higher food abundance influences breeding by many amphibians (Wilbur 1980).

Thus, for temporary pond breeders the probability of extinction is influenced by pond hydroperiod (Holomuzki 1986; Semlitsch 1987; Loman 1988; Pechmann et al. 1989; Laan and Verboom 1990; Kupferberg et al. 1994). In New Brunswick, the abundance of potential aquatic breeding habitat is greater in recently clearcut sites than in surrounding natural forested areas (Waldick 1994). Catchments are created by ditch building and soil excavation (leaving 'borrow-pits') and receive runoff from surrounding areas after snowmelt and rainfall. However, these ponds may be inaccessible to amphibians or they may be so shallow that their hydroperiod is too short to permit successful juvenile recruitment (Shoop 1974; Enge and Marion 1986; Semlitsch 1987; Pechmann et al. 1989; Waldick 1994).

I investigated the suitability of ephemeral ponds for breeding amphibians by monitoring adult and larval amphibians at ponds in a variety of habitats (Waldick 1994). Although a variety of species of amphibians were observed to breed at these ponds, the increased exposure in the clearcut areas lead to recruitment failure in *A. maculatum* and *Pseudacris crucifer*. Catastrophic mortality will occur if ponds dry before larvae reach the minimum size for metamorphosis (Semlitsch 1987; Waldick 1994). Even among those individuals that may be able to metamorphose, accelerated larval development leads to reduced size at metamorphosis which can ultimately lower future reproductive fitness (Semlitsch 1987; Semlitsch et al. 1988; Scott 1994) and increase desiccation risk (Shoop 1974; Wassersug 1975; Enge and Marion 1986). So despite the tendency for forestry-operations to increase the density of ephemeral ponds in an area, the reduced hydroperiod in these exposed areas can offset any overall benefit.

If these aquatic habitats manage to successfully rear larvae to metamorphosis, they must still be in close proximity to suitable terrestrial refuges and/or overwintering habitat. However, these ponds may act as 'stepping-stones' which facilitate recolonization of suitable habitats by allowing dispersal across intermittent patches of unfavourable habitat.

The presence of edge or submergent vegetation is important for egg deposition and as a refuge for larvae from predators and exposure to the elements. Vegetation can help slow evaporative losses, thereby extending hydroperiods. Forestry operations can affect the edge and submergent vegetation at aquatic habitats if the heavy machinery used during a harvest has destroyed the existing vegetation or compacted the soil. Consistent with these observations, survivorship of eggs, larvae, and newly metamorphosed individuals is greater at ponds when edge vegetation is present (Holomuzki 1986; Pechmann et al. 1989; Waldick 1994).

Forestry activities can directly affect aquatic habitat quality by the purposeful filling-in of wetlands during the construction of logging roads. Indirect degradation of aquatic habitats also occurs as erosion results in the accumulation of sediment in ponds. Streams in clearcuts receive increased sediment loads and organic inputs which can reduce the types and abundance of stream biota, including amphibians (Corn and Bury 1989a). Although studies of this sort have not been carried out for amphibians in eastern North America, it is probable that some stream dwellers, like *Desmognathus fuscus* and *Eurycea bislineata*, could be similarly affected.

Clearcutting

All studies of the effects of clearcutting on amphibians have observed lower amphibian densities and diversities in clearcut sites than in nearby, undisturbed forests (Blymyer and McGinnis 1977; Pough et al. 1987; Ash 1988; Raymond and Hardy 1991; Petranka et al. 1993). Scott and Ramotnik (1992) questioned the effectiveness of sampling methods and the correlations between amphibians and features of their habitat, but amphibian abundance is likely a function of habitat quality, because physiological limitations limit activity and survival in marginal or low-quality habitats (Jaeger 1978, 1980; Frisbie and Wyman 1992; Wyman and Jancola 1992). Statistically significant correlations between adult breeding populations at ponds in clearcuts and the distance to the nearest intact forest suggest that amphibians migrate from the surrounding forests to these available breeding ponds (Laan and Verboom 1990). Because increases in climatic extremes in clearcut sites reduce resource quality, some amphibians will necessarily be limited or excluded from these sites (Ash 1988; Buhlmann et al. 1988; Collins et al. 1988; Wyman 1988; Elmberg 1993). In northwestern Louisiana, Raymond and Hardy (1991) individually marked *Ambystoma talpoideum* in an area prior to its harvest. Following clearcutting, marked animals were discovered in adjacent, marginal habitats, and salamander mortality remained higher in the clearcut than in the surrounding marginal habitats for 3 yr.

In eastern North America, declines in *P. cinereus* density following clearcutting appear to be the combined result of direct mortality and emigration (Pough et al. 1987; Raymond and Hardy 1991; Petranka et al. 1993; Waldick 1994) and led Petranka et al. (1993) to project that 14 million salamanders are lost annually in western North Carolina as a direct result of forestry-associated mortality. Petranka et al. (1993) cited the territorial and philopatric behaviour of *P. cinereus* as evidence against emigration, but I found strong evidence of mass emigration from a clearcut by *P. cinereus* (Waldick 1994). I established drift fences in 6 habitats in southeastern New Brunswick, 1 of which was < 500 m from an area that was subsequently clearcut. The traps were opened early the following spring. At the clearcut site, I captured > 30 adults in a single day, coinciding precisely with the first warm spring rain. None of the other sites captured any salamanders during this interval. Moreover, the traps near the clearcut continued to capture small numbers of salamanders consistent with rainfall over the next 2 mo, and the other drift-fence sites failed to capture any animals.

The increased evaporative losses and reduced litter abundance at the soil-litter interface after clear--cutting may be the primary factor limiting the abundance of *P. cinereus* at clearcut sites (Pough et al. 1987; Ash 1988, 1994). The dry surface conditions force salamanders to remain below ground, where foraging opportunities are limited (Jaeger 1980; Pough et al. 1987; Feder and Burggren 1992). A similar avoidance response may occur in *A. maculatum*, which has been known to forego breeding in particularly dry years (Husting 1965) or experience complete reproductive failure (Semlitsch 1987; Semlitsch et al. 1988; Pechmann et al. 1989). In North Carolina, overall salamander densities were 5 × greater in mature, mixed-species forests than in clearcut areas (Petranka et al. 1993). In a north Florida Pine forest, the density and diversity of amphibians was reduced 8-fold following clearcutting (Enge and Marion 1986). Enge and Marion (1986) also noted that reduced amphibian abundance following clearcutting was, in part, the result of reduced reproductive success.

In Virginia, Blymyer and McGinnis (1977) found the density of salamanders in 60-to 100-yr old deciduous forests to be > 4 × higher than those of the regenerating cutover areas 2- to 7-yr old. The failure of amphibian populations to recover from tree-harvesting has also been observed by Pough et al. (1987) in central New York State, where recovery was not recorded during the 7 yr following clearcutting. However, 60-yr old, naturally regenerated stands contained amphibian densities indistinguishable from those in old-growth forests. The duration of the hiatus in amphibian recovery reflects the time required for habitat features to recover to levels that are acceptable for amphibians. Similar evidence of recovery lags have also been observed in other regions of eastern North America (Blymyer and McGinnis 1977; Pough et al. 1987; Petranka et al. 1993; Waldick 1994) as well as in redwood forests along the Pacific Coast of the United States (Bury 1983).

Plantation Establishment Versus Natural Regeneration

Some areas will be left to regenerate naturally following clearcutting activities. In eastern Canada, succession proceeds towards forests that typically include a mix of coniferous and broad-leaf deciduous trees and thus have greater levels of structural and ground surface complexity than similarly aged plantations (Freedman 1986). Thus, natural forests provide superior microenvironmental conditions than clearcut sites or conifer plantations (Blymyer and McGinnis 1977; Bury 1983; Pough et al. 1987; Waldick 1994). If natural regeneration is allowed to proceed, microhabitat features of relative humidity, leaf litter, and coarse woody debris (CWD) are able to recover to levels characteristic of mature, natural forests. If, however, these sites are reharvested before these features recover from the previous disturbance, recovery will be compromised.

Canopy closure and diverse structural and systematic floral composition create an environment with favourable ambient and ground surface conditions (i.e., moisture levels) and moderate temperature fluctuations. Because many habitat features fail to recover, plantations remain of poor quality for amphibians even after canopy closure (Pough et al. 1987; Waldick 1994). Drier microclimates and less stratified ground surface structure characteristic of conifer plantations (Table 1) reduce foraging and overwintering opportunities for amphibians in these sites (Heatwole 1962; Jaeger 1972, 1978, 1980; Pinder et al. 1992), undermining the suitability of plantations as year-round habitat. Nevertheless, certain species of amphibians are capable of recovering to densities similar to those found in natural forested stands (Pough et al. 1987; Bury and Corn 1988a; Petranka et al. 1993; Waldick 1994). *Notophthalmus viridescens*, for example, occupies closed canopy plantations and natural-forested stands (Pough et al. 1987; Waldick 1994). Once the vegetation recovers, recolonization of a site by amphibians may be partly dependent on the nature of the regeneration, features of the surrounding habitat, the presence of source populations for recolonization, and the tolerance ranges of the species (Andren 1994; Gulve 1994). Neither *N. viridescens* nor *P. cinereus* populations were found to be detrimentally affected by small-scale firewood harvesting (Pough et al. 1987). Bennett et al. (1980) have suggested that small-scale disturbances, like selective cutting, can be beneficial for some amphibians by creating additional niches. Slash piles in moist habitats, may serve as additional CWD habitat. Raphael (1988) documented such a trend in *Bufo boreas* and *Hyla regilla* in the Pacific Northwest. Both of these species are typically found in dry habitats and population sizes were found to increase by 45% and 160%, respectively. However, Raphael (1988) found that there were also amphibian species that declined as a result of the old-growth harvesting. Overall, those species which declined had restricted ranges or occupied threatened habitats while those that benefited had widespread distributions.

Most plantations are also subject to further habitat degradation due to successive tree harvesting at these sites. This further reduces the overall suitability of these sites for amphibians. Scott and Ramotnik (1992) described an exceptionally short-rotation cutting cycle and its probable detrimental impacts on amphibian populations. They concurred with Enge and Marion (1986) that the progressive depletion of leaf litter and large-dimension logs will reduce the suitability of these sites, thereby altering amphibian community structure.

Habitat Fragmentation

Forestry and agricultural land use is creating a patchwork landscape. The proximity of harvested sites to other types of habitat, and the size of these cut over areas, influence the distribution, densities, and community composition of amphibians (Enge and Marion 1986; Rosenburg and Raphael 1986). The regional persistence of a species depends directly on the relative rates of extinction and recolonization of subpopulations (Levins 1969). Habitats must be large enough to allow populations to survive environmental fluctuations and demographic stochasticity (Soulé 1983).

The dynamics of *N. viridescens* metapopulations illustrate the importance of the presence of multiple, connected populations (Gill 1978). This species requires relatively permanent aquatic breeding habitat for adults and larvae. However, the post-metamorphic, juvenile (i.e. red-eft) stage, which may last 1 to 7 yr, is entirely terrestrial. Different subpopulations at breeding ponds are either sources, sites with successfully reproducing local populations that provide new immigrants for recolonization of surrounding habitats, or sinks, where some or all of the life history stages are not being supported (Istock 1967; Wilbur 1980). It is the unpredictable nature of their aquatic habitats that dictates this metapopulation structure. Many amphibian species undergo dynamic patterns of local extinction followed by recolonization similar to that of *N. viridescens* (Corn and Fogleman 1984; Laan and Verboom 1990; Bogart 1992; Gulve 1994). The metapopulation structure, in which multiple breeding populations are present in a landscape, buffers a species against large-scale extirpation by increasing the probability that at least some 'source' populations will be are present at any given time (Gilpin 1987). However, individuals emigrating from 'source' populations must be able to gain access to these other habitats.

As natural forest patches continue to become less common, the distance between source patches and other suitable habitat patches will increase, which will reduce accessibility to new immigrants. Moreover, if the size of natural forest stands containing source populations becomes too small to support minimum viable populations, the existing source populations will also face increasing extinction threats. It is unclear what the minimum number of amphibians required to sustain a population is, or whether Simberloff's (1988) 50-500 rule will suffice for amphibians.

By reducing the area of clearcuts, dispersal across suitable areas would be improved. Juveniles and adults could cross unsuitable areas more readily, thereby improving their chances of locating favourable habitat. This is most important when colonization relies on dispersal by desiccation-sensitive juveniles. Enge and Marion (1986) proposed that amphibian recovery in relatively small cut-over areas will always occur more rapidly than in larger patches. However, overly small patches may be inadequate for supporting viable populations. Rosenburg and Raphael (1986) found that Douglas-fir forest patches of 10 ha supported lower densities of certain amphibian species. Such small patches may not contain self-sustaining populations and, therefore, represent sink populations.

The susceptibility of small, relatively isolated populations in small habitat patches to local extinction events such as disease, stochastic breeding failures, or catastrophic disturbances can be extremely high (Gilpin 1987; Lande 1987, 1988; Soulé et al. 1988; Blaustein et al. 1994b). Such periodic extinction events also occur more frequently and require longer periods for recolonization when patches also have low connectivity, and are distributed across a fragmented landscape (Soulé et al. 1988; Samson 1983; Andren 1994). The percentage of suitable area in a region is related to the patch size and density of available habitats over a landscape (Levins 1969; Lande 1987, 1988). The relatively high local extinction rates observed in small, isolated populations indicates that metapopulations composed of smaller subpopulations will be more susceptible to extirpation than those with larger, more stable subpopulations (Primack 1993).

Many amphibian species are limited to maximum dispersal distances of approximately 1 km, even under humid conditions (Dole 1965, 1971, 1972; Berven and Grudzien 1990; Sinsch 1990; Gibbs 1993). Conditions that are especially dry will limit dispersal across exposed areas, such as clearcuts, which may separate habitable patches. Some terrestrial salamanders disperse short distances (< 300 m; Semlitsch 1981; Parmelee 1993) while the more desiccation-tolerant *N. viridescens* has been observed as far as 1800 m from water (R. Waldick, pers. obs.). Dispersal distances of anurans have been recorded at around 1.5 km for a variety of species (Berven and Grudzien 1990; Sinsch 1990). When conditions for dispersal, or the distances across which dispersal must occur, are limiting, colonists from nearby populations may be unable to recolonize a suitable area, or else several years of habitat succession may be required to permit their migration across these disturbed areas (Berven and Grudzien 1990; Travis 1994).

The survivorship of individuals that successfully repopulate a site following a harvest has not been studied. The status of these populations may be tenuous (Dodd and Seigel 1991). Pough et al. (1987) found that adult *P. cinereus* were present in a 25 yr old conifer plantation, but the absence of juveniles suggested that immigration, not local reproduction, was sustaining the population. Bennett et al. (1980) and Pough et al. (1987) have similarly suggested that the presence of red-efts of *N. viridescens* in conifer plantations reflects habitat availability, not habitat suitability.

Conclusions

Habitat modifications associated with tree-canopy removal and the establishment of plantations have detrimental effects on amphibians. Whether the majority of forestry-associated population declines in amphibian species are the result of direct mortality (Petranka et al. 1993) or reduced survivorship (Enge and Marion 1986; Raymond and Hardy 1991; Waldick 1994) is unknown. Interspecific and intraspecific differences in the responses of amphibians suggest that a combination of anthropogenic and environmental factors is influencing these declines. In eastern North America, all forest-dwelling amphibian species appear to be acutely or chronically affected by forestry activities. While some species are only limited or excluded from disturbed areas until the canopy starts to regrow (eg. *N. viridescens*), others also require the re-establishment of additional habitat features (eg. *P. cinereus* and *Ambystoma* spp.). Leaf litter abundance, coarse woody debris availability, and microclimatic conditions are particularly important for the more susceptible species. Both plantations and clearcuts appear to be unsuitable for these species because these sites fail to provide these features in the same state as undisturbed forests. Moreover, silvicultural practices that progressively fragment natural, mixed-species forest or convert it to conifer plantations are reducing the total amount of available habitat for many amphibian species. Some degree of patch connectivity will offset the high local extinction rates that can occur in small, inadequate or highly isolated habitat patches.

Terrestrial salamanders appear the most susceptible to habitat disturbance by clearcutting, silviculture, and habitat fragmentation. Conserving the structure of amphibian communities requires that the habitat requirements of these most sensitive species be satisfied. This may be accomplished by leaving some relatively large (10 ha), undisturbed areas and by maintaining patches of naturally regenerated forest. Selective cutting, along with provisions directed at sustaining long-term inputs of coarse woody debris, will help mitigate the negative effects on these species.

Because the habitat requirements of amphibians at different life history stages can vary enormously, a species may respond to forestry in different ways at different times. Declines in *Scaphiopus holbrooki, Gastrophryne carolinensis,* and *Rana sphenocephala* in Florida were attributable to reduced juvenile recruitment following logging operations (Enge and Marion 1986), because the most severe effect of forestry on these species occurred at the larval and metamorphic stages of their development. The short hydroperiod and shallowness of artificially-created ponds used for breeding was insufficient to permit larvae to succesfully metamorphose. Massive juvenile mortalities caused by early pond drying may be common. To alleviate future recruitment failures, forestry operations should ensure that some shade trees are retained in cut-over areas to facilitate juvenile dispersal and metamorphosis. In New Brunswick, populations of larval and juvenile *P. crucifer* and *A. maculatum* in ponds with short hydroperiods (i.e., those in clearcuts) experienced high mortality rates (Waldick 1994). Concessions for retaining shade trees would also reduce evaporative losses at ponds, as would the retention of other types of edge vegetation. Perhaps the best case scenario is a situation where edge vegetation is retained and dugouts are purposely made deeper during road-building operations to help extend their hydroperiods. This sort of mitigation could substantially improve juvenile recruitment in exposed habitats.

Acknowledgments

Sincere thanks are extended to D. Green and 2 anonymous reviewers who provided excellent advice regarding the focus of the manuscript. Thanks also to R. J. Wassersug and B. Freedman for helping me by reviewing an early draft of this manuscript. Thanks are perhaps owed foremost to my friend Jack Gee, who made the study of amphibians an integral part of my life. My original research that appears, in part, within this manuscript was made possible through support from the Natural Sciences and Engineering Research Council of Canada.

Literature Cited

Allmon WD. 1991. A plot study of forest floor litter frogs, Central Amazon, Brazil. Journal of Tropical Ecology 7:503–522.

Andren H. 1994. Effects of habitat fragmentation on birds and mammals in landscapes with different proportions of suitable habitat: a review. Oikos 71:355–366.

Ash AN. 1988. Disappearance of salamanders from clearcut plots. Journal of the Elisha Mitchell Society 104:116–122.

Ash AN, Bruce RC. 1994. Impacts of timber harvesting on salamanders. Conservation Biology 8:300–301.

Aubry KB, Jones LLC, Hall PA. 1988. Use of woody debris by plethodontid salamanders in Douglas-Fir in Washington. In: Szaro RC, Severson KE, Patton DR, editors. Management of amphibians, reptiles, and small mammals in North America. Fort Collins, CO: USDA Forest Service. General Technical Report RM-166. p 32–37.

Beiswenger RE. 1988. Integrating anuran amphibian species into environmental assessment programs. In: Szaro RC, Severson KE, Patton DR, editors. Management of amphibians, reptiles, and small mammals in North America. Fort Collins, CO: USDA Forest Service. General Technical Report RM-166. p 109–128.

Bellis ED. 1962. The influence of humidity on wood frog activity. American Midland Naturalist 68:139–148.

Bennett SH, Gibbons JW, Glanville J. 1980. Terrestrial activity, abundance, and diversity of amphibians in differently managed forest types. American Midland Naturalist 103:412–416.

Berrill M, Bertram S, Wilson A, Louis S, Brigham D, Stromberg C. 1993. Lethal and sublethal impacts of pyrethroid insecticides on amphibian embryos and tadpoles. Environmental Toxicology and Chemistry 12:525–539.

Berven JA, Grudzien TA. 1990. Dispersal in the wood frog (*Rana sylvatica*): implications for genetic population structure. Evolution 44:2047–2056.

Bishop SC. 1941. The salamanders of New York. New York State Museum Bulletin 324:1–365.

Blaustein AR, Wake DB. 1990. Declining amphibian populations-a global phenomenon? Bulletin of the Ecological Society of America 71:127–128.

Blaustein AR, Hoffman PD, Hokit DG, Kiesecker JM, Walls SC, Hays JB. 1994a. UV repair and resistance to solar UV-B in amphibian eggs: a link to population declines? Proceedings of the National Academy of Sciences 91:1791–1795.

Blaustein AR, Wake DB, Sousa WP. 1994b. Amphibian declines: judging stability, persistence, and susceptibility of populations to local and global extinctions. Conservation Biology 8:60–71.

Blymyer MJ, McGinnis BS. 1977. Observations on possible detrimental effects of clearcutting on terrestrial amphibians. Bulletin of the Maryland Herpetological Society 13:79–83.

Bogart JP. 1992. Monitoring genetic diversity. In: Bishop CA, Pettit KE, editors. Declines in Canadian amphibian populations: designing a national monitoring strategy. Ottawa: Canadian Wildlife Service. Occasional Paper 76. p 50–52.

Bonin J. 1991. Effects of forest age on woodland amphibians and the habitat and status of stream salamanders in Southwestern Quebec [thesis]. Montréal: McGill University.

Bormann FH, Likens GE. 1979. Pattern and process in a forested ecosystem. New York: Springer--Verlag.

Brandon RA, Huheey JE. 1975. Diurnal activity, avian predation, and the question of warning and cryptic coloration in salamanders. Herpetologica 31:252–255.

Buhlmann KA, Pague CA, Mitchell JC, Glasgow RB. 1988. Forestry operations and terrestrial salamanders: techniques in a study of the Cow Knob salamander, *Plethodon punctatus*. In: Szaro RC, Severson KE, Patton DR, editors. Management of amphibians, reptiles, and small mammals in North America. Fort Collins, CO: USDA Forest Service. General Technical Report RM-166. p 4–10.

Burton TM, Likens GE. 1975. Energy flow and nutrient cycling in salamander populations in the Hubbard brook experimental forest, New Hampshire. Ecology 56:1068–1080.

Bury RB. 1983. Differences in amphibian populations in logged and old-growth Redwood forests. Northwest Science. 57:167–178.

Bury RB, Corn PS. 1988a. Douglas-Fir forests in the Oregon and Washington Cascades: relation of the herpetofauna to stand age and moisture. In: Szaro RC, Severson KE, Patton DR, editors. Management of amphibians, reptiles, and small mammals in North America. Fort Collins, CO: USDA Forest Service. General Technical Report RM-166. p 11–22.

Bury RB, Corn PS. 1988b. Responses of aquatic and streamside amphibians to timber harvest: a review. In: Raedeke KJ, editor. Streamside management: riparian wildlife and forestry interactions. Seattle: University of Washington Institute of Forest Resources. Contribution 59. p 165–181.

Cagle FR. 1942. Herpetological fauna of Jackson and Union counties, Illinois. American Midland Naturalist 28:164–200.

Canadian Forestry Service. 1992. Selected forestry statistics for Canada. Ottawa: Natural Resources Canada Policy and Economics Directorate. Information Report E-X-47.

Collins JP, Jones TR, Berna HJ. 1988. Conserving genetically distinctive populations: the case of the Huachuca tiger salamander (*Ambystoma trigrinum stebbinsi* Lowe). In: Szaro RC, Severson KE, Patton DR,editors. Management of amphibians, reptiles, and small mammals in North America. Fort Collins, CO: USDA Forest Service. General Technical Report RM-166. p 45–53.

Corn PS, Bury RB. 1989a. Logging in Western Oregon: Responses Of headwater habitats and stream amphibians. Forest Ecology and Management 29:39–57.

Corn PS, Bury RB. 1989b. Terrestrial amphibian communities in the Oregon Coast range. In: Ruggiero LF, Aubry KB, Carey AB, Huff MH, editors. Wildlife and vegetation of unmanaged Douglas-Fir forests. Portland, OR: USDA Forest Service. General Technical Report GTR-PNW-GTR-285. p 316–321.

Corn PS, Bury RB. 1991. Terrestrial amphibian communities in the Oregon Coast Range. In: Ruggiero LF, Aubry KB, Carey AB, Huff MH, editors. Wildlife and vegetation of unmanaged douglas-fir forests. Portland, OR: USDA Forest Service. General Technical Report PNW-GTR-285. p 305–317.

Corn PS, Fogleman JC. 1984. Extinction of montane populations of the northern leopard frog (*Rana pipiens*) in Colorado. Journal of Herpetology 18:147–152.

Dodd CK Jr, Seigel RA. 1991. Relocation, repatriation, and translocation of amphibians and reptiles: are they conservation strategies that work? Herpetologica 47:336–350.

Dole JW. 1965. Summer movements of the leopard frog (*Rana pipiens*) in Northern Michigan. Ecology 46:236–255.

Dole JW. 1971. Dispersal of recently metamorphosed leopard frogs, *Rana pipiens*. Copeia 1971:221–228.

Dole JW. 1972. Homing and orientation of displaced toads, *Bufo americanus*, to their home sites. Copeia 1972:151–158.

Dunson WA, Connell J. 1982. Specific inhibition of hatching in amphibian embryos by low pH. Journal of Herpetology. 16:314–316.

Elmberg J. 1993. Threats to boreal frogs. Ambio 22:254–255.

Enge KM, Marion WR. 1986. Effects of clearcutting and site preparation on herpetofauna of a North Florida flatwoods. Forest Ecology and Management 14:177–192.

Fauth JE, Crother BI, Slowinski JB. 1989. Elevational patterns of species richness, evenness, and abundance of the Costa Rican leaf-litter herpetofauna. Biotropica 21:178–185.

Feder ME, Burggren WW, editors. 1992. Environmental physiology of the amphibians. Chicago: University of Chicago Press.

Freda J, Dunson WA. 1984. Sodium balance of amphibian larvae exposed to low environmental pH. Physiological Zoology 57:435–443.

Freda J, Dunson WA. 1985. The effect of acidic precipitation on amphibian breeding in temporary ponds in Pennsylvania. Washington: USDI Fish and Wildlife Service. Biological Report 80(40.22).

Freedman B, Duinker PN, Morash R. 1986. Biomass and nutrients in Nova Scotian forests, and implications of intensive harvesting for future site productivity. Forest Ecology and Management 15:103–127.

Frisbie MP, Wyman RL. 1992. The effect of soil chemistry on sodium balance in the red-backed salamander: a comparison of two forest types. Journal of Herpetology 26:434–442.

Gibbs JP. 1993. The importance of small wetlands for persistence of local populations of wetland-associated animals. Wetlands 13:25–31.

Gill DE. 1978. Metapopulation ecology of the red-spotted newt, *Notophthalmus viridescens* (Ranfinesque). Ecological Monographs 48:145–166.

Gilpin ME. 1987. Spatial structure and population vulnerability. In: Soulé ME editor. Viable populations for conservation. Cambridge, UK: Cambridge University Press. p 125–139.

Graham RT, Harvey AE, Jurgensen MF, Jain TB, Tonn JR, Page-Dumroese DS. 1994. Managing coarse woody debris in forests of the Rocky Mountains. Ogden, UT: USDA Forest Service. Research Paper Int-RP-477.

Gulve PS. 1994. Distribution and extinction patterns within a northern metapopulation of the pool frog, *Rana lessonae*. Ecology 75:1357–1367.

Harmon ME, Franklin JF, Swanson FJ, Sollins P, Gregory SV, Lattin JD, Anderson NH, Cline SP, Aumen NG, Sedell JR, Lienkaemper GW, Cromack K Jr, Cummins KW. 1991. Ecology of coarse woody debris in temperate ecosystems. Advances in Ecological Research 15:133–302.

Heatwole H. 1962. Habitat selection and activity of the wood frog, *Rana sylvatica* Le Conte. American Midland Naturalist 66:301–313.

Heinen JT. 1993. Aggregations of newly metamorphosed *Bufo americanus*: tests of two hypotheses. Canadian Journal of Zoology 71:331–338.

Herrington RE. 1988. Talus use by amphibians and reptiles in the Pacific Northwest. In: Szaro RC, Severson KE, Patton DR, editors. Management of amphibians, reptiles, and small mammals in North America. Fort Collins, CO: USDA Forest Service. General Technical Report RM-166. p 216–221

Holomuzki JR. 1986. Effect of microhabitat on fitness components of larval tiger salamanders, *Ambystoma tigrinum nebulosum*. Oecologia 71:142–148.

Hurlbert SH. 1969. The breeding migrations and interhabitat wandering of the vermilion-spotted newt *Notophthalmus viridescens* (Rafinesque). Ecological Monographs 39:465–488.

Husting EL. 1965. Survival and breeding structure in a population of *Ambystoma maculatum*. Copeia 1965:352–361.

Istock CA. 1967. The evolution of complex life cycle phenomena: an ecological perspective. Evolution 21:592–605.

Jaeger RG. 1972. Food as a limited resource in competition between two species of terrestrial salamanders. Ecology 53:535–546.

Jaeger RG. 1978. Plant climbing by salamanders: periodic availability of plant-dwelling prey. Copeia 1978:686–691.

Jaeger RG. 1980. Fluctuations in prey availability and food limitation for a terrestrial salamander. Oecologia 44:335–341.

Jaeger RG. 1981. Birds as inefficient predators on terrestrial salamanders. American Naturalist 117:835–837.

Jaeger RG, Joseph R, Barnard DE. 1981. Foraging tactics of a terrestrial salamander: choice of diet in structurally simple environments. American Naturalist 117:639–664.

Jones KB. 1988a. Comparison of herpetofaunas of a natural and altered riparian ecosystem. In: Szaro RC, Severson KE, Patton DR, editors. Management of amphibians, reptiles, and small mammals in North America. Fort Collins, CO: USDA Forest Service. General Technical Report RM-166. p 222–227.

Jones KB. 1988b. Distribution and habitat associations of herpetofauna in Arizona: comparisons by habitat types. In: Szaro RC, Severson KE, Patton DR, editors. Management of amphibians, reptiles, and small mammals in North America. Fort Collins, CO: USDA Forest Service. General Technical Report RM-166. p 109–128.

Kimmins H. 1992. Balancing act: environmental issues in forestry. Vancouver, BC: University of British Columbia Press.

Kupferberg SJ, Marks JC, Power ME. 1994. Effects of variation in natural algal and detrital diets on larval anuran (*Hyla regilla*) life-history traits. Copeia 1994:446–457.

Laan R, Verboom B. 1990. Effects of pool size and isolation on amphibian communities. Biological Conservation 54:251–262.

Lande R. 1987. Extinction thresholds in demographic models of territorial populations. American Naturalist 130:624–635.

Lande R. 1988. Genetics and demography in biological conservation. Science 241:1455–1460.

Likens GE, Bormann FH, Pierce RS, Reiners WA. 1978. Recovery of a deforested ecosystem. Science 199:492–496.

Levins R. 1969. Some genetic and demographic consequences of environmental heterogeneity for biological control. Bulletin of the Entomological Society of America 15:237–240.

Loman J. 1988. Breeding by *Rana temporaria*: the importance of pond size and isolation. Memoranda Societatis pro Fauna et Flora Fennica 64:113–115.

Maiorana VC. 1978. Difference in diet as an epiphenomenon: space regulates salamanders. Canadian Journal of Zoology 56:1017–1025.

Parmelee JR. 1993. Microhabitat segregation and spatial relationships among four species of mole salamanders (genus *Ambstoma*). University of Kansas Museum of Natural History Occasional Papers 160:1–33.

Pechmann JHK, Scott DE, Gibbons JW, Semlitsch RD. 1989. Influence of wetland hydroperiod on diversity and abundance of metamorphosing juvenile amphibians. Wetland Ecology and Management 1:3–11.

Pechmann JHK, Scott DE, R.D. Semlitsch, J.P. Caldwell, L.J. Vitt, and J.W. Gibbons. 1991. Declining amphibian populations: the problem of separating human impacts from natural fluctuations. Science 253:892–895.

Petranka JW, Eldridge ME, Haley KE. 1993. Effects of timber harvesting on Southern Appalachian salamanders. Conservation Biology 7:363–370.

Pinder AW, Storey KB, Ultsch GR. 1992. Estivation and hibernation. In: Feder ME, Burggren WW, editors. Environmental physiology of the amphibians. Chicago: University of Chicago Press. p 250–274.

Pough FH, Wilson RE. 1977. Acid precipitation and reproductive success of *Ambystoma* salamanders. Water, Air, and Soil Pollution 7:307–316.

Pough FH, Smith EM, Rhodes DM, Collazo A. 1987. The abundance of salamanders in forest stands with different histories of disturbance. Forest Ecology and Management 20:1–9.

Primack RB. 1993. Essentials of conservation biology. Sunderland, MA: Sinauer Associates.

Raedeke KJ, editor. 1988. Streamside management: riparian wildlife and forestry interactions. Seattle: University of Washington Institute of Forest Resources. Contribution 59.

Raphael MG. 1988. Long-term trends in abundance of amphibians, reptiles, and mammals in Douglas-fir forests of northwestern California. In: Szaro RC, Severson KE, Patton DR, editors. Management of amphibians, reptiles, and small mammals in North America. Fort Collins, CO: USDA Forest Service. General Technical Report, RM-166. p 4–10.

Raymond LR, Hardy LM. 1991. Effects of a clearcut on a population of the mole salamander, *Ambystoma talpoideum*, in an adjacent unaltered forest. Journal of Herpetology 25:509–512.

Rosenburg KV, Raphael MG. 1986. Effects of forest fragmentation on vertebrates in Douglas-fir forests. In: Verner J, Morrison ML, Ralph CJ, editors. Wildlife 2000: modeling habitat relationships of terrestrial vertebrates. Madison, WI: University of Wisconsin Press. p 340–345.

Sadinski WJ, Dunson WA. 1992. A multilevel study of effects of low pH on amphibians of temporary ponds. Journal of Herpetology 26:413–422.

Samson FB. 1983. Minimum viable populations-a review. Natural Areas Journal 3:15–23.

Scott DE. 1994. The effect of larval density on adult demographic traits in *Ambystoma opacum*. Ecology 75:1383–1396.

Scott NJ Jr, Ramotnik CA. 1992. Does the Sacramento Mountain salamander require old-growth forests? In: Kaufmann MR, Moir WH, Bassett RL, editors. Old-growth forests in the Southwest and Rocky Mountain Regions: proceedings of a workshop. Fort Collins, CO: USDA Forest Service. General Technical Report RM-213. p 170–178.

Seale DB. 1980. Influence of amphibian larvae on primary production, nutrient flux, and competition in a pond ecosystem. Ecology 61:1531–1550.

Semlitsch RD. 1981. Terrestrial activity and summer home range of the mole salamander (*Ambystoma talpoideum*). Canadian Journal of Zoology 59:315–322.

Semlitsch RD. 1985. Analysis of climatic factors influencing migrations of the salamander *Ambystoma talpoideum*. Copeia 1985:477–489.

Semlitsch RD. 1987. Relationship of pond drying to the reproductive success of the salamander *Ambystoma talpoideum*. Copeia 1987:61–69.

Semlisch RD, Scott DE, Pechmann JHK. 1988. Time and size at metamorphosis related to adult fitness in *Ambystoma talpoideum*. Ecology 69:184–192.

Shoop CR. 1974. Yearly variation in larval survival of *Ambystoma maculatum*. Ecology 55:440–444.

Simberloff D. 1988. The contribution of population and community biology to conservation biology. Annual Review of Ecology and Systematics 19:473–511.

Sinsch U. 1990. Migration and orientation in anuran amphibians. Ethology, Ecology, and Evolution 2:65–79.

Soulé ME. 1980. Thresholds for survival: maintaining fitness and evolutionary potential. In: Soulé ME, Wilcox BA, editors. Conservation biology: an evolutionary ecological perspective. Sunderland, MA: Sinauer Associates. p 151–169.

Soulé ME. 1983. What do we really know about extinction? In: Schonewald-Cox C, Chambers S, MacBryde B, Thomas W, editors. Genetics and conservation: a reference for managing wild animal and plant populations. Menlo Park, CA: Benjamin/Cummings Publishing Company. p 111–124.

Soulé ME, Bolger DT, Alberts AC, Sauvajot RS, Wright J, Sorice M, Hill S. 1988. Reconstructed dynamics of rapid extinction of chaparral-requiring birds in urban habitat islands. Conservation Biology 2:75–92.

Spotila JR. 1972. Role of temperature and water in the ecology of lungless salamanders. Ecological Monographs 42:95–125.

Taub FB. 1961. The distribution of the red-backed salamander, *Plethodon c. cinereus*, within the soil. Ecology 42:681–698.

Travis J. 1994. Calibrating our expectation in studying amphibian populations. Herpetologica 50:104–108.

Wake DB. 1991. Declining amphibian populations. Science 253:860.

Waldick RC. 1994. Implications of forestry-associated habitat conversion on amphibians in the vicinity of Fundy National Park, New Brunswick [thesis]. Halifax: Dalhousie University.

Wassersug RJ. 1975. The adaptive significance of the tadpole stage with comments on the maintenance of complex life cycles in anurans. American Zoologist 15:405–417.

Wilbur HM. 1980. Complex life cycles. Annual Review of Ecology and Systematics 11:67–93.

Wyman RL. 1988. Soil acidity and moisture and the distribution of amphibians in five forests of southcentral New York. Copeia 1988:394–399.

Wyman RL. 1990. What's happening to the amphibians? Conservation Biology 4:350–352.

Wyman RL, Jancola J. 1992. Degree and scale of terrestrial acidification and amphibian community structure. Journal of Herpetology 26:392–401.

©1997 by the Society for the Study of Amphibians and Reptiles
Amphibians in decline: Canadian studies of a global problem. David M. Green, editor.
Herpetological Conservation 1:206–225.

Chapter 22

VULNERABILITY OF AMPHIBIANS IN CANADA TO GLOBAL WARMING AND INCREASED SOLAR ULTRAVIOLET RADIATION

Kristiina Ovaska

Renewable Resources Consulting Services Ltd, #214 Marine Technology Centre, 9865 West Saanich Road, Sidney, British Columbia V8L 3S1, Canada

Abstract.—Effects of climate warming on amphibians in Canada are likely to be both beneficial because of increases in temperature in winter and spring and deleterious because of decreased precipitation combined with elevated summer temperatures. Potential direct effects include increased physiological stress, decreased mobility, earlier reproduction in spring, more rapid development, shorter period of hibernation, and longer period of aestivation. Indirect effects, including habitat loss and fragmentation and changes in interactions with prey, competitors, predators and parasites, probably form the most serious adverse consequences of climate warming on amphibian populations. Potential direct effects of increased solar UV-radiation on amphibians consist of abnormal embryonic and larval development, damage to the eye and skin, and systemic effects through the suppression of the immune system. Indirect effects include changes in the relative abundance and species composition of competitors, predators and/or parasites, as well as toxic effects of chemicals produced or released as a result of photochemical reactions. Nocturnal and secretive habits of many amphibians protect them from exposure to solar UV. Pigmentation and an ability to repair UV-induced damage are likely to determine the sensitivity of those species that are regularly exposed to solar radiation at different phases of their life cycle. Most frog species in Canada are potentially vulnerable to increased solar UV, especially during the egg and larval stages, based on their potential exposure. About 50% of the salamander species in Canada are not exposed to solar radiation at early stages of their development. Species with relatively high vulnerability include *Bufo boreas, Rana pretiosa* complex, *B. hemiophrys, Ambystoma tigrinum, A. gracile*, and *Taricha granulosa*. Many amphibian species in Canada may be tolerant of the predicted changes associated with global warming, based on the results of a numerical scoring system. However, terrestrial plethodontid salamanders and *Ascaphus truei* had relatively high vulnerability scores.

Resumé.—Tout porte à croire que les effets du réchauffement climatique sur les amphibiens du Canada seront à la fois bénéfiques, puisqu'ils contribueront à la hausse des températures hivernales et printanières et délétères puisqu'ils auront pour effet de faire décroître les précipitations et d'augmenter parallèlement les températures estivales. Parmi les effets directs potentiels, mentionnons l'augmentation des stress physiologiques, une mobilité réduite, l'avancement de la date de reproduction au printemps, un développement plus rapide, une hibernation écourtée et une estivation prolongée. Parmi les effets indirects figurent la disparition et la fragmentation des habitats et des modifications dans les rapports avec les proies, les concurrents, les prédateurs et les parasites. Ce sont là, vraisemblablement, les conséquences les plus graves du réchauffement climatique sur les populations d'amphibiens. Les effets directs potentiels d'une augmentation des rayonnements solaires ultraviolets sur les amphibiens consistent dans des anomalies du développement larvaire et embryonnaire, des lésions oculaires et cutanées et des effets systémiques par suppression du système immunitaire. Les modifications dans l'abondance relative et la composition des espèces de concurrents, de prédateurs et(ou) de parasites ainsi que les effets toxiques des substances chimiques produites ou émises à l'issue de réactions photochimiques figurent au nombre des effets indirects. De nombreux amphibiens vivent la nuit et se cachent pendant le jour; ils sont donc protégés des rayonnements ultraviolets. La pigmentation de leur peau et leur pouvoir de guérison des lésions induites par les ultraviolets vont probablement influer sur la sensibilité des espèces régulièrement exposées aux rayonnements solaires, aux différents stades de leur cycle biologique. La plupart des grenouilles du Canada sont potentiellement vulnérables à l'augmentation des rayonnements ultraviolets, notamment au stade embryonnaire et larvaire; tout dépend du degré d'exposition. Environ 50% des espèces de salamandres du Canada ne sont pas exposées aux rayonnements solaires aux premiers stades de leur développement. Les espèces particulièrement vulnérables sont *Bufo boreas, Rana pretiosa* complexe, *B. hemiophrys, Ambystoma tigrinum, A. gracile*, et *Taricha granulosa*. Si l'on en croit les résultats d'un système de notation digital, de nombreuses espèces d'amphibiens du Canada devraient pouvoir tolérer les changements qui résulteront du réchauffement planétaire. D'après ce système, les salamandres pléthodontides terrestres et *Ascaphus truei* sont par contre très vulnérables.

Global warming and ozone depletion as a result of human activities are predicted to continue through at least the next 50 yr. The environmental effects of these processes can potentially be severe and widespread. Such effects are of particular concern in Canada, because increases in both temperature and solar ultraviolet (UV) radiation associated with ozone depletion are predicted to be amplified at middle and high latitudes in the northern hemisphere (Mitchell et al. 1990; Madronich 1993). Mitigation may be possible in some cases if these effects can be anticipated.

The Earth's surface and the lower atmosphere will warm as concentrations of CO_2 and other greenhouse gases (i.e. molecules that absorb infrared radiation thus preventing its escape into space) continue to increase (Mitchell et al. 1990; Government of Canada 1991; Gates 1993). In Canada, an increase of 3.5 to 4.2° C in average annual temperature is predicted if atmospheric CO_2 concentration, or its equivalent in other greenhouse gases, doubles (Government of Canada 1991). The anticipated increase in temperature is greatest at higher latitudes in winter and some models predict an elevation of 8 to 12° C in parts of the Canadian north during this season (Boer et al. 1992). The warming trend is predicted to be associated with changes in precipitation and evaporation patterns which, in turn, will influence soil moisture and hydrology at regional and local levels. The models predict a decrease in precipitation and soil moisture in summer, particularly in the interior of the continent. Although models differ with respect to the speed of the warming, the predicted change over decades or centuries is likely to be unprecedented, when compared to prehistoric climate changes of a similar magnitude over millennia during the last glacial episode (Davis 1989; Rizzo and Wiken 1992). A doubling of atmospheric CO_2 or its equivalent may occur as early as 2025 (Government of Canada 1991).

Concomitantly with climate warming, solar ultraviolet radiation within the UV-B range of the spectrum (280 to 320 nm) is predicted to continue to increase over the next several decades as a result of ozone depletion. Stratospheric ozone that shields the Earth's surface from biologically damaging UV-B radiation has declined at an alarming rate in recent years, possibly due to the release of ozone-destroying substances such as chlorofluorocarbons into the atmosphere by human activities. A decrease of 8% in ozone per decade occurred from 1979 to 1989 in the northern hemisphere (Madronich 1993) and up to a 35% increase in UV-B per year was recorded over Toronto in winter and early spring from 1989-1993 (Kerr and McElroy 1993). Although UV-B radiation constitutes only about 0.5% of the total solar radiation reaching the Earth's surface (Blumthaler 1993), it has a high potential to cause biological damage because the high-energy wavelengths can be absorbed by nucleic acids and proteins of living organisms (Tevini 1993). However, the consequences of increased UV-B radiation on natural populations and ecosystem processes are uncertain (Caldwell and Flint 1994).

Amphibians may be particularly vulnerable to atmospheric changes, due to the highly permeable skin of adults and eggs that are exposed directly to the environment (Duellman and Trueb 1986; Blaustein 1994). There is concern that those declines of amphibian populations that have taken place in relatively pristine habitats worldwide during the past few decades may be the result of subtle changes in atmospheric conditions, pollutants, or disease-causing organisms that are widely circulated by atmospheric or hydrological processes or by movements of animal vectors. Recent studies by Blaustein et al. (1994) indicate that the ambient levels of UV-B radiation affect hatching success of some amphibian species and may be a factor in the declines of their populations. However, hatching success of other species was not affected. These observations, together with the high variability in life history characteristics among different species, suggest that amphibians can be expected to differ greatly in both their sensitivity and vulnerability to global atmospheric changes.

This chapter presents an overview of potential effects of global warming and increased UV-B radiation on amphibians in Canada, based on a synthesis of existing information. In addition, I used a numerical scoring system to assess the relative vulnerability to the anticipated changes of all species of amphibians that occur regularly in Canada. My approach followed methodology of Herman and Scott (1992a,b, 1994), who used a similar system to assess the relative vulnerability of Nova Scotian vertebrates to global warming. The overall goal of this study is to provide background information and a framework for setting priorities in research, environmental monitoring, and conservation programs for amphibian species in Canada.

GLOBAL WARMING

Climate warming is likely to result in both beneficial and adverse effects on amphibians. Herman and Scott (1992a,b) concluded that the anticipated increases in temperature in winter are likely to be beneficial for Nova Scotian amphibians, whereas decreased precipitation in combination with elevated temperatures and evaporation in summer can be expected to be deleterious.

Elevated temperatures and associated changes in precipitation and soil moisture can affect amphibians directly, through increased physiological stress, decreased mobility, and changes in the timing and duration of reproduction, development, and periods of hibernation or aestivation, and indirectly through changes in habitat characteristics, food supply and interactions with competitors, predators, and parasites, including disease-causing microbial organisms. These direct and indirect pathways (Fig. 1) are likely to be interconnected through various feedback mechanisms. For example, the direct effects of temperature on the timing of spring migration may expose an amphibian population to a novel complement of predators or competitors. This effect will be superimposed on the independent responses of individual predators or competitors to climatic change.

The immediate responses of amphibians to climate warming are likely to consist of acclimation (phenotypic adjustments in tolerance or efficiency of performance following exposure to new environmental conditions) and changes in the timing of life history events. These adjustments may be sufficient to permit a population to persist within its former range. However, if the magnitude of the environmental change exceeds the ability of organisms to adjust phenotypically, persistence will depend on their ability to either shift their ranges along latitudinal or altitudinal gradients to areas where conditions are more favourable or to adapt genetically (Peters and Darling 1985; Cohn 1989; Fields et al. 1993). The anticipated speed of climate change will severely limit the degree of adaptation that is possible. Therefore, adaptation will most likely occur through variation that already exists within metapopulations rather than through the incorporation of novel mutations. Thus, the existing genetic and phenotypic variability within a species or population may largely govern their responses to climate warming (Fields et al. 1993).

Physiological Stress

Most northern and temperate species of amphibians can tolerate a broad range of temperatures and have a well-developed ability to acclimate to various temperature regimes (Brattstrom 1963; Zweifel 1977; Duellman and Trueb 1986). However, the range and limits in temperature tolerance vary according to species and stage of development. Amphibians are most susceptible to temperature extremes during early embryonic development, when the breadth of their temperature tolerance is relatively narrow (Zweifel 1977; Duellman and Trueb 1986). Abnormal development will occur if the limits of specific critical upper or lower thresholds are exceeded. However, the period of greatest sensitivity is brief, and acclimation within hours or days after laying can rapidly increase embryonic temperature tolerance of most species studied (Zweifel 1968, 1977).

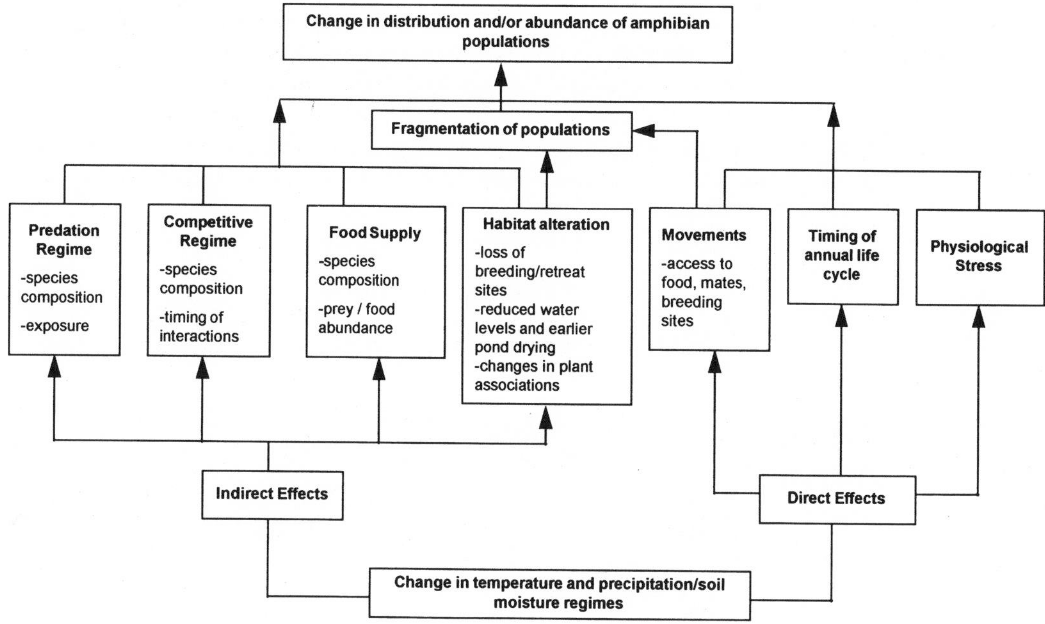

Figure 1. *Main pathways showing how global climate warming could affect amphibian populations.*

In general, elevated mean temperatures are unlikely to directly affect the survival of amphibian populations in Canada. The geographical ranges of most species extend well south into the United States, where temperatures are similar or higher than those predicted for Canada. However, if temperature extremes, in addition to temperature means, also increase under climate warming (Mitchell et al. 1990, Katz and Brown 1992), then adverse effects on embryonic development are a possibility. Cold-adapted species with narrow thermal tolerances such as *Ascaphus truei* (Brown 1975) or *Rana aurora* (Licht 1971), as well as all species that breed in early spring, are most likely to be affected by such changes.

Drier conditions in summer are more likely to result in adverse effects on amphibians in Canada than are elevated temperatures. Amphibians regulate water loss mainly by behavioural means and can avoid stress by changing their periods of activity or by selecting appropriate microhabitats. Morphological mechanisms, including sculpturing of the skin that increases rehydration rates, and physiological mechanisms such as production and storage of urea to increase water retention through osmosis, also play a role in regulating water balance and largely account for differences among species in their ability to tolerate dry conditions (Ruibal 1962; Lillywhite and Licht 1974; Lopez and Brodie 1977; Hillman 1980). Thus, species adapted to arid conditions, such as *Spea* spp. and *Ambystoma tigrinum*, may be better able to tolerate drier conditions associated with global warming.

Tailed frog, Ascaphus truei. *Photo by David M. Green.*

Although all terrestrial amphibians may be affected by drier summer conditions, the effects are likely to be most pronounced on lungless salamanders of the family Plethodontidae. Nine species are present in Canada, of which 5 are entirely terrestrial (Cook 1984). Because respiration takes place through their moist, highly vascularized skin, these salamanders are particularly susceptible to dehydration and their local distributions are constrained by the availability of moist microhabitats (Spotila 1972; Feder 1983). In addition to increased dehydration rates, the assimilation efficiency of plethodontid salamanders decreases with elevated temperatures. Jaeger (1981) concluded that *Plethodon cinereus* had difficulties in maintaining a positive energy balance during warm periods in summer. Physiological stress due to both susceptibility to dehydration and decreased assimilation efficiency of plethodontid salamanders at elevated temperatures are likely to be magnified in summer under global warming.

Mobility

Because the responses of amphibians to temperature and moisture conditions are largely behavioural, their activity and movement patterns may be affected by predicted climatic changes. Many amphibians respond to unfavourable conditions by retreating into underground burrows or sheltered microhabitats (Ruibal et al. 1969; Feder 1983). Low body temperatures and associated low metabolic rates of amphibians allow them to persist for months or even years in these retreats. This opportunistic strategy is particularly well developed in species such as *Spea* spp. and *A. tigrinum* inhabiting arid areas with unpredictable patterns of rainfall (Duellman and Trueb 1986). Due to metabolic rates that are the lowest of all vertebrates and low energy requirements, plethodontid salamanders also possess a remarkable ability to persist in environments with long periods of unfavourable conditions (Feder 1983).

Although survival of amphibians may not be affected directly, climate warming may constrain periods of surface activity that are necessary for reproduction and foraging. Dry conditions in summer decreased mobility and foraging success of *P. cinereus* (Jaeger 1980). Access to mates or reproductive sites may also be restricted by decreased mobility associated with dry conditions, because breeding activities of many species are linked to patterns of rainfall (Duellman and Trueb 1986). In contrast, some species increase rather than decrease their activity in response to dehydration, and this immediate response has been interpreted as an attempt to escape unfavourable conditions (Putnam and Hillman 1977). Increased movements could bring animals into contact with predators or competitors that differ from those normally encountered.

Timing of the Annual Life Cycle

Life history characteristics of many species of amphibians are highly responsive to local environmental conditions, as attested by intraspecific differences along geographical and altitudinal gradients (Duellman and Trueb 1986). Local differences in the timing of metamorphosis, growth rates, body size, or fecundity can have both a genetic and phenotypic component (Riha and Berven 1991). Within limits of their ability to adapt and acclimatize, amphibian populations are likely to respond to changing conditions by modifying their life history characteristics. Potential immediate changes to climate warming include earlier oviposition and more rapid larval development, earlier metamorphosis in response to both elevated temperatures and pond drying, longer periods of estivation, and later onset of hibernation.

Temperature and rainfall are the most important environmental cues for salamanders and anurans to initiate breeding (Duellman and Trueb 1986). Elevated temperatures, associated with winter rains under global warming, are likely to result in earlier breeding by those amphibians that oviposit in the spring. Present variation in the timing of breeding along latitudinal and altitudinal gradients of wide-ranging species can be used as a model for future expected changes. Variation spanning many months has been recorded for species at different altitudes or latitudes within their range (Eagleson 1976; Guttmann et al. 1991; Riha and Berven 1991). Guttmann et al. (1991) calculated an increase of 5.2 days in the mean date of breeding with every decrease of 1° in latitude for *R. sylvatica*. The timing of breeding also differs among years at the same location in response to changes in temperature, although such variation is usually of days or weeks rather than months (Licht 1973).

Changes in the timing of the annual life cycle with climate warming can have both beneficial and adverse effects on amphibians. Early breeding is thought to be beneficial for amphibians in northern and temperate climates because the time available for growth is increased (Hodge 1976, Duellman and Trueb 1986). Based on their studies on thermal adaptations of hylid treefrogs, John-Alder et al. (1988) suggested that the physical performance of late-breeding species at low temperatures, together with their embryonic temperature tolerances, curtail early breeding. Thus, climate warming would free late-breeding species from the constraints imposed by low temperatures at the northern part of their range. However, if the breeding phenology by different species is greatly compressed, increased temporal overlap among potential competitors or predators may result in adverse effects. Such a compression could take place if the magnitude of the temperature change is great and/or if physiological refractory periods prevent very early onset of breeding. Such refractory periods are pronounced in some amphibian species, such as *Ambystoma maculatum* (Sexton et al. 1990). Experiments in artificial ponds have demonstrated that intense competition can occur among anuran larvae of species that breed asynchronously under natural conditions (Wilbur and Alford 1985; Morin 1987). Decreased larval survival and mass at metamorphosis resulting from interspecific competition were linked to nutrient limitation. However, the inclusion of predatory newts, *Notophthalmus viridescens*, reduced the intensity of competition. Similar changes in interspecific interactions may occur under climate warming if the breeding phenology is altered. However, the changes in competition and predation regimes are likely to be more complex, reflecting the much greater number of interacting species under natural conditions.

Earlier breeding in spring and later onset of winter under global warming may increase time available for reproduction, development, growth, and assimilation of energy. However, these benefits may be offset by decreased precipitation and soil moisture in summer. Summer dryness may result in decreased activity and food availability, and pond drying may result in early metamorphosis of aquatic-breeding species. In many species, early metamorphosis is linked with a small body size, which in turn can decrease subsequent survival rates of juveniles (Wilbur 1980).

The ratio of benefits to costs associated with changes in the timing of life history is likely to vary among species of amphibians, based on their specific responses to the anticipated changes. Benefits are most likely to exceed costs at the northern limits of the distribution of those species that are limited by low temperatures and a short growing season.

Habitat Alteration and Fragmentation

Habitat alteration and fragmentation are probably the greatest threats that climate warming poses to Canadian amphibians. Decreased soil moisture, drying of ponds, and lowered water levels in summer, as well as changes in plant associations, can all alter the suitability of habitat to amphibians. Decreased soil moisture can result in fragmentation of habitat patches for most plethodontid salamanders and terrestrial phases of other amphibians, whereas a reduction in the number of seasonal and temporary ponds may limit breeding opportunities for species with aquatic eggs and larvae.

Many species of amphibians in Canada breed in shallow, seasonal or temporary ponds, and such water bodies are likely to be the first to disappear with increased aridity. These changes may be especially severe in the shallow wetlands of the prairies, where the intervals between periods of rainfall are highly unpredictable and where the expected changes in soil moisture are the greatest (Poiani and Johnston 1991). Species that breed in more permanent water bodies, such as *Rana catesbeiana*, *R. clamitans*, and *Ambystoma gracile*, may be somewhat buffered from the loss of breeding sites. However, if such events do occur, reduced survival is likely because larval development is prolonged and may span several seasons.

Northwestern salamander, Ambystoma gracile. *Photo by David M. Green.*

In addition to changes in moisture regimes, changes in plant associations, particularly in forest cover, may result in habitat loss or fragmentation for amphibians. In middle latitudes in the northern hemisphere, every 1° C increase in temperature is predicted to shift the temperature isotherms northwards by 100 to 125 km, and these shifts are likely to be accompanied with changes in forest distribution and species composition (Melillo et al. 1990; Government of Canada 1991). The long generation time of many forest trees limits their responses to climate change through dispersal, and forests in some areas may be susceptible to massive die-offs in response to changes in climate or associated invasions of herbivorous insects (Cohn 1989). Thus forest-dwelling amphibians are particularly vulnerable. Wild fires may also increase due to drier summers (Balling et al. 1992; Torn and Fried 1992), resulting in further fragmentation of habitats.

Populations of amphibians are inherently fragmented to various degrees due to their dependence on moist microhabitats, the presence of aquatic water bodies for breeding, and/or the presence of suitable diurnal or seasonal retreat sites. Climate warming is likely to amplify this fragmentation. Metapopulation dynamics are thought to be important in maintaining local adaptations and the persistence of amphibians in marginal habitats (Pechmann and Wilbur 1994). As a result of climate warming, reduced gene flow among subpopulations may lead to the extirpation of marginal or sink populations, as well as reducing genetic variance within subpopulations. Consequently, the potential of the metapopulation for adaptation may be reduced.

Shifts in Distribution Patterns

The number of amphibian species decreases along a latitudinal gradient in the northern hemisphere and many species reach the northern limits of their geographical distribution in Canada (Larsen and Gregory 1988). Short northern summers may provide insufficient time for reproduction, development and energy assimilation, and this and low winter temperatures are thought to contribute to the latitudinal pattern (Hodge 1976; Larsen and Gregory 1988). Species that are limited by these factors are likely to expand their ranges northward as the climate warms. However, similar northward expansion is unlikely by those species that require specialized conditions, such as certain soil types or vegetation associations, or whose population growth is limited by other factors.

Amphibians with narrow temperature tolerances may also exhibit altitudinal migrations in response to climate warming. These include *Dicamptodon tenebrosus, Gyrinophilus porphyriticus* and *Ascaphus truei*, all of which require cool, forested environments. As a result, shrinkage of the range and increased fragmentation of the populations are likely.

The rate of range shifts of animals in response to global warming is likely to be influenced by both the dispersal ability of particular species and the presence of natural and artificial barriers (Peters and Darling 1985). The dispersal movements of amphibians may be severely restricted by both drier summer conditions and by habitat fragmentation.

Changes in Communities and Ecosystems

Using a classification function model, Rizzo and Wiken (1992) concluded that changes in energy and moisture balances and productivity are likely to result in major shifts in the boundaries and areas of ecoclimatic provinces in Canada. The Arctic and Boreal Ecoclimatic Provinces were predicted to decrease in area whereas the Cool Temperate, Moderate Temperate, and Grassland Provinces were predicted to expand. In addition, they predicted that extensive arid regions not characterized by the present system of land classification would form in the continental interior. However, because of time lags due to reproductive cycles and dispersal of plants, these changes would not be realized for decades or centuries.

The predicted changes may be beneficial for those species that inhabit ecoclimatic regions that are expected to increase in size. *Acris crepitans, Ambystoma jeffersonianum*, and *A. texanum* have very limited geographical ranges in Canada associated with the Moderate Temperate Ecoclimatic Province. However, adverse immediate effects due to summer dryness or to high variability in temperature and moisture patterns may limit the realization of these potential long-term benefits. Conversely, the shrinkage of the boreal ecoclimatic province, in particular, may reduce habitat for northern species such as *Bufo hemiophrys*.

In contrast to the shifts in ecoclimatic zones, changes in plant and animal communities can occur relatively rapidly as a result of individual responses of different species and populations to climate change (Peters and Darling 1985; Cohn 1989; Melillo et al. 1990). Differential range shifts, adaptations, acclimation responses, and extirpations are likely to result in novel, largely unpredictable assemblages of communities. Consequently, a population is likely to be exposed to a different complement of competitors, predators and food sources. Changes in community structure are difficult to predict due to the large number of potential interactions, but some characteristics render a species or a population particularly susceptible to climate change. The most susceptible populations are those that are peripheral to the main distribution, geographically localized, highly specialized, poor dispersers, and/or inhabit montane, Arctic, or coastal areas (Peters and Lovejoy 1992). These same factors are likely to contribute to the vulnerability of amphibians.

Solar Ultraviolet Radiation

Increased UV-B fluxes can potentially expose living organisms to levels of UV radiation that are higher than those to which they are adapted. These relative increases, rather than the absolute levels of UV-B, may form the most important factor that governs the effects of ozone depletion and increased UV-B radiation on living systems. However, small increases in absolute values may also be important if certain critical thresholds are exceeded (Caldwell and Flint 1994). The potential direct effects of increased UV-B radiation on amphibians include abnormal embryonic and larval development, damage to the eye and skin, and systemic effects through the suppression of the immune system (Fig. 2). Indirect effects include changes in the relative abundance of predators, parasites and/or competitors, as well as toxic effects of chemicals produced or released from the sediments as a result of photochemical reactions.

Effects on Development

Amphibians may be most susceptible to UV-B radiation during early development (Worrest and Kimeldorf 1975, 1976; Blaustein et al. 1994). Most frogs in Canada and some salamanders, such as *Ambystoma* spp., oviposit in locations that are exposed to solar radiation and breed in the spring when the rate of increase in UV-B has been relatively high (Kerr and McElroy 1993). However, many species, including plethodontid salamanders, *Necturus maculosus*, and *A. truei*, lay their eggs in concealed sites that are not exposed to solar radiation.

The exposure and sensitivity of embryos of different species to solar UV-radiation can vary greatly even among those species that lay their eggs in exposed sites, due to the location of the eggs in the water column, shading provided by aquatic vegetation, turbidity of water, and the shape and size of the egg mass. Scully and Lean (1994) found that in lakes with high concentrations of dissolved organic carbon, only about 1% of UV-B wavelengths penetrated to the depth of 1 m, whereas in clear lakes the penetration was many m. Eggs on the surface of a globular mass may be affected, but those in the middle of the mass may be shielded by the surrounding eggs. Thus, eggs that are laid in thin sheets or strings, such as those of *R. catesbeiana* and *Bufo* spp., are probably more exposed than are eggs that are laid in large globular masses.

The sensitivity of different species can potentially be influenced by pigmentation and various repair mechanisms that operate after damage has occurred (Duellman and Trueb 1986; Tevini 1993). Melanin commonly occurs in eggs of those species that are exposed to intense sunlight (Duellman and Trueb 1986). Among European species, the heavily pigmented eggs of *Rana temporaria* were less susceptible to radiation damage than the paler eggs of *Rana esculenta* (Beudt 1930). However, little evidence is available on the role of pigments in protecting eggs and larvae from solar UV-induced damage.

Repair mechanisms for the DNA molecule, a primary locus of UV-B induced damage, include photoreactivation, excision, and post-replication repair (Tevini 1993). The relative importance of the different mechanisms can vary among groups of organisms. In mammals, light-independent excision repair and post-replication are the most important mechanisms. Photoreactivation repair occurs in eggs and larvae of some species of amphibians (Zimskind and Schisgall 1955; Worrest and Kimeldorf 1975, 1976; Blaustein et al. 1994), but the importance of the other 2 mechanisms is unknown.

Blaustein et al. (1994) showed that ambient levels of UV-B affect the hatching success of some species in the Oregon Cascade Mountains, whereas other species were more resilient. The hatching success of *Bufo boreas* and *Rana cascadae* was significantly lower in treatment groups exposed to solar UV-B radiation than in control groups where transmission of wavelengths < 315 nm was blocked by filters. The hatching success of *Hyla regilla* was unaffected. This difference among species was linked to differential activity of a photolyase that functions in the repair of UV-B induced damage to DNA through its action on cyclobutane pyrimidine dimers.

In contrast, ambient UV-B levels did not affect the hatching success of *Rana aurora* and *H. regilla* on Vancouver Island, British Columbia, in April 1995 (Ovaska et al. 1997). However, the hatching success of *R. aurora*, but not that of *H. regilla*, was reduced under levels of UV-B that were 15 to 30% above ambient levels at mid-day under clear skies. For *R. aurora*, the mean hatching success was 81.0% where UV-B was blocked by filters, 91.7% where ambient levels were maintained, and 56.0% where UV-B levels were enhanced by 15 to 30% (ANOVA, $F_{2,18} = 10.00$, $P = 0.001$). The corresponding values for *H. regilla* under the 3 treatments were 87.7, 71.8, and 87.1%, respectively ($F_{2,9} = 0.62$, $P = 0.559$). Under the enhanced UV-B treatment, the embryonic mortality of *R. aurora* was accompanied by the disintegration of the jelly that surrounds the ovum. The differences between these results and Blaustein et al. (1994) may have resulted from differences in ambient levels of UV-B due to elevation. The absolute levels of solar UV-B increase with altitude (Blumthaler 1993) and thus would have been greater at the high-elevation study sites of Blaustein et al. (1000 to 2000 m) than at the low elevation site on Vancouver Island (100 m). Geographic, seasonal and/or annual variation in levels of solar UV-B could also have contributed to the observed differences.

Intensive UV exposure in the laboratory can result in embryonic mortality and abnormal development in *Ambystoma mexicanum* (Sergeev and Smirnof 1939), *Triturus alpestris* (Kraft 1968), *Xenopus laevis* (Gurdon 1960), *Rana pipiens* (Higgins and Sheard 1926), *R. pipiens* and *R. catesbeiana* (Zimskind and Schisgall 1955), and *B. boreas* (Worrest and Kimeldorf 1975, 1976). However, the relationship between the observed damage and possible effects under natural conditions is not clear.

Amphibian larvae are also likely to be susceptible to UV-radiation. Worrest and Kimeldorf (1975, 1976) found that early survival before Gosner's (1960) developmental stage 30, which immediately precedes toe development, was not affected but that abnormalities became evident after this stage. The morphological changes in larvae included thickening and increased pigmentation of the cornea and abnormal curvature of the spine.

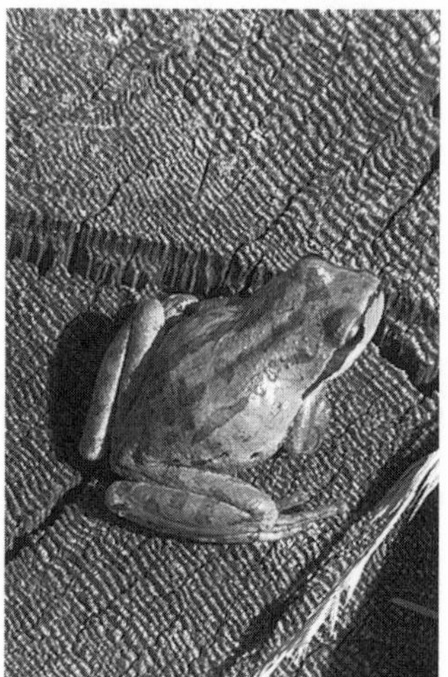

Pacific treefrog, Hyla regilla. *Photo by Stephen Corn.*

Effects on Tissues, Organs, and the Immune System

Because UV-B radiation penetrates only the superficial layers of the body, the primary effects on vertebrates are confined to the skin and eyes (van der Leun and Gruijl 1993). The relationship between non-melanoma skin cancers and UV-B exposure in humans is well established, and similar effects may occur in other vertebrates. The effects on eyes, such as acute inflammation of the superficial layers of the eye ("snow-blindness") and cataracts, are governed mainly by the level of exposure, as eyes have little resistance to UV radiation. The nocturnal and secretive behaviours of most salamanders and many frogs in Canada virtually eliminate their exposure to solar radiation. However, frogs such as *Rana septentrionalis*, *R. clamitans*, and *B. boreas* that regularly bask in the sun may be susceptible to UV damage to the skin and eyes.

Systemic effects of UV radiation could result in consequences throughout the body (van der Leun and Gruijl. 1993). In mammals, UV radiation can suppress components of the immune system, and increased susceptibility to some types of tumours and sensitization to certain allergens have been recorded. Van der Leung and Gruijl (1993) suggested that those infections or diseases that are mediated through the skin may increase with increased exposure to solar UV radiation. The moist and exposed skin of amphibians provides potential breeding grounds for fungal and bacterial infections, and amphibians produce a variety of antibiotic secretions to combat these organisms such as mycobacteria that can cause lesions of the skin (Duellman and Trueb 1986). Increased exposure to UV-B radiation could in theory alter this balance, either through its effects on the immune system of the host or on its independent effects on the disease-causing organisms.

Because many direct, physiological effects of UV-B radiation are cumulative (for example, development of cataracts and tumours, and immunosuppression), long-lived organisms are more likely to be affected than those with a short life span. Salamanders, in general, may live > 10 yr (Tilley 1977). However, the exposure of most salamanders to solar radiation is limited due to their activity and habitat use patterns. *Bufo boreas* can also live decades (Russell and Bauer 1993), whereas *A. crepitans* rarely live > 1 to 2 yr (Oldham and Campbell 1989). Cumulative damage to the reproductive cells through systemic effects is of particular concern, because such changes could affect many generations.

Water Chemistry

Increased UV radiation can affect amphibians indirectly through changes in water chemistry. The high energy wavelengths of UV radiation can result in the formation of a variety of free radicals through photochemical reactions (Lean et al. 1994). These reactions are mediated through dissolved organic carbon that is excited by UV radiation, leading to the formation of superoxide, hydrogen peroxide, hydroxyl radicals, and other highly reactive substances. Hydrogen peroxide can affect cycling of metals, particularly copper and iron, as well as the fate of pollutants in freshwater systems (Cooper and Lean 1992; Cooper et al. 1994). However, the biological consequences of these changes on freshwater ecosystems and on amphibians, in particular, are poorly known.

Ecological Interactions

In addition to effects on water chemistry, increased UV radiation may alter the environment of amphibians through its effects on prey or food organisms, predators or competitors. Caldwell and Flint (1994) concluded that the alteration of ecological interactions is the most likely consequence of increased solar UV-B on plant populations. Similarly, changes in ecological interactions are likely to be important for amphibians (Fig. 2).

Interactions Between Global Warming and Increased Solar Ultraviolet Radiation

Although global warming and increased UV-radiation result from separate human-induced changes in atmospheric processes, they can either re-enforce or ameliorate the effects of each other. Decreased cloud cover in summer, predicted under global warming, could increase the exposure of amphibians to solar UV. Elevated temperatures could either increase or decrease water turbidity, hence affecting the exposure of eggs and larvae of aquatic-breeding amphibians to UV radiation. In the Experimental Lakes Area in northwestern Ontario, the transparency of water has increased in response to a 2°C elevation in annual average water temperature over the past 20 yr (Schindler et al. 1990). The changes in water clarity were attributed to decreased import of dissolved organic matter from the surrounding areas. The concentration of dissolved organic carbon, in turn, greatly affects the transmission of UV radiation through the water column (Scully and Lean 1994).

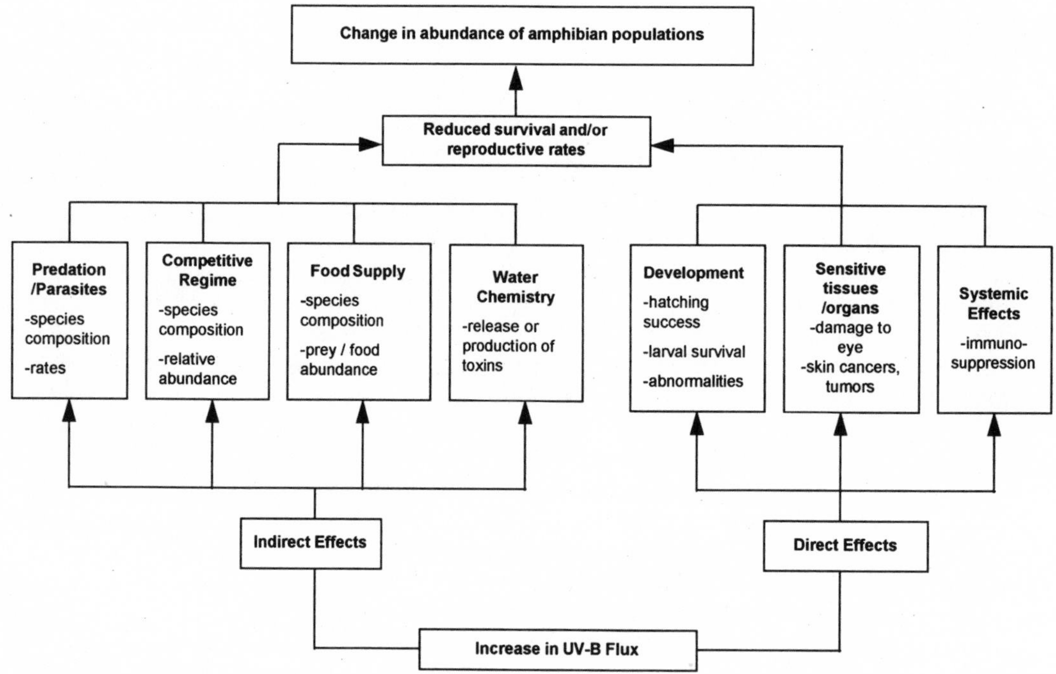

Figure 2. *Main pathways showing how increased UV-B radiation could affect amphibian populations.*

In addition to increasing exposure, interactions between elevated temperatures and UV radiation could change the rates of photochemical reactions in amphibian habitats. Both the production and decay of hydrogen peroxide by biological and chemical processes in freshwater systems occur more rapidly at elevated temperatures (Cooper et al. 1994). The consequences of these interactions to biogeochemical processes and living systems are poorly understood.

Analysis of Relative Vulnerability

To assess the relative vulnerability of the amphibians that occur in Canada to climate warming and to increased solar UV-radiation, I used a numerical scoring system modified from that developed by Herman and Scott (1992a,b, 1994) for Nova Scotia. I modified their categories to encompass the entire country and to consider amphibians specifically. I also added a 3rd category. Thus the new system consisted of population and distribution variables (= "biological variables" in the system of Herman and Scott 1992a,b), vulnerability to increased UV-radiation, and climatic sensitivity variables (Appendix).

Many frog species have declined or are suspected to be declining over a wide part of their range. Scores = 40 or 50 for the "population trend" variable were frequent (Fig. 3). In contrast, no widespread declines have been reported for salamanders, although many species have high vulnerability scores based on their other population/distribution variables (Appendix). The highest total scores (≥ 60) for the population/distribution variables were for *A. truei, A. crepitans, Spea intermontana, Spea bombifrons, R. pipiens, Bufo fowleri, B. cognatus, Plethodon idahoensis*, and *G. porphyriticus*. These scores represent vulnerability of the entire species in Canada, but local or regional differences may exist.

Most of the frog species, unlike the salamanders, are vulnerable to increased solar UV, with the egg and larval stages usually having higher scores than the adult stage (Fig. 4). *Bufo boreas, R. pretiosa* complex, and *B. hemiophrys* had the highest total UV scores. Other species with high total UV scores were *R. sylvatica, B. cognatus, B. fowleri, S. intermontana, A. tigrinum*, and *A. gracile*. Exposed eggs and larvae, diurnal activity of adults, and/or distribution patterns that extend to high altitudes or latitudes contributed to the high scores of these species.

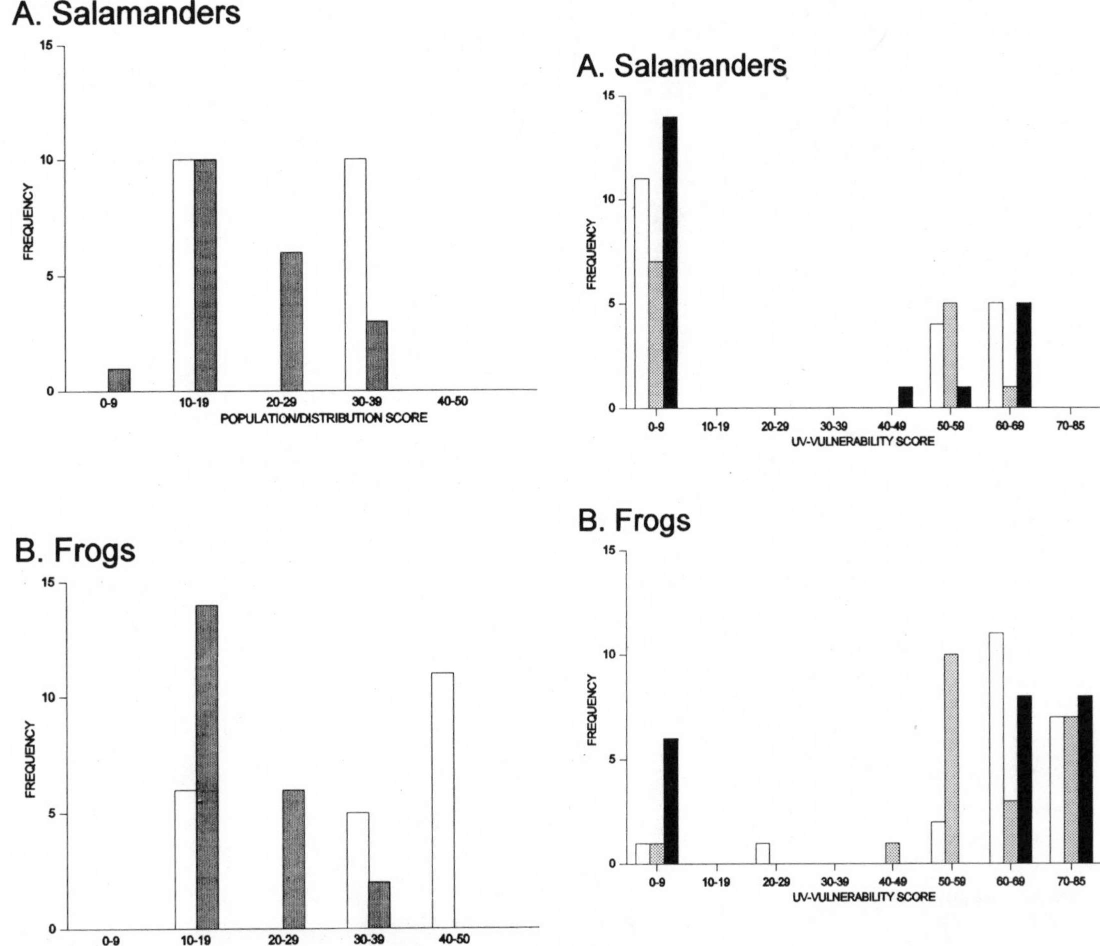

Figure 3. *Frequency distribution of "population and distribution" scores for salamander (top) and frog (below) species in Canada. High scores indicate increasing vulnerability. Open bar: "population trend" variable; closed bar: other "population and distribution" variables.*

Figure 4. *Frequency distribution of "UV vulnerability" scores for salamander (top) and frog (below) species in Canada. High scores indicate increasing vulnerability. Open bar: eggs; stippled bar: larvae; closed bar: adults.*

Unlike sensitivity to UV-radiation, which has been poorly studied, a large body of data exists on the sensitivity of amphibians to elevated temperatures and dehydration (Zweifel 1977; Spotila 1972; Feder 1983; Duellman and Trueb 1986), and I also considered habitat requirements and life history of the different species in assigning ratings. Most amphibian species may be relatively resistant to the predicted changes associated with climate warming (Fig. 5). However, all plethodontid salamanders as well as *A. truei* had high scores. For eggs and larvae, reduced summer soil moisture, lower summer water table, lower summer streamflow, increased spring/summer water temperatures, and increased winter/spring flooding all contributed to the total climatic sensitivity scores. Reduced summer soil moisture, lower summer water table, and reduced summer rainfall contributed to higher total scores for adults and metamorphosed juveniles.

The Grassland, Cordilleran, Interior Cordilleran, and Moderate Temperate Ecoclimatic Provinces (Ecoregions Working Group 1989) contained the greatest proportions of amphibian species with high vulnerability scores for the population/distribution variables (Table 1). The Grassland, Interior Cordilleran, and Cordilleran Ecoclimatic Provinces contained greatest proportions of species with high UV-sensitivity scores, whereas the climatic sensitivity scores were the highest in the Pacific Cordilleran,

Cordilleran, and Cool Temperate Ecoclimatic Provinces. Geographical ranges encompassing high elevation areas, together with other species-specific habitat use variables affecting the exposure of amphibians to solar radiation, account for the high UV vulnerability scores in the montane ecoclimatic zones. The pattern of high climatic sensitivity scores largely reflects the geographical distribution of terrestrial plethodontid salamanders which are likely to be among the most sensitive amphibians to global warming.

Discussion

The main advantage of the numerical scoring system is that it allows a systematic approach to the assessment of relative vulnerability among groups of organisms. The system also lends itself to the examination of a variety potential environmental threats on a wide range of spatial scales. The main disadvantage of the system is that the scores for the individual variables are subject to error depending on the accuracy of the available information. Because the scores are relative, data that are uneven in quality and level of detail among species could further bias the results. To avoid these potential sources of error, I attempted to maintain the scoring at a general level, while at the same time incorporating enough detail to achieve a reasonable resolution of vulnerability among species.

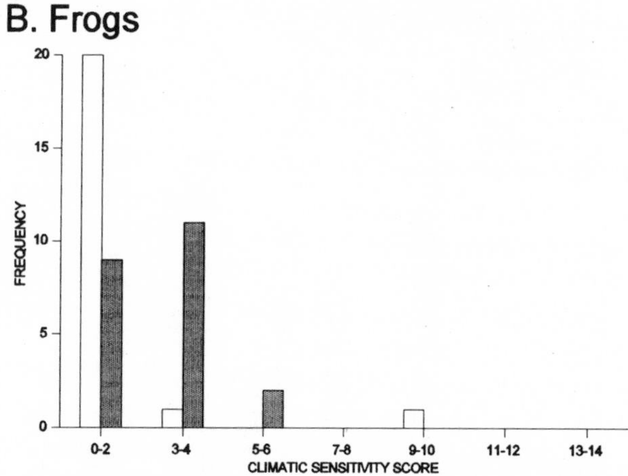

Figure 5. *Frequency distribution of "climatic sensitivity" scores for salamander (top) and frog (below) species in Canada. High scores indicate increasing vulnerability. Open bar: eggs and larvae; closed bar: adults.*

Table 1. *Amphibian species with high vulnerability scores in different ecoclimatic provinces in Canada.*

		Percent of species with high scores			
Ecoclimatic province	Number of species	Population and distribution variables (score ≥ 60)	UV vulnerability (score ≥ 200)	Climatic sensitivity (score ≥ 10)	Total (score ≥ 250)
Subarctic	7	14.3	14.3	0	14.3
Boreal	18	5.5	11.1	11.1	22.2
Grassland	8	37.5	25.0	0	50.0
Cordilleran	11	36.4	27.3	18.2	45.5
Interior Cordilleran	8	25.0	25.0	0	75.0
Pacific Cordilleran	11	9.1	18.2	27.3	18.2
Cool Temperate	18	5.6	11.1	16.7	11.1
Moderate Temperate	16	18.8	6.3	0	18.8

Paucity of data exists on population trends of different species of amphibians and on their sensitivity to current levels and predicted increases in solar UV-radiation at all stages of their life history. The scoring results can be used to focus efforts towards those species that are predicted to be most vulnerable. The variables scored can also be viewed as a set of hypotheses that can be experimentally tested. Because many potential effects of both climate warming and increased solar UV are indirect and occur through habitat or ecological relations, monitoring is also desirable. The anticipation and early detection of environmental effects of the predicted atmospheric changes are important to help shape policies and to facilitate the implementation of possible mitigative measures.

Acknowledgments

This study was made possible by funding from the Canadian Wildlife Service (Ottawa) and the Ecological Monitoring Coordination Office of Environment Canada, and I greatly appreciate their support. I also thank Ron Jakimchuk and Lennart Sopuck for their suggestions and comments on the scoring system and early drafts of the manuscript. Regional representatives of Declining Amphibians in Canada Task Force graciously supplied me with their unpublished manuscripts of studies on amphibian populations in Canada. I also wish to acknowledge the support that I have received from Renewable Resources Consulting Services Ltd. throughout the study.

Literature Cited

Balling RC Jr, Meyer G, Wells S. 1992. Climate change in Yellowstone National Park: is the drought-related risk of wildfires increasing? Climatic Change 22:35–45.

Beudt EL. 1930. Der Einfluss der Lichtes der Quarz-Queck-silber Lampe auf der Furschungs- und Larvenstadien verschieden Amphibien. Zoologische Jahrbucher Abteilung für Systematik, Okologie und Geographie der Tiere. 47:623–684.

Blaustein AR. 1994. Chicken Little or Nero's fiddle? A perspective on declining amphibian populations. Herpetologica 50:85–97.

Blaustein AR, Hoffman PD, Hokit DG, Kiesecker JM, Walls SC, Hays JB. 1994. UV repair and resistance to solar UV--B in amphibian eggs: a link to population declines? Proceedings of the National Academy of Science 91:1791–1795.

Blumthaler M. 1993. Solar UV measurements. In: Tevini M, editor. UV--B radiation and ozone depletion: effects on humans, animals, plants, microorganisms, and materials. Boca Raton, FL: Lewis Publishers. p 71–91.

Boer GJ, McFarlane NA, Lazare M. 1992. Greenhouse gas--induced climate change simulated with the CCC second--generation general circulation model. Journal of Climate 5:1045–1077.

Brattstrom BH. 1963. A preliminary review of the thermal requirements of amphibians. Ecology 44:238–255.

Brown HA. 1975. Temperature and development of the tailed frog, *Ascaphus truei*. Comparative Biochemistry and Physiology A 51:863–873.

Caldwell MM, Flint SD. 1994. Stratospheric ozone reduction, solar UV-B radiation and terrestrial ecosystems. Climatic Change 28:375–394.

Cohn JP. 1989. Gauging the biological impacts of the greenhouse effect. BioScience 39:142–146.

Cook FR. 1984. Introduction to Canadian amphibians and reptiles. Ottawa: National Museum of Natural Sciences.

Cooper WJ, Lean DRS. 1992. Hydrogen peroxide dynamics in marine and fresh water systems. Encyclopaedia Earth System Science 2: 527–535.

Cooper WJ, Shao C, Lean DRS, Gordon AS, Scully FE Jr. 1994. Factors affecting the distribution of H_2O_2 in surface waters. Advances in Chemistry Series 237:391–422.

Davis M. 1989. Lags in vegetation response to greenhouse warming. Climatic Change 15:75–80.

Duellman WE, Trueb L. 1986. Biology of amphibians. New York: McGraw--Hill.

Eagleson GW. 1976. A comparison of the life histories and growth patterns of populations of the salamander *Ambystoma gracile* (Baird) from permanent low--altitude and montane lakes. Canadian Journal of Zoology 54:2098–2111.

Ecoregions Working Group. 1989. Ecoclimatic regions of Canada: first approximation. Ottawa: Environment Canada, Canadian Committee on Ecological Land Classification. Ecological Land Classification Series 23.

Feder ME. 1983. Integrating the ecology and physiology of plethodontid salamanders. Herpetologica 39:291–310.

Fields PA, Graham JB, Rosenblatt RH, Somero GN. 1993. Effects of expected global climate change on marine faunas. Trends in Ecology and Evolution 8:361–366.

Gates DM. 1993. Climate Change and Its Biological Consequences. Sunderland, MA: Sinauer Associates. 280 p.

Gosner KL. 1960. A simplified table for staging anuran embryos and larvae with notes on identification. Herpetologica 16:183–190.

Government of Canada. 1991. The State of Canada's Environment. Ottawa: Environment Canada.

Gurdon JB. 1960. The effects of ultraviolet irradiation on uncleaved eggs of *Xenopus laevis*. Quarterly Journal of Microscope Science 101:299–305.

Guttman D, Bramble JE, Sexton OJ. 1991. Observations on the breeding immigration of wood frogs *Rana sylvatica* reintroduced in east-central Missouri. American Midland Naturalist 125:269–274.

Herman TB, Scott FW. 1994. Protected areas and global climate change: assessing the regional or local vulnerability of vertebrate species. Proceedings of the IVth world congress on national parks and protected areas, Caracas, Venezuela. Gland, Switzerland: IUCN—The World Conservation Union. p 13–27.

Herman TB, Scott FW. 1992a. Global change at the local level: assessing the regional or local vulnerability of vertebrate species to climatic warming. Developments in Landscape Management and Urban Planning 7:353–367.

Herman TB, Scott FW. 1992b. Assessing the vulnerability of amphibians to climatic warming. In: Bishop CA, Pettit KE, editors. Declines in Canadian amphibian populations: designing a national monitoring strategy. Ottawa: Canada Wildlife Service. Occasional Paper 76. p 46–49.

Higgins C, Sheard C. 1926. Effects of ultraviolet radiation on the early larval development of *Rana pipiens*. Journal of Experimental Zoology 46:333–343.

Hillman SS. 1980. Physiological correlates of differential dehydration tolerance in anuran amphibians. Copeia 1980:125–129.

Hodge RP. 1976. Amphibians and reptiles in Alaska, the Yukon and Northwest Territories. Anchorage, AK: Alaska Northwest Publishing Company.

Jaeger RG. 1981. Diet diversity and clutch size of aquatic and terrestrial salamanders. Oecologia 48:190–193.

Jaeger RG. 1980. Fluctuations in prey availability and food limitation for a terrestrial salamander. Oecologia 44:335–341.

John-Alder HB, Morin PJ, Lawler S. 1988. Thermal physiology, phenology, and distribution of tree frogs. American Naturalist 132:506–520.

Katz RW, Brown BG. 1992. Extreme events in a changing climate: variability is more important than averages. Climatic Change 21:289–302.

Kerr JB, McElroy CT. 1993. Evidence for large upward trends of ultraviolet--B radiation linked to ozone depletion. Science 262:1032–1034.

Kraft AV. 1968. Larvengestalt und Eigeveide-situs beim Alpenmolch (*Triturus alpestris*) nach Halbseiter UV-bestrahlung von Neurula- und Nachneurula-keime. Wilhelm Roux' Archiv für Entwicklungsmechanik der Organismen 160:259–297.

Larsen KW, Gregory PT. 1988. Amphibians and reptiles in the Northwest Territories. Occasional Papers Prince of Wales Northern Heritage Centre 3:31–51.

Lean DRS, Cooper WJ, Pick FR. 1994. Hydrogen peroxide formation and decay in lake waters. In: Helz GR, Zepp RG, Crosby DG, editors. Aquatic and surface photochemistry. Boca Raton, FL: CRC Press. p 207–214

Licht LE. 1973. Behavior and sound production by the northwestern salamander *Ambystoma gracile*. Canadian Journal of Zoology 51:1055–1056.

Licht LE. 1971. Breeding habits and embryonic thermal requirements of the frogs, *Rana aurora aurora* and *Rana pretiosa pretiosa* in the Pacific northwest. Ecology 52:116–124.

Lillywhite AP, Licht P. 1974. Movement of water over toad skin: functional role of epidermal sculpturing. Copeia 1974:165–171.

Lopez CH, Brodie ED Jr. 1977. The function of costal grooves in salamanders (Amphiuma, Urodela). Journal of Herpetology 11:372–374.

Madronich S. 1993. UV radiation in the natural and perturbed atmosphere. In: Tevini M, editor. UV-B radiation and ozone depletion: effects on humans, animals, plants, microorganisms, and materials. Boca Raton, LA: Lewis Publishers. p 17–61.

Melillo J, Callaghan TV, Woodward FI, Salati E, Sinha SK. 1990. Effects on ecosystems. In: Climate change, the ICPP scientific assessment. Cambridge, UK: Cambridge University Press. p 287–318.

Mitchell JFB, Manabe S, Meleshko V, Tokioka T. 1990. Equilibrium climate change - and its implications for the future. In Climate change: The IPCC scientific assessment. Cambridge UK: Cambridge University Press. p 139–173.

Morin PJ. 1987. Predation, breeding asynchrony, and the outcome of competition among treefrog tadpoles. Ecology 68:675–683.

Oldham MJ, Campbell CA. 1989. Status report on the Blanchard's cricket frog *Acris crepitans blanchardi* in Canada. Ottawa: Environment Canada, Committee on the Status of Endangered Wildlife in Canada.

Ovaska K, Davis TM, Novales Flamarique I. 1997. Hatching success and larval survival of the frogs *Hyla regilla* and *Rana aurora* under ambient and artificially enhanced solar UV radiation. Canadian Journal of Zoology (in press).

Pechmann JHK, Wilbur HM. 1994. Putting declining amphibian populations in perspective: natural fluctuations and human impacts. Herpetologica 50:65–84.

Peters RL, Darling JDS. 1985. The greenhouse effect and nature reserves. Bioscience 35:707–717.

Peters RL, Lovejoy TE, editors. 1992. Global warming and biological diversity. New Haven, CT: Yale University Press.

Poiani KA, Johnson WC. 1991. Global warming and prairie wetlands. BioScience 41:611–618.

Putnam RW, Hillman SS. 1977. Activity responses of anurans to dehydration. Copeia 1977:746–749.

Riha VF, Berven KA. 1991. An analysis of latitudinal variation in the larval development of the wood frog (*Rana sylvatica*). Copeia 1991:209–221.

Rizzo B, Wiken E. 1992. Assessing the sensitivity of Canada's ecosystems to climatic change. Climatic Change 21:37–55.

Ruibal R. 1962. The adaptive value of bladder water in the toad, *Bufo cognatus*. Physiological Zoology 35:218–223.

Ruibal R, Tevis L Jr, Roig V. 1969. The terrestrial ecology of the spadefoot toad *Scaphiopus hammondii*. Copeia 1969:571–584.

Russell AP, Bauer AM. 1993. The amphibians and reptiles of Alberta. A field guide and primer of boreal herpetology. Calgary, AB and Edmonton, AB: University Calgary Press and University of Alberta Press.

Sergeev AM, Smirnov KS. 1939. The color of eggs of Amphibia. Voprosy Ekologii i Biotsenologii 5:319–324.

Schindler DW, Beaty KG, Fee EJ, Cruikshank DR, Debruyn ER, Findlay DL, Linsey GA, Shearer JA, Stainton MP, Turner MA. 1990. Effects of climatic warming on lakes of the central boreal forest. Science 50:967–970.

Scully NM, Lean DRS. 1994. The attenuation of ultraviolet radiation in temperate lakes. Archiv für Hydrobiologie Beiheft 43:135–144.

Sexton OJ, Phillips C, Bramble JE. 1990. The effects of temperature and precipitation on the breeding migration of the spotted salamander (*Ambystoma maculatum*). Copeia 1990:781–787.

Spotila JR. 1972. Role of temperature and water in the ecology of lungless salamanders. Ecological Monographs 42:95–125.

Tevini M. 1993. Molecular biological effects of ultraviolet radiation. In: Tevini M, editor. UV-B radiation and ozone depletion: effects on humans, animals, plants, microorganisms, and materials. Boca Raton, FL: Lewis Publishers. p 1–15.

Tilley SG. 1977. Studies of life histories and reproduction in North American plethodontid salamanders. In: Taylor DH, Guttman SI, editors. Reproductive biology of amphibians. New York: Plenum Press. p 1–41.

Torn M, Fried JS. 1992. Predicting the impacts of global warming on wildland fire. Climatic Change 21:257–274.

Van der Leun JC, de Gruijl FR. 1993. Influences of ozone depletion on human and animal health. In: Tevini M, editor. UV-B radiation and ozone depletion: effects on humans, animals, plants, microorganisms, and materials. Boca Raton, LA: Lewis Publishers. p 95–119.

Wilbur HM. 1980. Complex life cycles . Annual Review of Ecology and Systematics 11:67–93.

Wilbur HM, Alford RA. 1985. Priority effects in experimental pond communities: responses of *Hyla* to *Bufo* and *Rana*. Ecology 66:1106–1114.

Worrest RC, Kimeldorf DJ. 1975. Distortions in amphibian development induced by ultra-violet-B enhancement (290-315 NM) of a simulated solar spectrum. Photochemistry and Photobiology 24:377–382.

Worrest RC, Kimeldorf DJ. 1976. Photoreaction of potentially lethal, UV--induced damaged to boreal toad (*Bufo boreas*) tadpoles. Life Science 17:1545–1550.

Zimskind PD, Schisgall RM. 1955. Photorecovery from ultraviolet-induced pigmentation changes in anuran larvae. Journal of Cellular and Comparative Physiology 45:167–175.

Zweifel RG. 1968. Reproductive biology of anurans of the arid southwest, with emphasis on adaptation of embryos to temperature. Bulletin of the American Museum of Natural History 140:1–64.

Zweifel RG. 1977. Upper thermal tolerances of anuran embryos in relation to stage of development and breeding habits. American Museum Novitates 2617:1–21.

APPENDIX

Variables Used to Score Vulnerability of Canadian Amphibian Species to Environmental Perturbations

Criterion . . . . . Score

I. Population and Distribution Variables (adapted from Herman and Scott 1992a,b, 1994)

A. Population trend. Overall, persistent change in abundance or distribution within the past 20 years. Includes both recorded population trends and indirect habitat loss or alteration.

1. Decrease recorded over a wide area (> 20% of the taxon's range in Canada) 50
2. Trend unknown but a decrease is suspected, or declines with unknown causes recorded but the population has recovered. . . 40
3. Local decreases or habitat loss recorded (affecting < 20% of the taxon's range in Canada), or no information on trend is available 30
4. Population trend unknown but taxon appears to be widespread and abundant over a large part of its range 15
5. Population size known to be stable or increasing 0

B. Other population and distribution variables.

1. Range size in Canada, delineated by outermost locations. Does not consider patchiness of distribution due to habitat suitability.

a. ≤ 5000 km² 10
b. 5001–50,000 km² 6
c. 50,001–500,000 km² 2
d. > 500,000 km² 0

2. Fragmentation of the overall range in Canada including dispersion of habitats, habitat requirements, and vagility. Not including fine-scale fragmentation due to habitat suitability.

a. Fragmentation of overall distribution. May indicate historical shrinkage of range.

i. ≥ 2 isolated parts or population isolated from rest of the range 2
ii. contiguous. 0

b. Patchiness of suitable large-scale habitat units, eg. due to forest or substrate types.

i. greatly fragmented 10
ii. moderately fragmented 6
iii. lower degree of fragmentation 3
iv. limited fragmentation. 0

3. Number of distinct habitats occupied during different phases of life history.

a. 3. 2
b. 2. 1
c. 1. 0

4. Population concentration.

a. Seasonal concentration.

i. large seasonal concentrations occur due to mating, oviposition or hibernation 3
ii. limited seasonal concentration 0

b. Length of oviposition season.

i. explosive breeder (< 1 month) 2
ii. prolonged breeder (≥ 1 month) 0

5. Reproductive specialization: conditions required for mating, oviposition or brooding.

a. Highly specialized 4
b. Moderately specialized 2
c. Limited specialization 0

6. Habitat specialization. Includes dependence on particular moisture regimes, plant or animal communities, substrate types and/or habitat structures.

a. Highly specialized 4
b. Moderately specialized 2
c. Limited or no specialization 0

7. Dietary specialization. Considers foraging behaviour, space, and food items.

a. Highly specialized 4
b. Moderately specialized 2
c. Limited or no specialization 0

8. Reproductive potential for recovery.

a. Average clutch size

i. < 30 5
ii. 30–99 4
iii. 100–999 2
iv. 1000–4999 1
v. ≥ 5000 0

b. Average age of female at first reproduction

i. ≥ 6 yr 4
ii. 4–5 yr 2
iii. 1–3 yr 0

II. Vulnerability to Solar UV Radiation

A. Exposure. Habitat and activity related variables that affect the exposure of amphibians to predicted increases in solar UV radiation at different stages of their life-cycle.

1. Eggs.

a. Microhabitat: The location of eggs in relation to their potential exposure to solar radiation. For aquatic eggs, turbidity of water and shading provided by aquatic vegetation is included.

i. floating on water surface 30

ii. eggs submerged in water column (scores are additive)

a) turbidity low 20
b) turbidity moderate or highly variable 10
c) turbidity high 5
d) shielding by vegetation low 10
e) shielding by vegetation moderate 5
f) shielding by vegetation high 0

iii. eggs not exposed to solar radiation, e.g. under rocks or inside logs 0

b. Egg mass characteristics: Includes shape of egg masses and the number of eggs/mass. These affect the proportion of eggs/clutch that are potentially exposed to solar radiation, as eggs in the center of a mass may be shielded by the surrounding eggs. Only scored if II.A.1.a > 0.

i. eggs laid singly, in strings or in flattened masses 5
ii. eggs laid in small clusters 3
iii. eggs laid in large globular masses 0

c. Shading provided by forest cover: Only scored if II.A.1.a > 0.

i. low (open habitats) 5
ii. moderate (semi-open, or found in both forested and open habitats) 3
iii. high (habitat densely forested) 0

d. Peak period oviposition. The percent increase in UV is predicted to be the greatest in January to April at latitudes above 45°N. Only scored if II.A.1.a > 0.

i. includes January to April 5
ii. includes May to June 3
iii. includes June but continues into July or later 1
iv. as above but does not include June 0

e. Proportion of range in Canada at high altitudes. UV levels increase with altitude from sea level. Scored only if Point II.A.1.a > 0.

i. > 10% above 1000 m; occurs above 2000 m 5
ii. > 10% above 1000 m; does not occur above 2000 m 3
iii. < 10% above 1000 m 0

2. Larvae.

a. Microhabitat of aquatic larvae: Includes turbidity of water, shading by aquatic vegetation and diel activity/exposure of larvae. Scores for turbidity and shading are additive.

i. frequently active/exposed during day

a) turbidity low 25
b) turbidity moderate or highly variable 15
c) turbidity high 5
d) shading by aquatic vegetation low 10
e) shading by aquatic vegetation high 0

ii. occasionally active/exposed during day

a) turbidity low . . . 15
b) turbidity moderate or highly variable . . . 10
c) turbidity high . . . 5
d) shading by aquatic vegetation low . . . 10
e) shading by aquatic vegetation high . . . 0

iii. secretive, shelter in bottom debris or under cover during day . . . 0

b. Shading provided by forest cover: Only scored if II.A.2.a > 0.

i. low (open habitats) . . . 5
ii. moderate (semi-open, or found in both forested and open habitats) . . . 3
iii. high (habitat densely forested) . . . 0

c. Seasonal pattern of larval period. The % increase in UV-B is predicted to be the greatest in January to April at latitudes above 45° N. Only scored if II.A.2.a > 0.

i. larval period includes January to April . . . 5
ii. larval period does not include January to April but includes May to June . . . 3
iii. larval period does not include January to June; 10% of range is above 55°N . . . 1
iv. larval period does not include January to June; % of range is above 55°N . . . 0

d. Proportion of range in Canada at high altitudes. UV levels increase with altitude from sea level. Scored only if II.A.2.a > 0.

i. > 10% above 1000 m; occurs above 2000 m . . . 5
ii. > 10% above 1000 m; does not occur above 2000 m . . . 3
iii. < 10% is above 1000 m . . . 0

3. Adults and metamorphosed juveniles.

a. Habitat use: Includes diel activity pattern and behaviour (e.g. fossorial). For species using both aquatic and terrestrial habitats, only the habitat with the highest use is scored. In the case of equal use, the habitat with the highest score is included.

i. diurnal surface activity in terrestrial habitats

a) frequent . . . 35
b) moderate . . . 20
c) infrequent or absent . . . 0

ii. diurnal activity in aquatic habitats

a) frequent, turbidity low . . . 35
b) frequent, turbidity high . . . 20
c) moderate, turbidity low . . . 20
d) moderate, turbidity high . . . 10
d) infrequent or absent . . . 0

b. Shading provided by forest cover. Only scored if II.A.3.a > 0.

i. low (open habitats) . . . 5
ii. moderate (semi-open, or found in both forested and open habitats) . . . 3
iii. high (habitat densely forested) . . . 0

c. Proportion of range in Canada at high latitudes: The rate of change in UV-levels in summer (mid-June to mid-August) is predicted to increase with latitude above 45°N. Scored only if II.A.3.a > 0.

i. > 10% above 55° N; occurs above 65° N. . . . 5
ii. > 10% above 55° N; does not occur above 65° N. . . . 3
iii. ≥ 90% below 55° N . . . 0

d. Proportion of range in Canada at high altitudes. UV levels increase with altitude from sea level. Calculations are based on the proportion of overall range at specific altitude ranges and known occurrence of the taxon at these altitude ranges.

i. > 10% above 1000 m; occurs above 2000 m . . . 5
ii. > 10% above 1000 m; does not occur above 2000 m . . . 3
iii. < 10% above 1000 m . . . 0

B. Sensitivity to UV radiation.

1. Experimental evidence of mortality or physiological damage: Measured in response to current levels or predicted increases in UV-radiation. Scored separately for eggs, larvae and adults/metamorphosed juveniles if II.A.(1–3).a > 0.

a. High . . . 40
b. Unknown . . . 30
c. Moderate . . . 20
d. Low . . . 0

2. Efficiency of repair mechanisms: Function in repair of molecular damage to DNA caused by UV-B exposure, including excision repair and photoreactivation mechanisms. Scored separately for eggs, larvae and adults/metamorphosed juveniles if II.A.(1–3).a > 0.

a. Low . . . 10
b. Unknown . . . 8
c. Moderate . . . 5
d. High . . . 0

III. Climatic Sensitivity (according to Herman and Scott 1992a,b, 1994).

The 5 direct effects below (A–E) were scored for the following 9 variables: reduced summer soil temperature, lower summer water table (drying of ponds and streams, lower water levels in larger water bodies), reduced summer rainfall, 4) lower summer stream flow, increased spring/ summer water surface temperature, increased summer soil temperature, increased winter water bottom temperature, reduced snow and ice cover, and increased winter/spring flooding.

A. Food supply or access to food.

1. High 3
2. Moderate 1
3. Low. 0

B. Dispersal mobility.

a. High 3
b. Moderate 1
c. Low. 0

C. Habitat reduction or loss of quality.

a. High 3
b. Moderate 1
c. Low. 0

D. Exposure to predation.

a. High 3
b. Moderate 1
c. Low. 0

E. Physiological stress.

a. High 3
b. Moderate 1
c. Low. 0

©1997 by the Society for the Study of Amphibians and Reptiles
Amphibians in decline: Canadian studies of a global problem. David M. Green, editor.
Herpetological Conservation 1:226–232.

Chapter 23

THE EFFECTS OF TEMPERATURE AND ACIDITY ON SPAWNING OF THE SPOTTED SALAMANDER, *AMBYSTOMA MACULATUM*, IN FUNDY NATIONAL PARK

DOUGLAS CLAY

Resource Conservation, Fundy National Park, P.O.Box 40, Alma, New Brunswick E0A 1B0, Canada

ABSTRACT.—Spotted salamanders (*Ambystoma maculatum*) were surveyed in Fundy National Park to monitor possible effects of acid precipitation. Egg mass counts were recorded in conjunction with environmental and physical parameters. Acidity of the monitored water bodies and precipitation did not appear to have changed significantly during the decade of monitoring although egg mass counts changed up to 5 fold. Spawning of *A. maculatum* fell within a narrow time period in early spring, the mean date of spawning in Fundy National Park was 9 May. A strong relationship ($r^2 = 0.76$) was found between spawning date and maximum daily air temperature during the spawning season.

RÉSUMÉ.—Des salamandres maculées (*Ambystoma maculatum*) ont été observées dans le Parc national Fundy afin de déterminer les effets possibles des précipitations acides. Le décompte des masses d'oeufs a été effectué en même temps que l'enregistrement des paramètres environnementaux et physiques. Aucun changement significatif n'a été observé dans le pH des précipitations et des masses d'eau surveillées pendant la période visée par l'étude alors que le décompte des masses d'oeufs a parfois changé du simple au quintuple. La fraie de *A. maculatum* a diminué pendant une courte période au début du printemps, la date moyenne de la fraie dans le Parc national Fundy étant fixée au 9 mai. Nous avons observé un rapport très marqué ($r^2 = 0,76$) entre la date de la fraie et la température ambiante maximale pendant la saison de la fraie.

In the 1970s there was increasing concern (Anonymous 1980) about acidification of Canadian lakes and rivers, especially in Nova Scotia and Ontario. Low pH values were correlated with reduced populations of fish, stressed plants, and possibly more widespread effects on both aquatic and terrestrial ecosystems. Vernal ponds are the aquatic ecosystems most likely to be sensitive to atmospheric acidification (Pierce 1985). Permanent ponds and lakes have varying amounts of inorganic buffers that are taken up from the soil and provide some protection against increasing acidity. In contrast, temporary ponds form when shallow depressions collect rainfall and snowmelt. This surface runoff does not have the same opportunity to absorb inorganic buffers from the soil. Thus, the acidity of water in temporary ponds more directly reflects the acidity of precipitation in a given locality. As the acidity of ponds increases, either the amphibian eggs fail to hatch or the larvae are deformed (Pough and Wilson 1977). Some amphibian species are more tolerant of acid conditions than others. The natural lakes in Fundy National Park are mostly < 4-ha in area and < 1-m deep. Kettle ponds and isolated woodland pools provide much of the habitat for spring-breeding amphibians which include the spring peeper (*Pseudacris crucifer*), wood frog (*Rana sylvatica*), green frog (*Rana clamitans*), American toad (*Bufo americanus*), spotted salamander (*Ambystoma maculatum*), blue-spotted salamander (*Ambystoma laterale*), and red-spotted newt (*Notophthalmus viridescens*).

Ambystoma maculatum was identified by Brooks (1991) and Clark (1991) as likely to be one of the prime candidate species for testing responses to changes in pH in Canada and by Glooschenko et al. (1992) as extremely sensitive to acid deposition in the Canadian Shield. Pierce (1985) recorded embryo mortality approaching 50% at pH slightly > 5, however Dale et al. (1985) found hatching success in controlled laboratory conditions ranging from 25 to 40% between pH 5 and pH 7. This species breeds in the temporary ponds in Fundy National Park, together with other, presumably more acid-tolerant, species. If counts of egg masses are assumed to provide an index of adult abundance, they would provide an indication of population change. Thus *A. maculatum* has become the "indicator of choice" for the monitoring program in Fundy National Park. Park staff monitored the reproductive output of *A. maculatum* in small ponds, expecting that this would provide an early, sensitive indicator of any negative effects of atmospheric acidification on the aquatic ecosystems of the park.

Spotted salamander, Ambystoma maculatum. *Photo by Jacques Brisson.*

MATERIALS AND METHODS

Study Area

Fundy National Park occupies 207 km² of southeastern New Brunswick on the upper Bay of Fundy coast (Fig. 1). The soils are mainly sandy, gravelly glacial till derived from andesitic and basaltic lavas (Aalund and Wicklund 1950). These soils contain a higher proportion of calcium than soils derived from granitic rock, leading to a "moderate" rating of acid sensitivity of the surface water and soils (Anonymous 1988). This calcium provides some buffering against any increase in acid deposition.

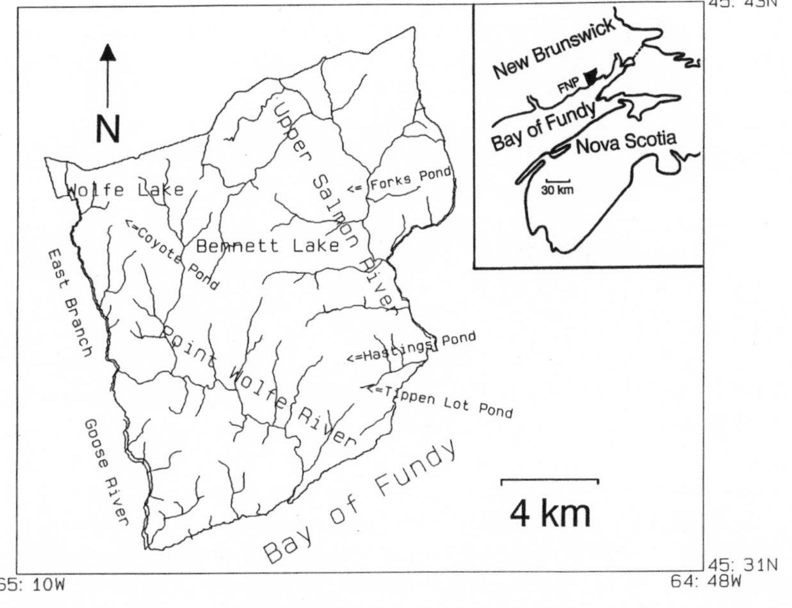

Figure 1. *Locations of 4 ponds monitored for breeding of* Ambystoma maculatum *in Fundy National Park, New Brunswick, 1985 to 1991.*

The 4 study ponds (Fig. 1) were selected because they fill primarily from precipitation and surface runoff. Three ponds were located inland from the coast on the rolling plateau (Fig. 1) at elevations between 305 m and 320 m above sea level. The 4th pond was located on a glaciofluvial terrace about 1.5 km from the coast at an elevation of 91 m. The forest cover in the park was predominantly spruce and fir with maple-birch hardwood stands on drier, inland ridges. Tippen Lot pond (725 m² area) is an isolated kettle depression and Coyote pond (660 m² area) is a borrow pit adjacent to Highway 114. In most years both hold water year round. In contrast, Hastings pond (30 m^2 area) and Forks pond (330 m^2 area) are both seasonal and dry up in late spring or early summer. None of the 4 ponds contains fish nor supports a permanent aquatic macrophyte community, however, mortality of amphibian eggs can be caused by caddis fly larvae, aquatic mites, or terrestrial vertebrate predators.

Sampling Protocol

After the ice melt each spring, at about 3-day intervals, the total number of *A. maculatum* egg masses was recorded. Sampling continued until all egg masses had hatched or disappeared. All egg masses were counted while circling the pond and returning to a designated starting point. No attempt was made to record mortality.

Beginning in 1988, in addition to circling the 2 larger ponds on foot, an inflatable boat was used to count egg masses in deeper water (McAlpine 1991). No attempt was made to calibrate this alteration in technique with the previous work. Therefore, this change must be considered when analyzing the egg mass counts for Coyote and Tippen Lot ponds over the 1st 7 yr of monitoring.

Water temperatures and pH were recorded at the time of each sample count. Water temperatures were taken near the surface and usually 50 cm from shore. Water samples were collected in 200 ml glass bottles which were thoroughly rinsed in pond water prior to taking the samples. pH of the water samples was measured using a Corning 145 pH meter. For comparison pH was measured in 4 permanent water bodies in the park (Wolfe Lake, Bennett Lake, and above the head of tide on the Point Wolfe and Upper Salmon rivers). The acidity of precipitation at the Alma Climate Station during the study period was measured from rainfall collected in a standard plastic rain gauge.

Daily air temperature records from the Alma Climate Station were combined for the 7-yr period (1985 to 1991) to provide a proxy for water temperature. The daily maximum air temperature of each year was compared to the 7-yr daily mean air temperature and the 7-yr minimum and 7-yr maximum air temperatures that occurred on that day (Fig. 2). The mean spawning dates of *A. maculatum* were overlaid on these figures to identify any patterns that might exist. Considering that rainfall will influence amphibian spawning, monthly rainfall from the Alma Climate Station was reviewed.

The mean spawning date, the date at which 50% of the egg masses have been laid, was determined for each pond in each year but the 2 seasonal ponds receiving low and irregular egg deposition were not used in the final estimate. Regression analysis was used to test that daily maximum air temperature was more significant in defining mean spawning date than either the mean or minimum daily air temperatures. The 7-yr mean daily maximum air temperature was taken as the base from which to determine warm from cool years. The daily deviations of the maximum air temperature from this base were summed for the pre-spawning period 1 April to 15 May. The resulting value was the deviation in degree-days from the base (mean) value.

Results

Pond Acidity

All 4 ponds were acidic with a mean annual pH of 5.5 ± 0.2 (Table 1). The mean pH for individual ponds changed little during the study. Hastings and Forks ponds tended to be more acidic than the other 2 ponds (mean pH = 5.3 compared to 5.6).

Number of Egg Masses

The number of egg masses laid in any given pond was highly variable between years (Table 2). Tippen Lot pond was the most variable with changes 60% occurring between successive years. Coyote pond was relatively stable in comparison; most changes in egg mass numbers were 20% between years. The variability observed here was not correlated with the pH of the pond water and was not consistent from pond to pond. Only in 1989, did all 3 ponds show a similar change in *A. maculatum* egg mass counts.

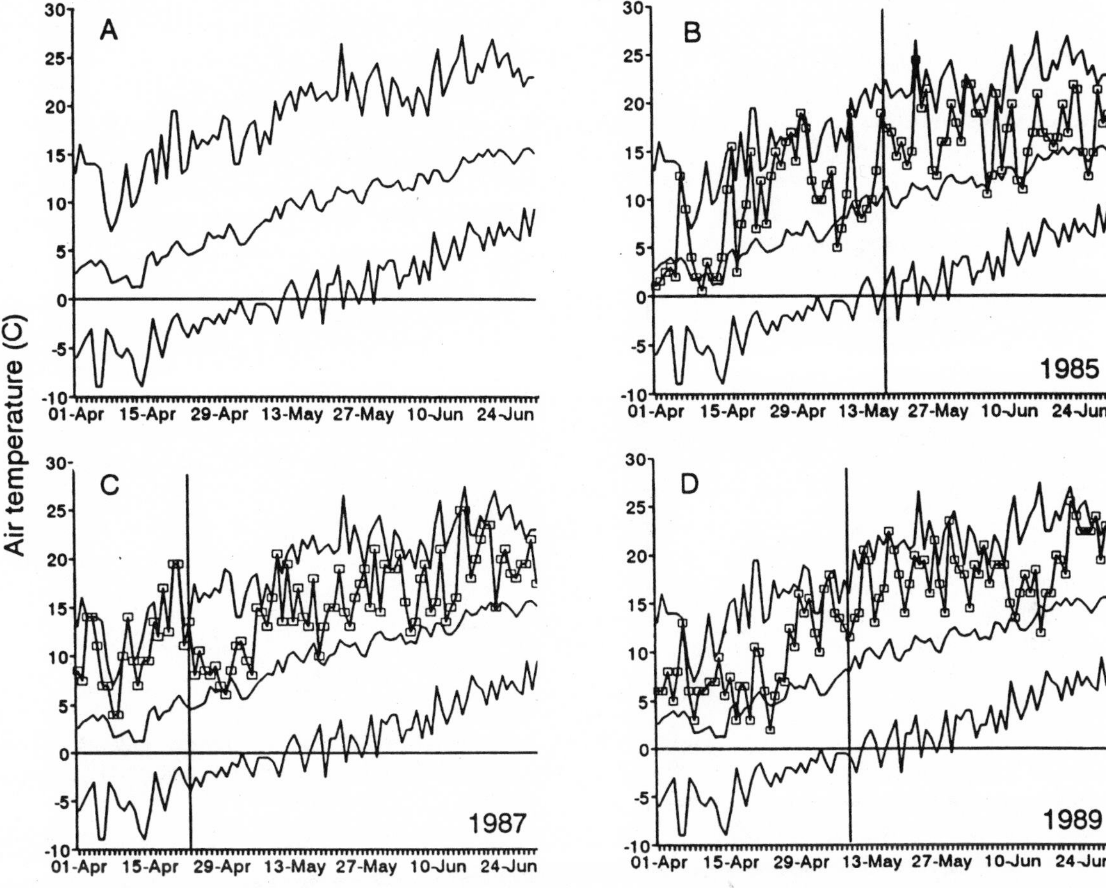

Figure 2. *A) The 3 solid lines are respectively (top to bottom), 7-yr maximum daily, mean daily, and 7-yr minimum daily air temperature. The date of mean spawning of* Ambystoma maculatum *(50% of egg masses laid) in Fundy National Park, New Brunswick, is represented by the vertical bar. The line with the open box symbols is the daily maximum air temperature for the respective year. B) year with late spawning, C) year with early spawning, D) year with average date of spawning.*

Table 1. *Mean annual pH of water in 4 ponds in Fundy National Park during April to June, 1985 to 1989.*

Pond	Mean pH 1985	1986	1987	1988	1989	Pond mean
Forks	–[a]	5.47	5.37	5.27	5.12	5.31
Hastings	5.32	5.42	5.52	5.25	5.35	5.37
Coyote	5.58	5.51	5.56	5.06	5.78	5.50
Tippen Lot	6.61	5.77	5.71	5.81	5.68	5.72
Yearly mean	5.50	5.54	5.54	5.35	5.48	5.48

[a]not sampled

Table 2. *Number of* Ambystoma maculatum *egg masses observed at 4 ponds in Fundy National Park.*

Year	Pond: Forks	Hastings	Coyote[a]	Tippen Lot[a]
1985	–[b]	2	232	406
1986	15	1	297	336
1987	12	0	238	438
1988	30	0	625	1065
1989	20	0	520	850
1990	3	0	680	234
1991	20	0	653	380

[a]boat used for counts after 1987

[b]not sampled

Mean Spawning Date

The average date of 1st spawning was 4 May with mean spawning on 9 May (Table 3). Although the major spawning activity took place over a relatively short time period, *A. maculatum* exhibited a protracted spawning season, often still arriving at the pond to lay eggs in mid-May. The latest egg-laying date observed was 29 May 1985. A regression of each year's mean spawning date expressed as Julian days on the season's total deviation from the base temperature resulted in coefficient of determination of 0.76 (Fig. 3).

Table 3. Dates of 1st spawning and mean spawning (50% of egg masses laid) of Ambystoma maculatum *in Fundy National Park.*

	Spawning date	
Year	1st	Mean
1985	20 May	22 May
1986	27 April	2 May
1987	18 April	27 April
1988	11 May	20 May
1989	8 May	13 May
1990	1 May	4 May
1991	4 May	7 May
Mean	4 May	9 May

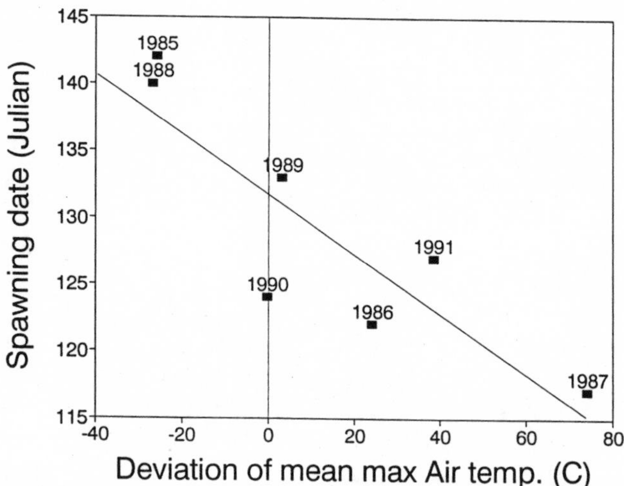

Figure 3. *The date of mean spawning of* Ambystoma maculatum *in Fundy National Park, New Brunswick, and the seasonal sum deviation, in degree-days, of the maximum daily air temperature from the 7-yr mean maximum.*

Discussion

Tippen Lot and Coyote ponds, being permanent, provide 2 of the more dependable amphibian breeding sites in Fundy National Park, secure from desiccation and predation by fish. Consequently, large numbers of adult frogs and salamanders use these 2 ponds, laying hundreds of egg masses in each (Table 2).

Despite the expectation that a decline in the number of egg masses in a given pond would be correlated with an increase in acidity, the results show the number of egg masses per pond can fluctuate 5 fold between years and between ponds, apparently independent of pH values of pond water. Many unknown factors affect the number of amphibian eggs in a given pond (Pechmann et al. 1991). In a recent study of *A. maculatum* in Massachusetts (Portnoy 1990), the most acidic ponds with pH < 4.5 contained no egg masses. Ponds with pH ≥ 4.5 were used as egg laying sites by *A. maculatum,* yet the number of egg masses per pond showed no relationship to the level of acidity. The critical value of acidity was pH 4.5, below which there was no breeding activity and above which breeding abundance and embryonic survival did not appear affected. These results suggest that there may be no advance indication from declining egg mass counts that the ponds in Fundy National Park are approaching a critical pH. This monitoring program concentrated on egg mass counts and did not monitor actual hatching success of amphibian larvae. Thus, adults may continue to return to an acidic pond for several years to mate and lay eggs, even if those eggs have poor hatching success. If so, a better indication of the first negative effects of increasing acidification would be the detection of interrupted development of the embryos and/or the presence of deformed larvae.

In Massachusetts, where spring is about a month earlier than in southern New Brunswick, 1st spawning of *A. maculatum* occurred 1 April, and mean date of egg deposition was 2 wk later (Stangel 1988), a month earlier than observed in New Brunswick. The mean date of spawning is determined mainly by temperature, however other factors can affect the relationship. The high rainfall in April and May 1990 provides one explanation for that apparent anomalous year in the date of mean spawning, about 10 days earlier than predicted from air temperature alone (Fig. 3).

Spotted salamander, Ambystoma maculatum, *larva. Photo by John Mitchell.*

Although bio-monitoring for effects of acid precipitation has long been advocated, a more direct approach of pH measurements of precipitation and surface waters might be more effective and provide a more timely warning of threats from acid deposition. Bio-monitoring, however, does provide more than just a warning of acid precipitation. It provides long-term biological data that otherwise would not be available and it provides an early warning of problems not yet popularized, or even articulated. UV-B radiation is now of global concern (MacIver et al. 1994), but in 1985, when this study was first proposed, it was not a major consideration and was not considered as a major threat by Parks Canada. Yet amphibians are likely to be affected by this aspect of atmospheric change.

For a given species, tolerance to acidity varies with geographic location; populations breeding where long-term natural acidity has existed seem to be more tolerant to acidity and continue to breed at pH < 5.0 (Dale et al. 1985; Portnoy 1990). The larger aquatic ecosystems of Fundy National Park are reasonably well buffered due to the relatively high calcium ion concentrations present in the lakes and rivers, ranging from 10 to 40 mg/l. Our survey ponds, however, have lower values, in the range of 3 to 8 mg/l. Because our amphibian populations have not been acclimatized to naturally acidic conditions below pH 5, they may become stressed at pH > 4.5. However despite this potential sensitivity, there does not appear to be cause for immediate concern as the pH values in Fundy National Park are nearly 6 and have not changed over the past decade.

Acknowledgments

I am continuing the work begun by previous Fundy National Park staff. The initial work and design was implemented by Resource Conservation staff, Steven Woodley, Eugene Taylor, later Laurie Collingwood, and presently Thane Watts. Their effort and foresight are appreciated. Stephen Flemming, Conservation Biologist, Fundy National Park and Donald McAlpine, Curator of Zoology, New Brunswick Museum, Saint John provided helpful critiques of an earlier version of this manuscript. Two unidentified reviewers offered much useful advice to provide direction to this manuscript.

Literature Cited

Aalund H, Wicklund RE. 1950. Soil survey report of southeastern New Brunswick. Third report of the New Brunswick Soil Survey. Fredericton, NB: Experimental Farm Service, Dominion Department of Agriculture and New Brunswick Agriculture.

Anonymous. 1980. The acid earth: soil and acidity. Harrowsmith 4:32–41.

Anonymous. 1988. Acid rain: a national sensitivity assessment. Ottawa: Inland Waters and Lands Directorate, Environment Canada. Environmental Fact Sheet 88-1.

Brooks RJ. 1991. Intensive monitoring: biology of amphibians in Canada. In: Bishop CA, Pettit KE, editors. Declines in Canadian amphibian populations: designing a national monitoring strategy. Ottawa: Canadian Wildlife Service. Occasional Paper 76. p 106–108.

Clark KL. 1991. Monitoring effects of acidic deposition on amphibian populations in Canada. In: Bishop CA, Pettit KE, editors. Declines in Canadian amphibian populations: designing a national monitoring strategy. Ottawa: Canadian Wildlife Service. Occasional Paper 76. p 63–66.

Dale J, Freedman B, Kerekes J. 1985. Experimental studies of the effects of acidity and associated water chemistry on amphibians. Proceedings of the Nova Scotia Institute of Science 35:35–54.

Glooschenko V, Weller WF, Smith PGR, Alvo R, Archbold JHG. 1992. Amphibian distribution with respect to pond water chemistry near Sudbury, Ontario. Canadian Journal of Fishes and Aquatic Science (Supplement 1):114–121.

MacIver D, Wheaton EE, Craine I, Scott P. 1994. Biodiversity and atmospheric change. In: Biodiversity in Canada: a science assessment for Environment Canada. Ottawa: Environment Canada. p 181–198.

McAlpine DF. 1991. The status of New Brunswick amphibian populations. In: Bishop CA, Pettit KE, editors. Declines in Canadian amphibian populations: designing a national monitoring strategy. Ottawa: Canadian Wildlife Service. Occasional Paper 76. p 26–29.

Pechmann HJ, Scott DE, Semlitsch RD, Caldwell JP, Vitt LJ, Gibbons JW. 1991. Declining amphibian populations: the problem of separating human impacts from natural fluctuations. Science 253:892–895.

Pierce BA. 1985. Acid tolerance in amphibians. BioScience 35:239–243.

Portnoy JW. 1990. Breeding biology of the spotted salamander *Ambystoma maculatum*, Shaw in acidic temporary ponds at Cape Cod USA. Biological Conservation 53:61–76.

Pough FH, Wilson RE. 1977. Acid precipitation and reproductive success of *Ambystoma* salamanders. Water, Air, and Soil Pollution 7:307–316.

Stangel PW. 1988. Premetamorphic survival of the salamander *Ambystoma maculatum*, in eastern Massachusetts. Journal of Herpetology 22:347–348.

Amphibians in decline: Canadian studies of a global problem. David M. Green, editor.
Herpetological Conservation 1:233–245.

Chapter 24

EFFECTS OF PESTICIDES ON AMPHIBIAN EMBRYOS AND LARVAE

MICHAEL BERRILL AND SUSAN BERTRAM

Biology Department, Trent University, Peterborough, Ontario K9J 7B8, Canada

BRUCE PAULI

Environment Canada, Canadian Wildlife Service, National Wildlife Research Centre, 100 Gamelin Blvd., Hull, Québec K1A 0H3, Canada

ABSTRACT.—We assessed the sensitivity of a variety of amphibian species to some of the more commonly used insecticides and herbicides sprayed for pest control in Canadian forests and croplands. We exposed embryos and newly hatched tadpoles of 1 or more of the 5 anurans, *Rana sylvatica, R. pipiens, R. clamitans, R. catesbeiana*, and *Bufo americanus*, and newly hatched larvae of the salamander *Ambystoma maculatum* to low concentrations of 3 insecticides permethrin, fenvalerate, and fenitrothion and 6 herbicides hexazinone, triclopyr, glyphosate, bromoxynil, triallate, and trifluralin. Embryos were unaffected by the experimental exposures, exhibiting the same high rate of hatching success and low rate of morphological abnormalities as the controls. However, newly hatched tadpoles were characteristically paralysed by the exposures, recovering gradually after termination of exposure periods. Concentrations of pesticides at levels higher than those inducing paralysis often resulted in death of the tadpoles. Some species were consistently more sensitive to the exposures than were others. We concluded that aquatic stages of amphibians are generally comparable to fresh-water fish in their vulnerability to exposure to low levels of pesticides.

RÉSUMÉ.—Nous avons évalué la sensibilité de plusieurs espèces d'amphibiens à quelques-uns des insecticides et herbicides les plus souvent utilisés pour la lutte antiparasitaire dans les forêts et cultures canadiennes. Nous avons exposé des embryons et des têtards qui venaient d'éclore d'une ou de plusieurs espèces d'anoures (*Rana sylvatica, R. pipiens, R. clamitans, R. catesbeiana*, et *Bufo americanus*) ainsi que des larves nouvellement écloses de salamandres (*Ambystoma maculatum*) à de faibles concentrations de 3 insecticides (perméthrine, fenvalérate et fénitrothion) et de 6 herbicides (hexazinone, triclopyr, glyphosate, bromoxynil, triallate et trifluraline). Les embryons n'ont pas été affectés par les expositions expérimentales et ont affiché le même taux élevé d'éclosion et le même faible taux d'anomalies morphologiques que leurs témoins. Toutefois, les têtards qui venaient d'éclore ont été paralysés par les expositions, récupérant graduellement leur mobilité lorsque l'exposition a pris fin. Des concentrations de pesticides supérieures à celles qui induisent la paralysie ont souvent entraîné la mort des têtards. Certaines espèces affichent une plus grande sensibilité aux expositions que d'autres. Nous en concluons qu'au stade aquatique, la vulnérabilité des amphibiens aux expositions de faibles concentrations de pesticides est généralement comparable à celle des poissons d'eau douce.

Application of pesticides in forestry and agriculture results in unavoidable contamination of adjacent water bodies. Amphibian communities of small ponds in the application areas are particularly vulnerable. For most of the pesticides in use or proposed for use in Canada, there may be some information on their effects on fish but rarely is there any on amphibians. We therefore assessed the sensitivity of the aquatic stages of a variety of amphibian species, mostly ranid frogs, to some of the more commonly used insecticides and herbicides sprayed for pest control in Canadian forests and croplands. In the process, we developed a standardized exposure test to compare the effects of pesticides on embryos and newly hatched tadpoles (Berrill et al. 1993, 1994, 1995).

Sublethal effects of pesticides on amphibians may influence the survival and success of exposed aquatic stages. If they are not directly lethal, the exposures may result in increased predation, reduced feeding, and delayed growth, and tadpoles may fail to reach metamorphosis at an appropriate time or size (Werner 1986). Although we have determined lethal levels for some of the pesticides we tested (Berrill et al. 1993, 1994), we attempted to keep exposure levels and durations within the range that could occur following standard applications. Therefore, we have focused on sublethal effects such as abnormal morphology, delayed growth, and abnormal behavior (Little 1990).

The characteristic avoidance response of tadpoles is to dart away rapidly when prodded, and we used it as an indicator of health. A tadpole not killed by exposure to pesticides may be weak and swim away only very slowly and for a short distance when prodded, or else be paralysed and unable to swim even a short distance when prodded. As a tadpole recovers from the effects of exposure, its avoidance response improves until it may again appear to be normal.

We selected pesticides of widely different applications and biological effects (Table 1). Our most intense work involved the pyrethroid insecticides permethrin and fenvalerate, the organophophorus insecticide fenitrothion, and the herbicide triclopyr. We also tested the herbicide hexazinone and have completed preliminary experiments using 4 other herbicides: glyphosate, bromoxynil, triallate, and trifluralin.

Although standardized experiments comparing the sensitivity of eggs and tadpoles of different species of amphibians to these pesticides have not been done prior to ours, extensive work on the effects of low pH exposure, usually exposing embryo stages, indicates that different species may differ considerably in sensitivity. For example wood frog, *Rana sylvatica*, embryos are relatively more tolerant of low pH exposure than those of American toads, *Bufo americanus*, and spotted salamanders, *Ambystoma maculatum* (Clark and LaZerte 1985; Freda 1986). One of our objectives was to continue this work, comparing the sensitivity of different amphibian species to different pesticides in order to determine whether species are consistently different in their sensitivity and what the range in sensitivity might be.

For many of our tested pesticides, reasonable data exist concerning the sensitivity of fresh-water fish to lethal and sublethal exposures. Fresh-water fish are often, but not always, more sensitive to pesticide exposures than birds and mammals and, in some cases, may even be as sensitive as the target insects (Edwards et al. 1986). Aquatic stages of amphibians are also likely to be similarly sensitive (Berrill et al. 1993). Furthermore, different developmental stages of fresh-water fish may differ in response to pesticide exposures (Holdway and Dixon 1988) and the same may be true of amphibians. We concentrated on a comparison of the sensitivity of the embryos and newly hatched tadpole stages of various amphibian species, but the sensitivity of other stages in development, such as premetamorphic tadpoles, should also be assessed.

Northern leopard frog, Rana pipiens. *Photo by Martin Ouellet.*

Assuming that amphibians are sensitive to low concentrations of pesticides and that lethal and sublethal effects are measurable, we tested several hypotheses: 1st, that amphibians are approximately as sensitive as fresh-water fish to exposure to low levels of pesticides; 2nd, that amphibian species differ from each other in sensitivity to pesticide exposure in consistent and predictable ways; and

3rd, that amphibian developmental stages, embryos and tadpoles, differ in sensitivity to pesticides. Our major objective was to predict the sensitivity of the amphibian community to levels of pesticides that are likely to occur following standard operational applications of the compounds.

Materials and Methods

Egg Collections and Tadpole Culture

We collected freshly laid egg masses of *R. sylvatica*, northern leopard frogs (*Rana pipiens*), *B. americanus*, *A. maculatum*, green frogs (*Rana clamitans*), and bullfrogs (*Rana catesbeiana*). These species lay their eggs during the spring and summer in the ponds and lakes of south central Ontario. *Rana sylatica* breeds earliest in spring and *R. catesbeiana* breeds last. Eggs were cultured in the lab until embryos reached mid-neurula stage for embryo exposure tests, or until 1 to 2 days after hatching, for exposure tests of newly hatched tadpoles or larvae. Surviving tadpoles were cultured for a further 1 to 2 wk recovery period, fed with small amounts of boiled lettuce.

Experimental Conditions

Embryos and tadpoles were exposed to 3 insecticides, permethrin, fenvalerate (both pyrethroids), and fenitrothion (an organophosphorus insecticide), and 6 herbicides, hexazinone, triclopyr ester, triallate, trifluralin, glyphosate, and bromoxynil. Exposure levels, with the exception of hexazinone, were intended to overlap with those that could occur in natural water bodies (Tables 1 and 2). Experiments with the pyrethroids were carried out during the breeding seasons of 1989 and 1990, and experiments with fenitrothion and the herbicides were carried out during the breeding seasons of 1992 and 1993. All experimental exposures occurred in static water. Nominal concentrations for the various treatments were made by adding desired amounts of stock solutions to hard, filtered water from the Otonabee River in Peterborough Co., Ontario.

Table 1. *The major Canadian use and effects of the pesticides used in amphibian exposure experiments.*

Pesticide	Major use	Action	Water solubility* (maximum levels observed following spraying)	Aquatic fate (half life)
Insecticides:				
Pyrethroids	cereals, fruit	affects nervous system, sodium channels	low (sprayed with solvent and synergist), 0.1 ppb[8]	1–3 days in clear water, longer in pond water
Fenitrothion	spruce budworm, hemlock looper	cholinesterase inhibitor	low (sprayed with solvent and emulsifier), 0.2–2.5 ppm[2,3]	24–48 hr in warm, shallow sunlit water, longer in cool or shaded ponds
Herbicides:				
Hexazinone	forestry	probable photosynthesis inhibitor	high, 14 ppm[1]	degrades slowly
Triclopyr ester	forestry, industrial sites	plant growth inhibitor	high, 2.5 - 4.0 ppm[7]	6–24 hr in warm sunlit water
Triallate	cereals	mitotic poison, electron-transport inhibitor	low (4 mg/l) sprayed with solvent, 0.1 ppb[5]	3–15 days
Trifluralin	canola, root crops	inhibits cell elongation; neurotoxin	low (1 mg/l) sprayed with solvent, 0.24 ppb[4]	1 hr in shallow, sunlit water
Glyphosate	cereals, forestry, industrial sites	inhibits aromatic amino acids	high (12 g/l)	10 wk in clear water, 6 days in pond water
Bromoxynil	cereals, corn	electron transport inhibitor	low (1.3 mg/l) sprayed with solvent, 0.1 - 0.2 ppb[6]	15 days in shallow water

*References: [1]Bouchard et al., 1985; [2]Ernst et al., 1991; [3]Fairchild et al., 1989; [4]Kent et al., 1992a; [5]Kent et al., 1992b; [6]Waite et al., 1992; [7]Wan et al., 1987; and, [8]Zitco and McLeese, 1980.

Table 2. *Experimental exposures of amphibian embryos (e) and larvae (l) to low pesticide concentrations.*

	Insecticides			Herbicides					
	Permethrin	Fenvalerate	Fenitrothion	Hexazinone	Triclopyr	Triallate	Trifluralin	Glyphosate	Bromoxynil
Concentration (ppm)	0.01–0.1	0.01–0.1	0.5–8.0	100	0.6–4.8	0.5–8.0	0.5–8.0	0.5–8.0	0.01–1.0
Exposure duration	22 hr	22 hr	24 hr	7 days	48 hr	96 hr	96 hr	96 hr	96 hr
Rana sylvatica	e		l						
Rana pipiens	e, l	l	e, l	e, l	e, l				
Rana clamitans	e, l	e, l	e, l	e, l	e, l	e, l	e, l	e, l	e, l
Rana catesbeiana			e, l	e, l	e, l				
Bufo americanus	e		l						
Ambystoma maculatum	l	l	l						

Newly hatched tadpoles of *R. pipiens* and *R. clamitans* were exposed to 0.01 and 0.1 ppm permethrin and fenvalerate, and newly hatched *A. maculatum* larvae were exposed to 0.01 ppm of both pyrethroids for 22 to 24 hr. Each pyrethroid was first dissolved in a 1:1 acetone:ethanol mixture that was subsequently added to 100 ml deionized water to make a stock solution. Newly hatched tadpoles of *R. sylvatica, R. pipiens, R. clamitans, R. catesbeiana*, and *B. americanus,* and newly hatched larvae of *A. maculatum,* were exposed to 0.5, 1.0, 2.0, 4.0, and 8.0 ppm nominal concentrations of fenitrothion for 24 hr. Mid-neurula embryos of *R. pipiens, R. clamitans*, and *R. catesbeiana* were also exposed for the same duration. Because fenitrothion is only very weakly soluble in water, a 99% technical grade formulation of 1260 g active ingredient (g AI) per l was dissolved in the solvent Dawanol and the emulsifier Atlox was then added. Water extractions were analysed on a gas chromatograph equipped with a flame photometric detector.

Mid-neurula embryos and newly hatched tadpoles of *R. clamitans* were exposed to all 6 of the herbicides that we tested (Table 2). The same stages of *R. pipiens* and *R. catesbeiana* were also exposed to triclopyr and hexazinone. Preliminary experiments indicated that hexazinone was harmless at levels that might occur in natural water bodies and so further exposures lasted 8 days at the nominal concentration of 100 ppm. A 97% technical grade formulation of 240 g AI/l was used. A solvent was unnecessary due to the high water solubility of the herbicide. The triclopyr butoxyethyl ester, at 96% technical grade, 480 g AI/l, is also relatively soluble in water and carrier solvents were similarly unnecessary. Embryos and tadpoles were exposed to nominal concentrations of 0.6, 1.2, 2.4, and 4.8 ppm for 48 hr. Residue analysis for triclopyr acid indicated that the ester underwent little hydrolysis to triclopyr acid, which is relatively non-toxic, during the 1st 24 hr of the experimental exposures.

Only *R. clamitans* mid-neurula embryos and newly hatched tadpoles were exposed to the other 4 herbicides. End-use formulations of all four herbicides were used: triallate as Avadex B.W.® at 400 g AI/l, trifluralin as Rival EC® at 500 g AI/l, glyphosate as Roundup® at 70 g AI/l, and bromoxynil as Pardner® at 280 g a.i./l. All four therefore contained unknown amounts of unknown solvents.

All exposures were carried out at 15° C in darkness, except for the pyrethroid exposures which were repeated at 20° C, because the enzymes that detoxify pyrethroids are temperature sensitive (Coats et al. 1989). Embryos and tadpoles were exposed in 1-l beakers containing 500 ml of water. Embryos were exposed in groups of 20 per treatment beaker and tadpoles were exposed in groups of 10. Because newly hatched tadpoles are 4 to 8 mm total length, they were a consistently small load in each treatment beaker. All experiments were run in triplicate, with appropriate controls including solvents and emulsifier carriers.

Measurements

Embryos and tadpoles were examined daily for mortality and for evidence of abnormalities. Both hatching success and timing of hatching were determined for embryos following exposure. All tadpoles were prodded daily for assessment of their avoidance response. Any tadpole that darted away when prodded was considered to be normal or to have recovered from the effects of exposure to the pesticide. Abnormal avoidance behavior in response to prodding was highly variable and ranged from weak, slow swimming for a distance of several body lengths to complete, unresponsive immobility or paralysis.

Results

Sensitivity of Embryos

Embryos of all species exposed to all the pesticides hatched at the same time as control samples, and with the same hatching success (≥ 95%; Table 2). Hatched tadpoles did not possess any higher degree of abnormalities than the controls, with only 1 to 4% exhibiting characteristic tadpole deformities such as bent backs or heads larger or smaller than normal.

Sensitivity of Tadpoles and Salamander Larvae

Most newly hatched *R. pipiens, R. clamitans*, and *A. maculatum* larvae were initially weak or paralysed in response to prodding following exposure to the pyrethroids permethrin or fenvalerate (Figs. 1 and 2). All of those exposed as controls to uncontaminated water, or to water containing only the solvent, exhibited normal avoidance behavior throughout the experimental period. Their mortality rate ranged from 0 to 3% per treatment. Tadpoles of both *R. pipiens* and *R. clamitans* recovered normal avoidance responses gradually over several days following exposure to 0.01 ppm and recovered more quickly at 20°C than at 15°C. After exposure to 0.1 ppm of either pyrethroid at 15° C, most *R. pipiens* tadpoles remained weak or paralysed throughout the post-exposure period, and all *R. clamitans* tadpoles failed to recover. At 20° C following exposure to 0.1 ppm of either pyrethroid, most tadpoles recovered but more slowly than at 15° C (Figs. 1 and 2). Newly hatched larvae of *A. maculatum* were considerably more sensitive to both pyrethroids than either anuran, remaining paralysed 10 days after exposure to 0.01 ppm concentrations at 15° C and recovering much more slowly than the anuran tadpoles at 20° C (Figs. 1 and 2). Despite the extensive paralysis, mortality of tadpoles or larvae following exposures to either pyrethroid did not exceed 7% in any treatment.

For fenitrothion, newly hatched *B. americanus, R. sylvatica, R. pipiens, R. clamitans*, and *A. maculatum* were unaffected by 24-hr exposures to 0.5, 1.0, or 2.0 ppm, and their avoidance responses remained similarly normal in controls containing the solvent and emulsifier. Newly hatched *R. catesbeiana* were also unaffected by the same 3 low levels of fenitrothion but up to 30% were initially sensitive to the control containing the solvent and emulsifier, recovering normal avoidance behavior within 48 hr. However, following exposure to 4 or 8 ppm fenitrothion, tadpoles or larvae of all 6 species were affected, exhibiting varying degrees of abnormality, paralysis, and mortality (Fig. 3). Tadpoles of *B. americanus* were the most tolerant. *Rana sylvatica, R. pipiens, R. clamitans*, and *A. maculatum* were more sensitive, and tadpoles of *R. catesbeiana* were the most sensitive.

The tested herbicides varied greatly in their effects (Figs. 4 and 5). Hexazinone, at environmentally unrealistic levels (7-day exposures at 100 ppm), had no effect on newly hatched *R. pipiens, R. clamitans*, or *R. catesbeiana*. In contrast, bromoxynil was particularly toxic, with extensive mortality (though little paralysis) of *R. clamitans* tadpoles occurring at 0.01 ppm. Triallate, trifluralin, and glyphosate all induced weak abnormal avoidance responses of 5 to 42% of *R. clamitans* tadpoles at 4 ppm. At exposure levels < 4 ppm, tadpoles were unaffected, but at 8 ppm almost all died or became paralysed without recovering (Fig. 4). Of the 3 species exposed to the triclopyr ester, *R. clamitans* and *R. catesbeiana* tadpoles were more sensitive than were *R. pipiens* tadpoles, with complete mortality occurring during exposure to 4.8 ppm (Fig. 5). The avoidance behavior of tadpoles exposed to uncontaminated water in control treatments for the herbicide experiments remained normal, and tadpole mortality rate in control treatments ranged from 0 to 3%.

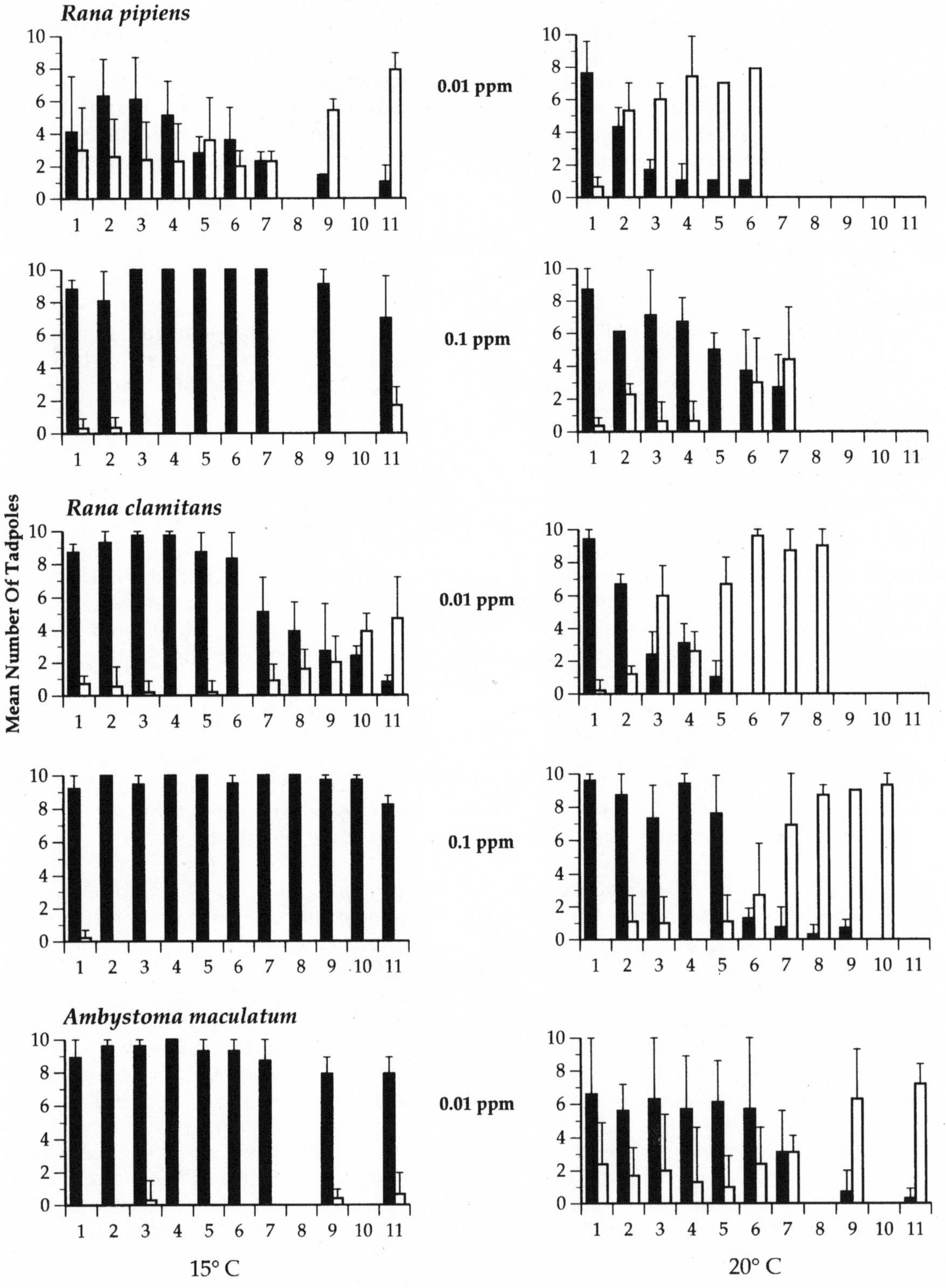

Figure 1. *Sensitivity of* Rana pipiens, R. clamitans, *and* Ambystoma maculatum *larvae to the pyrethroid insecticide permethrin. Black bars indicate mean numbers of tadpoles paralysed or with weak avoidance responses. Clear bars indicate mean numbers of normal (recovered) tadpoles. Error bars indicate standard deviations of triplicate samples.*

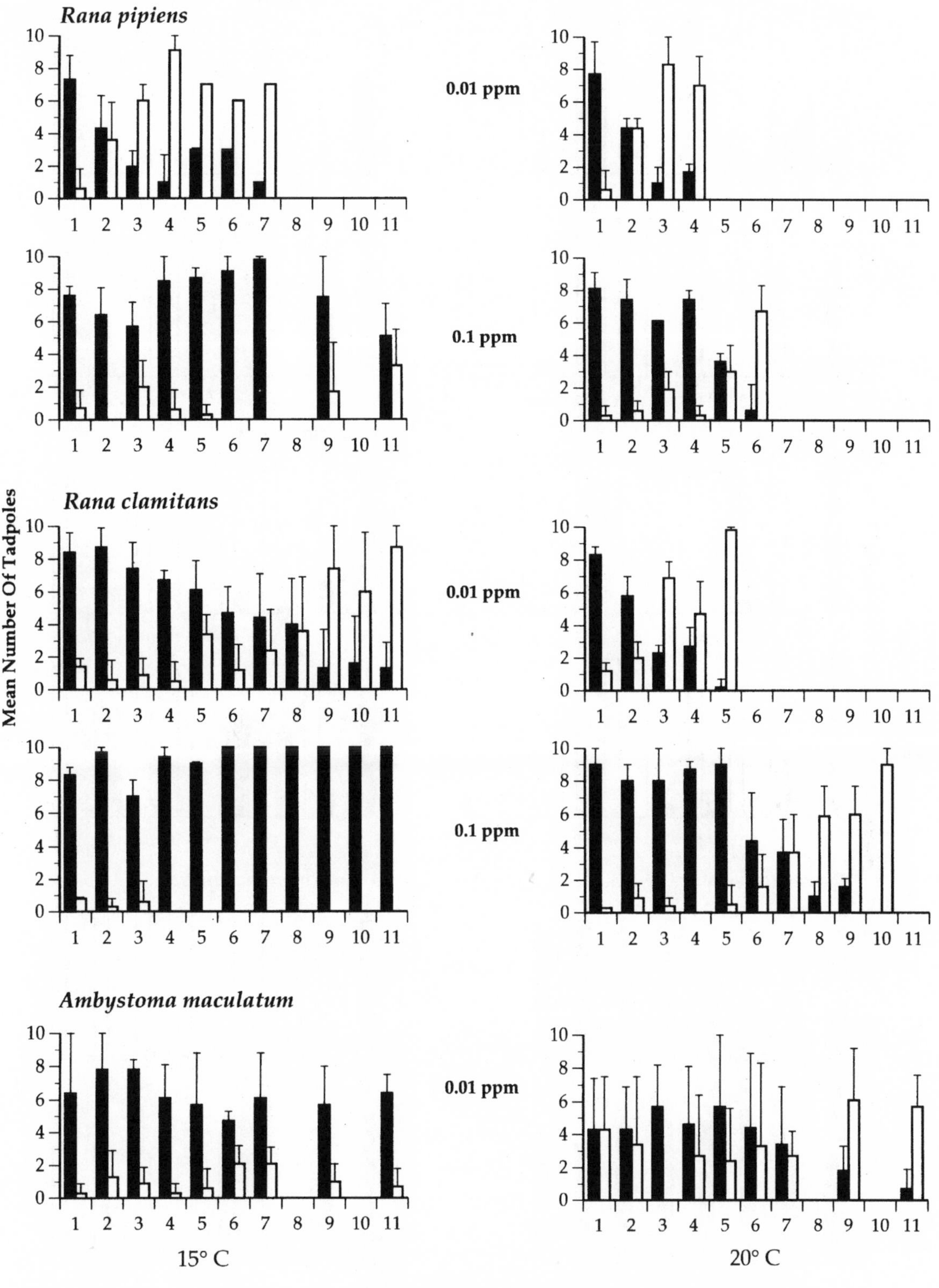

Figure 2. *Sensitivity of* Rana pipiens, R. clamitans, *and* A. maculatum *larvae to the pyrethroid fenvalerate. Legend is as in Fig. 1.*

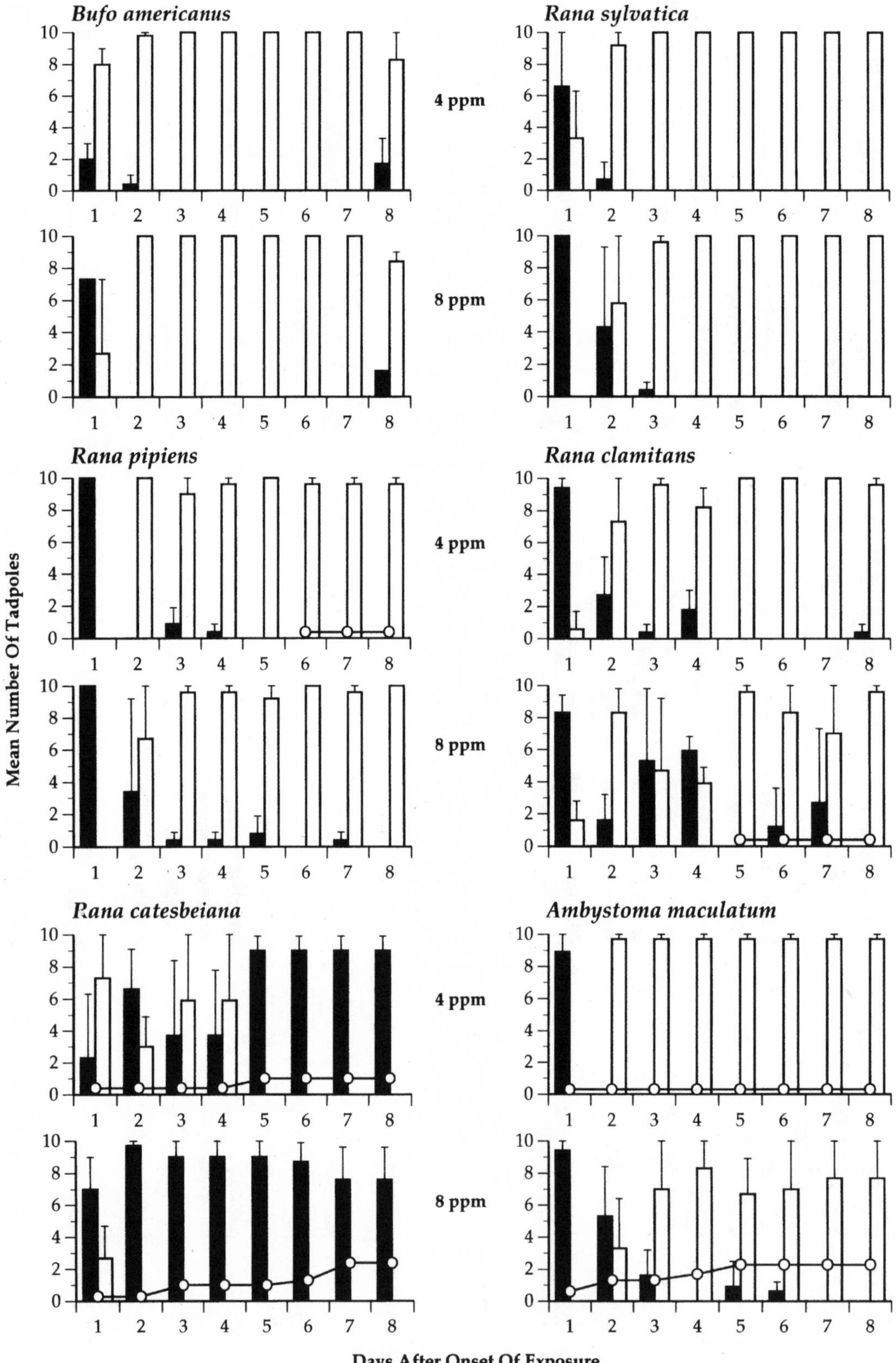

Figure 3. *Sensitivity of* Rana sylvatica, R. pipiens, R. catesbeiana, R. clamitans, Bufo americanus, *and* Ambystoma maculatum *larvae to exposure to the insecticide fenitrothion. Solid line indicates accumulated mortality, other symbols are as as in Fig. 1.*

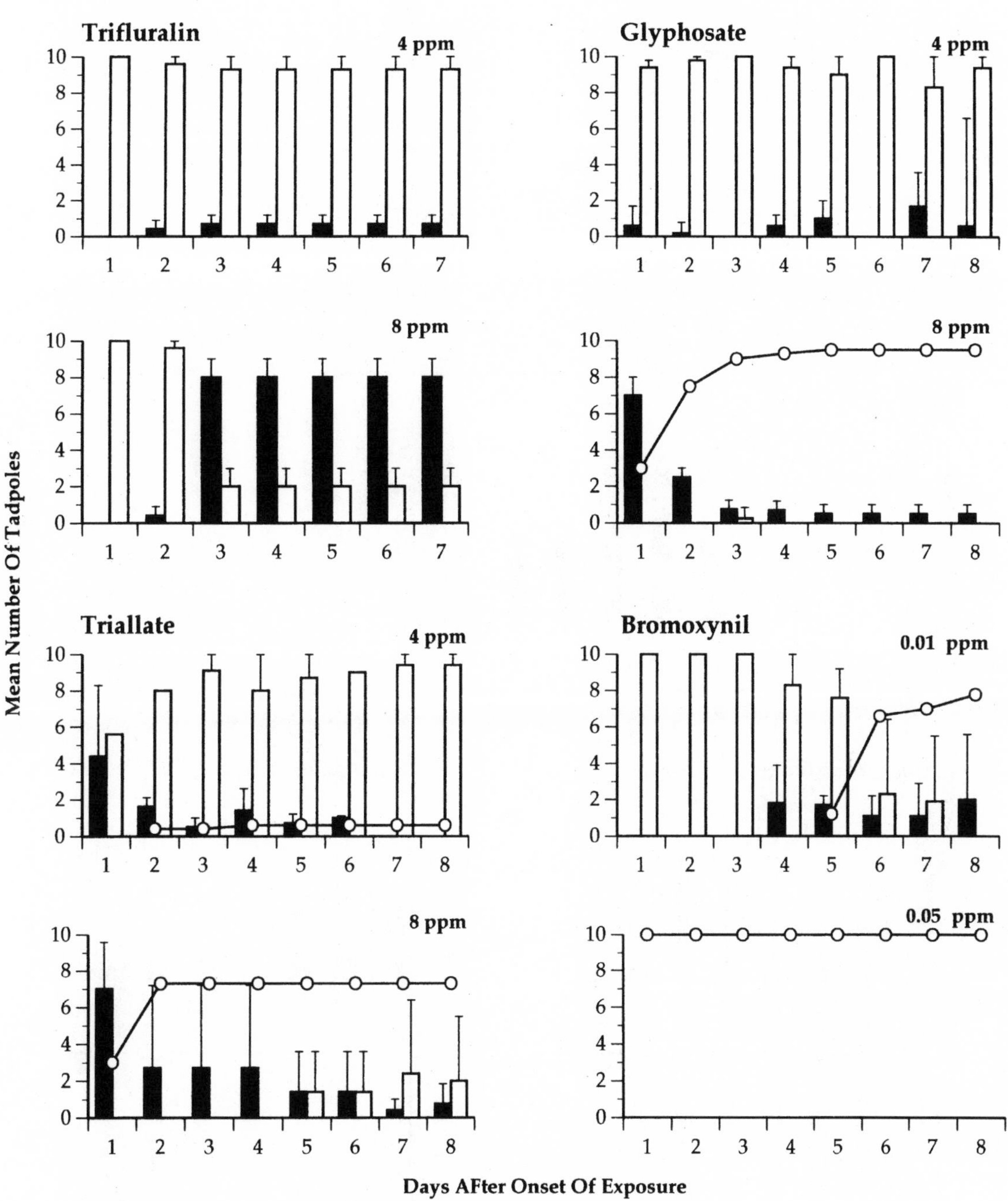

Figure 4. *Sensitivity of* Rana clamitans *to 4 herbicides. Solid lines indicates accumulated mortality, other symbols are as as in Fig. 1.*

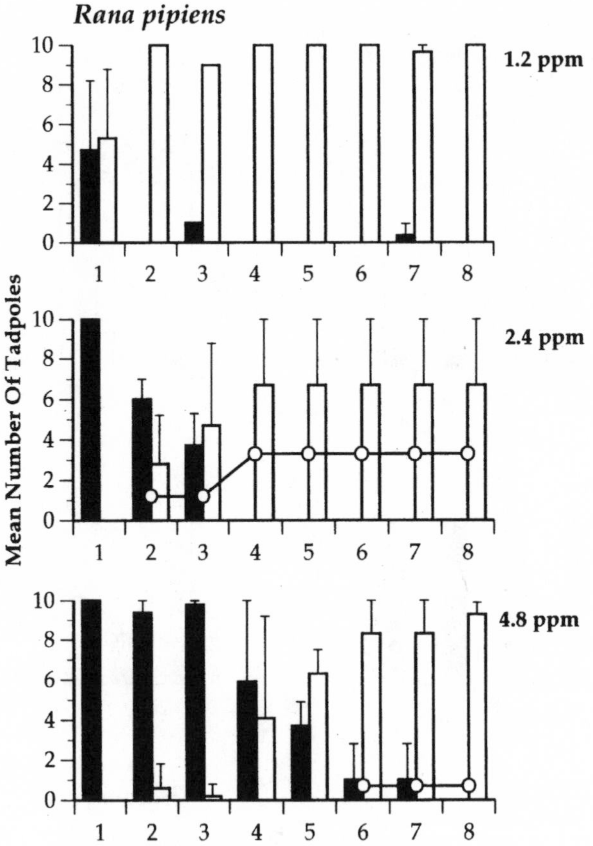

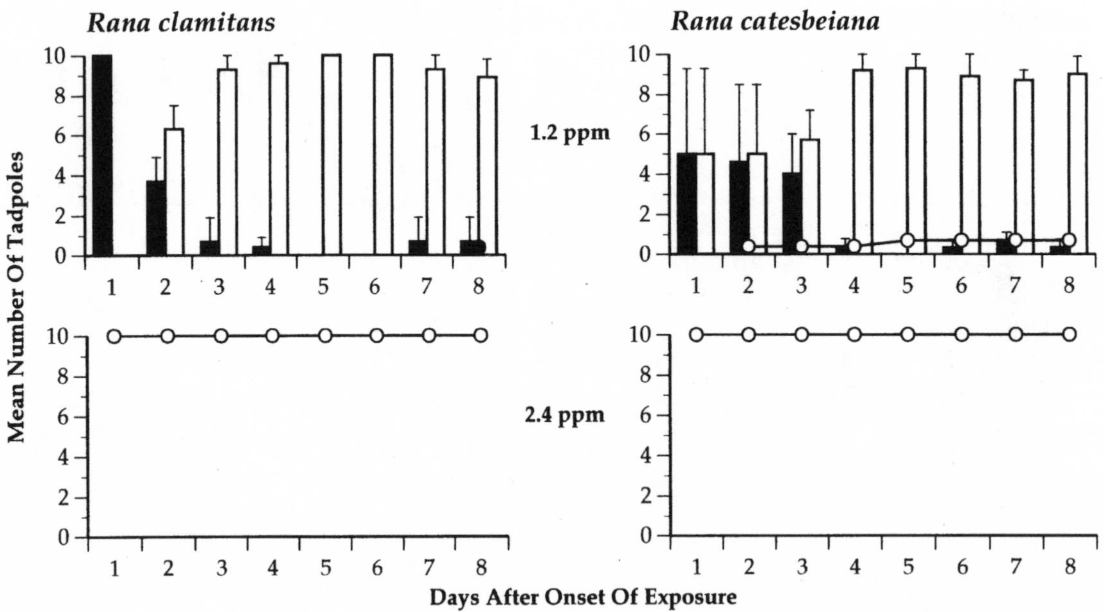

Figure 5. *Sensitivity of* Rana pipiens, R. clamitans, *and* R. catesbeiana *to the herbicide triclopyr. Solid lines indicates accumulated mortality, other symbols are as as in Fig. 1.*

DISCUSSION

Given the exposure levels and durations that we used in our experiments, mid-neurula amphibian embryos are unlikely to be affected by the low concentrations of the pesticides that we used. Other research has shown that eggs that are dejellied may be quite sensitive to pesticide exposure (Elliott-Feeley and Armstrong 1982), and it is likely that the jelly layer provides considerable protection to the embryos.

Tadpoles of the amphibians we tested had sensitivities to the pesticides relatively similar to those of fresh-water fish, such as various species of trout and salmon (Table 3). Strict comparisons are impossible to make to make, because lethal levels may be determined from 24 or 96 hr exposures and sublethal effects may range from ability to recover normal behaviour, as with our experiments, to ability to detect and avoid contaminants. It is also likely that some of the sensitivity reported may be due to solvent and carrier chemicals that are used in end-use formulations. Our experiments indicated that only tadpoles of *R. catesbeiana* were affected by exposure to the solvent/emulsifier used to carry fenitrothion.

Like some fresh-water fish species, the newly hatched stages of the amphibians we tested are likely to be sensitive to levels of contamination that could occur as a result of spray drift and run-off. Exposure of the tadpoles to the higher concentrations of each of the pesticides resulted either in direct mortality, or abnormal avoidance behavior (including paralysis), or both, irrespective of the chemical or its mode of action. More subtle effects, at lower concentrations, may exist and longer exposures would certainly lead to greater mortality. We also did not assess the impact on tadpoles of multiple exposures to > 1 chemical. We therefore consider our results to be conservative assessments of amphibian sensitivity. Furthermore, the level of sensitivity to different chemicals indicates that although the amphibian community may not be affected by most low level contamination events that typically occur following standard applications, the community is likely to be very sensitive to any incidents that result in contamination at concentrations even slightly higher than usual. Several of the pesticides we studied are particularly toxic and warrant close attention.

Although newly hatched tadpoles of the species we tested were affected in approximately the same way and at the same levels of exposure to the various pesticides, some species were consistently more sensitive than others. Among the ranid frogs, where comparisons are possible, *R. clamitans* and *R. catesbeiana* tadpoles were more sensitive than *R. pipiens* or *R. sylvatica* tadpoles. Further comparative data are limited, but *B. americanus* tadpoles appear to be the most tolerant and *A. maculatum* larvae the least tolerant of the 6 species tested.

The sensitivity of aquatic stages of these amphibians to pH stress has also been determined, with similar conclusions. *Rana sylvatica* is the most tolerant of low pH stress, *A. maculatum* is the most sensitive, and the other ranids and *B. americanus* lie somewhere in between (Freda 1986). The stresses provoked by low pH and pesticide ex-

Table 3. *Summary of tadpole sensitivity results and comparison with sensitivity of fish to the same pesticides (concentrations expressed as ppm).*

	Fish*		Amphibian tadpole	
Pesticide	lethal	sublethal	lethal	sublethal
Pyrethroids	0.002–7.0[3,4]	0.01[4]		0.01
Fenitrothion	1–3[10]	< .1[8]	> 8	2–8
Hexazinone	> 200[5]		> 100	> 100
Triclopyr (ester)	1–4[11]	.7[5]	2–4	1–2
Triallate	1.3[1]	.2[2]	8	4
Trifluralin	0.1–1.0[12]		> 8	4–8
Glyphosate	1.3–29.0[4,7]		8	4
Bromoxynil	0.05[9]		.01–.05	.01

*References: [1]Grover and Cessna, 1988; [2]Kent et al., 1992b; [3]Hansen et al., 1983; [4]Holdway and Dixon, 1988; [5]Janz et al., 1991; [6]Kennedy, 1984; [7]Mitchell et al, 1987; [8]Morgan et al., 1990; [9]Muir et al., 1991; [10]Takimoto et al., 1987; [11]Wan et al., 1987; and, [12]Worthing and Hance, 1991.

posures may be too different to expect a species that is more sensitive to one to also be more sensitive to the other. However, we propose, and our data seem to suggest, that the early spring-breeding ranids *R. sylvatica* and *R. pipiens* as well as the toad *B. americanus* are likely to be less sensitive to pesticide exposures than the later breeding ranids *R. clamitans* and *R. catesbeiana*. This trend may be the result of the size and developmental stage at which the tadpoles hatch. The greater sensitivity to fenitrothion of newly hatched tadpoles of *R. clamitans* and *R. catesbeiana*, in comparison with that of *R. pipiens* and *R. sylvatica* tadpoles, is correlated with the smaller size of their eggs. *Rana clamitans* and *R. catesbeiana* tadpoles hatch at smaller sizes and earlier developmental stages, suggesting that the difference in species sensitivity may in part be a result of greater sensitivity of earlier developmental stages.

We know nothing about the relative sensitivity of later tadpole stages, but predict that they should be less sensitive, because developing fish also appear to become less sensitive with age (Holdway and Dixon 1988). But unlike fish, tadpoles develop toward metamorphosis, a time of developmental and metabolic stress, and it is likely that premetamorphic tadpoles will also be relatively sensitive.

The tadpole and larval stages of the amphibians we have tested are vulnerable to low-level pesticide contamination. As a result, decline and even extinctions of local populations are likely at any sites in close proximity to heavy or frequent use of pesticides. As the timing of pesticide applications often overlaps that of the tadpole stage of many amphibian species, and the timing, frequency, and rates of the pesticide applications are determined by the pests to be controlled, such declines might be difficult to prevent. Fortunately, if water quality were to improve, it takes only a few colonizers and 1 or 2 fertilized egg masses to repopulate a pond. If pesticides are playing a more general role in amphibian decline, then it should be possible to identify that role and propose ways to eliminate or reduce their impact.

Acknowledgments

We wish to thank Deneen Brigham, Donna Coulson, Mark Kolohon, Sharon Louis, Lise McGillivray, Dean Ostrander, Christina Stromberg, Graeme Taylor, and Anne Wilson who helped with various aspects of the research, which was funded by the Natural Sciences and Engineering Council of Canada, the Canadian Wildlife Service, and the World Wildlife Toxicology Fund.

Literature Cited

Berrill M, Bertram S, Wilson A, Louis S, Brigham D, Stromberg C. 1993. Lethal and sublethal impacts of pyrethroid insecticides on amphibian embryos and tadpoles. Environmental Toxicology and Chemistry 12:525–539.

Berrill M, Bertram S, McGillivray L, Kolohon M, Pauli B. 1994. Effects of low concentrations of forest-use pesticides on frog embryos and tadpoles. Environmental Toxicology and Chemistry 13:657–664.

Berrill M, Bertram S, Pauli B, Coulson D, Kolohon M, Ostrander D. 1995. Comparative sensitivity of amphibian tadpoles to single and pulsed exposures of the forest-use insecticide, fenitrothion. Environmental Toxicology and Chemistry 14:1011–1018.

Bouchard C, Lavy TL, Lawson ER. 1985. Mobility and persistence of hexazinone in a forest watershed. Journal of Environmental Quality 14:229–233.

Coats JR, Symonik DM, Bradbury SP, Dyer SD, Timson LK, Atchison GJ. 1989. Toxicology of synthetic pyrethroids in aquatic organisms: An overview. Environmental Toxicology and Chemistry 8:671–679.

Clark KL, Lazerte BD. 1985. A laboratory study of the effects of aluminum and pH on amphibian eggs and tadpoles. Canadian Journal of Fisheries and Aquatic Sciences 42:1544–1551.

Edwards R, Millburn P, Hutson DH. 1986. Comparative toxicity of cis-cyprmethrin in rainbow trout, frog, mouse and quail. Toxicology and Applied Pharmacology 84:512–522.

Elliot-Feeley E, Armstrong JB. 1982. Effects of fenitrothion and carbaryl on *Xenopus laevis*. Toxicology. 22:319–335.

Ernst W, Julien G, Hennigar P. 1991. Contamination of ponds by fenitrothion during forest spraying. Bulletin of Environmental Contamination and Toxicology 46:815–821.

Fairchild WL, Ernst WR, Mallet VN. 1989. Fenitrothion effects on aquatic organisms. In: Ernst WR, Pearce PA, Pollock TL, editors. Environmental effects of fenitrothion use in forestry. Halifax, NS: Environment Canada. p 109–160.

Freda J. 1986. The influence of acidic pond water on amphibians: A review. Water, Air, and Soil Pollution 30:439–450.

Grover R, Cessna AJ. 1988. Environmental chemistry of herbicides. Vol. 2. Boca Raton, FL: CRC Press Inc.

Hansen DL, Goodman LR, Moore JC, Higdon PK. 1983. Effects of the synthetic pyrethroids AC 222,705, permethrin and fenvalerate on sheepshead minnows in early life stage toxicity tests. Environmental Toxicology and Chemistry 2:251–258.

Holdway DA, Dixon DG. 1988. Acute toxicity of permethrin or glyphosphate pulse exposure to larval white sucker (*Catostomus commersoni*) and juvenile flagfish (*Jordanella floridae*) as modified by age and ration. Environmental Toxicology and Chemistry 7:63–68.

Janz DM, Farrell AP, Morgan JD, Vigers GA. 1991. Acute physiological stress responses of juvenile coho salmon (*Oncorhynchus kisutch*) to sublethal concentrations of Garlon 4, Garlon 3A and Vision herbicides. Environmental Toxicology and Chemistry 10:81–90.

Kennedy GL, Jr. 1984. Acute and environmental toxicity studies with hexazinone. Fundamental and Applied Toxicology 4:603–611.

Kent RA, Tache M, Caux PY, Pauli BD. 1992a. Canadian water quality guidelines for trifluralin. Ottawa: Environment Canada. Scientific Series 190.

Kent RA, Tache M, Caux PY, De Silva S, Lemky K. 1992b. Canadian water quality guidelines for triallate. Ottawa: Environment Canada. Scientific Series 195fs.

Little EE. 1990. Behavioral toxicology: Stimulating challenges for a growing discipline. Environmental Toxicology and Chemistry 9:1–2.

Mitchell DG, Chapman PM, Long TL. 1987. The acute toxicity of Round-up and Rodeo herbicides to rainbow trout, chinook and coho salmon. Bulletin of Environmental Contamination and Toxicology 37:1028–1035.

Morgan MJ, Fancey LL, Kiceniuk JW. 1990. Response and recovery of brain acetylcholinesterase activity in Atlantic salmon (*Salmo solar*) exposed to fenitrothion. Canadian Journal of Fisheries and Aquatic Sciences 47:1652–1654.

Muir DCG, Kenny DF, Grift NP, Robinson RD, Titman RD, Murkin HR. 1991. Fate and acute toxicity of bromoxynil esters in an experimental prairie wetland. Environmental Toxicology and Chemistry 10:395–406.

Takimoto T, Ohshima M, Miyamoto J. 1987. Comparative metabolism of fenitrothion in aquatic organisms. I. Metabolism in the euryhaline fish, *Oryzias latipes* and *Mugil cephalus*. Ecotoxicology and Environmental Safety 13:104–117.

Waite DT, Grover R, Westcott ND, Sommerstad H, Kerr L. 1992. Pesticides in ground water, surface water and spring runoff in a small Saskatchewan watershed. Environmental Toxicology and Chemistry 11:741–748.

Wan MT, Moul DJ, Watts RG. 1987. Acute toxicity to juvenile Pacific salmonids of Garlon 3A, Garlon 4, triclopyr, triclopyr ester, and their transformation products: 3,5,6-Trichloro-2-pyridinol and 2-methoxy-3,5,6-trichloropyridine. Bulletin of Environmental Contamination and Toxicology 39:721–728

Werner EE. 1986. Amphibian metamorphosis: Growth rate, predation rate and optimal size at transformation. American Naturalist 128:319–341.

Worthing CR, Hance RJ, editors. 1991. The pesticide manual. 9th ed. Farnham, U.K.: The British Crop Protection Council.

Zitco V, McLeese DW. 1980. Evaluation of hazards of pesticides used in forest spraying to the aquatic environment. Canadian Technical Report of Fisheries and Aquatic Sciences 985:1–27.

©1997 by the Society for the Study of Amphibians and Reptiles
Amphibians in decline: Canadian studies of a global problem. David M. Green, editor.
Herpetological Conservation 1:246–257.

Chapter 25

MEASURING THE HEALTH OF FROGS IN AGRICULTURAL HABITATS SUBJECTED TO PESTICIDES

JOËL BONIN[1], MARTIN OUELLET[1], JEAN RODRIGUE, AND JEAN-LUC DESGRANGES

Canadian Wildlife Service, Québec Region, P.O. Box 10100, 1141 Route de L'Église, Ste-Foy, Québec G1V 4H5, Canada

FRANÇOIS GAGNÉ

Environment Canada, St. Lawrence Centre, 105 McGill St., Suite 400, Montréal, Québec H2Y 2E7, Canada

TIMOTHY F. SHARBEL[2] AND LESLIE A. LOWCOCK[3]

Redpath Museum, McGill University, 859 Sherbrooke St. W., Montréal, Québec H3A 2K6, Canada

ABSTRACT.—A number of clinical methods were used to determine the health of anurans living in agricultural habitats in southern Québec to identify the most useful methods for detecting sublethal effects of agricultural pollutants. In 1993, metamorphosed and metamorphosing *Rana clamitans* were collected from ponds and ditches in 2 potato fields, 3 sweet corn fields, and 3 uncultivated areas having similar habitats. Physical and postmortem examinations revealed diseases and hindlimb deformities mainly in metamorphosing individuals from potato fields. Blood smears indicated comparable white cell counts, hemoparasite presence, and micronucleus frequencies in all sites. Hematological and blood biochemical analyses, for adults only, appeared consistent with normal health at all locations. Brain acetylcholinesterase activity levels were also normal. Flow cytometry revealed genomic disruption in adult and metamorphosing individuals from all cultivated areas. Deformed individuals from the potato field had increased genome size variability. Similar genomic effects were also found in pesticide-exposed individuals showing no apparent physical or physiological changes. Analyses showed correspondingly high genotoxicity values for water samples taken from the cultivated sites. Clinical examination and flow cytometry are practical techniques for examining the sublethal effects of environmental contaminants on frogs. This investigation indicates a general mutagenic effect of agricultural pollutants on anurans, in addition to potential teratogenic and pathogenic effects in some cases.

RÉSUMÉ.—Différentes méthodes cliniques ont été employées afin de vérifier l'état de santé d'anoures vivants dans les milieux agricoles du sud du Québec. L'objectif était de reconnaître les méthodes les plus utiles pour détecter les effets subléthaux de polluants agricoles. En 1993, des individus métamorphosés et en métamorphose de *Rana clamitans* ont été échantillonnés dans des étangs et fossés de 2 champs de pommes de terre, 3 champs de maïs sucré et 3 sites non-cultivés présentant des habitats similaires. Les examens physique et post-mortem ont révélé des maladies principalement chez les in-

1 Present Address: *Redpath Museum, McGill University, 859 Sherbrooke St. W., Montréal, Québec H3A 2K6, Canada.*

2 Present Address: *Arbeitsgruppe Michiels, Max-Planck-Institut für Verhaltensphysiologie, D--82319 Seewiesen (Post Starnberg), Germany.*

3 Present Address: *c/o Centre for Biodiversity and Conservation Biology, Royal Ontario Museum, 100 Queen's Park, Toronto, Ontario, M5S 2C6 Canada.*

dividus en métamorphose provenant d'un champ de pommes de terre. L'analyse des frottis sanguins indiquait des valeurs comparables dans tous les sites pour le différentiel des globules blancs, la prévalence des hémoparasites et la fréquence des micro-noyaux. Les valeurs hématologiques et biochimiques obtenues chez les adultes seulement, représentaient probablement des valeurs normales dans toutes les stations. Les niveaux d'activité de l'acétylcholinestérase étaient également normaux. La cytométrie en flux a révélé des modifications génétiques chez les adultes et les individus en métamorphose dans tous les sites cultivés. Les individus malformés du champ de pommes de terre avait une variabilité génétique accrue. Des effets génétiques similaires ont été également trouvés chez des individus exposés aux pesticides qui ne présentaient aucunes anormalités physiques ou physiologiques. Les analyses ont aussi démontré une génotoxicité élevé de l'eau dans les sites cultivés. L'examen clinique et la cytométrie en flux sont donc des méthodes utiles pour étudier les effets subléthaux des contaminants environnementaux sur les anoures. Cette première investigation indique un effet mutagénique général des polluants agricoles sur les anoures, en plus d'effets tératogènes et pathogènes possibles dans certains cas.

Amphibians may be especially sensitive to environmental contaminants due to their biphasic life histories and semi-permeable skin (Hall and Henry 1992). In farming areas, where wetlands adjacent to crops become contaminated with pesticides (Wauchope 1978; Giroux and Morin 1992; Berryman and Giroux 1994; Giroux 1994), anuran breeding populations may be affected (Hazelwood 1970; Cooke 1981). There exist few comprehensive field studies of the effects of environmental contaminants on amphibian populations (Bishop 1992; Hall and Henry 1992), and little comparative information for the symptomatic diagnosis of frog population health is available (Crawshaw 1992a, this volume). Most toxicological tests examine the lethal effects of various compounds on amphibians under controlled laboratory conditions but virtually no light is thereby shed upon their transient or permanent sublethal effects in the natural milieu (Harfenist et al. 1989; Hall and Henry 1992).

Diverse degrees of mutagenic, teratogenic, and physiological effects on animals have been reported for agricultural pesticides (Hayes and Laws 1991). Documented effects on amphibians include increased mortality, lowered hatching success of eggs, retarded growth, abnormal development, behavioural changes, alteration in thermal resistance, and hematological variation (Harfenist et al. 1989; Bishop 1992; Berrill et al., this volume). The inhibition of brain acetylcholinesterase activity has been seen as indicative of poisoning by organophosphate and some carbamate pesticides (Martin et al. 1981), but amphibians were found to be very resistant (Wang and Murphy 1982). However, a common occurrence after chemical exposure in amphibians is the clastogenic generation of micronuclei (Fernandez et al. 1993; Krauter 1993).

The objective of this study was to identify health assessment techniques that could be useful for detecting sublethal effects of agricultural pollutants on frog populations in the wild. A number of clinical methods were used to determine the health of green frogs, *Rana clamitans*, living in aquatic habitats adjacent to potato and sweet corn fields. These crops are farmed extensively in southern Québec, and the numerous pesticides applied during their cultivation contaminate watercourses (Giroux and Morin 1992; Giroux 1993, 1994; Berryman and Giroux 1994). From the exposure to agricultural pollutants, we expected differences in physical and postmortem conditions of frogs, including hematology and blood biochemistry profiles, brain acetylcholinesterase activity, and genomic characteristics (Lowcock et al. 1997; Ouellet et. al. 1997).

Materials and Methods

We concentrated our study on *R. clamitans* because of its sedentary lifestyle (Martof 1953) and omnipresence in southern Québec. Frog populations from 2 potato fields (sites 1 and 2), 3 sweet corn fields (sites 3, 4 and 5) and 3 control sites (sites 6, 7 and 8) from uncultivated open areas were sampled in southern Québec. The corn fields were located within 2 km of each other, while the 2 potato fields were 150 km apart and separated by the St. Lawrence River. The control sites were spread over a distance of 150 km, in the vicinity of the corn and potato fields. Frogs were collected from ponds and ditches exposed to pesticide contamination through direct application and runoff (Table 1) and from similar habitats at the control sites. Land use at all sites had been the same for > 10 yr. Adult frogs (those ≥ 1-yr post-metamorphosis) were caught by hand or dipnet in July 1993, 1 to 5 days after application of the insecticides carbofuran (carbamate) on the sweet corn and azinphos-methyl (organophosphate) on the potatoes. Metamorphosing or newly metamorphosed juveniles were collected by dipnet in late August 1993. Field examination of the general appearance and behaviour of each frog was conducted immediately after capture. The frogs were then placed in separate plastic jars containing pond water and stored in a cool dark room pending laboratory analysis.

Within 12 hr of capture, physical examination and tissue collection were completed in the laboratory. After external examination, each frog was decerebrated following a rapid decapitation at the first cervical vertebra (Canadian Council on Animal Care 1984) and its brain was placed in liquid nitrogen for cholinesterase activity analysis. A necropsy was performed and a blood sample (0.5 or 1 ml) was collected from the exposed heart using a syringe. A few drops of blood were taken for blood smears and for flow cytometric DNA analysis (Sharbel et al., this volume). If possible, a greater volume of blood was also collected for hematology and for biochemical profiles. Each frog was then measured (snout to vent length, and length of the right tibia), weighed, sexed and given a complete physical and postmortem examination during which evident parasites and/or parasitic cysts were collected from the body surface, the oral cavity, beneath the skin, or from the surfaces of the viscera. In selected cases, tissues were collected and preserved in 10% buffered formalin for histopathological examination.

Blood smears were made from both juveniles and adults and were immediately fixed with absolute methanol. Smears were then stained using a modified Wright-Giemsa colorant ("Diff-Quik", Baxter). Differential white cell counts, thrombocyte estimates, mitotic indices and screenings for hemoparasites, micronuclei, and other cellular abnormalities were performed manually on 82 apparently healthy adults and 7 juve-

Table 1. *Pesticides applied to potato and sweet corn fields in southern Québec in 1993. For potato fields, the 1st application of pesticides was in mid-May, the 2nd was from late June to mid-August, and the 3rd was at the end of August. For sweet corn fields, the 1st application of pesticides was during May and the 2nd (twice) was in July through early August. Common names of pesticides are from Worthing and Hance (1991).*

		Pesticide applications		
Site	Crop	1st	2nd	3rd
1	potato	linuron[H]	azinphos-methyl[I] cypermethrin[I] oxamyl[I] mancozeb[F] chlorothalonil[I]	diquat[H]
2	potato	metribuzin[H] phorate[I]	same as site 1	diquat[H]
3	sweet corn	atrazine[H]	carbofuran[I]	
4	sweet corn	atrazine	carbofuran	
5	sweet corn	atrazine glyphosate[H] butylate[H]	carbofuran	
		Crop rotation and adjacent crops		
		Period	Crop	Pesticide applications
1	potato	3 yr	potato	same as other potato fields
2	potato	2 yr	potato	same as other potato fields
3	sweet corn	2 yr	cucumber cantaloupe	endosulfan[I] chlorothalonil maneb[F]
4	sweet corn	2–3 yr	cabbage raspberry strawberry	endosulfan cypermethrin captan[F] metamidophos[I] benomyl[F] mancozeb
5	sweet corn	none	cabbage	metamidophos mancozeb

[H]herbicide, [I]insecticide, [F]fungicide

niles in various states of health as determined through physical and postmortem examination. The number of micronuclei found in 1000 mature erythrocytes was recorded twice for each slide (Krauter et al. 1987). Only micronuclei clearly separated from the nucleus were recorded. Before the blood smears were scored, all slides were coded and then scored blind by a single observer.

Eight hematological values and 23 blood biochemical parameters were measured in samples from adults that had a sufficient volume of blood (0.6 ml for hematology and > 0.6 ml for biochemistry). Hematology was performed on 78 adults (> 19 g). Fresh blood samples kept in Microtainer (Becton Dickinson) with lithium heparin were analyzed within 24 hr at Vita-tech Canada Inc., Markham, Ontario, using a Technicon H*1 Hematology analyzer. Complete biochemical analysis was performed on 32 adults (> 30 g). Plasma samples kept frozen (−80°C) were analyzed at the Université de Montréal, Faculty of Veterinary Medicine, St. Hyacinthe, Québec, using a Synchron Cx Systems apparatus (Beckman Instruments).

Acetylcholinesterase activity was measured on brain samples from 104 frogs to determine if this enzyme was inhibited in frogs exposed to pesticides. Analyses were performed at the National Wildlife Research Centre, Hull, Canada. Activity was estimated by the rate of formation of thionitrobenzoic acid (measured spectrophotometrically at 405 nm), a product of the reaction between dithiodinitrobenzoic acid and thiocholine, which originates from the hydrolyzation of acetylthiocholine iodine by the cholinesterase present in the brain sample (Ellman et al. 1961; Hill and Fleming 1982). To verify inhibition by organophosphorus pesticides, samples having a low enzymatic activity were reactivated chemically using pyridine 2-aldoxime methiodide (2-PAM; Martin et al. 1981).

The DNA content of erythrocyte nuclei in samples of 45 juveniles and 53 adults was measured using flow cytometry. For each individual, more than 15,000 cells were plotted. Abnormal profiles (aneuploid mosaic, polyploid or other) were identified and, in the case of a normal profile, the c-value (in pg DNA / haploid nucleus) was determined and the coefficient of variation (CV) was calculated as a measure of intra-individual genome size variability (Bickham et al. 1988; Licht and Lowcock 1991).

Water samples for toxicity analyses were collected from sites 1 to 6 at the same time as the adult frogs were collected (Table 2). In order to evaluate the spatiotemporal variability of toxicity, the samples were collected from different areas of ponds 1 and 2, and pond 2 was resampled 3 days later. Two *in vitro* genotoxicity bioassays using rainbow trout (*Oncorhynchus mykiss*) hepatocytes were performed (Gagné and Blaise 1995). The 1st test was a nick translation assay (Snyder and Matheson 1985), which estimates small-scale chromatin damage by the relative amounts of labelled nucleotide uptake following DNA repair. The 2nd test was a DNA alkaline precipitation assay (Olive et al. 1988), an estimate of large-scale DNA damage via fluorometric detection of DNA strands. In addition, cellular viability was estimated by the propidium iodide exclusion test (Gagné and Blaise 1995). The results of 4 replicates of 5 different dilutions (0.1, 1, 10, 25, and 50% v/v) were compared by analysis of variance (ANOVA or the non-parametric Kruskal-Wallis test), followed by a comparison test (Dunnett test or the non-parametric Dunns test).

Results

Disease was found mostly in juveniles from site 2, a potato field which had the greatest density of metamorphosing frogs (Table 2). It was also the only site where dead frogs were found; 7 dead metamorphosing individuals were collected, all showing evidence of red leg (Crawshaw 1992b; this volume). Most of the sick frogs showed red leg symptoms such as erythema and cutaneous hemorrhages (n = 9), ocular lesions (panophthalmitis; n = 8), cutaneous ulcerations (n = 4), and subcutaneous edema (n = 1). Hepatic lipidosis, a degenerative change in the liver characterized by a fatty and yellowish appearance, was detected in 11 juveniles with red leg and in 5 others with no external signs of illness. Kidneys sometimes showed an abnormal yellowish coloration in individuals with hepatic lipidosis. Histologically, hepatocytes were pale and swollen with lipid vacuoles, and their nuclei were displaced to the periphery of the cells. At the other locations, only 1 juvenile (at control site 8) had an ocular lesion (possibly traumatic), and 1 adult (at sweet corn site 3) had a cutane-

Table 2. *Study sites description and number of* Rana clamitans *with disease or deformity collected from farmlands in southern Québec in 1993. Juveniles = metamorphosing individuals (mean mass = 4.9 ± 2.5 g, range 1.7 to 16.3 g). Adults were ≥ 1-yr post-metamorphosis (mass = 43.5 ± 15.7 g, 18.9 to 113.2 g).*

			Juveniles			Adults		
Site	Habitat	Crop	n	Disease	Deformity	n	Disease	Deformity
1	pond	potato	68		5	8		
2	pond	potato	203	27	15	9		
3	large ditch	sweet corn	3			12	1	1
4	pond	sweet corn	21			14		
5	ditch	sweet corn	0			10		
6	pond	none	1			21		1
7	pond	none	1			6		
8	ditch	none	33	1		0		
Total			330	28	20	80	1	2

ous mycosis (chromomycosis). Scars were found on 4 healthy adults from potato and sweet corn fields. No parasites were found during the physical examination except for 1 adult (at control site 6) with a leech on a hindlimb.

Deformities were found in 20 juveniles (7.4%) from the 2 potato sites (Table 2). In general, the deformities consisted of conspicuous skeletal and muscular defects of the hindlimbs which did not appear to be caused by mechanical amputation. One individual also had a deformed forelimb. Noticeable lesions included agenesis or segmental hypoplasia of 1 or more long bones (ectromelia) and absence of all or part of a digit (ectrodactyly). No individuals with similar deformities were found at the other sites but 1 adult male had abnormal muscular tissue on 1 hindlimb (sweet corn site 3) and another had a single enlarged testis (control site 6).

All adult frogs providing blood samples were diagnosed as normal based on the physical and post-mortem examinations. Hematological and biochemical values varied considerably among individuals from a given population. Differential white cell counts and about ¼ of the biochemical parameters had coefficients of variation > 50%. Part of the variation was related to sex and body size: hematocrit values, red blood cell counts, hemoglobin, and mean corpuscular hemoglobin concentrations differed by sex (Mann Whitney $P < 0.05$); hematocrit values, Na^+, Cl^- and CO_2 were correlated to body weight (Kendall's τ, $P < 0.05$). Small differences (Kruskal-Wallis and Mann-Whitney, $P < 0.05$) between sites, whether for the same or different crops, were obtained for a few parameters. However, these variations were often related to differences in body size, and thus seemed to fall within the range of normal health status. Differential white cell counts did not vary among crops or sites (ANOVA, n = 3 replicates of 82 adults, $P > 0.05$ for each cell type), and were

Metamorphosing green frog, Rana clamitans, *with a deformed right hind leg. Photo by Martin Ouellet.*

comparable between sick juveniles (red leg and/or hepatic lipidosis, n = 6) and other healthy individuals.

Most frogs (n = 89) carried intraerythrocytic parasites. Thirty-seven percent had *Lankesterella minima*, 27% had the rickettsia *Aegyptianella ranarum*, and 22% had *Hepatozoon* sp. (= *Haemogregarina* sp.). Three species of *Trypanosoma* sp., including *T. rotatorium* in 21% of the individuals and *T. ranarum* in 11%, were frequent in the plasma. These frequencies were similar between crop and control sites. Microfilariae were found in the blood of 2 individuals.

Although micronuclei were seen in some erythrocytes, cellular abnormalities were infrequent. The mean number of micronuclei in erythrocytes from the 82 adults and 7 juveniles was 0.6 ± 1.0‰ and 0.4 ± 0.5‰ respectively. Frequencies were also similar among crop and control sites. The mitotic index was 0.2 ± 0.5‰ for both juveniles and adults.

Acetylcholinesterase activity levels were similar regardless of crops (Kruskall-Wallis $P > 0.05$). The mean levels (in μmole/min/g ± SD) were 22.2 ± 5.3 for potato (n = 41), 21.2 ± 6.0 for sweet corn (n = 36), and 22.4 ± 3.6 for the controls (n = 27). Activity levels were not correlated to brain or body mass (Kendall's τ, $P > 0.05$ in both cases), and were thus similar in juveniles and adults (Mann-Whitney, $P > 0.05$). Lower acetylcholinesterase activity levels were measured in some of the samples for all crop scenarios. A test for enzyme reactivation was possible with only a few samples (4 for potato, 5 for sweet corn, and 1 for the controls) because the other brains were too small. Only 1 sample from potato site 1 gave a positive response with acetylcholinesterase activity increasing from 14.9 to 31.7 μmole/min/g after chemical reactivation by 2-PAM, suggesting prior inhibition by an organophosphate pesticide.

DNA content analysis indicated significant differences between the crop and control sites. Abnormal profiles (including aneuploids) were more frequent at the sweet corn than at the control sites (Fisher's exact test, $P = 0.03$). DNA content values were consistent with those known for *R. clamitans* (6.0 to 6.6 pg: L. Lowcock, pers. data), but there were differences in intra-individual genome size variability (CV values) between the crop and control sites as well as between juveniles and adults. In juveniles, CVs at the sweet corn sites were greater than at the control sites (Mann-Whitney, $P < 0.05$). In adults, CVs at both the potato and the sweet corn sites were greater than at the control sites (Mann-Whitney, $P < 0.01$ in both tests). Deformed juveniles (n = 3) from site 2 had greater CVs than normal ones (n = 18) from the same pond (Mann-Whitney, $P < 0.05$). There was no difference in CVs between healthy and sick juveniles (Mann-Whitney, $P > 0.05$) and the prevalence of abnormal DNA profiles and high CVs was not related to higher micronuclei and intraerythrocytic parasite frequencies.

A significant genotoxic effect was detectable in all the water samples. The samples from croplands appeared more genotoxic than that those from control site 6 (Fig. 1). The highest values of cellular toxicity were measured in water samples from potato fields. Large spatial and temporal variability in water sample toxicity was revealed by our tests (sites 1 and 2 in Fig. 1). The highest values of genotoxicity were comparable with industrial effluents of intermediate toxicity (Gagné and Blaise 1995).

Discussion

Physical and postmortem examinations provide an immediate assessment of frog health. However, diagnosis of disease often requires extensive use of microbiology and histopathology (Ouellet et al. 1994). Histopathology will also identify neoplasms (Rose and Harshbarger 1977; Mizgireuv et al. 1984). Although tumors were not diagnosed in this study, they are known to occur sporadically in frogs and would be expected to develop more frequently under the influence of certain chemicals (Rose and Harshbarger 1977; Mizgireuv et al. 1984; Crawshaw 1992a). Die-offs are particularly difficult to study because remaining bodies are often rotten.

Hepatic lipidosis has been linked with chemical poisoning in mammals (Jones and Hunt 1983), and red leg in amphibians is known to break out following environmental stressors (Crawshaw 1992a, this volume). However, causative factors for these diseases in our study still remain unknown. Hepatic lipidosis, red leg, and mortality encountered at potato site 2 has been also observed in remote habitats (Ouellet et al. 1994), which suggests that these diseases are not specific to cases of pesticide intoxication. Site 2 had a relatively high density of metamorphosing frogs. Three species of fish (*Culaea inconstans, Notemigonus crysoleucas*, and *Pimephales promelas*) and a leech (*Macrobdella decora*) were also found dying in this pond. High biological oxygen demand, water temperature increases, inputs of organic matter from cropland runoff, and lowered water levels resulting from pumping for irrigation are possible environmental stressors, independent of pesticide contamination, that might have contributed to this disease outbreak. Such conditions may be less than optimal for the frogs and may change the pond's microbial balance in favour of pathogenic organisms (Schotts et al. 1972).

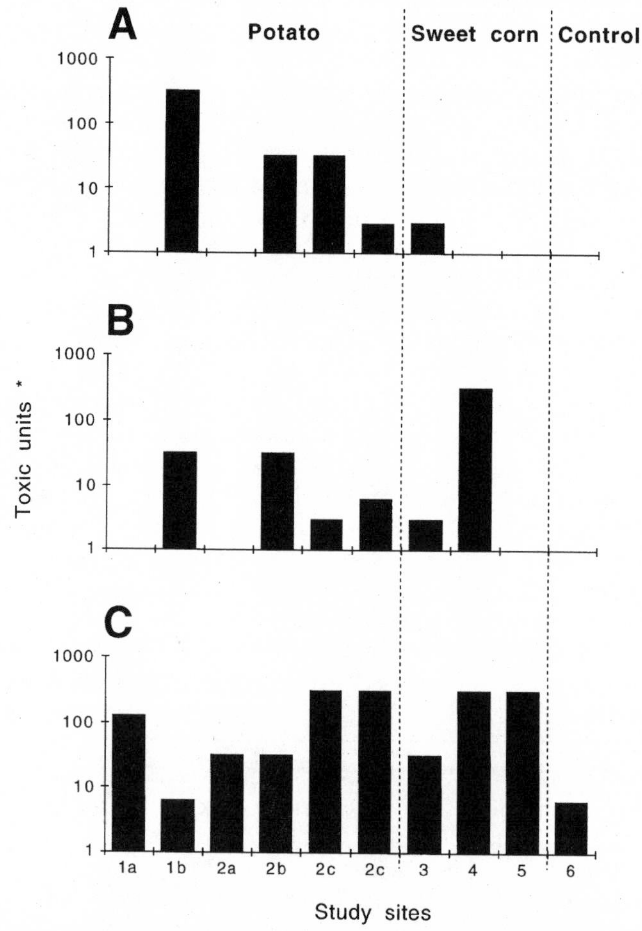

Figure 1. *Relative water toxicity values of 10 samples taken from potato fields, sweet corn fields, and a control site using 3 different bioassays: a) cellular viability, b) genotoxicity - DNA alkaline precipitation assay, c) genotoxicity - nick translation assay. Toxic unit = 100% / geometric mean of the no-observable-effect concentration and the lowest-observable-effect concentration in % v/v (Costan et al., 1993). Study sites are listed in Table 2. Samples "a" and "b" were from 2 different areas in the same pond while samples "c" were collected 3 days later.*

Limb deformities in metamorphosing frogs are uncommon events that have only rarely been studied in North America (Ouellet et al. 1997). The types of limb deformities encountered here contrast with those most commonly reported in the literature (polymely and polydactyly), but their frequency and unpredictable occurrence were comparable to what has been reported elsewhere (Rostand 1971; Dubois 1979; Mizgireuv et al. 1984; Borkin and Pikulik 1986; Vershinin 1989). Developmental defects could originate from genetic, abiotic (xenobiotic pollutants or environmental conditions), biotic (biological products or parasites [Sessions and Ruth 1990]) and nutritional factors. At the potato sites, putative genetic damage in the frogs, together with the genotoxicity of the water samples, suggest the action of environmental mutagens. The known teratogenic effect of various pesticides and fertilizers on eggs and tadpoles of frogs (Hazelwood 1970; Cooke 1981; Harfenist et al. 1989; Bishop 1992) raises a serious possibility that agricultural pollutants are the primary cause of the observed developmental defects.

Disease and deformity were not observed in all situations. It may be that they occurred only when toxic events were synchronous with specific sensitive stages of anuran development. Temporal variation in toxicity was suggested by repeated water sampling at site 2, and is reported in more detailed surveys (Berryman and Giroux 1994). Moreover, the prevalence of deformities in hindlimbs over other organs suggests a peculiar sensitivity of the hindlimb buds during their early development stage. Our results also indicate the susceptibility of frogs to disease and mortality during metamorphosis. During the resorption of the tail at metamorphosis, bioaccumulated pesticides are mobilized and can cause the death of young frogs (Cooke 1970).

The prevalence of disease and deformity may also be related to a greater contamination risk at some sites. Improper disposal of pesticide containers is a risk factor and the location of a pond may be important as well. For example, site 2, where diseases and deformities were frequent, is an isolated pond located downhill from a field, with few plants along the shore to stop runoff. A large amount of sediment was reaching this pond, and white foam were seen floating along the shore after a rain. This site was considered to be the most exposed to contamination and water tests indicated a correspondingly high toxicity.

Blood sampling in the field proved to be very difficult with tadpoles and moribund juveniles. In these debilitated animals, blood samples were often clotted and contaminated by lymph fluid, and as a result, hematological values were inaccurate. Complete hematology and biochemistry data were thus obtained only from apparently healthy adults, which limits the usefulness of the blood parameters in detecting the sublethal effects of agricultural pollutants.

Kaplan and Glaczenski (1965) observed decreased white blood cell counts and variations in the differential white cell counts following laboratory exposure to pesticides in the northern leopard frog, *Rana pipiens*. Our results do not confirm such clinical findings. The variation in many of the hematological and biochemical parameters with sex and body size in our data, and with other factors such as reproductive state, season, body temperature, nutritional state, desiccation, disease, and stress from recent capture (Hutchisson and Szarski 1965; Harris 1972; Crawshaw 1992b) make it difficult to interpret the results. In addition, there are few reference data for comparative purposes. A reliable diagnosis cannot be obtained from hematological data if sample size is small, and if the normal variability for a given species is unknown (Hutchisson and Szarski 1965). Consequently, controlled

Improper disposal of empty pesticide containers. Photo by Martin Ouellet.

laboratory tests or analyses of large samples from populations in the wild are necessary to determine whether pesticide treatments affect hematological and biochemical values.

Blood smears enabled evaluation of parasitic load. Hemoparasites were frequent in aquatic frogs and seemed to be well tolerated in the individuals examined. This appears to be a normal condition since Barta and Desser (1984) found similar frequencies of *Haemogregarina* sp. (in 49% of the individuals, n = 57), *Lankesterella minima* (14%), *Trypanosoma rotatorium* (44%), and *T. ranarum* (11%) in a population of *R. clamitans* from an unpolluted lake in Ontario.

Consistent acetylcholinesterase activity levels in our results probably represent normal activity levels for the frogs during the summer. The values were comparable for all crop sites and evidence of inhibition was rare. This is in agreement with previous studies (Andersen and Mikalsen 1978; Wang and Murphy 1982), suggesting that anurans are very resistant to acetylcholinesterase inhibition associated with pesticide exposure. However, for this assessment, the collection of frogs immediately after intoxication remains an essential condition which was hardly verifiable in the field. Furthermore, enzyme reactivation could not be performed on brains of juveniles because they were too small (50 mg) for sub-sampling.

The micronucleus test has been proposed as a simple and reliable method for evaluating genotoxic effects of freshwater pollutants (Fernandez et al. 1993). In the wild animals studied here, the frequency of micronuclei was low (< 1‰) compared to what has been observed in laboratory exposition tests with larvae of ranid frogs (Krauter et al. 1987; Krauter 1993) and other amphibians (Fernandez et al. 1993). Frequencies were not higher in the pesticide-exposed populations than in the controls. This suggests that these scores represent spontaneous frequencies of micronuclei. The spontaneous frequency of micronuclei in circulating erythrocytes of metamorphosing tadpoles of bullfrogs, *Rana catesbeiana*, (Krauter et al. 1987) was comparable to our scores in juvenile and adult *R. clamitans*. We also obtained comparable low values for adult *R. catesbeiana* collected from our study sites (unpubl. data). Changes from tadpole to adult erythropoietic centers and the possibility of selective removal of micronucleated erythrocytes from the peripheral blood of frogs are important consideration for the micronucleus assay (Krauter et al. 1987). Investigation of these aspects is required before we decide on the value of this test for wild metamorphosed anurans.

Flow cytometric analysis of DNA provided evidence of hidden damage which could not be evaluated using the other techniques in our study. Comparison of micronuclei scores and flow cytometric results indicated no relations between the 2 assessments of genotoxic effects. The correlation between putative DNA damage (high CVs) and hindlimb deformity does not necessarily indicate a causal link, although this remains a possibility. The use of flow cytometry in the study of environmental mutagens is in its infancy and ours has been its 1st application in assessing genomic disruption in wild amphibian populations (Lowcock et al. 1997; Sharbel et al., this volume). Comprehensive interpretation of the flow cytometric results will require further controlled investigations of the mechanisms and cells (somatic vs gonadal) which are involved.

The prevalence of genetic damage and the genotoxicity of water samples indicate that agricultural practices are affecting habitat quality and frog health. However, contamination of habitats or carcasses by pesticides was not determined and we cannot incriminate specific pesticides or crops. Both the potato and the sweet corn sites, which were exposed to different pesticides, had frog populations with similar genetic defects. Diverse degrees of mutagenicity have been reported for some of the pesticides used at the test sites, including azinphos-methyl, carbofuran, cypermethrin, endosulfan and diquat (Hayes and Laws 1991). Controlled *in vivo* toxicological studies will be required to identify their specific effects. Though pesticide residues in watercourses (Wauchope 1978; Giroux et Morin 1992) are usually at lower concentrations than those used in laboratory tests (Harfenist et al. 1989; Bishop 1992), actual contamination processes under field conditions (i.e. cumulative exposure, multiplicity of products, synergistic effects, etc.) might increase the risk of genotoxicity.

In evaluating the value of health assessment methods in detecting effects of agricultural pollutants, consideration should be given to the fact that pesticides and other agricultural pollutants might produce a broad spectrum of sublethal and lethal effects that are still poorly understood. The various methods can supplement each other for this task. Physical examination is easily applied in the field, and provides valuable information on disease and deformity. Blood samples for DNA, blood smears, and selected hematological or biochemical parameters can also be obtained without sacrificing the animal. Flow cytometry has proven to be a potentially useful tool for the measurement of the effects of environmental mutagens (Sharbel et al., this volume). Other approaches used here require sacrificing the animal. In this study, cholinesterase analysis, complete hematology, and blood biochemistry profiles were not feasible on juveniles, and gave few additional indications of health problems associated with agriculture. On the other hand, postmortem examination was crucial in the diagnosis of disease, and should include histopathological examination and bacterial cultures. A complete physical and postmortem examination, including blood, tissue and parasite collection, took a veterinarian and a technician approximately 45 min for an adult frog.

Health assessment proved to be a practical approach to detecting the effects of environmental stressors on amphibians in the wild. This approach is attractive in its simplicity of sampling (frogs are caught at random in their habitat), as compared to population studies where survivorship, recruitment and density need to be determined. However, population level studies are required to understand the ultimate impact of diseases on populations in the wild. Health is usually determined from survivors, and it is not known whether sick frogs are easier or harder to find than healthy ones. If sublethal responses to some contaminants are skewed toward lethality, and dead eggs, tadpoles, or frogs are not readily found, the impact of agricultural pollutants might be underestimated.

Acknowledgments

We wish to thank all the property owners and conservation officers involved for their help. For their professional advice in the identification of organisms or diagnosis of diseases, we thank J. Bergeron (Ministère de l'Environnement et de la Faune), G. J. Crawshaw (Metropolitan Toronto Zoo), I. K. Barker and J. R. Barta (University of Guelph), and S. Lair and D. Martineau (Université de Montréal). For their support as field technicians, we thank Y. Bachand, J. Boulé, R. Dauphin, and also D. Fontaine and E. Barten who, in addition, were excellent laboratory technicians. We also thank S. Trudeau for the acetylcholinesterase analysis and D. M. Green, F. Halwani, and C. Smith for their help in the genetic analysis. This study was funded by the Canadian Wildlife Service (Québec Region). We are also grateful for additional funding from Agriculture and Agri-Food Canada.

Literature Cited

Andersen RA, Mikalsen A. 1978. Substrate specificity, effect of inhibitors and electrophoretic mobility of brain and serum cholinesterase from frog, chicken and rat. General Pharmacology 9:177–181.

Barta JR, Desser SS. 1984. Blood parasites of amphibians from Algonquin Park, Ontario. Journal of Wildlife Diseases 20:180–189.

Berryman D, Giroux I. 1994. La contamination des cours d'eau par les pesticides dans les régions de culture intensive de maïs au Québec. Québec: Ministère de l'Environnement et de la Faune du Québec, Direction des écosystèmes aquatiques.

Bickham JW, Hanks BG, Smolen MJ, Lamb T, Gibbons JW. 1988. Flow cytometric analysis of low-level radiation exposure on natural populations of slider turtles (*Pseudemys scripta*). Archives of Environmental Contamination and Toxicology 17:837–841.

Bishop CA. 1992. The effects of pesticides on amphibians and the implications for determining causes of declines in amphibian populations. In: Bishop CA, Pettit KE, editors. Declines in Canadian amphibian populations: designing a national monitoring strategy. Ottawa: Canadian Wildlife Service. Occasional Paper 76. p 67–70.

Borkin LJ, Pikulik MM. 1986. The occurrence of polymely and polydactyly in natural populations of anurans of the USSR. Amphibia-Reptilia 7:205–216.

Canadian Council on Animal Care. 1984. Guide to the care and use of experimental animals, Vol. 2. Ottawa: Canadian Council on Animal Care.

Cooke AS. 1970. The effects of pp'-DDT on tadpoles of the common frog (*Rana temporaria*). Environmental Pollution 1:57–71.

Cooke AS. 1981. Tadpoles as indicators of harmful levels of pollution in the field. Environmental Pollution Series A 25:123–133.

Costan G, Bermingham N, Blaise C, Ferard JF. 1993. Potential ecotoxic effects probe (PEEP): a novel index to assess and compare the toxic potential of industrial effluents. Environmental Toxicology and Water Quality 8:115–140.

Crawshaw G J. 1992a. The role of disease in amphibian decline. In: Bishop CA, Pettit KE, editors. Declines in Canadian amphibian populations: designing a national monitoring strategy. Ottawa: Canadian Wildlife Service. Occasional Paper 76. p 60–62.

Crawshaw GJ. 1992b. Amphibian medicine In: Kirk RW, editor. Kirk's Current Veterinary Therapy XI. Philadelphia: W.B. Saunders. p 1219–1230.

Dubois A. 1979. Anomalies and mutations in natural populations of the *Rana "esculenta"* complex (Amphibia, Anura). Mitteilungen aus dem Zoologischen Museum in Berlin 55:59–87.

Ellman GL, Courtney KD, Andres V, Jr, Featherston RM. 1961. A new and rapid colorimetric determination of acetylcholinesterase activity. Biochemistry and Pharmacology 7:88–95.

Fernandez M, L'Haridon J, Gauthier L, Zoll-Moreux C. 1993. Amphibian micronucleus test(s): a simple and reliable method for evaluating in vivo genotoxic effects of freshwater pollutants and radiations. Initial assessment. Mutation Research 292:83–99.

Gagné T, Blaise C. 1995. Genotoxicity evaluation of environmental contamination to rainbow trout hepatocytes. Environmental Toxicology and Water Quality 10:217–229.

Giroux I. 1993. Contamination de l'eau souterraine par l'aldicarbe dans les régions de culture intensive de pommes de terre—1984 à 1991. Québec: Ministère de l'Environnement, Direction du milieu agricole et du contrôle des pesticides.

Giroux I. 1994. Contamination de l'eau souterraine par les pesticides et les nitrates dans les régions de culture de pommes de terre. Québec: Ministère de l'Environnement et de la Faune, Direction des écosystèmes aquatiques.

Giroux I, Morin C. 1992. Contamination du milieu aquatique et des eaux souterraines par les pesticides au Québec. Québec: Ministère de l'Environnement, Direction du milieu agricole et du contrôle des pesticides.

Hall RJ, Henry PFP. 1992. Assessing effects of pesticides on amphibians and reptiles: status and needs. Herpetological Journal 2:65–71.

Harfenist A, Power T, Clark KL, Peakall DB. 1989. A review and evaluation of the amphibian toxicological literature. Canadian Wildlife Service Technical Report Series 61:1–222.

Harris JA. 1972. Seasonal variation in some hematological characteristics of *Rana pipiens*. Comparative Biochemistry and Physiology 43:975–989.

Hayes WJ Jr, Laws ER Jr. 1991. Handbook of pesticide toxicology. San Diego, CA: Academic Press.

Hazelwood E. 1970. Frog pond contaminated. British Journal of Herpetology 41:177–184.

Hill EF, Fleming WJ. 1982. Anticholinesterase poisoning of birds: field monitoring and diagnosis of acute poisoning. Environmental Toxicology and Chemistry 1:27–38.

Hutchisson VH, Szarski H. 1965. Number of erythrocytes in some amphibians and reptiles. Copeia 1965:373–375.

Jones TC, Hunt RD. 1983. Veterinary pathology. 5th ed. Philadelphia: Lea and Febiger.

Kaplan HM, Glaczenski SS. 1965. Hematological effects of organophosphate insecticides in the frog (*Rana pipiens*). Life Sciences 4:1213–1219.

Krauter PW. 1993. Micronucleus incidence and hematological effects in Bullfrog tadpoles (*Rana catesbeiana*) exposed to 2-acetylaminofluorene and 2-aminofluorene. Archives of Environmental Contamination and Toxicology 24:487–493.

Krauter PW, Anderson SL, Harrison FL. 1987. Radiation-induced micronuclei in peripheral erythrocytes of *Rana catesbeiana*: an aquatic animal model for in vivo genotoxicity studies. Environmental and Molecular Mutagenesis 10:285–296.

Licht LE, Lowcock LA. 1991. Genome size and metabolic rate in salamanders. Comparative Biochemistry and Physiology B 100:83–92.

Lowcock LA, Sharbel TF, Bonin J, Ouellet M, Rodrigue J, DesGranges J-L. 1997. Flow cytometric assay for *in vivo* effects of pesticides in green frogs (*Rana clamitans*). Aquatic Toxicology (in press).

Martin AD, Norman G, Stanley P, Westlake GE. 1981. Use of reactivation techniques for the differential diagnosis of organophosphorus and carbamate pesticide poisoning in birds. Bulletin of Environmental Contamination and Toxicology 26:775–780.

Martof BS. 1953. Home range and movements of the green frog, *Rana clamitans*. Ecology 34:529–543.

Mizgireuv IV, Flax NL, Borkin LJ, Khudoley VV. 1984. Dysplastic lesions and abnormalities in amphibians associated with environmental conditions. Neoplasma 31:175–181.

Olive RL, Chan APS, Cu CS. 1988. Comparison between the DNA precipitation and alkali unwinding assays for detecting DNA strand breaks and cross links. Cancer Research 48:6444–6448.

Ouellet M, Bonin J, Rodrigue J, DesGranges J-L. 1994. Diseases investigation, pathological findings and impact on anuran populations in southern Québec. In: Proceedings of the 4th annual meeting of the Task Force on Declining Amphibian Populations in Canada. Winnipeg, MB: Manitoba Museum of Man and Nature. p 85–89.

Ouellet M, Bonin J, Rodrigue J, DesGranges J-L, Lair S. 1997. Hindlimb deformities (ectromelia, ectrodactyly) in free-living anurans from agricultural habitats. Journal of Wildlife Diseases 33:95–104.

Rose FL, Harshbarger JC. 1977. Neoplastic and possibly related skin lesions in neotenic tiger salamanders from a sewage lagoon. Science 196:315–317.

Rostand J. 1971. Les étangs à monstres. Histoire d'une recherche (1947–1970). Paris: Stock.

Schotts EB, Gains JL, Martin L, Prestwood AK. 1972. *Aeromonas*-induced deaths among fish and reptiles in a eutrophic inland lake. Journal of American Veterinary Medical Association 161:603–607.

Sessions SK, Ruth SB. 1990. Explanation for naturally occurring supernumerary limbs in amphibians. Journal of Experiemental Zoology 254:38–47.

Snyder RD, Matheson DW. 1985. Nick translation - A new assay for monitoring DNA damage and repair in cultured human fibroblasts. Environmental Mutagenesis. 7:267–279.

Vershinin VL. 1989. Morphological anomalies in urban amphibians. Ékologiya 3:58–66. (in Russian)

Wang C, Murphy SD. 1982. Kinetic analysis of species difference in acetylcholinesterase sensitivity to organophosphate insecticides. Toxicology and Applied Pharmacology 66:409–419.

Wauchope RD. 1978. The pesticide content of surface water draining from agricultural fields: a review. Journal of Environmental Quality 7:459–472.

Worthing CR, Hance RJ. 1991. The pesticides manual, 9th ed. Farnham, U.K.: The British Crop Protection Council.

©1997 by the Society for the Study of Amphibians and Reptiles
Amphibians in decline: Canadian studies of a global problem. David M. Green, editor.
Herpetological Conservation 1:258–270.

Chapter 26

DISEASE IN CANADIAN AMPHIBIAN POPULATIONS

Graham J. Crawshaw

Metropolitan Toronto Zoo, 361A Old Finch Avenue, Scarborough, Ontario M1B 5K7, Canada

ABSTRACT.—It has not yet been established that any Canadian amphibian species has suffered decline as a result of infectious or non-infectious disease alone, but there is mounting evidence from several countries, including Canada, that viral and bacterial infections are directly or indirectly involved in episodes of mortality in amphibian populations. Die-offs of leopard frogs in Manitoba in the 1970s and 1980s were probably caused by bacterial infections secondary to oxygen depletion or other environmental stressors. Viruses have been found in Ontario frogs but have not yet been implicated in die-offs, as they have in other parts of the world. A variety of blood parasites, particularly trypanososmes, have been recognized in amphibians in Ontario and Québec. Leeches have been incriminated as vectors for some hemoparasites. Protozoan and metazoan parasites, although common, are unlikely to be mediators of amphibian decline. There is growing evidence that environmental toxins have caused morphologic abnormalities, and even death, in amphibians in some localities.

RÉSUMÉ.—Nous n'avons pas encore pu déterminer avec certitude si les espèces canadiennes d'amphibiens diminuaient à cause de maladies infectieuses ou non infectieuses mais de plus en plus de preuves tendent à démontrer, dans plusieurs pays, incluant le Canada, que les infections virales et bactériennes sont directement ou indirectement liées à des épisodes aigus de mortalité dans les populations d'amphibiens. La disparition rapide des grenouilles léopards du Manitoba dans les années 1970 et 1980 a probablement été causée par des infections bactériennes secondaires à l'appauvrissement en oxygène ou à d'autres stress environnementaux. Des virus ont été découverts dans des grenouilles de l'Ontario mais rien ne permet encore d'affirmer qu'ils sont responsables de cette mortalité massive, comme c'est le cas dans d'autres pays. Plusieurs parasites sanguins, notamment des trypanosomes, ont été identifiés dans des amphibiens de l'Ontario et du Québec. Les sangsues semblent jouer un rôle de vecteur de certains parasites sanguins. Les parasites protozoaires et métazoaires, quoique courants, ne semblent pas jouer un rôle de médiateurs dans le déclin des amphibiens. De plus en plus de preuves tendent à démontrer que les toxines environnementales sont à l'origine d'anomalies morphologiques, voire de mortalité massive chez les amphibiens de certaines localités.

As evidence is collected for declines of certain populations of amphibians worldwide, there is increased awareness of the effect of infectious and non-infectious disease on fitness, reproductive success, and survival (Bradford 1991; Carey 1993). Most of the reports of disease in Canadian free-living amphibians to date have been either anecdotal or are records of the distribution or prevalence of parasites. The effect of the infectious agent on the host has mostly been ignored or has been a secondary consideration. However, there is some evidence of an association of infectious disease with die-offs in at least 1 amphibian species in Canada (Koonz 1992).

Defining amphibian diseases and measuring their effects are not easy. Microorganisms and parasites that produce demonstrable pathological effects on an individual may have no impact on longevity, overall fitness, or reproductive success of a population. Neither can disease be evaluated without consideration of the interactions of environmental and ecological factors with the physiology and behaviour of the animal. Many published reports of disease outbreaks in wild amphibian populations are based on findings from small and statistically insignificant numbers of animals. In other reports diagnosis has been based on isolation of bacteria alone without histopathological examination of tissues. Identification or isolation of any agent must be correlated with clinical findings and histologic changes to substantiate a disease claim. Experimental duplication of the disease with the purified agent will provide definitive diagnostic evidence.

INFECTIOUS AGENTS

Viruses

An increasing number of viruses are being identified in amphibians worldwide, but, for the most part, studies of their effect upon the host have been limited. There is mounting evidence, however, that die-offs in local amphibian populations have been caused, or been precipitated, by viral infections. Viruses should be considered in any disease investigation even if more apparent causes, such as bacteria, are identified. Viruses may themselves suppress immune function leading to death from bacterial disease, and viral infections may be overlooked unless thorough investigations are performed.

Some examples of viruses identified in amphibian populations include a DNA-virus (given the name Bohle iridovirus, or BIV) that was the cause of death in young frogs (*Lymnodynastes ornatus)* brought into captivity in Australia (Speare and O'Shea 1989; Speare and Smith 1992). More recent evidence indicates that a similar virus may be responsible for precipitous decline of several tropical frog species in northern Australia (Laurance et al. 1996). The Lucké herpesvirus can cause significant abnormalities (renal tumours) in infected leopard frogs, *Rana pipiens*, in the United States (McKinnell 1984), and there was a high correlation with a poxvirus-like infection in die-offs of common frogs, *R. temporaria*, in England (Cunningham et al. 1993). Mortalities from viral infections have also been seen in batches of anuran tadpoles brought into captivity in the United States. Of particular interest is the report of a virus isolated from an amphibian that has been shown experimentally to cause high mortality in certain species of fish (Moody and Owens 1994), which suggests that agents might be sought beyond usual taxonomic boundaries.

In Canada, an intraerythrocytic virus was identified in bullfrogs (*Rana catesbeiana*), mink frogs (*R. septentrionalis*), and green frogs (*R. clamitans*) in Ontario in the 1980s (Desser and Barta 1984a; Gruia-Gray et al. 1989). In some *R. catesbeiana*, ≥ 90% of erythrocytes were found to be infected. Most of these heavily affected frogs were weak and listless, and some died in captivity. Electron microscopy revealed the viral particles were consistent with the icosahedral cytoplasmic virus group, and the virus has been designated as frog erythrocytic virus (FEV). Subsequent studies on FEV have associated the virus with anemia and increased mortality in young frogs, although adult *R. catesbeiana* were refractory to infection (Gruia-Gray and Desser 1992). A similar virus was seen in the white blood cells (monocytes and heterophils) of mink and green frogs in Ontario (Desser 1992). Viruses of this type have been incriminated as the cause of disease in a number of amphibian species around the world, usually affecting tadpoles and recently metamorphosed juveniles (Wolf et al. 1968; Speare and O'Shea 1989; Speare and Smith 1992; Speare et al. 1991, unpublished). However, many amphibians harbour these viruses without evidence of disease or pathological lesions. There is huge scope for research into the natural history of amphibian viruses, their origin, and transmissibility to other animal groups.

Rickettsia

Non-viral intraerythrocytic inclusions were seen in blood smears from *Rana clamitans* in Ontario that on electron microscopy were characteristic of organisms of the Order Rickettsia. No evidence of pathogenicity was given (Desser and Barta 1984a). Similar bodies, identified as *Aegyptianella*, have been seen in European frog species (Desser and Barta 1989).

Chlamydia

The intracellular organism *Chlamydia* has caused mortality in zoo and laboratory frogs but there have been no reports of infection in free-living amphibians.

Bacteria

There are an increasing number of reports of mortality associated with bacteria in North American amphibians as well as from other parts of the world. In some cases mortality moves in waves through an area suggesting that a causative agent is dispersing (Laurance et al. 1996). Studies are needed to determine whether these incidents represent particularly virulent strains of bacteria, or whether viruses, chemicals, environmental stressors, or other agents are increasing susceptibility to bacterial disease. Changes in environmental conditions, particularly cold spells, can precipitate die-offs but if that were the only causative factor, it would be expected that populations would rebound in years of normal weather conditions. It appears that, in many areas, populations are not returning to previous levels, and that mortalities continue in the residual populations (Koonz 1992).

Because most pathogenic bacteria associated with disease in amphibians may be found in the environment, or even within apparently healthy animals, finding a causal relationship between microorganism, disease, and mortality is not easy. However, there is now enough confirmed and anecdotal evidence to suggest that bacteria are the ultimate, if not the proximate, causes of die-offs in certain amphibian populations. The bacterial disease red leg, usually associated with the bacterium *Aeromonas hydrophila,* has on several occasions been recognised in wild amphibians, but primary stressors have not been identified, although they are frequently implicated (Nyman 1986; Koonz 1992; Carey 1993).

In the United States red leg was incriminated in a mass mortality in a population of the mountain yellow-legged frog, *Rana muscosa.* Affected animals showed the typical signs of lethargy, poor coordination, with visible congestion and hemorrhage on the limbs and ventral surfaces (Bradford 1991). However, the carcasses were also emaciated, a feature that suggests a more chronic underlying disease or nutritional process that predisposed the frogs to a terminal *Aeromonas* septicemia. Bacterial infections were recognised in reductions in frog populations in Wisconsin (Hine et al. 1981), and the serious decline of the endangered Wyoming toad, *Bufo hemiophrys baxteri*, has also been associated with a bacterial disease which has continued to cause morbidity and mortality in captive animals.

Similar die-offs have been identified in other species elsewhere in North America, including wood frogs, *Rana sylvatica*, in Rhode Island (Nyman 1986), Tarahumara frogs, *Rana tarahumarae*, in Arizona, Yosemite toads, *Bufo canorus,* in California, and ranids in New Mexico (Scott 1993). In many cases mortality occurred in post-metamorphic animals prompting the term "Post-metamorphic Death Syndrome", PDS, to be proposed by Scott (1993).

In northern Arizona and Utah, salamander populations were found to experience mass mortality in which most larvae died within a week, with signs of lethargy and reddening of the body, especially as they approached metamorphosis (Pfennig et al. 1991). High levels of clostridial bacteria were present in the livers. Similarly, the disappearance of boreal toads, *Bufo boreas*, from the mountains of western Colorado has been blamed at least in part on bacterial infections (Carey 1993). Again, bacterial septicemia was considered secondary to immune suppression caused by unknown factors, such as pH or temperature. Amphibians remain healthy by virtue of the competence of a variety of protective mechanisms including the immune system. When these protective mechanisms are compromised by environmental stresses, competition, poor nutrition, viruses, and a variety of other factors, bacterial disease may ensue.

Dead bullfrog, Rana catesbeiana, tadpoles. Photo by Martin Ouellet.

Bacterial infections were seen in individual *R. pipiens* in some locations in Alberta from 1976 to 1979, though not affecting entire populations (Roberts 1992). In Manitoba in the late 1970s and early 1980s, large die-offs of leopard frogs occurred in the shallow lakes of the Interlake region where they had normally existed in large enough numbers to attract collection for educational and scientific purposes (Koonz 1992). Piles of dead frogs were found, up to a metre deep at the "frog holes" and along shorelines. Die-offs occurred in relatively remote and pristine areas unaffected by human development, encroachment, or agriculture. Many of the frogs appeared to die from red leg (Koonz 1992). Although such catastrophes have not recurred in more recent years, smaller die-offs continue. The populations are still reduced in those areas previously affected by the outbreaks and yet frogs are present in reasonable numbers in surrounding areas (Koonz 1992). It appears that the factors which precipitate these epidemics remain in the environment. One possible factor is a reduction in available oxygen to hibernating and emerging frogs as a result of depletion by organic material and nitrogenous compounds (eutrophication), which may result in abnormally high levels of pathogenic organisms. Such conditions have been associated with mortality in reptiles and fish (Schotts et al. 1972). Amphibians may be even more susceptible in view of their permeable skins and susceptibility to environmental change.

Symptoms of red leg in wild amphibians include pinpoint hemorrhages on the skin, especially of the abdomen, hind legs, and tail. Edema, skin ulceration, and ocular lesions may also been seen. Affected animals may not be active feeders, and tadpoles may lie on the bottom of the pond or remain unresponsive close to the surface, rendering them more susceptible to predation.

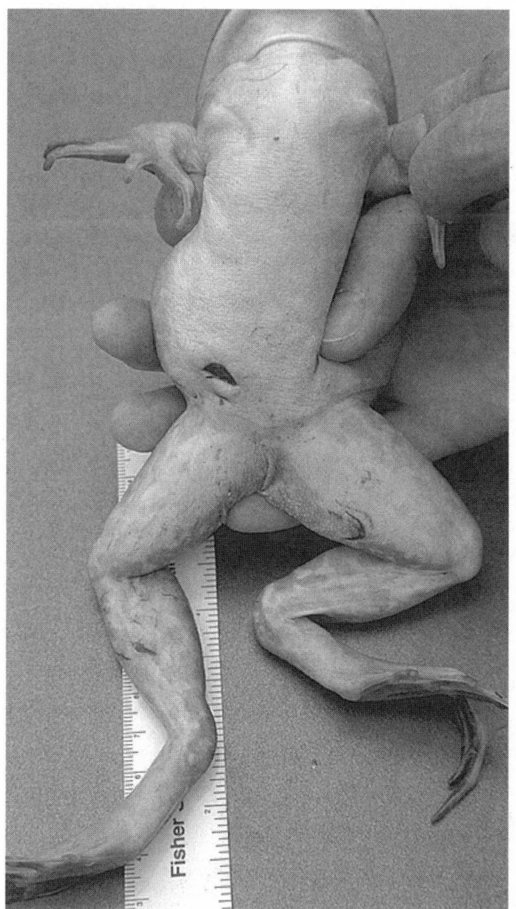

Bullfrog, Rana catesbiana, *with puncture wounds caused by attempted predation. Photo by Martin Ouellet.*

Isolation of particular pathogen from a carcass does not confirm the organism as the cause of death. Similar findings need to be observed in a number of animals, and correlated with the presence or absence of other organisms, and histopathologic findings. For definitive conviction, the disease should be reproduced using the purified agent.

Aeromonas is common but is an inconsistent pathogen. Other bacteria may cause septicemia characterised by erythema and hemorrhages, grossly and histologically indistinguishable from red leg. *Pseudomonas, Citrobacter*, and *E. coli* have also been associated with septicemia in Canadian amphibians examined in Toronto and Saskatoon. *Acinetobacter, Streptococcus*, and many other organisms have been incriminated elsewhere (Glorioso et al. 1973; Worthylake and Hovingh 1989). *Salmonella, Leptospira*, and other potentially pathogenic bacteria have been isolated from healthy amphibians worldwide without signs of disease (Everard et al. 1979, 1983, 1988).

Another intriguing possibility is that die-offs in certain amphibian populations could be caused by toxins produced by cyanobacteria. This group of bacteria, formerly known as blue-green algae, are found in fresh and salt water environments. They produce several types of potent toxins that have been known to kill animals as large as cattle. Increases in nitrogen and phosphorus in water, in association with warm temperatures and favourable pH, promote blooms of cyanobacteria. Other toxins are produced by the marine algae that cause "red tides" which result in mortality in marine invertebrates and secondary poisoning in animals and humans (Carmichael 1994).

Fungi

Fungal infections are frequently reported in wild, and captive tropical amphibians. Temperate species appear affected to a lesser degree. Fungal spores may be found dormant in the tissues and may develop into fulminating disease if the host's immune function is compromised. One fungus, *Basidiobolus*, has been a cause of mortality in captive *B. hemiophrys baxteri,* which has experienced catastrophic decline in recent years (T. Thorne, pers. comm.). Aquatic fungi such as *Saprolegnia* are usually considered to be secondary invaders of the skin and underlying musculature of aquatic forms and larvae. The mudpuppy, *Necturus maculosus*, is particularly susceptible. Recently the fungus has attracted attention as a cause of mortality of *B. boreas* eggs (Blaustein et al. 1994). There are no published reports of fungi causing disease in Canadian amphibians but they would not be expected to cause primary disease in wild populations.

PARASITES

There have been a number of studies in wild vertebrates attempting to correlate the level of parasitism on mate selection and hence success, in reproductive terms, of individuals (Hamilton and Zuk 1982). However, studies in amphibians such as the gray tree frog, *Hyla versicolor,* and Couch's spadefoot toad, *Spea couchii*, have failed to show any effect on reproductive success in animals with low or moderate parasite loads (Hausfater et al. 1990). Considering the diversity of host species and the nature of parasite life cycles, it is highly unlikely that parasitic diseases are having a serious impact on Canadian amphibians. In most environments host animals and their parasites exist in relative equilibrium. Most parasites do not benefit by killing their hosts, which would reduce the ability of the parasite itself to survive and reproduce. Heavy parasite loads certainly can affect growth rates and survivability, but the effects are usually seen in young and more susceptible subgroups. Extensive surveys for hemoprotozoal infections in Canadian amphibians have been conducted in southern Ontario and in Algonquin Park in central Ontario (Woo 1969; Desser and Barta 1984b). The prevalence of infection increased with age of the host, and the relative abundance of parasites was related to the presence of suitable vectors (Desser and Barta 1984b). For instance, aquatic frogs showed a higher incidence of infection with some trypanosomes than terrestrial species, suggesting that aquatic organisms such as leeches are important vectors in aquatic environments. Other trypanosomes were only seen in mature amphibians, which suggests that aerial vectors could be responsible for transmission in those species. Studies such as this still encompass a small fraction of the geographical range of some species. In most areas of the country there have been few attempts even to identify parasites or pathogenic organisms affecting amphibians. Surveys in United States complement Canadian studies in the same species (Levine and Nye 1977).

Protozoans

Many of the familiar protozoan parasites of other vertebrates have also been recognised in amphibians. Flagellates are commonly found within the intestinal tract but some, notably the diplomonads, may invade the blood and other organs. Heavy infections may be pathogenic. One such parasite, *Brugerolleia*, was first described from a Canadian *R. clamitans* in Ontario (Desser and Jones 1985; Desser et al. 1993).

Ciliates and the multinucleated opalinids may be found in the gastrointestinal tract of almost all amphibians but symptoms of disease are rare. Nyctotherids and balantidia are the most frequently found intestinal ciliates. *Trichodina* is an aquatic ciliate that can be pathogenic on the skin and gills of amphibian larvae. There are no reports of studies in Canadian amphibians but a similar organism, *Tetrahymena*, caused heavy mortality in larval spotted salamanders, *Ambystoma maculatum*, obtained from Michigan ponds (Ling and Werner 1988).

Coccidia.—The intestinal coccidians of Amphibia have been poorly described and the only published record in Canada is 1 survey of *R. catesbeiana, R. septentrionalis, R. sylvatica*, and *R. clamitans* by Chen and Desser (1989). Two *Eimeria* species were identified in the feces. Young adult *R. catesbeiana* were the most heavily infected but few adults harboured these parasites, which is consistent with this genus in other vertebrates. This trend was less apparent in *R. septentrionalis*. No histopathological studies were performed. Coccidia of all species including amphibians are usually very host-specific and pathogenic only in heavy infections. They are unlikely candidates for agents of disease or mass mortality in amphibians in Canada.

Hematozoa and trypanosomes.—The blood protozoans, or hematozoa, *Hepatozoon* (= *Haemogregarina*), *Thrombocytozoons, Lankesterella*, and *Babesiasoma* have been seen in Canadian amphibians (Fantham et al. 1942; Barta and Desser 1984, 1986; Desser and Barta 1984b; Bonin et al., this volume). In a survey for the erythrocytic parasite *Lankesterella* in *R. catesbeiana* in Ontario, 54.8% of tadpoles and 29.4% of adults were infected (Desser et al. 1989). The parasite was also found in *R. clamitans* and *R. septentrionalis*. Leeches appear to be responsible for transmission of the organism, because forms were also found within leeches parasitizing the amphibian hosts. However no development of the parastite occurs in the vector (Tse et al. 1986).

In recent years there have been numerous studies on the taxonomy, distribution, and biology of trypanosomes, the motile blood protozoans. Five species of trypanosome were identified in the blood of ranids in Ontario (Woo 1969). Two additional trypanosome species were found in *H. versicolor* in Ontario and southern Manitoba (Reilly and Woo 1982; Woo and Bogart 1984), while other species had been identified earlier in American toads, *Bufo americanus*, in Quebec (Fantham 1942). Some trypanosomes may be quite host-specific; for example, species obtained from hylids were not infective for five species of ranids, *Xenopus*, or *Notophthalmus* (Reilly and Woo 1982). This would make them poor candidates as causes of disease across taxonomic boundaries. Other trypanosome species have proven less selective in their choice of hosts. A single species *Trypanosoma ambystomae*, was found in five species of *Ambystoma* salamanders in southern Ontario (Woo et al. 1980). Trypanosomes were found in up to 26% of salamanders, and in *B. americanus* and several ranids in other studies in Ontario (Barta and Desser 1984; Woo and Bogart 1986) and Québec (Bonin et al., this volume). Trypanosomes are generally considered to be non-pathogenic in their natural hosts.

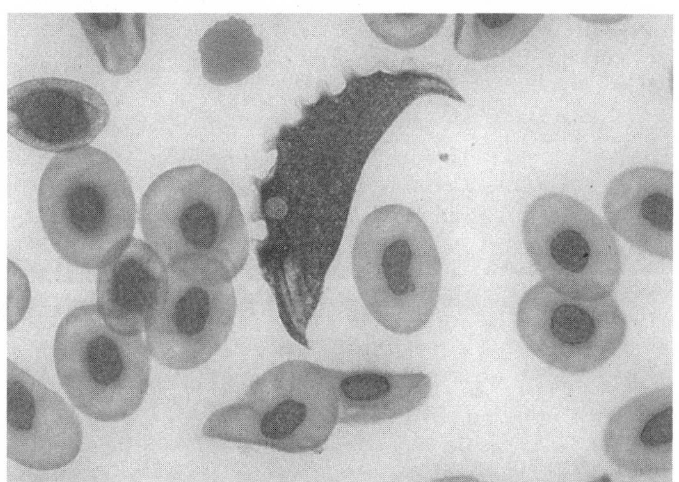

Trypanosome and red blood cells. Photo by Martin Ouellet.

Myxosporea.—Developmental and mature stages of the myxosporean *Sphaerospora ohlmacheri* were seen in the tubules and glomerular spaces of the kidneys of *R. catesbeiana* tadpoles in Ontario (Desser et al. 1986). Quite common and widely distributed, the myxosporeans are usually considered non-pathogenic to the host. However, the parasites in these bullfrogs were associated with extensive dystrophic changes in the renal tubular epithelium and cell necrosis. Although no definitive correlation was made between the parasite and the lesions, it is conceivable that heavy infections could affect renal function and hence survivability.

Nematodes

Nematodes of amphibians comprise 5 main groups. Many species of parasite have evolved concurrent with their hosts while some have remained little changed for millions of years (Baker 1984). Genera which have been identified in Canadian amphibians include the ubiquitous lungworm *Rhabdias* (Baker 1978b), *Ozwaldocruzia* (Baker 1987a), *Cosmocercoides* (Vanderburgh and Anderson 1987) and *Megalobactrachonema* (Richardson and Adamson 1988). Three species of *Rhabdias* were found in Ontario amphibians (Baker 1978b). The life cycles of *Rhabdias americanus* in *B. americanus*, in which it is very common, and *Rhabdias ranae* in *R. sylvatica* were elucidated from specimens obtained in Ontario (Baker 1979a,b). These parasites are common in the lungs of anurans and some caudates. Larva may be found in various tissues. Most infections cause no adverse effect upon the host. Seasonal changes in the population density of *Ozwaldocruzia pipiens* in *R. sylvatica* in Ontario were measured (Baker 1978a). *Megalobatrachonema*, a kathlaniid nematode was reported from larval and adult northwestern salamanders, *Ambystoma gracile*, in British Columbia (Richardson and Adamson 1988).

Microfilaria, presumed to be a *Foleyella* sp., were seen in the blood of *B. americanus, R. clamitans, R, sylvatica*, and *R. septentrionalis* in Ontario (Barta and Desser 1984). The adults of these worms are usually found in the fascial planes and subcutis of amphibians, with the larval microfilaria circulating in the bloodstream.

Trematodes

Amphibians are host to a large variety of monogenetic (single host) trematodes as well as both larvae and adults of digenetic (2 or more hosts) species. Intermediate stages (metacercariae) may be found in the skin, musculature, intestinal wall, kidneys, and other tissues of frogs and tadpoles throughout the world including Canada. Examination of *R. clamitans* collected in the St. Lawrence River drainage in northern New York State, showed a high prevalence of echinostomatid metacercarial cysts in the kidneys (Martin and Conn 1990). Similar infections have been found in adjacent areas of Canada (Bonin et al., this volume). Some cysts were surrounded by inflammatory cells and fibrosis. In heavily infected frogs, there was considerable reduction of functional renal tissue, although the authors were unable to evaluate whether the lesions had a deleterious effect on the host. Flukes have also been found in the lungs, musculature, and other tissues of frogs in Canada.

Monogenean flukes are common parasites of the skin of amphibians and fish. Flukes resembling *Gyrodactylus*, the fish gill fluke, were found on *R. catesbeiana* tadpoles obtained in Algonquin Park, Ontario. Significant mortality attributed to the parasite occurred in a group of wild-caught tadpoles being maintained in captivity (P. Wright, pers. comm.).

Leeches

The glossiphoniid leech, *Desserobdella* (formerly *Batrachobdella/Clepsine*) *picta* commonly feeds on many amphibians in North America. Heavy infestations may cause dermal ulceration with hemorrhage around the point of attachment to the skin (Barta and Sawyer 1990). Such lesions may become the portal of entry of bacteria or other microorganisms. Leeches are frequently found on adults and tadpoles of *R. clamitans, R. septentrionalis, R. catesbeiana*, and other amphibians in Ontario (Barta and Desser 1984). Aquatic leeches are vectors of blood-borne parasites of amphibians. Trypanosomes, *Lankesterella, Babesiasoma*, and other hematozoa are transmitted by leeches and it is also possible that viral infections are spread by the same method.

Other Conditions

In view of their permeable skins and their developmental stages, amphibians are particularly susceptible to the effects of environmental alterations and toxins. Both field and laboratory studies are underway to evaluate the effects of pH, ultraviolet light, and environmental contaminants, such as pesticides, on amphibian development, reproductive capacity, and health (Bishop 1992; Clark 1992; Bonin et al., this volume; A. Gendron, pers. comm.). A high prevalence of neoplastic and non-neoplastic abnormalities has been seen in salamanders developing in contaminated ponds in Texas (Rose and Harshbarger 1977). Continuing work in Canada suggests that environmental pollutants are having an adverse impact on physiology and normal development of amphibians (Ouellet et al. 1997). Dead *R. clamitans* were found after pesticide application on fields in Québec (Boninet al., this volume).

Methods of Investigation

The importance of a complete investigation for disease in cases of population decline or mass mortality cannot be stressed too strongly. This will include the collection of appropriate specimens from both live and freshly dead, or preferably, euthanised sick animals, and analysis of environmental factors. Gross and histological examination of tissues should be complemented with attempts at isolation of pathogens and toxins from both affected animals and the environment. It is possible to collect specimens, such as blood samples, from live animals allowing their release unharmed. Anesthesia can be readily accomplished in the field or the laboratory, avoiding the unnecessary sacrifice of healthy animals. It is essential to involve specialists in pathology, microbiology, toxicology, genetics, and other disciplines in order to obtain reliable data.

The Disease and Pathology Subgroup of the Declining Amphibian Population Task Force has prepared a protocol for specimen collection to promote consistency and thoroughness in examinations and specimen collection (Table 1). A network of Canadian veterinary pathology laboratories and other laboratories with expertise in wildlife diseases has also recently been established. The Canadian Cooperative Wildlife Health Centre (CCWHC) is a joint project among the 4 veterinary colleges in Canada, Environment Canada, and provincial and territorial governments to improve the state of knowledge of the health of free-living vertebrate wildlife, and to provide expertise in the investigation of disease occurrences (Table 2). This resource is available to provide advice and training for scientists in other disciplines who encounter disease in free-living wildlife. Researchers are encouraged to contact one of the institutions in the event they encounter, or wish to document, disease in wild amphibians. Further information on the proper techniques for sample collection may be obtained from the veterinary diagnostic laboratories (Table 2) and relevant references (Crawshaw 1992; Green 1993).

Table 1. *Important diagnostic tests and samples to be taken in cases of decline or mortality. More specific tests may require other samples or specialised methods of handling (see Green, 1993). "Heathy" signifies apparently healthy, live animals.*

Test	Condition	Sample	Life stage	Preservation	Container
Voucher specimen	Healthy, sick	Whole	Egg, larva, adult	Formalin[a], ethanol[b]	Bottle[c]
Electrophoresis	Healthy, sick	Whole, tissue	Larva, adult	5° C, freeze	Bottle, cryotube
Necropsy	Sick, recently dead	Whole	Egg, larva, adult	5° C	Bottle, plastic bag
Histology	Sick, recently dead	Whole, tissue	Egg, larva, adult	5° C, formalin	Bottle
Serology	Healthy, sick	Whole, serum	Larva, adult	5° C, freeze	Blood tube, cryotube
Hematology	Healthy, sick	Whole, blood[d]	Larva, adult	5° C	Blood tube, glass slide
Virus culture	Sick, recently dead	Whole, tissue	Larva, adult	Freeze	Plastic bag
Bacterial culture	Sick, recently dead	Tissue, swab	Egg, larva, adult	5° C, freeze	Bacterial swab, plastic bag
Fungal culture	Sick	Tissue, swab	Egg, larva, adult	5° C, freeze	Bacterial swab, plastic bag
Protozoology	Healthy, sick	Whole, tissue	Egg, larva, adult	5° C, formalin	Plastic bag
Helminths, ectoparasites	Recently dead	Whole	Larva, adult	Ethanol	Bottle, blood tube
Toxicology: animal tissue	Dead, decomposed	Whole, tissue	Egg, larva, adult	Freeze	Plastic bag
Toxicology: soil	Soil			Freeze	Plastic bag, glassware[e]
Toxicology: water	Water			Room temperature	Glassware

[a]10% formalin solution

[b]70% ethanol concentration

[c]Bottle or vial in either glass or plastic

[d]Either whole blood in heparin or other anticoagulant, or blood smear on clean microscope slide

[e]Acid-cleaned, opaque glass

Table 2. *The Cooperative Wildlife Health Centre.*

Office	Address
Headquarters, Western and Northern Region	Department of Pathology, Western College of Veterinary Medicine, University of Saskatchewan, Saskatoon, Saskatchewan S7N 0W0 Tel: (306) 966-5099, Fax: (306) 966-8747 *National Information Line* — Tel: 1 (800) 567-2033
Atlantic Region	Department of Veterinary Pathology, Atlantic Veterinary College, University of Prince Edward Island, 550 University Avenue, Charlottetown, Prince Edward Island C1A 4P3 Tel: (902) 566-0667, Fax: (902) 566-0958
Québec	Département de Pathologie, Faculté de Médecine Vétérinaire, Université de Montréal, C.P. 5000, 3200 rue Sicotte, Saint-Hyacinthe, Québec J2S 7C6 Tel: (514) 773-8521, Fax: (514) 773-2161
Ontario	Department of Pathobiology, Ontario Veterinary College, University of Guelph, Guelph, Ontario N1G 2W1 Tel: (519) 823-8800 ext. 4616, Fax: (519) 824-5930
Environment Canada, Ontario Region	Canada Centre for Inland Waters, P.O. Box 5050, 867 Lakeshore Road, Burlington, Ontario L7R 4A6

There are several considerations in collecting specimens for diagnostic purposes to optimize results: the nature of the specimen itself, the quality of the specimen, and the way in which it is collected and preserved. Amphibians decompose very rapidly after death resulting in bacterial and fungal overgrowth, and loss of tissue integrity. For worthwhile results, necropsies must be performed on freshly dead animals. Euthanized, sick or moribund animals provide the best material diagnostically in cases of fatal disease.

If submission of live, or recently dead, animals to a diagnostic facility cannot be guaranteed within 3 days, field autopsies and submission of preserved specimens are acceptable, although it is preferable for experienced pathologists to examine the tissues both grossly and microscopically. Live adult amphibians should be carried in dark insulated containers on moist substrate. Aquatic forms are best transported in jars or bags containing a large volume of pond water with a large air interface, placed in insulated boxes.

Data may also be obtained from live animals in the field, allowing their release unharmed. Blood, lymph, skin scrapings, and other samples can be collected. Anesthesia may be indicated and is essential for any painful procedure. Amphibians can be anesthetised by immersion in 0.5 to 2 mg/l MS-222 (3-aminobenzoic acid ethyl ester; Sigma Chemical, St. Louis, Missouri, USA), and euthanized by overdose with the same drug followed by pithing or exsanguination, or overdose with barbiturates. Further details on amphibian care and handling may be found in the Canadian Council on Animal Care (1984) publications. As much material as possible should be collected in the 1st instance, because the opportunity for follow-up samples may be quickly lost. If numbers are sufficient, animals should be divided into at least 5 groups for various diagnostic and documentary purposes. These categories follow in decreasing order of importance.

Necropsy, Histology, Cultures, and Other Diagnostic Testing

Ideally, live, sick animals should be submitted to a diagnostic laboratory. Alternatively, recently dead animals may be submitted for histology but are of reduced value for cultures. If diagnostic facilities are not immediately available, thin blood smears on microscope slides should be made and serum collected; carcasses should be placed on wet ice and transported to the laboratory within 3 days. Avoid freezing carcasses intended for necropsy or histology. Carcasses for bacterial and viral culture may be frozen.

If diagnostic facilities are remote, preserve some carcasses immediately for histology, and freeze others for culture. Small carcasses are best placed whole in 10% buffered formalin; larger specimens (> 30g) should have their abdomens opened prior to placement in formalin. Where facilities permit, field necropsies can be performed. Pieces of all organs should be preserved in formalin for histopathology. Swabs or small pieces of major organs and lesions should be taken and placed individually in appropriate transport media, or frozen, for bacterial and viral culture.

Toxicology and Water Chemistry

All life stages are acceptable. Freezing tissues or carcasses in bags is the preferred method. Soil samples may be collected in plastic bags; water samples or samples for analysis of petroleum products should be collected in acid-cleaned, opaque glassware. If collecting water from a stream, consider collecting mutliple samples at intervals upstream and downstream from the casualty site.

Blood Tests (Serology, Hematology, Serum Chemistries, and Blood Parasites)

These necessitate the collection of samples from live animals (normal and sick). Animals destined for sacrifice should be anesthetised, the heart exposed by dissection, and then blood samples taken directly from the heart prior to euthanasia. For survival animals, blood may be taken from the heart or accessible vessels by fine needle aspiration. Anesthesia is required in most species, and reduces the degree of trauma and stress. Blood should be placed in plain blood tubes for serology (detection of antibodies against infectious disease) and biochemistry, and in heparinized tubes for blood cell counts and plasma collection for biochemistry. At least 3 blood smears should be made at the time of blood collection, and blood samples submitted as soon as practical to a laboratory. Whole blood should be refrigerated but should never be frozen.

Parasite Collection and Identification

Very fresh carcasses, or live, sick animals are required for the diagnosing protozoan infections. Helminths identified upon dissection of fresh, euthanized specimens should be preserved whole in 70% ethanol.

Voucher Specimens for Identification, Preservation, and Biochemical Genetics

This requires live or very recently dead animals. Whole specimens for identification may be frozen or preserved in 10% formalin (McDiarmid 1994). Because speciation without morphologic change is common in many species of salamanders proper identification of species and hybrids may require isozyme electrophoresis (Rye et al., this volume; Galbraith, this volume). Specimens for electrophoresis should preferably be purged of ingesta by holding them alive for 3 to 7 days, euthanized, and frozen in dry ice or liquid nitrogen (Jacobs and Heyer 1994). For biochemical analysis, tissues should be frozen individually.

Conclusions

Our knowledge of disease processes in wild amphibians in Canada and elsewhere in the world is woefully deficient compared with other vertebrates. There are still large gaps in our understanding of amphibian pathogens and pathophysiology. We are attempting to catch up but it will take long and careful investigation to develop a picture of how wild amphibians are affected by disease, and how environmental and intrinsic factors interact. It is encouraging that within Canada, there now exist both the mechanisms and the enthusiasm to address some of these deficiencies.

Acknowledgments

I am grateful to Jo Welch of the Canadian Cooperative Wildlife Health Centre for providing autopsy data and reference material. Special thanks are due Dr. D. Earl Green, Coordinator of the DAPTF Disease and Pathology Subgroup for the inclusion of his sampling protocol, and for review of the manuscript.

Literature Cited

Baker MR. 1978a. Development and transmission of *Oswaldocruzia pipiens* Walton, 1929 (Nematoda: Trichostrongylidae) in amphibians. Canadian Journal of Zoology 56:1026–1031.

Baker MR. 1978b. Morphology and taxonomy of *Rhabdias* spp. (Nematoda: Rhabdiasidae) from reptiles and amphibians of southern Ontario. Canadian Journal of Zoology 56:2127–2141.

Baker MR. 1979a. The free-living and parasitic development of *Rhabdias* spp. (Nematoda: Rhabdiasidae) in amphibians. Canadian Journal of Zoology 57:161–178.

Baker MR. 1979b. Seasonal population changes in *Rhabdias ranae* Walton, 1929 (Nematoda: Rhabdiasidae) in *Rana sylvatica* of Ontario. Canadian Journal of Zoology 57:179–183.

Baker MR. 1984. Nematode parasitism in amphibians and reptiles. Canadian Journal of Zoology 62:747–757.

Barta JR, Desser SS. 1984. Blood parasites of amphibians from Algonquin Park, Ontario. Journal of Wildlife Diseases 20:180–189.

Barta JR. 1986. Light and electronic microscopic observations on the intraerythrocytic development of *Babesiasoma stableri* (Apicomplexa, Dactylsomatidae) in frogs from Algonquin Park, Ontario. Journal of Protozoology 33:359–368.

Barta JR, Sawyer RT. 1990. Definition of a new genus of glossiphoniid leech and a redescription of the type species, *Clepsine picta* Verrill, 1872. Canadian Journal of Zoology 68:1942–1950.

Bishop CA. 1992. The effects of pesticides and the implications for determining causes of declines in amphibian populations. In: Bishop CA, Pettit KE, editors. Declines in Canadian amphibian populations: designing a national monitoring strategy. Ottawa: Canadian Wildlife Service. Occasional Paper 76. p 67–70.

Blaustein AR, Hokit DG, O'Hara RK, Holt RA. 1994. Pathogenic fungus contributes to amphibian losses in the Pacific-Northwest. Biological Conservation 67:251–254.

Bradford DF. 1991. Mass mortality and extinction in a high-elevation population of *Rana muscosa*. Journal of Herpetology 25:174–177.

Canadian Council on Animal Care. 1984. Guide to the care and use of experimental animals. Volume 2. Ottawa: Canadian Council on Animal Care.

Carey C. 1993. Hypothesis concerning the disappearance of boreal toads from the mountains of Colorado. Conservation Biology 7:355–362.

Carmichael WW. 1994. The toxins of cyanobacteria. Scientific American 1994(1):78–86.

Chen GJ, Desser SS. 1989. The Coccidia (Apicomplexa: Eimeriidae) of frogs from Algonquin Park, with descriptions of two new species. Canadian Journal of Zoology 67:1686–1689.

Clark KL. 1992. Monitoring the effects of acidic deposition on amphibian populations in Canada. In: Bishop CA, Pettit KE, editors. Declines in Canadian amphibian populations: designing a national monitoring strategy. Ottawa: Canadian Wildlife Service. Occasional Paper 76. p 63–66.

Crawshaw GJ. 1992. The role of disease in amphibian decline. In: Bishop CA, Pettit KE, editors. Declines in Canadian amphibian populations: designing a national monitoring strategy. Ottawa: Canadian Wildlife Service. Occasional Paper 76. p 19–20.

Cunningham AA, Langton TES, Bennett PM, Drury SEN, Gough RE, Kirkwood JK. 1993. Unusual mortality associated with poxvirus-like particles in frogs (*Rana temporaria*). Veterinary Record 133:141–142.

Desser S. 1992. Ultrastructural observations on an icosahedral cytoplasmic virus in leukocytes of frogs from Algonquin Park, Ontario. Canadian Journal of Zoology 70:833–836.

Desser S, Barta JR. 1984a. An erythrocytic virus and rickettsia of frogs from Algonquin Park, Ontario. Canadian Journal of Zoology 62:1521–1524.

Desser S, Barta JR. 1984b. *Thrombocytozoons ranarum* Tchacarof 1963, a prokaryotic parasite in thrombocytes of the mink frog *Rana septentrionalis* in Ontario. Journal of Parasitology 70:454–456.

Desser S, Barta JR. 1989. The morphological features of *Aegytianella bacterifera*: an intraerythrocytic rickettsia of frogs from Corsica. Journal of Wildlife Disease 25:313–318.

Desser S, Hong H, Siddall ME, Barta JR. 1993. An ultrastructural study of *Brugerolleia algonquinensis* gen. nov., sp. nov. (Diplomonadina: Diplomonadida), a flagellate parasite in the blood of frogs from Ontario, Canada. European Journal of Protistology 29:72–80.

Desser S, Jones S. 1985. *Hexamita intestinalis* Dujardin in the blood of frogs from Southern and Central Ontario. Journal of Parasitology 71:841.

Desser S, Lom J, Dykova I. 1986. Developmental stages of *Sphaerospora ohlmacheri* (Whinery, 1893) n. comb. (Myxozoa: Myxosporea) in the renal tubules of bullfrog tadpoles, *Rana catesbeiana* from Lake of Two Rivers, Algonquin Park, Ontario. Canadian Journal of Zoology 64:2344–2347.

Desser S, Siddall ME, Barta JR. 1989. Ultrastructural observations on the developmental stages of *Lankesterella minima* (Apicomplexa) in experimentally infected *Rana catesbeiana* tadpoles. Journal of Parasitology 76:97–103.

Everard COR, Tota B, Basset D, Ali C. 1979. *Salmonella* in wildlife from Trinidad and Grenada, W. I. Journal of Wildlife Diseases 15:213–219.

Everard COR, Carrington D, Korver H, Everard JD. 1988. Leptospires in the marine toad (Bufo marinus) on Barbados. Journal of Wildlife Diseases 24:334–338.

Everard COR, Fraser-Chanpong GM, Bagwandin LJ, Race MW, James AC. 1983. Leptospires in wildlife from Trinidad and Granada. Journal of Wildlife Diseases 19:192–199.

Fantham HB, Porter A, Richardson LR. 1942. Some haematozoa observed in vertebrates in eastern Canada. Parasitology 34:199–226.

Glorioso JC, Amborski GF, Culley DD. 1973. Microbiological studies on septicemic bullfrogs (*Rana catesbeiana*). American Journal of Veterinary Research 335:1241–1245.

Green DE. 1993. Diagnostic assistance for investigating amphibian declines and mortalities. Froglog 5:2.

Gruia-Gray J, Desser SS. 1992. Cytopathological observations and epizootiology of frog erythrocytic virus in bullfrogs (*Rana catesbeiana*). Journal of Wildlife Diseases 28:34–41.

Gruia-Gray J, Petric M, Desser SS. 1989. Ultrastructural, biochemical and biophysical properties of an erythrocytic virus of frogs from Ontario, Canada. Journal of Wildlife Diseases 25:497–506.

Hamilton WD, Zuk M. 1982. Heritable true fitness and bright birds: a role for parasites. Science 218:384–387.

Hausfater G, Gerhardt HC, Klump GM. 1990. Parasites and mate choice in gray treefrogs, *Hyla versicolor*. American Zoologist 30:299–311.

Hine R, Les B, Helnich B. 1981. Leopard frog populations and mortality in Wisconsin, 1974–1976. Wisconsin Department of Natural Resources Technical Bulletin 122:1–39.

Jacobs JF, Heyer WR. 1994. Collecting tissue for biochemical analysis. In: Heyer WR, Donnelly MA, McDiarmid RW, Hayek LC, Foster MS, editors. Measuring and monitoring biological diversity. Standard methods for amphibians. Washington: Smithsonian Institution Press. p 299–301.

Koonz W. 1992. Amphibians in Manitoba. In: Bishop CA, Pettit KE, editors. Declines in Canadian amphibian populations: designing a national monitoring strategy. Ottawa: Canadian Wildlife Service. Occasional Paper 76. p 19–20.

Laurance WF, McDonald KR, Speare R. 1996. Epidemic disease and the catastrophic decline of Australian rain forest frogs. Conservation Biology 406–413.

Levine ND, Nye RR. 1977. A survey of blood and tissue parasites of leopard frogs *Rana pipiens* in the United States. Journal of Wildlife Diseases 13:17–32.

Ling RL, Werner JK. 1988. Mortality in *Ambystoma maculatum* larvae due to *Tetrahymena infection*. Herpetological Review 19:26.

Martin TR, Conn DB. 1990. The pathogenicity, localization, and cyst structure of echinostomatid metacercariae (Trematoda) infecting the kidneys of the frogs *Rana clamitans* and *Rana pipiens*. Journal of Parasitology 76:414–419.

McDiarmid RW. 1994. Preparing amphibians as scientific specimens. In: Heyer WR, Donnelly MA, McDiarmid RW, Hayek LC, Foster MS, editors. Measuring and monitoring biological diversity. Standard methods for amphibians. Washington: Smithsonian Institution Press. p 289–297.

McKinnell RG. 1984. Lucké tumor of frogs. In: Hoff GL, Frye FL, Jacobson ER, editors. Diseases of amphibians and reptiles. New York: Plenum Press. p 581–605.

Moody NJG, Owens L. 1994. Experimental demonstration of the pathogenicity of a frog virus, Bohle iridovirus, for a fish species, barramundi *Lates calcarifer*. Diseases of Aquatic Organisms 18:95–102.

Nyman S. 1986. Mass mortality in larval *Rana sylvatica* attributable to the bacterium, *Aeromonas hydrophila*. Journal of Herpetology 20:196–201.

Ouellet M, Bonin J, Rodrigue J, DesGranges J-L, Lair S. 1997. Hindlimb deformities (ectromelia, ectrodactyly) in free-living anurans from agricultural habitats. Journal of Wildlife Diseases 33:95–104.

Pfennig DW, Loeb MLG, Collins JP. 1991. Pathogens as a factor limiting the spread of cannibalism in tiger salamanders. Oecologia 88:161–166.

Reilly BO, Woo PTK. 1982. The biology of *Trypanosoma andersoni* n. sp. and *Trypanosoma grylli* Nigrelli, 1944 (Kinetoplastida) from *Hyla versicolor* LeConte, 1825 (Anura). Canadian Journal of Zoology 60:116–123.

Richardson JPM, Adamson ML. 1988. *Megalobatrachonema (Chabaudgolvania) waldeni* n.sp. Nematoda: Kathlaniidae) from the intestine of the northwestern salamander, *Ambystoma gracile* (Baird). Canadian Journal of Zoology 66:1505–1505.

Roberts W. 1992. Declines in amphibian populations in Alberta. In: Bishop CA, Pettit KE, editors. Declines in Canadian amphibian populations: designing a national monitoring strategy. Ottawa: Canadian Wildlife Service. Occasional Paper 76. p 14–16.

Rose FL, Harshbarger JC. 1977. Neoplastic and possibly related skin lesions in neotenic salamanders from a sewage lagoon. Science 208:315–317.

Schotts EB, Gains JL, Martin L, Prestwood AK. 1972. *Aeromonas*-induced deaths among fish and reptiles in a eutrophic inland lake. Journal of the American Veterinary Medical Association 161:603–607.

Scott NJ Jr. 1993. The postmetamorphic mortality syndrome in ranid frog declines in the American West. Froglog 7:1.

Speare R. 1990. The influence of parasite infection on mating success in spadefoot toads, *Scaphiophus couchii.* American Zoologist 30:313–324.

Speare R, O'Shea P. 1989. The marine toad, *Bufo marinus* and the search for a killer disease. In: Olsen JH, Eisenacher M, editors. Proceedings of the American Association of Zoo Veterinarians annual meeting 14–19 October 1989; Greensboro, N.C. Philadelphia: American Association of Zoo Veterinarians. p 166–172.

Speare R, Smith JR. 1992. An iridovirus-like agent isolated from the ornate burrowing frog (*Lymnodynastes ornatus*) in northern Australia. Diseases of Aquatic Organisms 14:51–57.

Speare R, Freeland WJ, Bolton SJ. 1991. A possible iridovirus in erythrocytes of *Bufo marinus* in Costa Rica. Journal of Wildlife Diseases 27: 457–462.

Tse B, Barta JR, Desser SS. 1986. Comparative ultrastructural features of the sporozoite of *Lankesterella minima* (Apicomplexa) in its anuran host and leech vector. Canadian Journal of Zoology 64:2344–2347.

Vanderburgh DJ, Anderson RC. 1987. The relationship between nematodes of the genus *Cosmocercoides* Wilkie, 1930 (Nematoda: Cosmocercoidea) in toads (*Bufo americanus*) and slugs (*Deroceras laeve*). Canadian Journal of Zoology 65:1650–1661.

Wolf K, Bullock GL, Dunbar CE, Quimby WC. 1968. Tadpole edema virus: a viscerotropic pathogen for anuran amphibians. Journal of Infectious Diseases 118:253–262.

Woo PTK. 1969. Trypanosomes in amphibians and reptiles in southern Ontario. Canadian Journal of Zoology 47:981–988.

Woo PTK. 1986. Trypanosome infection in salamanders (order: Caudata) from eastern North America with notes on the biology of *Trypanosoma ogawai* in *Ambystoma maculatum*. Canadian Journal of Zoology 64:121–127.

Woo PTK, Bogart JP. 1984. *Trypanosoma* spp. (Protozoa: Kinetoplastida) in Hylidae (Anura) form eastern North America with notes on their distributions and prevalences. Canadian Journal of Zoology 62:820–824.

Woo PTK, Bogart JP, Servage DL. 1980. *Trypanosoma ambystomae* in *Ambystoma* spp. (order Caudata) in southern Ontario. Canadian Journal of Zoology 58:466–469.

Worthylake K M,Hovingh E. 1989. Mass mortality of salamanders (*Ambystoma tigrinum*) by bacteria (*Acinetobacter*) in an oligotrophic seepage mountain lake. Great Basin Naturalist 49:364–372.

Amphibians in decline: Canadian studies of a global problem. David M. Green, editor.
Herpetological Conservation 1:271–281.

Chapter 27

ASSESSING AND CONTROLLING AMPHIBIAN POPULATIONS FROM THE LARVAL PERSPECTIVE

RICHARD J. WASSERSUG

Department of Anatomy and Neurobiology, Sir Charles Tupper Medical Building, Dalhousie University, Halifax, Nova Scotia B3H 4H7, Canada

ABSTRACT.—For assessing the status of amphibian populations, larval surveys are both difficult and unreliable, and therefore of questionable value. The accuracy of techniques (such as traps) currently available for sampling aquatic amphibians from anything but the smallest of ponds has not been validated for tadpoles. Techniques that are believed to be effective for tadpoles (e.g., seine netting) are environmentally disruptive and thus undesirable. In any case, the typical, biphasic amphibian life history is characterized by high mortality during or shortly after metamorphosis, which means that tadpoles can be abundant even when there is zero recruitment into the adult population. Tadpoles are efficient swimmers capable of high linear acceleration and small turning radii and are highly sensitive to waterborne chemicals that signal the presence of potential aquatic predators. They respond to these chemicals by reducing their activity level and moving into refugia. Thus tadpoles should not be easy to observe or to collect. When amphibian larvae remain in the shallows of otherwise large and structurally complex ponds, it is because they are threatened by predators in other parts of the pond. Thus an abundance of tadpoles within easy reach of the herpetologist's net is as likely a sign that they are in a malevolent as a benevolent environment. The chemical compounds that tell amphibian larvae that a predator is near should be identified and synthesized. These natural, species-specific repellents drive amphibian larvae into suboptimal habitats and lead to reduced overall fitness. These compounds have great potential in the control of undesirable exotic amphibians, such as *Rana catesbeiana*, where they threaten indigenous amphibians.

RÉSUMÉ.—L'évaluation de situation des populations d'amphibiens, à partir de leurs larves est à la fois difficile et manque de fiabilité; l'intérêt de cette méthode est par conséquent discutable. La précision des techniques disponibles actuellement (comme les pièges) pour l'échantillonnage des amphibiens aquatiques dans des habitats les plus variés n'a pas pu être validée pour les têtards. Les techniques considérées comme efficaces pour les têtards (l'emploi de seines par exemple) perturbent l'environnement et sont par conséquent peu recommandables. Quoi qu'il en soit, le cycle biphasique type des amphibiens se caractérise par une mortalité élevée pendant ou peu de temps après la métamorphose ce qui en d'autres termes signifie que les têtards peuvent être abondants même lorsque le recrutement est nul dans la population adulte. Les têtards sont par ailleurs d'excellents nageurs capables d'accélérations linéaires et angulaires importantes. Ils sont également très sensibles aux substances chimiques d'origine hydrique qui leur signalent la présence de prédateurs aquatiques potentiels. Ils réagissent à ces substances chimiques en réduisant leur niveau d'activité et en cherchant refuge. Il est par conséquent difficile d'observer ou de capturer les têtards. Lorsque les larves des amphibiens restent dans les eaux peu profondes de mares par ailleurs grandes et structurellement complexes c'est parce qu'elles sont menacées par des prédateurs ailleurs dans la mare. L'abondance de têtards à portée du filet de l'herpétologue est sans doute le signe que leur environnement est malveillant. Les substances chimiques qui signalent la proximité des prédateurs devrait être identifiées et synthétisées. Ces substances naturelles spécifiques d'espèces entraînent les larves d'amphibiens dans des habitats sous-optimaux et altèrent leur potentiel global. Ces substances affichent par contre un excellent potentiel dans le contrôle des amphibiens exotiques indésirables, comme *Rana catesbeiana*, dans les habitats où ils menacent les espèces indigènes.

It is not easy to assess changes in amphibian populations from larval surveys. The scarcity of studies in this volume that focus on larvae supports that assertion. There are 3 reasons. First is the simple fact that the most common amphibians in Canada are tadpoles. Given their ubiquitous nature, tadpoles may be present even when a population is otherwise vulnerable or threatened. Except in the extreme, when larvae are clearly absent, their presence is little assurance that a species is not in local decline. Second, it is difficult to accurately sample amphibian larvae because of the way that they use microhabitats and respond to changes in their biotic and abiotic environments. Most techniques for surveying ponds are likely to be misleading and inaccurate unless they are massively intense or environmentally destructive, and therefore undesirable. And finally, tadpoles should *not* be easy to find and capture. When they are, this is more often than not an indication that they are in a hostile rather than a benevolent environment. This counter-intuitive fact further indicates that local larval abundance tells us little about the overall health of an amphibian population. However, aspects of larval biology point to new ways of controlling exotic amphibians, such as the Bullfrog (*Rana catesbeiana*) and the Marine toad (*Bufo marinus*), where they have taken hold in parts of the world and now threaten native species. My focus is clearly on anurans but the general conclusions apply to salamander larvae as well.

Tadpoles Can Be Common Even When Their Species Is Locally in Trouble

In the grand scheme of things, at any moment in North America, the commonest amphibian life stage is the tadpole. Obviously, during the full life cycle, eggs are more numerous than tadpoles, tadpoles are more numerous than juveniles, and juveniles are more numerous than adults. But if we multiply the length of time that anurans spend in each of those 4 life stages by the average number of individuals at that stage, there is no question that the tadpole should be the most common life stage encountered in the field.

Mortality for anurans is severe at metamorphosis (Arnold and Wassersug 1978) and survivorship for *Rana* larvae in ponds generally is less than 10%(Duellman and Trueb 1985). If the steepest part of the mortality curve normally occurs at or shortly after metamorphosis (cf. Berven 1990), then tadpole abundance gives little assurance that a species is successfully maintaining itself. In populations that are crashing, tadpoles may continue to be numerous even when juvenile recruitment is nil. Waldick's study (this volume) of amphibians in a clear-cut forest in New Brunswick nicely illustrates this point. A logging operation actually increased the number of ponds in which adults could, and did, deposit eggs. At first glance that would appear to be a positive sign for the local amphibians; larvae were indeed found in abundance in those new aquatic habitats. But despite the abundance of the aquatic larvae, intense sampling for postmetamorphic individuals proved that there was no successful recruitment of amphibians in the disturbed landscape. The plethora of larvae was an unreliable measure of the overall ability of that disturbed environment to support or maintain amphibian species. Lots of larvae is not necessarily a positive sign for the population overall. Seburn et al. (this volume) reach a similar conclusion in their study of the leopard frogs (*Rana pipiens*) in Alberta.

It Is Difficult to Census Tadpoles Accurately

It is important to distinguish between surveying to establish the simple presence or absence of a species versus surveying to obtain a population census. The presence of amphibian larvae can be assessed by visual inspection (Dodd 1979), by netting, by use of traps (Griffiths 1985a; Richter 1995), or by sampling techniques, which are essentially a combination of traps and netting (Shaffer et al. 1994). All of these approaches have strengths and weaknesses (Griffiths and Raper 1994; R.A. Griffiths, S.J. Raper, and L.D. Brody, unpublished), but visual inspection is the easiest and quickest technique simply to know if tadpoles are present or not.

Unfortunately, visual surveys, as a basis for population enumeration, are unreliable for anything but the smallest and shallowest bodies of water. In ponds with much microhabitat variation, amphibian larvae use different parts of the pond over a single diel cycle (Anderson and Graham 1967; Ashby 1969; Holomuzki 1986a; Griffiths et al. 1988; Warkentin 1992a; Peterson et al. 1992). Different species may have different diurnal activity cycles within the same pond. Temperature, light level, and O_2 concentration all vary over a 24-hr cycle in natural aquatic habitats, and tadpoles are known to react to all of these factors (Noland and Ultsch 1981; Griffiths 1985b). It is obvious then that visual surveys for the purpose of tallying the number of individuals of even a single species in a single pond may give very different results depending on the time of day, light levels, or temperature when the survey is made (cf. Cooke 1995).

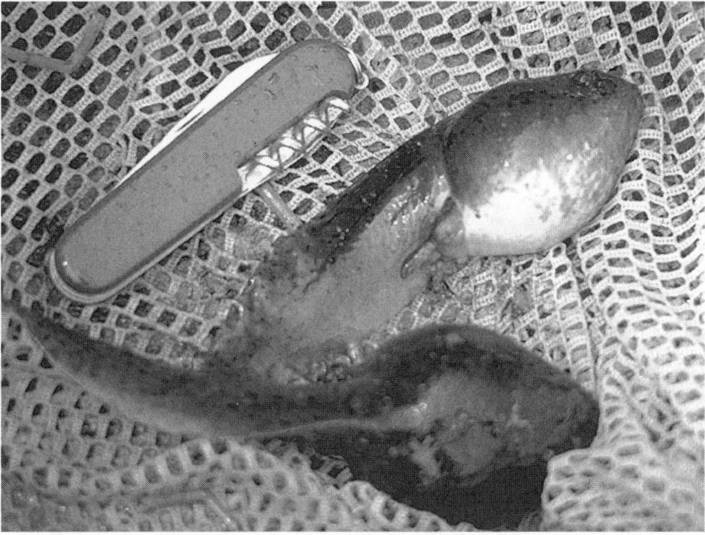

Bullfrog, Rana catesbeiana, *tadpoles. Photo by Claude Daigle.*

Even when one knows the best time of day to survey for a particular species, aquatic amphibians change their microhabitat utilization patterns in response to predator pressure, competition, water level, or other factors, all of which change throughout the growing season. Tadpoles of the same species change their microhabitat preferences as they grow (Alford 1986; Crawshaw et al. 1992). The result is that one cannot expect a visual survey of one part of a pond to reflect accurately the overall abundance of species in the pond across days, weeks, or months.

It is because of the problems of visual surveys that more intensive surveying techniques are often attempted. In principle, it is possible to survey a whole pond, including both its conspicuous and inconspicuous microhabitats, with traps. This approach has been extensively evaluated in studies with European newts, *Triturus* (Griffiths and Raper 1994; Cooke 1995) and is now considered the best overall technique for both detecting species presence and the number of individuals actually sampled. Unfortunately, it can be expensive and time consuming to survey large ponds this way. More to the point, there are few, if any, studies that evaluate the effectiveness of traps for both benthic and pelagic anuran larvae.

There are no tried and true techniques for sampling larval anuran populations with traps, although some trapping procedures appear reasonably effective with particular species in particular habitats (Griffiths 1985a; Richter 1995). It is not clear what one should or could use as bait in traps for tadpoles, if anything at all (Richter 1995). Most of Canada's tadpoles are dietary generalists that will eat a variety of food stuffs. In the one study that has attempted to relate *in situ* feeding activity of tadpoles with their diel activity cycle (Warkentin 1992b), tadpole movements did not directly correlate with feeding activity. This does not bode well for surveying amphibian larvae with baited traps.

Griffiths and Raper (1994) considered baiting unnecessary when trapping for newts, and it is not uncommon for unbaited minnow traps to be used to catch larvae of other amphibians (Shaffer et al. 1994; Leclair and Caetano, this volume). Unfortunately, it is also not uncommon to find that tadpoles, which wander into such traps, end up themselves as bait for newts and other predators. Amphibian larvae are at increased risk of both predation and drowning in traps. In this regard, Griffiths and Raper (1994) report, from a survey of > 40 herpetologists involved with surveying amphibians, that traps were far more likely to harm amphibians than any other survey technique. Furthermore, a negative result with traps does not establish the absence of a species. Leclair and Caetano (this volume) abandoned using minnow traps in certain microhabitats, not because the amphibians were necessarily absent, but because trapping proved unsuccessful.

Concerning dipnetting, Shaffer et al. (1994) asserted that all microhabitats must be sampled so that species with restricted distributions are not missed. This is laudable in principle but impossible in much of northern North America. In the post-glacial terrain of Canada, pond margins and bottoms range from water-logged peat to glacial erratics. Dipnets do not work well, and enclosure sampling (for example, "stove pipe" sampling) similarly falters where the bottom is either particularly soft, irregular, or heavily vegetated (Warkentin 1992b; Wassersug 1992).

In my estimation, the most thorough and reliable way to sample larval anurans is with a seine net drawn through as much of the pond as possible. In shallow vegetated areas it helps to have a row of marching assistants to drive the amphibians out of cover and into the net. But even this is ineffective in dense vegetation or deep water. Worse still, in order to be most effective the bottom must be raked by the net and the vegetation thoroughly trampled, massively disturbing the environment. In sum there is no established, easy, reliable, and environmentally friendly way to survey amphibian larvae in anything but the smallest of ponds.

TADPOLES SHOULD *NOT* BE EASY TO CAPTURE BECAUSE...

They Aren't Slow...

Tadpoles are not designed for sustained swimming over large distances (Wassersug and Hoff 1985), but that does not mean they are poor swimmers when it really matters. Tadpoles do not need marathon capabilities in the small bodies of water where they commonly reside. What matters most to a tadpole trying to evade a pursuing predator is fast acceleration and burst swimming. In that regard, tadpoles compare well with fishes. Medium-sized *Rana* tadpoles, for example, can quickly reach speeds in excess of 1 m/s, equal to almost 30 body lengths/sec (Wassersug 1989). Hoff (1987) reported an average acceleration of 3.9 m/s^2 for an assortment of Nova Scotian tadpoles just 60 msec after they had begun to move. The highest acceleration that she recorded at 60 msec into a fast–start was an astonishing 11.8 m/s^2 for a 10-cm *Rana catesbeiana*. However, by 200 msec most tadpoles in her study had ceased to accelerate. Such short but explosive bursts of activity may not move a tadpole very far. But, in many cases, they are all that is necessary for a tadpole to escape into vegetation and away from incoming fish jaws or heron beaks.

Tadpoles, with their thin tails conjoined to their globose head-bodies, are admittedly not the epitome of grace. However, those thin tails are highly flexible. That flexibility allows for large amplitude movements and, consequently, extremely short turning radii, which can be as important as maximum swimming speed in evading predators (Feder 1983; Gatten et al. 1984). Hoff (1987) reported that tadpoles rotate between 40° and 180° during fast–starts, with 90% of all fast–starts involving turns between 60° and 140°.

Even when they are moving in a straight line, tadpoles have high amplitude oscillations at the snout and tail tip—they wobble a lot. These high amplitude movements produce large diameter trailing vortices which, for fish at least, mean poor mechanical efficiency (Triantafyllou et al. 1993). But computational fluid dynamic (CFD) analyses suggest that the tadpole's wobbly gait is not nearly as inefficient as it subjectively appears (Liu et al. 1996). The kinematics for tadpoles closely matches their unique body form. Using computational models, Liu et al. examined the efficiency of both tadpole-shaped and fish-shaped objects swimming in a tadpole mode (i.e., high amplitude movements at the snout) and in a fish-like mode (i.e., with no lateral deflections at the snout). Efficiency was indeed low when their "fish" model swam like a tadpole. Conversely, efficiency was also low when their model "tadpole" swam like a fish; in fact it was much lower than when the tadpole swam like a tadpole. Liu et al. extended their analyses to fish- and tadpole-shaped objects with exposed hind limbs. They concluded that tadpoles with hind limbs protruding just caudal to the crease between their head-body and tail (i.e., where the hind limbs naturally develop) incur little cost in locomotor efficiency compared to what they would experience if they were either built like or swam like fishes. Thus, tadpole shape and locomotion are not inefficient *per se*. Rather the tadpoles' shape and swimming style are coadapted and appropriately fit their ultimate destiny: to develop limbs and transform into frogs.

And They Aren't Stupid.

A decade or 2 ago, the general belief was that tadpole survival depended greatly, if not wholly, on the wisdom of their parents. In effect, tadpoles survived because the adults knew where to lay their eggs, as has been documented, for example, by Crump (1991) for *Hyla pseudopuma* and by Hopey and Petranka (1994) for *Rana sylvatica*. The "preferred" habitat for a typical North American anuran was a pool or pond, permanent enough to allow the larvae enough time to reach metamorphosis, but not so permanent that the pond acquired a large resident population of predators (Wassersug 1975; Smith 1983; Woodward 1983; Roth and Jackson 1987; Pearman 1995). In other words, a tadpole's fate was thrust upon it.

Against this traditional view of tadpole survivability, there has been a growing body of literature in behavioral ecology showing that aquatic amphibians are capable of highly discriminatory behavioral choices (Table 1). We now know, for example, that tadpoles can distinguish their own from other amphibian species based on chemical signals (Hews and Blaustein 1985; Hews 1988; Kiseleva 1993). Furthermore, many species can distinguish siblings from non-siblings and some can even distinguish half sibs from full sibs (Waldman 1991; Blaustein and Waldman 1992; Pfennig et al. 1993, 1994). We also know that amphibian larvae can detect the presence of both potential competitors (Morin 1983; Alford 1986, 1989; Waringer-Löschenkohl 1988; Griffiths 1991; Griffiths and Denton 1992; Werner 1994; Werner and McPeek 1994) and predators (Table 1). Most importantly, amphibian larvae alter their behavior in response to these biotic factors. Some species may even alter their morphology as a defensive response to predator pressure (Smith and Van Buskirk 1995).

The most common reaction of amphibian larvae to aquatic predators is immobility (Table 1). But once under attack, aquatic amphibians flee, altering their flight path in response to perceived predator pressure elsewhere in the pond. One of the earliest field reports of this phenomenon was Taylor's (1983) observation that startled aquatic *Ambystoma gracile* were likely to flee into deep water in a lake without piscine predators, but into shallow water in lakes with those predators. I have personally witnessed this same variable fright response with tadpoles. In an artificial pond on a trout farm in Nova Scotia, *Rana* tadpoles rushed into the vegetation at the margin of the pond when disturbed by my approach. In many cases the tadpoles actually swam toward me seeking refuge in vegetation along the shore rather than into the deeper open water inhabited by fish. In contrast, in a natural trout-free pool less than 50-m away, tadpoles of the same size, stage, and species swam directly away from me as soon as I approached the shore. The 2 bodies of water were separated by a dirt road only a few yr old. This close proximity suggests that the tadpoles in the artificial pond were derived from the natural resident population. The young age of the trout pond made it unlikely that the tadpole's ability to assess predatory risk and adjust their behavior accordingly was the product of natural selection in that single pond. Rather, it suggests that the discriminatory skill was innate for the wild population.

Northwestern salamander, Ambystoma gracile. *Photo by David M. Green.*

Implications For Assessment

Because tadpoles are neither slow nor stupid, we would not expect them to be easy to find or to catch. The few exceptions in Canada are *Bufo* spp. tadpoles and *Rana catesbeiana*. These tadpoles appear to be unpalatable to many piscine predators (Wassersug 1973; Kruse and Francis 1977; Werner and McPeek 1994) and need not rely solely on speed or stealth for survival. Notably they are among the more conspicuous Canadian tadpoles found in shallow open water.

Excluding *Bufo* spp. and *R. catesbeiana*, what are we to conclude about the status of an amphibian population when we come across tadpoles in a situation where they are both abundant and easy to capture? The most intuitive conclusion—that the microhabitat is beneficial for the larvae and therefore for the species—is probably dead wrong. I suggest that when tadpoles are easy to catch, then the environment is probably *not* a friendly one (see also Wassersug 1992). Local abundance, in a particularly vulnerable microhabitat, is not what one expects from quick and clever animals. Tadpoles should only restrict themselves to microhabitats that are easy for herpetologists to sample (such as shallow, open water) when there are more serious risks of mortality in neighboring microhabitats. In the most common situation where tadpoles are found in great numbers and are easy to collect, namely when they are entrapped in a drying pool, the risk of death by desiccation is obvious. But the pattern also holds for larger, more structurally complex aquatic habitats. When tadpoles make themselves vulnerable to aerial predators (including the herpetologist's net), they are probably avoiding life-threatening forces elsewhere in the pond (see also Wassersug 1992). When tadpoles do not flee the approaching herpetologist, it is not because they are flightless dodos, adapted to a predator-free environment. A better metaphor is that they are too petrified to move. Tadpoles may not have an elaborate behavioral repertoire, but they are not suicidal.

In sum, when tadpoles are easy to find and to catch they are probably in great jeopardy. This leads to a paradox: if tadpoles are hard to find, their populations may still be doing fine; but when tadpoles are easy to find, they may in fact be in trouble. In either case, we are hard pressed to arrive at a correct conclusion about the health of an anuran population by sampling larvae alone. So what is the take-home message? A sad one—tadpole assessments may tell us little about declining amphibian populations until it is too late.

Implications For Control

There is, nevertheless, a positive aspect to this story of amphibian larval behaviour. But it is in the realm of amphibian control rather than amphibian assessment. There are naturally occurring, species-specific repellents to larval amphibians, released by aquatic predators (Table 1) and their prey (Pfeiffer 1966). These water soluble compounds have the effect of driving amphibian larvae into suboptimal microhabitats—usually shallow water and out of the reach of larger aquatic predators. The repellant compounds appear to be effective at very low concentrations and I suggest that they could be used as natural stressors on pest species such as *Bufo marinus* in Australia (Tyler 1976) and *Rana catesbeiana* in western North America (Bury and Whelan 1984; Hayes and Jennings 1986). They could be useful agents for depressing the population growth rates of undesirable amphibians wherever they occur.

By driving amphibian larvae into marginal habitats, these chemical repellants increase local density and decrease the access of individual larvae to food patches (Horat and Semlisch 1994). We would thus expect lower growth rates when larvae are exposed to the chemical signatures of their natural predators, as has been reported for *Bufo americanus* (Skelley and Werner 1990), *Ambystoma tigrinum* (Skelley 1992), *Pseudacris triseriata*, *P. crucifer* (Skelley 1995), and *Rana clamitans* (Werner 1992, 1994). It has been repeatedly demonstrated that amphibian larvae raised at high density have depressed larval growth rates, metamorphose at smaller sizes, and take longer to reach metamorphosis (Petranka and Sih 1986; Alford and Harris 1988; Berven 1990; Scott 1990; Brönmark et al. 1991; Warner et al. 1991; Tejedo and Reques 1992; Gascon and Travis 1992; Leips and Travis 1994; Schmuck et al. 1994). Merely by taking longer to reach metamorphic size, the larvae that react to

Table 1. *Evidence of discriminatory behaviour by amphibian larvae.*

Larval response	References
Detection of predators	Holomuzki (1986a), Semlitsch (1987), Hews (1988), Lawler (1989), Fauth (1990), Skelly and Werner (1990), Semlitsch and Reyer (1992), Skelly (1992, 1994, 1995), Azevedo--Ramos et al. (1992), Sih and Kats (1994), Relyea (1995)
Immobility as a reaction to predators	Werschkul and Christensen (1977), Caldwell et al. (1980), Lawler (1989), Azevedo-Ramos et al. (1992), Chovanec (1992), Peterson et al. (1992), Semlitsch and Reyer (1992), Sih et al. (1992), Lefcort and Eiger (1993), Tejedo (1993), Wilson and Lefcort (1993), Hawkins (1994), Horat and Semlitsch (1994)
Detection of predators by chemical signals	Kats (1988), Britton (1991), Lefcort and Eiger (1993), Jackson and Semlitsch (1993), Stauffer and Semlitsch (1993), Tejedo (1993), Wilson and Lefcort (1993), Hawkins (1994), Horat and Semlitsch (1994), Sih and Kats (1994), Relyea (1995)
Movement into refugia and marginal microhabitats in the presence of predators or chemical signals from predators	Holomuzki (1986a,b), Petranka et al. (1987), Kats (1988), Kats et al. (1988), Fauth (1990), Werner (1991), Peterson et al. (1992), Sih et al. (1992), Jackson and Semlitsch (1993), Stauffer and Semlitsch (1993)

these compounds are at increased risk of their ponds drying up before they can transform (Scott 1990). Thus these compounds could work indirectly, but naturally, as larvicidal agents.

Metamorphosing leopard frog, Rana pipiens, *tadpole. Photo by John Mitchell.*

Even in situations where larvae, exposed to these compounds, manage to metamorphose, we would expect them to be of a smaller size than unstressed conspecifics. Amphibian larvae that metamorphose at a small size either take longer than normal to reach reproductive size or are smaller when they reproduce, and have smaller clutches (Semlitsch et al. 1988; Smith 1987). *Ambystoma opacum* that are raised at high density before metamorphosis have reduced lipid stores and are prone to starvation (Scott 1994). *Bufo woodhousii* tadpoles that are raised at high density metamorphose into toadlets that are smaller, cannot jump as far, and have less stamina than siblings raised at low density (John-Adler and Morin 1990). *Bufo marinus* individuals that metamorphose at a small size, in particular, are at greater risk of desiccation and are less likely to disperse than larger individuals (Cohen and Alford 1993). All of these factors reduce individual fitness and should help reduce population size overall. A concerted effort should be made to identify the molecular structures of these water soluble repellents as a first step toward their commercial synthesis. Considering how many studies have documented the existence of these compounds (Table 1), it is surprising how little we know about their chemistry.

Acknowledgments

This work was supported by Natural Sciences and Engineering Research Council of Canada. Susan Hall and Nathalie Major helped with the literature review and manuscript production. Helpful comments on drafts of the manuscript were provided by Susan Hall, Melissa Meeker, Monika Fejtek, and Ruth Waldick.

LITERATURE CITED

Alford RA. 1986. Habitat use and positional behavior of anuran larvae in a northern Florida temporary pond. Copeia 1986:408–423.

Alford RA. 1989. Variation in predator phenology affects predator performance and prey community competition. Ecology 70:206–219.

Alford RA, Harris RN. 1988. Effects of larval growth history on anuran metamorphosis. American Naturalist 131:91–106.

Anderson JD, Graham RE. 1967. Vertical migration and stratification of larval *Ambystoma*. Copeia 1967:371–374.

Arnold SJ, Wassersug RJ. 1978. Differential predation on metamorphic anurans by garter snakes (*Thamnophis*): Social behavior as a possible defense. Ecology 59:1014–1022.

Ashby KR. 1969. The population ecology of a self-maintaining colony of the common frog (*Rana temporaria*). Journal of Zoology (London) 158:453–474.

Azevedo-Ramos C, Van Sluys M, Hero JM, Magnusson WE. 1992. Influence of tadpole movement on predation by odonate naiads. Journal of Herpetology 26:335–338.

Berven KA. 1990. Factors affecting population fluctuations in larval and adult stages of wood frog (*Rana sylvatica*). Ecology 71:1599–1608

Blaustein AR, Waldman B. 1992. Kin recognition in anuran amphibians. Animal Behaviour 44:207–221.

Britton CM. 1991. Effects of environmental history, sibship, and age on tadpole anti-predator responses. American Zoologist 31:112A.

Brönmark C, Rundle SD, Erlandsson A. 1991. Interactions between freshwater snails and tadpoles: competition and facilitation. Oecologia 87:8–18.

Bury RB, Whelan JA. 1984. Ecology and management of the bullfrog. Washington: USDI Fish and Wildlife Service. Resource Publication 155.

Caldwell JP, Thorp JH, Jervey TO. 1980. Predator-prey relationships among larval dragonflies, salamanders, and frogs. Oecologia 46:285–289.

Chovanec A. 1992. The influence of tadpole swimming behavior on predation by dragonfly nymphs. Amphibia-Reptilia 13:341–349.

Cohen MP, Alford RA. 1993. Growth, survival and activity patterns of recently metamorphosed *Bufo marinus*. Wildlife Research 20:1–13.

Cooke AS. 1995. A comparison of survey methods for crested newts (*Triturus cristatus*) and night counts at a secure site, 1983-1993. Herpetological Journal 5:221–228.

Crawshaw LI, Rausch RN, Wollmuth LP, Bauer EJ. 1992. Seasonal rhythms of development and temperature selection in larval bullfrogs, *Rana catesbeiana* Shaw. Physiological Zoology 65:346–359.

Crump ML. 1991. Choice of oviposition site and egg load assessment by a treefrog. Herpetologica 47:308–315.

Duellman WE, Trueb L. 1985. Biology of Amphibians. New York: McGraw-Hill.

Dodd CK. 1979. A photographic technique to study tadpole populations. Brimleyana 2:131–136.

Fauth JE. 1990. Interactive effects of predators and early larval dynamics of the treefrog *Hyla chrysocelis*. Ecology 71:1609–1616.

Feder ME. 1983. The relation of air breathing and locomotion to predation on tadpoles, *Rana berlandieri*, by turtles. Physiological Zoology 56:522–531.

Gatten RE Jr, Caldwell JP, Stockard ME. 1984. Anaerobic metabolism during intense swimming by anuran larvae. Herpetologica 40:164–169.

Gascon C, Travis J. 1992. Does the spatial scale of experimentation matter? a test with tadpoles and dragonflies. Ecology 73:2237–2243.

Griffiths RA. 1985a. A simple funnel trap for studying newt populations and an evaluation of trap behavior in smooth and palmitate, *Triturus vulgaris* and *T. helveticus*. Herpetological Journal 1:5–10.

Griffiths RA. 1985b. Diel pattern of movement and aggregation in tadpoles of the common frog *Rana temporaria*. Herpetological Journal 1:10–13.

Griffiths RA. 1991. Competition between common frog, *Rana temporaria*, and natterjack toad, *Bufo calamita*, tadpoles: the effect of competitor density and interaction level on tadpole development. Oikos 61:187–196.

Griffiths RA, Denton J. 1992. Interspecific association in tadpoles. Animal Behaviour 44:1153–1157.

Griffiths RA, Raper SJ. 1994. A review of current techniques for sampling amphibian communities. Peterborough, UK: Joint Nature Conservation Committee. Report 210.

Griffiths RA, Getliff JM, Mylotte VJ. 1988. Diel patterns of activity and vertical migration in tadpoles of the common toad, *Bufo bufo*. Herpetological Journal 1:223–226.

Hawkins CP. 1994. Larval tailed frogs alter their feeding behavior in response to non-visual cues from four aquatic predators. Northwest Science 68:129.

Hayes MP, Jennings MR. 1986. Decline of ranid frog species in western North America: Are bullfrogs (*Rana catesbeiana*) responsible. Journal of Herpetology 20:490–509.

Hews DK. 1988. Alarm response in larval western toads, *Bufo boreas*: Release of larval chemicals by a natural predator and its effect on predator capture efficiency. Animal Behaviour 36:125–133.

Hews DK, Blaustein AR. 1985. An investigaton of the alarm response in *Bufo boreas* and *Rana cascadae* tadpoles. Behavioral and Neural Biology 43:47–57.

Hoff K. vS. 1987. Morphological determinants of fast-start performance in anuran tadpoles [dissertation]. Halifax, NS: Dalhousie University.

Holomuzki JR. 1986a. Predator avoidance and diel patterns of microhabitat use by larval tiger salamanders. Ecology 67:737–748.

Holomuzki JR. 1986b. Effect of microhabitat on fitness components of larval tiger salamanders, *Ambystoma tigrinum nebulosum*. Oecologia 71:142–148.

Hopey MF, Petranka JW. 1994. Restriction of wood frogs to fish-free habitats: How important is adult choice? Copeia 1994:1023–1025.

Horat P, Semlitsch RD. 1994. Effects of predation risk and hunger on the behaviour of two species of tadpoles. Behavioral Ecology and Sociobiology 34:393–401.

Jackson ME, Semlitsch RD. 1993. Paedomorphosis in the salamander *Ambystoma talpoideum*: effects of a fish predator. Ecology 74:342–350.

John-Adler HB, Morin PJ. 1990. Effects of larval density on jumping ability and stamina in newly metamorphosed *Bufo woodhousei*. Copeia 1990:856–860.

Kats LB. 1988. The detection of certain predators via olfaction by small mouthed salamander larvae (*Ambystoma texanum*). Behavioral and Neural Biology 50:126–131.

Kats LB, Petranka JW, Sih A. 1988. Antipredator defenses and the persistence of amphibian larvae with fishes. Ecology 69:1865–1870.

Kiseleva EI. 1993. Chemical interaction of common toad tadpoles (*Bufo bufo*, Anura, Amphibia) with tadpoles of other anurans dwelling in the same reservoirs. Zhurnal Obshchei Biologii 54:311–319. [In Russian]

Kruse KC, Francis MG. 1977. A predation deterrent in larvae of the bullfrog, *Rana catesbeiana*. Transactions of the American Fisheries Society 106:248–252.

Lawler SP. 1989. Behavioural responses to predators and predation risk in four species of larval anurans. Animal Behaviour 38:1039–1047.

Lefcort H, Eiger SM. 1993. Antipredatory behaviour of feverish tadpoles: implications for pathogen transmission. Behaviour 126:13–27.

Leips J, Travis J. 1994. Metamorphic responses to changing food levels in two species of hylid frogs. Ecology 75:1345–1356.

Liu H, Wassersug RJ, Kawachi K. 1996. A computational fluid dynamics study of tadpole swimming. Journal of Experimental Biology 199:1245–1260.

Morin PJ. 1983. Predation, competition, and the composition of larval anuran guilds. Ecological Monographs 53:119–138.

Noland R, Ultsch GR. 1981. The role of temperature and dissolved oxygen in microhabitat selection by the tadpoles of a frog (*Rana pipiens*) and a toad (*Bufo terrestris*). Copeia 1981:645–652.

Pearman PB. 1995. Effects of pond size and consequent predator density on two species of tadpoles. Oecologica 102:1–8.

Peterson AG, Bull CM, Wheeler LM. 1992. Habitat choice and predator avoidance in tadpoles. Journal of Herpetology 26:142–146.

Petranka JW, Sih A. 1986. Environmental instability, competition, and density-dependent growth and survivorship of a stream-dwelling salamander. Ecology 67:729–736.

Petranka JW, Kats LB, Sih A. 1987. Predator-prey among fish and larval amphibians: use of chemical cues to detect predatory fish. Animal Behaviour 35:420–425.

Pfeiffer W. 1966. Die verbreitung der schrekreation bei kaulquappen und die herkunft des schrechstoffes. Zeitschrift für Vergleichende Physiologie 52:79–98.

Pfennig DW, Reeve HK, Sherman PW. 1993. Kin recognition and cannibalism in spadefoot toad tadpoles. Animal Behaviour 46:87–94.

Pfennig DW, Sherman PW, Collins JP. 1994. Kin recognition and cannibalism in polyphenic salamanders. Behavioral Ecology 5:225–232.

Relyea RA. 1995. Predator discrimination by six species of larval anurans. Bulletin of the Ecological Society of America 76(supplement):224.

Richter KO. 1995. A simple aquatic funnel trap and its application to wetland amphibian monitoring. Herpetological Review 26:90–91.

Roth AH, Jackson JF. 1987. The effect of pool size on recruitment of predatory insects and on mortality in a larval anuran. Herpetologica 43:224–232.

Schmuck R, Geise W, Linesmair KE. 1994. Life cycle strategies and physiological adjustments of reedfrog tadpoles (Amphibia, Anura, Hyperolidae) in relations to environment conditions. Copeia 1994:995–1007.

Scott DE. 1990. Effect of larval density in *Ambystoma opacum*: an experiment in large-scale field enclosures. Ecology 71:296–306.

Scott DE. 1994. The effect of larval density on adult demographic traits in *Ambystoma opacum*. Ecology 75:1383–1396.

Semlitsch RD. 1987. Interactions betwen fish and salamander larvae. Oecologia 72:481–486.

Semlitsch RD, Reyer HU. 1992. Modification of anti-predator behavior in tadpoles by environmental conditioning. Journal of Animal Ecology 61:353–360.

Semlitsch RD, Scott DE, Pechmann JHK. 1988. Time and size at metamorphosis related to adult fitness in *Ambystoma talpoideum*. Ecology 69:184–192.

Shaffer HB, Alford RA, Woodlawn BD, Richards SJ, Altig RG, Gascon C. 1994. Quantitative sampling of amphibian larvae. In: Heyer WR, Donnelly MA, McDiarmid RW, Hayek LAC, Foster MS, editors. Measuring and monitoring biological diversity. Standard methods for amphibians. Washington: Smithsonian Institution Press. p 130–141.

Sih A, Kats LB. 1994. Age, experience, and the response of streamside salamander hatchlings to chemical cues from predatory sunfish. Ecology 96:253–259.

Sih A, Kats LB, Moore RD. 1992. Effects of predatory sunfish on the density, drift, and refuge use of stream salamander larvae. Ecology 73:1418–1430.

Skelly DK. 1992. Field evidence for a cost of behavioral antipredator response in a larval amphibian. Ecology 73:704–708.

Skelly DK. 1994. Activity level and the susceptibility of anuran larvae to predation. Animal Behaviour 47:465–468.

Skelly DK. 1995. A behavioral trade-off and its consequences for the distribution of *Pseudacris* treefrog larvae. Ecology 76:150–164.

Skelly DK, Werner EE. 1990. Behavioral and life-historical responses of larval American toads to an odonate predator. Ecology 71:2313–2322.

Smith DC. 1983. Factors controlling tadpole populations of the chorus frog (*Pseudacris triseriata*) on Isle Royale, Michigan. Ecology 64:501–510.

Smith DC. 1987. Adult recruitment in chorus frogs: effects of size and date at metamorphosis. Ecology 68:344–350.

Smith DC, Van Buskirk J. 1995. Phenotypic design, plasticity, and ecological performance in two tadpole species. American Naturalist 145:211–233.

Stauffer HP, Semlitsch RD. 1993. Effects of visual, chemical and tactile cues of fish on the behavioral responses of tadpoles. Animal Behaviour 46:355–364.

Taylor J. 1983. Orientation and flight behavior of a neotenic salamander (*Ambystoma gracile*) in Oregon. American Midland Naturalist 109:40–49.

Tejedo M. 1993. Size-dependent vulnerability and behavioral responses of tadpoles of two anuran species to beetle larvae predators. Herpetologia 49:287–294.

Tejedo M, Reques R. 1992. Effects of egg size and density on metamorphic traits in tadpoles of the natterjack toad (*Bufo calamita*). Journal of Herpetology 26:146–152.

Triantafyllou MS, Triantafyllou GS, Grosenbaugh MA. 1993. Optimal thrust development in oscillating foils with application to fish propulsion. Journal of Fluids and Structures 7:205–224.

Tyler MJ. 1976. Frogs. Sydney: The Australian Naturalist Library, William Collins (Australia).

Waldman B. 1991. Kin recognition in amphibians. In: Hepper P, editor. Kin Recognition. Cambridge, U.K.: Cambridge University Press. p 162–219.

Warkentin KM. 1992a. Effects of temperature and illumination on feeding rates of green frog tadpoles (*Rana clamitans*). Copeia 1992:725–730.

Warkentin KM. 1992b. Microhabitat use and feeding rate variation in green frog tadpoles (*Rana clamitans*). Copeia 1992:731–740.

Waringer-Löschenkohl A. 1988. An experimental study of microhabitat selection and microhabitat shifts in European tadpoles. Amphibia-Reptilia 9:219–236.

Warner SS, Dunson WA, Travis J. 1991. Interaction of pH, density, priority effects on the survivorship and growth of two species of hylid tadpoles. Oecologia 88:331–339.

Wassersug RJ. 1973. Aspects of social behavior in anuran larvae. In: Vial JL, editor. Evolutionary biology of the anurans. Columbia, MO: University of Missouri Press. p 273–297.

Wassersug RJ. 1975. The adaptive significance of the tadpole stage with comments on the maintenance of complex life cycles in anurans. American Zoologist 15:405–417.

Wassersug RJ. 1989. Locomotion in amphibian larvae (or "Why aren't tadpoles built like fishes?"). American Zoologist 29:65–84.

Wassersug RJ. 1992. On assessing environmental factors affecting survivorship of premetamorphic amphibians. In: Bishop CA, Pettit KE, editors. Declines In Canadian amphibian populations: designing a national monitoring strategy. Ottawa: Canadian Wildlife Service. Occasional Paper 76. p 53–59.

Wassersug RJ, Hoff KvS. 1985. The kinematics of swimming in anuran larvae. Journal of Experimental Biology 119:1–30.

Werner EE. 1991. Nonlethal effects of a predator on competitive interactions between two anuran larvae. Ecology 72:1709–1720.

Werner EE. 1992. Individual behavior and higher-order species interactions. American Naturalist 140:S5–S32.

Werner EE. 1994. Ontogenetic scaling of comprtitive relations: size-dependent effects and responses in two anuran larvae. Ecology 75:197–213.

Werner EE, McPeek MA. 1994. Direct and indirect effects of predators on two anuran species along an environmental gradient. Ecology 75:1368–1382.

Werschkul DF, Christensen MT. 1977. Differential predation by *Lepomis macrochirus* on the eggs and tadpoles of *Rana*. Herpetologica 33:237–242.

Wilson DJ, Lefcort H. 1993. The effect of predator diet on the response of red-legged frog, *Rana aurora*, tadpoles. Animal Behaviour 46:1017–1019.

Woodward BD. 1983. Predator-prey interactions and breeding-pond use of temporary-pond species in a desert anuran community. Ecology 64:1549–1555.

©1997 by the Society for the Study of Amphibians and Reptiles
Amphibians in decline: Canadian studies of a global problem. David M. Green, editor.
Herpetological Conservation 1:282–290.

Chapter 28

THE ROLE OF MOLECULAR GENETICS IN THE CONSERVATION OF AMPHIBIANS

DAVID A. GALBRAITH[1]

Redpath Museum, McGill University 859 Sherbrooke Street W. Montréal, Québec H3A 2T5, Canada

ABSTRACT.—Conservation genetics is a rapidly developing field that seeks both to explain the dynamics of genetic variation within small populations and to provide tools for population managers to monitor demographic processes. Few applications of conservation genetics have been made to amphibian populations or to the questions posed by amphibian population declines. In general, the methodologies implemented in birds and mammals have been successfully applied to amphibians. However, the unique characteristics of amphibian life histories and breeding systems can be expected to introduce variance in effective population size and other fundamental population attributes. These in turn may affect the patterns of variation in rapidly-evolving genetic markers of interest to conservation geneticists. At present, theoretical foundations for the understanding of the demographics and genetic variation of populations undergoing decline are poorly developed relative to simpler stochastic models of the persistence of small populations.

RÉSUMÉ.—La génétique de conservation est une discipline qui connaît une expansion rapide et qui cherche à la fois à expliquer la dynamique des variations génétiques au sein de petites populations et à fournir des outils qui permettent de surveiller les différentes tendances démographiques. La génétique de conservation a rarement été appliquée aux Amphibiens ou utilisée pour élucider le déclin de leurs populations. En général, les méthodes utilisées pour les oiseaux et les mammifères s'appliquent très bien aux amphibiens. Toutefois, il convient de bien évaluer les caractéristiques uniques de leur vie et de leurs systèmes reproducteurs pour déterminer leurs effets vraisemblables sur les structures réticulaires des variations génétiques, dans un contexte de conservation. À l'heure actuelle, les fondements théoriques de la démographie et de la génétique des populations en voie de disparition sont peu développés, par rapport aux modèles stochastiques plus simples qui permettent de prédire la survie de petites populations. L'application de la génétique de conservation permettra de réunir les données démographiques nécessaires sur les populations soupçonnées d'être en voie de disparition.

Growing attention is being brought to amphibian population dynamics and population ecology because of concern over population declines. Since 1990, this has coincided with the accelerated development of population genetics concepts and methods applied to conservation (recently reviewed by Avise 1994). However, few studies have examined amphibian populations from the perspective of conservation genetics, and little or no mention has been made of the possible role of genetic factors in the decline and recovery of amphibian populations.

For other taxa, molecular genetics is having a growing influence on population-oriented conservation biology at many levels. Conservation genetics comprises 2 parallel conceptual fields. Population genetics using molecular methods can be used to study the demographics of populations, with the ultimate objective of determining effective population size, deme structure and patterns of allele flow. Alternatively, models of population dynamics and studies of the effects of inbreeding and hybridization may yield practical management recommendations for stabilizing and recovering declining populations.

1 Present address: *Royal Botanical Gardens, P.O. Box 399, Hamilton, Ontario L8N 3H8, Canada*

Molecular genetics has had a longer influence on the systematics of amphibians than it has on conservation, with significant contributions being derived from allozyme electrophoresis, mitochondrial DNA (mtDNA) studies, and cytogenetics. Molecular systematics is important in conservation genetics to recognize cryptic species and resolve the phylogenetic position or validity of taxa, and in forensic identification of the source of tissues (Blackett and Keim 1992; Guglich et al. 1993). The objective of this chapter is to examine some areas in which genetics can provide practical advice to conservation programs for amphibian populations.

Genetic Markers and Methods

A genetic marker is any discrete genetic variation which can be detected and studied. Any heritable character, from a colour phase to a DNA sequence can be a genetic marker, but molecular genetic markers are preferable in many cases because the basis of the variation being detected is understood (Avise 1994). The current "tool box" of techniques available to study molecular genetic markers ranges from allozyme electrophoresis to the many DNA sequence methods and presents a bewildering array of alternative nomenclature and acronyms. All molecular markers can be classified into 3 fundamental groups on the basis of inheritance patterns: single locus markers, multiple locus markers, and mitochondrial DNA (mtDNA) markers. Hoelzel (1992) provides excellent introductions to the variety of molecular genetic markers in use for population studies.

Single Locus Markers

Genetic markers that reveal variation at single loci are those most often used for population studies. Variant alleles of single genes can be detected by using allozyme electrophoresis, restriction fragment length polymorphisms (RFLPs), or more recently developed techniques employing hypervariable DNA microsatellites (simple sequence repeats). The principle characteristic of single locus information is that it consists of a genotype—an identifiable set of alleles for each individual at each locus studied.

A well-developed body of theory concerning allele frequency and allele flow (often called gene frequency and gene flow for historic reasons) has been developed around phenotypic characteristics displaying Mendelian inheritance patterns, and was elaborated and widely applied through the use of allozyme electrophoresis (Nei 1987). Fifteen to 50 genetic loci, representing genes coding for 10 to 30 enzymes, will usually be typed for each individual in studies using single locus allozyme methods (Avise 1994). Allozyme electrophoresis is inexpensive and reliable and continues to be a valuable method for assessing genetic variation and for taxonomic studies (May 1992). However, allozyme electrophoresis is limited by 2 factors. Specimens must be preserved at ultracold temperatures or else processed fresh to ensure that the proteins under study are not denatured. And, because the molecules under study are active enzymes, they may be under the influence of natural selection and thus violate the assumption that the genetic variation is neutral (Avise 1994).

Two types of single locus DNA markers, RFLPs and microsatellites, are now commonly used to detect DNA sequence variation in nuclear genes. RFLPs detect base pair substitutions in DNA which is usually non-coding and thus selectively neutral (Aquadro et al. 1992). Nuclear DNA microsatellites are now assuming a dominant role in demographic studies (Taylor et al. 1994). The method is fast, inexpensive once developed for a particular taxon, and reliable (Schlotterer et al. 1991). The variation detected by microsatellite markers is single-locus and based on *in vitro* amplification using the polymerase chain reaction (PCR). Therefore, very small samples of tissue, even degraded tissue, can be used. Usable DNA has already been extracted from tissues as ephemeral as butterfly wings (Rose et al. 1994), and shed snake skins. Microsatellite markers should be detectable from shed amphibian skin if the shed can be recovered before it is eaten.

All single locus markers have common Mendelian characteristics. They may include alleles related to obvious phenotypic traits such as colour morphs or differential allozyme expression or they may be neutral differences in DNA sequence at non-coding sites. Single locus methods all rely on the identification of a locus, whether it is coding or non-coding. For coding loci, variation is expected to be influenced to a greater or lesser extent by natural selection. Non-coding loci, such as most RFLPs, variable numbers of tandem repeats (VNTRs), or microsatellites, are neutral with respect to phenotype and are therefore the best markers for use in studies of demography (Amos and Hoelzel 1992). Single locus markers can also be used to estimate relatedness among individuals (Brookfield and Parkin 1993).

Multiple Locus Markers

Multiple locus DNA methods identify genetic variation at neutral, non-coding DNA sequences but do not attempt to identify specific loci (Bruford et al. 1992). DNA fingerprints, for example, simultaneously detect alleles at many loci using DNA probes that hybridize to loci that share a common, "core" repetitive DNA motif (Jeffreys et al. 1985). The DNA fingerprint profile of an individual is a non-Mendelian composite of the patterns of each of its parents. Random amplification of polymorphic DNAs (RAPDs), a 2nd form of multiple locus system, has has emerged from PCR technology (Williams et al. 1990). The presence or absence of individual DNA fragments represents variation in the underlying DNA sequence in both DNA fingerprinting and RAPDs. Although the results of DNA fingerprints and RAPDs are similar, they are based on detecting DNA sequence variation at completely different kinds of loci. Multi-locus profiles are powerful tools for detecting multiple paternity (Galbraith et al. 1993), assessing relatedness in populations (Lang et al. 1993; Geyer et al. 1993; Waldman et al. 1992), and even for species identification (Masters 1995). The alleles at the many loci are distinguishable from each other as different bands on an autoradiogram, but the identify of each locus is lost. Thus, both DNA fingerprints and RAPD profiles are phenotypes, not genotypes.

In a detailed study of the behaviour of different types of molecular genetic markers in amphibian populations, Scribner et al. (1994) examined several populations of the common toad (*Bufo bufo*) using 4 different types of genetic markers for nuclear DNA. With samples drawn from the same specimens, they examined allozyme variability, PCR-amplified microsatellites, VNTR single locus markers, and multiple locus DNA fingerprints. Their objective was to directly compare the estimates of heterozygosity and allele frequency derived from these 4 different classes of nuclear DNA sequence in real, semi-isolated populations. In general, all measurements provided similar estimates of genetic divergence among populations. As expected, Scribner et al. (1994) found allozyme loci were less variable than the hypervariable DNA classes of markers. Mutation rates of hypervariable loci were a significant contributing factor to interpopulation genetic differentiation relative to migration, indicating that these markers are very useful for demographic studies over short periods. These markers are less useful for studying larger scale demographic features or biogeographic phenomena for the same reason: the high mutation rate quickly obscures older, larger-scale information.

Mitochondrial DNA

Mitochondrial DNA, passed as a matriline from mothers to their offspring, is inherited in a very different mode from nuclear DNA. The small size of the mtDNA molecule, the completeness with which it has been characterised, and its lack of recombination, make it a powerful tool for taxonomy, demography and behavioral ecology. Several methodologies can access mtDNA variation. Most common are sequence information, usually generated by PCR sequencing (Hoelzel and Green 1992), and RFLP profiles (Tegelstrom 1992; Routman 1993). Sequences produced using PCR technology have largely replaced RFLP methods over the past 5 yr, but RFLPs remain an effective and sensitive method (Tegelstrom 1992). Moritz (1994) has differentiated 2 different uses for mtDNA data. In gene conservation, molecular genetic information is used to identify and direct the management of genetic diversity. In molecular ecology, genetic markers are used to study and manage demographic processes. Moritz (1994) points out that these 2 types of applications have differing goals and theory and address biological processes on different time scales.

Mitochondrial DNA sequences can be used to study processes occurring over a variety of time scales. At the taxonomic level, mtDNA can be used to estimate divergence times between sister taxa. This procedure has been used in gynogenetic salamanders to estimate the ages of clonal groups. For example, some clones of gynogenetic *Ambystoma* hybrids were estimated to be as much as 4 or 5 million yr old. However, the age of such hybrids estimated through nuclear DNA sequence divergence is not as great, and occasional nuclear introgression can take place (Spolsky et al. 1992, Hedges et al.

1992). The introgression of sperm-carried DNA is unusual among clonal species (Avise 1994). Such results emphasize that divergence times and phylogenetic relationships estimated for molecules cannot be assumed to be the same as the species carrying them (Avise 1994).

Molecular markers are one of the few tools that can resolve the complex patterns of hybridization which often arise in amphibians. In the case of *Ambystoma* [2] *laterale-jeffersonianum*, (LLJ; Rye et al., this volume) the haplotype of the mitochondrial DNA does not match that of either assumed parental species (Kraus and Miyamoto 1990). *Ambystoma* [2] *laterale-jeffersonianum* is a triploid which arose from hybridization between *A. laterale* and *A. jeffersonianum*, but the mtDNA in this hybrid is from *A. texanum*. The explanation for this condition involves 3 sequential hybridization events and remains to be confirmed.

Mitochondrial DNA has also been used to study more proximate relationships in amphibians. Adult *Bufo americanus* tend to return to their natal ponds for breeding and so it is likely that sibs could be encountered during mating (Waldman 1991). The mtDNA haplotypes of 86 amplexed pairs were analyzed by Waldman et al. (1992). They found significantly fewer pairs of toads had matching haplotypes, indicating close relatedness, than would be expected under a model of random mate choice given the distribution of haplotypes of the local populations. Waldmann et al. (1992) explained this as some form of kin recognition and subsequent avoidance of sib matings. They also found a positive correlation between relatedness as measured by the similarity of DNA fingerprint patterns and similarity of the characteristics of male calls, suggesting that the characteristics of the calls could contain kin-recognition signals that females might be able to process (Waldman et al. 1992).

Molecular Genetics as a Demographic Tool

Molecular genetic data can generate useful inferences about demographic factors affecting population persistence (Milligan et al. 1994). Most of these inferences concern the detection of significant levels of inbreeding on a population-wide level through the application of statistical methods to detect the partitioning of genetic variance among individuals, local deme-groups, and across a population as a whole (Parker and Whiteman 1993).

Sub-division by habitat fragmentation is an almost ubiquitous feature of populations affected by human land use patterns (Belovsky et al. 1994). The principle effects of subdivision are demographic, including restricted breeding groups or the disruption of previous migration and dispersal patterns. As a result, considerable stochastic variation may be introduced into fundamental demographic processes (Lande 1993). These demographic effects, in turn, may result in accelerated genetic drift or inbreeding because of restrictions to effective population size or allele flow. Genetic variation in recently isolated populations of snakes in Ontario (Gibbs et al. 1994) provides an example of how such studies should be implemented. Multiple locus RAPDs did not detect significant genetic differentiation among isolated populations of black rat snakes (*Elaphe obsoleta*) and eastern massasauga rattlesnakes (*Sistrurus catenatus*). However, mean heterozygosity in populations of *E. obsoleta* in Ontario may be significantly reduced relative to the larger, still contiguous populations farther south (K. Prior, pers. com.). This suggests that population fragmentation and isolation may have resulted in inbreeding, which increases homozygosity but does not alter allele frequency in isolated populations.

Populations that have encountered severe restrictions on their census or effective population sizes and have recovered to some extent bear genetic evidence of the bottleneck (Menotti and O'Brien 1993). A wide variety of vertebrate species that have suffered significant bottlenecks have now been subjected to molecular genetic investigation to measure the effects of genetic drift, inbreeding and erosion of genetic variation in the context of ecological and physiological studies (O'Brien 1994). The long-term effects of such events are unclear, and seem to vary considerably (Avise 1994).

Theoretical treatments of small population persistence have relied on the development of models of stochastic behaviour of a few key factors which are applicable across taxonomic or geographic boundaries (Lande 1993). However, declining populations are affected by a great many separate and contingent factors that are case-specific and cannot yet be modelled (Caughley 1994). In amphibian populations exhibiting dramatic fluctuations in population size (Bertram and Berrill, this volume; Green, this volume; Shirose and Brooks, this volume), a simple model can be constructed to predict the effect of demographics on the expected level of genetic variation present. Heterozygosity is strongly influenced by effective population size over time scales of generations. The expected relationship between change in heterozygosity at a single locus per generation and effective population size is: $\Delta H \cong -1/2\, N_e$, where ΔH is the change in heterozygosity per generation and N_e is the effective population size (Nei 1987). This model assumes that effective population size remains consistent over time.

The rapid change in census population size exhibited by amphibians (Green, this volume) may render such a simplistic model invalid. Pond-breeding amphibians are expected to experience repeated founder effects and bottlenecks. If a small proportion of the tadpole population reaches maturity, a bottleneck event is inferred. However, increasing numbers rapidly under favourable conditions can prevent the loss of genetic variation which would otherwise occur if the population remained small.

In the case of small amphibian populations in which the effective population size varies greatly from generation to generation, population genetic theory predicts that rare alleles will be very quickly lost, and alleles should be driven to fixation. Repeated bottleneck events should further reduce effective population size. At the same time, the basic life history of pond-breeding amphibians, in which large numbers of eggs are produced by most adult females, lends itself to stochastic recruitment and high variances in individual reproductive success. This could also be expected to result in a small effective population size relative to the census population. The mortality of developing embryos and offspring is spread over time from oviposition through metamorphosis to maturity, exposing the offspring to a series of risks. Of course, these range from catastrophic events with random effects with respect to genetic variation, such as the drying of an entire pond, to potential selective regimes such as predation attempts.

These risks suggests that the population genetics of even small, undisturbed amphibian populations may exhibit chaotic fluctuations in allele frequencies. Genetic drift, a change in allele frequencies caused by imperfect sampling of the available gametes with each generation, is slow in large populations and more pronounced in small populations. As the variance in reproductive success predicted in amphibian populations means a small effective population size relative to peak population census, genetic drift is expected to be high. The consequences of this situation cannot be predicted with certainty. On the average, small populations are expected to lose genetic variation because of genetic drift. This should result in the loss of rare alleles and an increase in homozygosity. At the same time, small populations are prone to inbreeding. This should result in a loss of deleterious rare alleles from the population over time, but may also depress reproductive performance relative to populations with higher levels of genetic variation.

Inbreeding depression and genetic drift, combined with increasing probabilities of extinction due to demographic fluctuations in small populations, have been collectively termed an extinction vortex (Gilpin and Soulé 1986). These effects are thought to exert a strong positive feed-back on each other and may increase the probability that the population will be extinguished.

Observations of allele frequency alone are often insufficient to extract demographic information from populations. However, detecting the partitioning of genetic variation within and among demographic units may be an effective way to detect demographic trends which could result in extinction (Milligan et al. 1994). An important example of how apparently conflicting results from different molecular markers can yield important insights into population processes is provided by *Cryptobranchus alleganiensis*. Allozyme electrophoresis revealed very little variation among 12 *C. alleganiensis* populations in the eastern USA., but variation was found among mtDNAs (Routman 1993). Within isolated populations, *C. alleganiensis* mtDNA did not vary greatly, but populations differed strongly from each other in mtDNA types (Routman 1993). Because mtDNA evolves quickly relative to allozymes, Routman (1993) inferred that *C. alleganiensis* went through a bottleneck in the past, which reduced allozyme variation. The population then expanded into relatively isolated populations which do not frequently exchange individuals. The homogeneity of mtDNA sequences within populations is therefore thought to be due to limited flow of genes among populations, and the variation between populations is due to mtDNA sequences evolving within each population (Routman 1993).

Metapopulation Concepts

Metapopulations are collections of smaller demographic or geographic sub-populations which can each be characterized in terms of genetic variation and demographic parameters, but which are linked by dispersal between the local populations. A considerable body of theory on the erosion of genetic variation in small groups has been developed and attempts to develop computer simulations of extinction to include the effects of inbreeding, random demographic changes, and catastrophic events of varying severity have been made for application to small population conservation (Amos and Hoelzel 1992). Amphibian populations may be prone to non-linear population dynamics to a great extent, because most amphibians exhibit a bimodal life history with strong, stochastic fluctuations in recruitment. These considerations are to some extent mitigated in wild populations where there is any significant flow of alleles or individuals from population to population. Some sub-populations may function as sources of individuals, while others may be sinks, and these roles may change with time. The demographic fate of individual sub-populations within metapopulations may not be greatly affected by their linkage to neighbours, but the probability of extinction of the whole metapopulation is greatly reduced relative to that of any single sub-population. Even if single sub-populations are lost to demographic fluctuations, dispersal from neighbouring sub-populations can recolonize the vacated habitat. If all of the sub-populations remain under pressure from external sources then the persistence of the metapopulation remains tenuous. Although theory exists for the persistence of small populations, there has been less development of theory dealing with the factors contributing to population declines and how to alleviate them (Caughley 1994).

Management Implications of Conservation Genetics

If conservation genetics is to have any relevance to declining amphibians it must have practical consequences, as a source of background information either on population declines or for management recommendations where intervention is necessary.

Genetic Screening

Population genetics has a role in efforts to support populations through supplementation or assisted breeding (Mistretta 1994). In such programs, release of individuals to a particular location should only be made when there is a good correspondence between the genetic material of the release group and the existing background population. Various types of genetic screening have been proposed to ensure the suitability of the implant group. Although the term genetic screening suggests a defined procedure, in fact it is a general idea rather than a lab test. It is necessary to ensure that the wild animals at the release site are going to be able to have offspring which will integrate with the offspring of the implant. It is also important to be sure that members of the implant group do not carry traits which would give their offspring a competitive advantage over the offspring of native individuals. The purpose of a reintroduction program is to alter the demographic situation of a population, but genetic manipulation of a wild population is a different objective, and accidental genetic manipulation should be stringently avoided (Mistretta 1994).

Genetic screening to prepare a group of candidates for implant may be as simple as looking at the historical records and pedigree information, if captive bred, to ensure that the released animals are not prone to known problems. Similarly, genetic screening for habitat suitability can be as simple as knowing where the predecessors of the original stock came from and ensuring that captivity has not imposed a breeding regime likely to have eliminated important alleles.

A list of information involved in genetic screening should include information on the origin of specimens involved, cytogenetic information to ensure obvious problems such as inversions are not present in the release group, and single or multi-locus genetic information to characterize both native and release groups. Ideally, genetic information should have a major role in reintroduction decisions (Templeton 1991).

Conservation management recommendations based on molecular genetic data should always be presented with clear warnings that the interpretation of the results of genetic studies may not be unambiguous. Predictions based on genetic information cannot be taken as certainties any more than predictions made on the basis of other types of biological information (Mistretta 1994).

Although there is no question that severe inbreeding and other genetic effects can pose serious problems for the survival of small populations, is must be recognized that these conditions are indications of demographic problems rather than problems that must themselves be solved. Lande (1988) pointed out that by the time genetic problems such as inbreeding depression become serious threats to population survival, the population in question is almost certainly doomed by the demographic factors leading to the expression of genetic problems. Therefore, examination of population genetic parameters should be mindful of demographic factors which may be eroding variation. Unfortunately, there have been very few, if any, studies of amphibians in which the long-term demographic history of a population has been known at the same time as the history of the population's genetic characteristics have been documented.

Museum Specimens as Population Samples

Museum reference collections are important resources in conservation genetics studies since genetic information may be retrieved from many types of specimens. The only specimens which are wholly unsuitable for molecular genetics studies are those which have been stored for > 1 wk in formalin or have been tanned (Mortilla et al. 1994). Otherwise, specimens which have been frozen, dried or preserved in alcohol can yield useful information through in-vitro DNA amplification procedures. For example, museum specimens were recently used to generate both mitochondrial DNA data and microsatellite genotypes of red wolves, adding further evidence to the theory that this species was generated by a hybridization event between grey wolves and coyotes (Roy et al. 1994a, b). Amphibian specimens are usually fixed in formalin and stored in alcohol. The short-term exposure to formalin required for fixation does not cause significant damage to DNA in most cases. Therefore, protection and assessment of such reference material should be a high priority for all museums. In addition, all destructive collection of live specimens, regardless of the intended purpose of the collection, should include preparation and deposition of a preserved tissue sample with a suitable museum collection.

Conclusions

The large body of information and theory on the persistence of small populations is of little direct application in populations undergoing decline, because small population theory is predicated on the assumption that small population size is itself the main factor putting the population at risk of extinction (Caughley 1994). On the other hand, populations in decline are being influenced by unpredictable and case-specific factors that can only be addressed by ecological studies and thus have attracted little interest by population theorists (Caughley 1994). Perhaps the present focus on single factors to explain global amphibian population declines is, in part, due to frustration generated by the intractability of what Caughley (1994) has identified as the "declining population paradigm".

Conservation genetics is a useful tool to augment long-term demographic studies. Unfortunately, the current vogue for genetics has been at some cost to long-term population studies. The tools and concepts of molecular genetics will greatly improve our understanding of amphibian population dynamics and may help to explain population declines. However, without long-term studies providing fundamental demographic information for single populations which can then be examined in the light of population genetics, causal associations cannot be made between amphibian population fluctuations and genetic mechanisms, or any other postulated extrinsic factor (Milligan et al. 1994).

Pilot research should be initiated on the genetics of amphibian populations already under long-term study in which both allele frequencies and heterozygosity should be measured by a variety of types of genetic markers. Such studies should expressly investigate the role of metapopulations. Detailed modelling should be undertaken to predict the consequences of amphibian life histories on hypervariable genetic markers. Theoretical modelling of the general phenomenon of declining populations should be encouraged, especially where such modelling includes realistic, multiple caused declines, and complex life histories. Extant populations could be compared to suitable museum specimens to find suitable samples of pre-decline and mid-decline populations to examine for effects of decline on genetic markers.

Acknowledgments

I thank the students and staff of the Durrell Institute of Conservation and Ecology, University of Kent, Canterbury, United Kingdom, and D. M. Green for many discussions of genetics applied to conservation. This chapter is dedicated to the memory of W. D. Galbraith (1926–1995).

Literature Cited

Amos B, Hoelzel AR. 1992. Applications of molecular genetic techniques to the conservation of small populations. Biological Conservation 61:133–144.

Aquadro CF, Noon WA, Begun DJ. 1992. RFLP analysis using heterologous probes. In: Hoelzel AR, editor. Molecular genetic analysis of populations: a practical approach. New York: IRL Press. p 115–157.

Avise JC. 1994. Molecular markers, natural history and evolution. New York: Chapman and Hall.

Belovsky GE, Bissonette JA, Dueser RD, Edwards TC, Luecke CM, Ritchie ME, Slade JB, Wagner FH. 1994. Management of small populations—concepts affecting the recovery of endangered species. Wildlife Society Bulletin 22:307–316.

Blackett RS, Keim P. 1992. Big game species identification by deoxyribonucleic acid (DNA) probes. Journal of Forensic Science 37:590–596.

Brookfield JFY, Parkin DT. 1993. Use of single-locus DNA probes in the establishment of relatedness in wild populations. Heredity 70:660–663.

Bruford MW, Hanotte O, Brookfield JFY, Burke T. 1992. Single-locus and multilocus DNA fingerprinting. In: Hoelzel AR, editor. Molecular genetic analysis of populations: a practical approach. New York: IRL Press. p 226–269.

Caughley G. 1994. Directions in conservation biology. Journal of Animal Ecology 63:215–244.

Galbraith DA, White BN, Brooks RJ, Boag PT. 1993. Multiple paternity in clutches of snapping turtles (*Chelydra serpentina*) detected using DNA fingerprints. Canadian Journal of Zoology 71:318–324.

Geyer CJ, Rryder OA, Chemnick LG, Thompson EA. 1993. Analysis of relatedness in the California Condors, from DNA fingerprints. Molecular Biology and Evolution 10:571–589.

Gibbs HL, Prior KA, Weatherhead PJ. 1994. Genetic analysis of populations of threatened snake species using RAPD markers. Molecular Ecology 3:329–337.

Gilpin ME, Soulé ME. 1986. Minimum viable populations: processes of species extinction. In: Soulé ME, editor. Conservation biology: the science of scarcity and diversity. Sunderland, MA: Sinauer Associates. p 19–34.

Guglich EA, Wilson PJ, White BN. 1993. Application of DNA fingerprinting to enforcement of hunting regulations in Ontario. Journal of Forensic Science 38:48–59.

Hedges SB, Bogart JP, Maxson LR. 1992. Ancestry of unisexual salamanders. Nature 356:708–710.

Kraus F, Miyamoto MM. 1990. Mitochondrial genotype of a unisexual salamander of hybrid origin is unrelated to either of its nuclear haplotypes. Proceedings of the National Academy of Sciences of the United States of America 87: 2235–2238.

Hoelzel AR. editor. 1992. Molecular genetic analysis of populations: a practical approach. New York: IRL Press.

Hoelzel AR, Green A. 1992. Analysis of population-level variation by sequencing PCR-amplified DNA. In: Hoelzel AR, editor. Molecular genetic analysis of populations: a practical approach. New York: IRL Press. p 297–305.

Jeffreys AJ, Wilson V, Thein SL. 1995. Hypervariable 'minisatellite' regions in human DNA. Nature 317:818–819.

Lande R. 1988. Genetics and demography in biological conservation. Science 241:1455–1460.

Lande R. 1993. Risks of population extinction from demographic and environmental stochasticity and random catastrophes. American Naturalist 142:911–927.

Lang JW, Aggarwal RK, Majumdar KC, Singh L. 1993. Individualization and estimation of relatedness in crocodilians by DNA fingerprinting with a BKM-derived probe. Molecular and General Genetics 238:49–58.

Masters BS. 1995. The use of RAPD markers for species identification in Desmognathine salamanders. Herpetological Review 26:92–95.

May B. 1992. Starch gel electrophoresis of allozymes. In: Hoelzel AR, editor. Molecular genetic analysis of populations: a practical approach. New York: IRL Press. p 1–27.

Menotti RM, O'Brien SJ. 1993. Dating the genetic bottleneck of the African cheetah. Proceedings of the National Academy of Sciences of the United States of America 90:3172–3176.

Milligan BG, Leebensmack J, Strand AE. 1994. Conservation genetics—beyond the maintenance of marker diversity. Molecular Ecology 3:423–435.

Mistretta O. 1994. Genetics of species reintroduction—applications of genetic analysis. Biological Conservation 3:184–190.

Moritz C. 1994. Applications of mitochondrial DNA analysis in conservation—a critical review. Molecular Ecology 3:401–411.

Mortilla M, Vaula G, St.-George-Hyslop PG. 1994. Assessment of genetic polymorphisms in DNA from formalin-fixed neurological tissues. Canadian Journal of Neurological Sciences 21:248–251.

Nei M. 1987. Molecular evolutionary genetics. New York: Columbia Univ. Press.

O'Brien SJ. 1994. A role for molecular genetics in biological conservation. Proceedings of the National Academy of Sciences of the United States of America 91:5748–5755.

Parker PG, Whiteman HH. 1993. Genetic diversity in fragmented populations of *Clemmys guttata* and *Chrysemys picta marginata* as shown by DNA fingerprinting. Copeia 1993:841–846.

Rose OC, Brookes MI, Mallet JLB. 1994. A quick and simple nonlethal method for extracting DNA from butterfly wings. Molecular Ecology 3:275.

Routman E. 1993. Mitochondrial DNA variation in *Cryptobranchus alleganiensis*, a salamander with extremely low allozyme diversity. Copeia 1993: 407–416.

Roy MS, Geffen E, Smith D, Ostrander EA, Wayne RK. 1994. Patterns of differentiation and hybridization in North American wolflike canids, revealed by analysis of microsatellite loci. Molecular Biology and Evolution 11:553–570.

Roy MS, Girman DJ, Taylor AC, Wayne RK. 1994. The use of museum specimens to reconstruct the genetic variability and relationships of extinct populations. Experientia 50:551–557.

Schlotterer C, Amos B, Tautz D. 1991. Conservation of polymorphic simple sequence loci in cetacean species. Nature 354:63–65.

Scribner KT, Arntzen JW, Burke T. 1994. Comparative analysis of intrapopulation and interpopulation genetic diversity in *Bufo bufo*, using allozyme, single-locus microsatellite, minisatellite, and multilocus minisatellite data. Molecular Biology and Evolution 11:737–748.

Spolsky CM, Phillips CA, Uzzell T. 1992. Antiquity of clonal salamander lineages revealed by mitochondrial DNA. Nature 356:706–708.

Taylor AC, Sherwin WB, Wayne RK. 1994. Genetic variation of microsatellite loci in a bottlenecked species—the northern hairy-nosed wombat *Lasiorhinus krefftii*. Molecular Ecology 3:277–290.

Tegelstrom H. 1992. Detection of mitochondrial DNA fragments. In: Hoelzel AR, editor. Molecular genetic analysis of populations: a practical approach. New York: IRL Press. p 89–113.

Templeton AR. 1991. The role of genetics in captive breeding and reintroduction for species conservation. Endangered Species Update 8:14–17.

Waldman B. 1991. Kin recognition in amphibians. In: Hepper PG, editor. Kin recognition. Cambridge, U.K.: Cambridge University Press. p 162–219.

Waldman B, Rice JE, Honeycut RL. 1992. Kin recognition and incest avoidance in toads. American Zoologist 32:18–30.

Williams JGK, Kubelik AR, Livak KJ, Rafalski JA, Tingey SV. 1990. DNA polymorphisms amplified by arbitrary primers are useful as genetic markers. Nucleic Acids Research 18:6531–6335.

©1997 by the Society for the Study of Amphibians and Reptiles
Amphibians in decline: Canadian studies of a global problem. David M. Green, editor.
Herpetological Conservation 1:291–308.

Chapter 29

PERSPECTIVES ON AMPHIBIAN POPULATION DECLINES: DEFINING THE PROBLEM AND SEARCHING FOR ANSWERS

DAVID M. GREEN

Redpath Museum, McGill University, 859 Sherbrooke St. W., Montréal, Québec H3A 2K6 Canada

ABSTRACT.—Impressions of amphibian population decline are rarely backed with historical or demographic data and are expressed either as a concern over declining sizes of populations or as a concern over declining numbers of populations. All instances in Canada where amphibians are considered to be in decline are due to population number losses, as is true elsewhere in the world. A research program broadly aimed at identifying demographic declines in population size is not practicable and monitoring methods designed to estimate population sizes are largely unreliable. Populations normally fluctuate in size over time periods ranging from years to decades; thus, rarity and population size reductions are not equivalent to declines and are not good predictors of population extinction. To study the problem, declines must be defined in terms of population numbers, not size. These declines may be ascertained by examining relative gains and losses of populations from year to year using verified presence/absence data from monitoring surveys. Current absence does not imply former presence at any particular site, and the loss of populations is not related to the sizes of the populations that remain. Due to the potential of environments to recover and of amphibians to disperse, local habitat destruction over the short term does not necessarily imply permanent amphibian population loss. There is a general paucity of information about natural amphibian populations under normal conditions and this information is needed to understand why declines in population number occur. Population changes are influenced by local conditions that may be possible to identify and are, in any case, usually specific and local. Amphibian populations are not universally declining. A single cause, or multiple synergistic causes, for global declines is elusive because it is difficult to prove that a particular factor involved in population loss is a primary or secondary agent. Amphibians' frequently complex life-histories, high reproductive potential, and high mortality rates require that conservation efforts concentrate on habitat preservation, including all stages of ecological succession, as well education and informed land use practices.

RÉSUMÉ.—Il est rare que des données historiques ou démographiques confirment l'hypothèse d'un déclin des populations d'amphibiens; par déclin on entend soit une décroissance des effectifs soit une décroissance du nombre de populations. Au Canada comme ailleurs dans le monde, le déclin des populations d'amphibiens prend la forme d'une déperdition au niveau du nombre de populations. Le programme de recherche qui vise essentiellement à identifier les déclins démographiques dans les effectifs n'est pas réaliste et les méthodes de surveillance conçues pour évaluer les effectifs manquent globalement de fiabilité. Les effectifs connaissent normalement des fluctuations qui peuvent durer plusieurs années voire plusieurs dizaines d'années. Par conséquent, la rareté et la réduction des effectifs ne sont pas nécessairement le signe d'un déclin de la population en question et ne peuvent pas permettre de prédire sa disparition. Pour étudier le problème, le déclin doit être défini en termes de nombre de populations et non d'effectifs. Ces déclins doivent être évalués en examinant les gains et les pertes relatives de populations d'années en années sur la base de données sur les présences/absences dûment vérifiées dans le cadre d'études de surveillance. L'absence d'une espèce ne signifie pas nécessairement sa présence à une date antérieure et le déclin démographique n'est pas nécessairement lié aux effectifs qui restent. Étant donné le potentiel des environnements à se régénérer et celui des amphibiens à se disperser, la destruction des habitats locaux à court terme n'entraîne pas nécessairement la disparition permanente des populations d'amphibiens. Les données sur les populations naturelles d'amphibiens dans des conditions normales sont d'une manière générale aussi pauvres qu'elles sont nécessaires pour comprendre les causes du déclin démographique observé dans certaines populations. Les changements démographiques dépendent des conditions locales qu'il est possible d'identifier et qui sont en règle générale à caractère très spécifique et localisé. Les populations d'amphibiens ne sont pas toutes en voie de disparition. Une cause unique ou des causes synergiques multiples ne suffisent pas à expliquer les phénomènes de déclin car il est très difficile de prouver qu'un facteur en particulier de déclin démographique est un agent primaire ou secondaire. Le cycle de vie des amphibiens est souvent très complexe et dans la mesure où leur potentiel reproducteur et leur taux de mortalité sont très élevés, il est essentiel que les efforts de conservation soient axés sur la protection de l'habitat, à toutes les étapes de la succession écologique, ainsi que sur l'éducation et l'adoption de pratiques raisonnables d'utilisation des sols.

When we discuss declines in amphibian populations, what do we mean? Is it a worry over the number of toads in a Toronto garden or the number of species of toads in Canada? the presence of salamanders across a wide region of Canadian prairie or the presence of a particular salamander in a particular Québec bog? population change in a common frog in British Columbia or preservation of an endangered frog in Southern Ontario? Are our concerns local, regional, or global? Can we identify a decline when it is real and avoid labelling something a decline when it is not? And how can we identify and remedy the conditions that lead to declines when we see them?

Seventeen species of Canadian amphibians have been identified as having suffered losses of populations (Weller and Green, this volume). Among them, *Ambystoma tigrinum* has declined in south central British Columbia (Orchard 1992) and losses of *Bufo hemiophrys* populations in Alberta have occurred over a decade (Roberts 1992), with lowered numbers of sightings of these toads reported from Manitoba as well (Koonz 1992). *Acris crepitans* and *Bufo fowleri* populations have been extirpated in Ontario from Pt. Pelee and Pelee Island in western Lake Erie (Oldham and Weller 1992; Green 1989). Drastic declines in *Rana pipiens* abundance have been noted throughout the western portion of its range from British Columbia through to northern Ontario (Roberts 1992; Seburn 1992; Didiuk, this volume; Koonz 1992; Oldham and Weller 1992). *Pseudacris triseriata* is now rare throughout the St. Lawrence Valley in southwestern Québec, where it was common in the 1950s (Bonin 1992; Daigle, this volume). At least 8 species are losing population numbers in British Columbia alone, considered to be threatened by forestry practises and land conversion. This echoes the greater numbers of declines seen in the western U. S. (Corn 1994) compared to the eastern U. S. (Pechmann et al. 1991). The population losses do not include those in habitats that are obviously disturbed by urbanization or intensive agricultural use. No one is considering the eradication of habitat in downtown Montréal, Toronto, or Vancouver, for instance, although amphibians surely once lived there. But neither has anyone seriously tested if the apparent declines are a significant departure from random or from normal levels of extinction.

There has been mixed opinion on precisely what is a decline and what is not. It is unfortunately easy to misinterpret biodiversity data of the sort that has so far been available (Rodda 1993) and a nagging problem for the Declining Amphibian Populations Task Force (DAPTF) initiative has been to define these terms of reference (Pechmann and Wilbur 1994; McCoy 1994). For example, declines in numbers of populations of *Dicamptodon tenebrosus* and *Ascaphus truei* in southwestern British Columbia have justifiably been attributed to habitat destruction, largely by logging (Orchard 1992), yet Farr (1989) expressed doubt that there was sufficient evidence to make a firm assessment or conclude that the declines were permanent. Although populations of some Canadian species have certainly been lost, there is no firm evidence that the sizes any remaining Canadian populations have permanently declined. It may appear to be paradoxical but population extinction is not necessarily preceeded by prior decline in population size (Blaustein et al. 1994a). As reported in this volume, the Canadian DAPTF Working Group undertook both wide-scale censusing to search for population loss and in-depth studies of particular populations to try to assess demographic decline. The results of these studies have shown how to define the problem more precisely and suggest the manner in which declining amphibian populations may be approached.

Types of Decline

The Demographic Approach

There are really 2 sorts of declines to consider: declines in the *size* of populations and declines in the *number* of populations. Declines in population size may be addressed with quantitative, demographic data and involve the estimation of N, the census population size, or even more importantly, N_e, the effective population size. Heightened mortality, lowered fecundity, and lowered recruitment may all lead to lowered abundance. In this vein, Vial and Saylor (1993) defined a decline, pertaining to amphibian populations, as:

> *"...a definite downward trend in numbers over a span of time appropriate to the species' life history, shown to be in excess of the normal fluctuations in population size."*

This is a demographic definition requiring us to know amphibian life histories, understand the extent of normal fluctuations in abundance, be able to accurately census population size, and be able to model the parameters of viable populations. It further requires charting the change in population size over time, ΔN, merely to establish what is the normal range of fluctuation while neglecting the natural processes of local population extinction and recolonization.

But declines fitting this demographic definition are not entirely what have inspired the DAP initiative (Barinaga 1990; Blaustein and Wake 1990; Wyman 1990; Wake 1991). The phenomenon most at issue is not necessarily demographic population size reductions but decrease in the numbers of populations due to extinctions over a broad scale. For instance, high abundance in a few populations does not mean that a species is holding its own if those are the only populations that remain. Therefore the Vial and Saylor definition does not encompass the whole of the phenomenon of concern regarding declining amphibian populations.

The Diversity Approach

The other concern regarding declining amphibians is the decrease in population number, as seen among the 17 identified declining species in Canada. Local extinctions, extirpations, and range reductions are at issue, rather than demographic parameters. These are qualitative and discrete phenomena, best assessed with good presence/absence data, in contrast to the quantitative data at the level within populations. Because of the stochastic nature of within-population parameters, the relationship between demographic variability and the probability of population extinction is not at all obvious (Goodman 1987; Schoener and Spiller 1992). Under normal, unstable conditions, and mindful of some amphibians' high fecundities, a reduction in N cannot be considered to lead inevitably to population loss or to be the singular hallmark of impending population loss. Thus another definition, complimentary to that of Vial and Saylor (1993), needs to be proposed:

> *A decline is the condition whereby the local loss of populations across the normal range of a species so exceeds the rate at which populations may be established, or re-established, that there is a definite downward trend in population number.*

This prescribes a much different research program. To assess losses in the number of populations, we need to understand normal extinction and recolonization rates and be concerned, at this interpopulational level, with landscape variability and the connections within metapopulation structures. All populations inevitably go extinct and the stability of individual populations can be very difficult to judge (Goodman 1987), so this implies assessing the persistence of populations and relative rates of loss and gain over large areas. The 2 different research programs engendered by the 2 different definitions of population declines each may contribute to understanding the phenomen.

The Demographic Research Program: Charting Fluctuating Population Size

There is considerable scope for increasing our understanding of amphibian population biology. Many of the models used to assess and conserve mammalian or avian populations, for instance, may be of limited use with the high fecundity, high mortality, and iteroparity of pond-breeding anurans. Different species of amphibians have very different life history strategies, and thus different reactions to perturbations affecting their numbers. They may also react quite differently to impacts at different times of the year (Berven 1990). For a short-lived species with high fecundity and high mortality, such as *Rana sylvatica* or *P. triseriata*, large losses of tadpoles and juveniles are always expected. The animals of greatest importance to the continuance of the population are the breeding adults in spring. So long as there is production and survival of breeding adults, the population is likely to persist. Therefore, the critical stage in the life history, and the one most critical to protect, is the adult in early spring at the onset of the breeding season. Their chances of breeding a 2nd time are often so low that losses of adults at any other time of year will have only minimal impact. However, for a long-lived, low-fecundity, low-mortality species, like *Plethodon cinereus*, losses of any individuals, especially the dispersing juveniles, may be more severely felt by the population as a whole. Nevertheless, both reproducers and dispersers are important for population persistence and the colonization of new populations.

Pacific tree frog, Hyla regilla. *Phot by David M. Green.*

Amphibians also confound standard population models where they exhibit biphasic life histories. Many ranid frogs overwinter as tadpoles and can survive as populations despite losses of adults (Bradford 1991). Contrarily, total reproductive failures due to massive die-offs of tadpoles may be masked by the presence of dispersed juveniles from other localities (Seburn et al., this volume). Thus amphibian population demographics need to be studied in rigorous fashion (Shirose and Brooks, this volume; Green, this volume) to discern in more detail amphibian survivorship, effective population size, minimum viable population size, and inbreeding, and determine how these demographic parameters affect metapopulation structure (Lande and Barrowclough 1987; Meffe and Carroll 1994). Many effects are density dependent. Although die-offs of tadpoles or metamorphs may be spectacular and alarming, epidemics of disease can only happen when population size is large enough for a pathogen to spread (Crawshaw, this volume).

Estimating Population Size

Although counting numbers is in itself insufficient for understanding population viability, the fundamental parameter remains how many individuals there are in a population. Capture-recapture methods and removal sampling may be sufficient to produce analysable data (Donnelly and Guyer 1994). There are many methods and much literature on the subject. A Lincoln/Peterson index based on 2 samples is both simple and unreliable. All animals captured in the 1st sample are marked and released. The 2nd sample assumes that the ratio of recaptured marked animals to the whole of the 2nd sample is proportional to the ratio of the 1st sample to whole of the population (i.e., N). The assumptions are rarely met and there are better means, but all require greater sampling effort. The program CAPTURE, (available from Colorado Cooperative Fish and Wildlife Research Unit, 201 Wager Bldg., Colorado State University, Fort Collins, CO, 80523 USA) can yield more reliable estimates of total numbers in closed populations based on repeated capture-recapture data (Rexstad and Burnham 1991; White et al. 1982). A closed population, where there is no immigration, emigration, birth or death, seems like a mere abstraction yet, for practical purposes, an anuran breeding chorus may approach the "closed" condition for the brief time the chorus endures. The Jolly-Seber method embodied in the program JOLLY (available from Dr. James E. Hines, Patuxent Wildlife Research Center, Laurel, MD 20708,USA) can apply to open populations (Pollack at al. 1990) but may not return the required total population size if used for mark-recapture within a single breeding season. It is valid for periodic sampling but may be hampered by low year to year recapture rates if survivorship is limited.

Sampling methods only aim to estimate the census size of the population, N_c, the absolute total number of animals present. The accuracy of estimating N_c depends upon the quality of the data, the model used, and the assumptions taken. But for judging the viability of a population and its real trends, N_c, is not entirely adequate. The effective population size, N_e, is more pertinent to the continuation of a population because not all individuals share equally in the total reproductive effort (Lande and Barrowclough 1987). But N_e is notoriously difficult to measure as it is governed by numerous factors, including bias in the effective sex ratio, variance in clutch size, and relative survival of offspring to maturity. Only a small proportion of individuals breed, for instance, and N_e contains only those which genetically contribute to the next generation. The formula

$$N_e = (4n_m \times N_f) / (N_m + N_f)$$

(Meffe and Carroll 1994) might be applied to the data of Bertram and Berrill (this volume) on breeding success in *Hyla versicolor*. Although the census size of the population was 87, 34, and 83 individuals in each of 3 yr, the numbers of amplectant pairs was considerably less than this—10, 4, and 19, respectively, and therefore, the effective population size was at most only 20, 8, and 38 individuals. But this reckoning still fails to account for a further reduction in N_e when some males breed more than once.

The variance in the numbers of successfully raised offspring per female, σ^2, also affects N_e. The greater the variance, the lower the value of N_e. In pond-breeding anurans, this may be significant, because the variance in clutch sizes may be considerable. According to Gilhen (1984), clutch sizes ranged from 396 to 1581 eggs (n = 87) for *Pseudacris crucifer* and from 1401 to 5289 eggs (n = 10) for *Rana clamitans* in Nova Scotia. In contrast, *P. cinereus* was recorded with clutch sizes only from 4 to 17 eggs (n= 58). However, lacking data on survivorship, calculating the variance in the distribution of progeny among females that is needed for this estimation of N_e is not possible.

Effective population size estimates calculated from demographic data complement those derived using genetic information. Fluctuations in population size decrease genetic N_e, because each crash to a small population size produces a bottleneck and a reduction in the genetic diversity available for the subsequently expanding population. Over time, the harmonic mean of populations sizes estimates genetic N_e, so a single population crash can greatly reduce its value. Many pond-breeding anurans experience such crashes (Bradford 1991; Pechmann et al. 1991). Nevertheless, this allows genetic methods to also be employed to estimate population parameters (Galbraith, this volume). Because heterozy-

Rough-skinned newt, Taricha granulosa. *Photo by David M. Green*

gosity is influenced by genetic effective population size, change in heterozygosity per generation (ΔH) allows an indirect estimation of N_e using allozyme or molecular data. The importance of such genetic analyses should not be underestimated. Until sufficient demographic and genetic data are at hand for any species, it will not be possible to determine the true vulnerability of populations, because we will be unable to estimate when a population may be approaching the critical minimum effective number of individuals required for population viability. We have no information on minimum viable population size from any Canadian amphibian.

Monitoring Population Demographics

The one effective way to establish population demographic parameters is via intensive, long-term study. Yet which demographic variables to monitor depends upon the questions being asked. According to Blaustein et al. (1994a), demographic variables need monitoring on a landscape scale for metapopulations. Demographic information such as age/stage distributions, natality and mortality schedules, and immigration and emigration are seldom stable in real populations and rarely known with accuracy. Yet these must be monitored on a real-time basis to have any predictive value. For example, it now appears that the number of calling adult male *B. fowleri* is dependent to an important but varying extent upon the numbers of yearling males joining the chorus (Green, this volume). Therefore, to predict chorus size for the following year, it may be best to monitor growth rate of juveniles during the summer. The number of calling males in 1 yr has proven to be a poor predictor of the number calling the next yr.

For many reasons, extensive monitoring of anuran choruses, as tested by numerous researchers in Canada, is too crude an estimation method of population size to enable meaningful evaluation of within-population parameters. Road call-counts are not useful as estimators of absolute animal abundance (Bonin et al., this volume Chapter 15; Bishop et al., this volume; Lepage et al., this volume). The scheme adopted experimentally by several Canadian monitoring studies used a scoring system for frog chorus intensity: 0 = no calls, 1= individuals heard plainly, 2 = countable numbers but overlapping calls, 3 = full chorus with uncountable numbers. In truth, the score of 0 is a negative datum that often merely indicates a non-observation. Negative data require confirmation before they can be accepted as probably true. Even scores of 1 or 2, under sub-optimal weather conditions, may be considered negative data indicating *absence* of a full chorus of the males present. Unless all possible males are calling, these scores do not really quantify the numbers of animals even within earshot. The subjectivity of these 2 categories in practise leaves considerable latitude for observer bias. Inadequately motivated or overwhelmed volunteers may not rigorously adhere to complicated protocols for data collection and may provide unreliable data. Anecdotal reportage almost always consists only of occurrence records, not reports of absences or losses. Repeated observations at precisely the same locations are vital to verify the truth of "0" data.

A call-count score of 3 is not quantitative. Anuran choruses saturate with callers as densities increase (Arak 1983). Once the air is full of calls, additional animals may be satellites or active searchers, depending upon species, and are in either case undetectable by ear. The relationship of the calling adult population to the total population also varies according to age structure, which may change from year to year. Furthermore, what constitutes a saturated chorus varies with species. A small number of toads of a species with a long call duration, such as *Bufo cognatus* or *Bufo americanus*, can easily fill the air near a listener whereas species with short call durations, such as *P. crucifer* or *Hyla regilla* may be more easily counted even in moderately large choruses. All monitoring groups operating in eastern Canada reported difficulty using the scoring scheme with *R. pipiens*, which does not form tight choruses of closely spaced individuals. A full chorus of *R. clamitans* is unlikely ever to be scored 3.

Time of night, relative humidity, and temperature also affect both the calling attitude of individual frogs or toads and the acoustic carrying capacity of the air. Weather conditions absolutely control whether or not there will be calls heard on any particular night and the proportion of animals calling, even at the height of a breeding season. In 1995, my students and I scored chorus strength on the 0 to 3 scale for both *B. americanus* and *B. fowleri* and estimated numbers of toads by ear (unpublished data) in conjunction with the Ontario Marsh Monitoring Program (see Bishop et al, this volume). We then waded into the chorus and counted all toads by intensive visual search. The aural estimates consistently underrated the number of toads observed to be present. Also in 1995, Shirose et al. 1997 found no relationship between aural index and population size for both *B. fowleri* and *Rana catesbeiana*. Within limits, call-count intensity data may be useful as a relative measure of population size from year to year, but listening for calls is likely to work best as a source of presence data for monitoring population loss or gain, not for demographic estimation of within-population parameters.

For amphibians that do not call, or do not call loudly, other methods must be used to obtain a census. Terrestrial salamanders present particular problems for determining population size, but use of artifical cover objects (Bonin and Bachand, this volume; Davis, this volume) may allow capture-recapture studies to proceed with reduced impact on microhabitats. However, assessing numbers of tadpoles and salamander larvae by any means is both difficult and largely inapplicable to monitoring the health or persistence of a population (Wassersug, this volume).

The Inherent Unpredictability of Population Size

The research program specified by the Vial and Saylor (1993) definition of declines is, for many species of amphibians, impossible to fulfill. The size of an animal population N at some time t is always a balance between gains of individuals by birth (B) and immigration (I) against losses by death (D) and emigration (E):

$$N_t = N_{t-1} + (B + I) - (D + E).$$

In species with the Type III survivorship curves that are typical of most North Temperate anurans, where both fecundity and mortality rates are very high, population parameters B and D are so inflated as to render it very difficult to establish population size based upon previous years' data. For long-lived species with low fecundities, like mammals, birds, or some terrestrial plethodontid salamanders and tropical frogs, populations may be stable enough that population growth models may be more predictive. But many species, such as *B. fowleri* (Breden 1988; Kellner and Green 1995), *B. americanus* (Acker et al 1986; Kalb and Zug 1990), *R. pipens* (Leclair and Castanet 1987) or *R. sylvatica* (Bastien and Leclair 1992), mature in only 1 or 2 yr and have very short natural life spans. *Bufo fowleri*, for example, requires an early successional habitat of dunes and ponds that are free from overgrowing vegetation (Green, this volume). These conditions often occur after major storms that also wreak catastrophic mortality upon the population, yet are nevertheless necessary for population growth. Like other amphibians (Wassersug 1975), the toads, then, are trapped in a perpetually and unpredictably fluctuating environment and respond in kind. Other species, such as *H. versicolor*, also require early or mid-succession habitats and find disturbed environments advantageous (McAlpine, this volume). Predator populations are subject to weather-mediated stochasticity. Variable growth rates of amphibians between individuals and between years influence maturation rates. Thus, unpredictability is inherent, and population viability theory predicts that small and fluctuating populations are more prone to random extinction (Soulé 1987). There is little chance within a human lifetime to identify what is the normal range of fluctuation in population size for many species of amphibians, and therefore no way to know if a population has deviated from this norm.

THE DIVERSITY RESEARCH PROGRAM: CHARTING DECLINES IN POPULATION NUMBER

Despite the demographic stochasticity of many amphibian populations, especially those suffering declines, a diversity research program based upon population number may be feasible. The number of species or populations present in a region over time depends upon a dynamic balance between their gain and loss (MacArthur and Wilson 1967; Levins 1969; Hanski and Gilpin 1991). Levins' (1969) metapopulation model considered the proportion of habitat patches occupied by a species at any time, modelling that to depend upon colonization of new patches (m) minus extinctions from occupied patches (e). But this concept can be simplified for our purposes by considering only numbers of populations rather than occupancy of habitat patches. Cast in the style of the previous equation describing change in population size, the numbers of populations (P) at time (t) can be seen to depend upon gains of new populations (G) minus losses of established populations (L):

$$P_t = P_{t-1} + (G - L).$$

Obviously, stable numbers of populations are achieved where $G - L = 0$, declining numbers occur where $G - L < 0$, and increases occur where $G - L > 0$. The flux, or amount of turnover (T) among populations is:

$$T = G + L,$$

and the rate of change in population number ΔP is:

$$\Delta P = \Delta G - \Delta L,$$

which reflects the rates of colonization (ΔG) and extinction (ΔL). This is clearly distinct from changes in population size, ΔN, because ΔP is not necessarily tightly coupled with the chance of extinction (Blaustein et al. 1994a; Schoener and Spiller 1992).

All populations go extinct eventually, and the single loss of a population may or may not be meaningful, depending upon the number of populations to begin with and the rate of population turnover. Demographic uncertainty, environmental uncertainty, genetic uncertainty, and catastrophe all lead to local extinctions. But similar uncertainties also lead to local recolonizations. Over a landscape scale, minimum viable population size is not so readily estimated as ΔP and T, and not so vital as the probability that all populations will go extinct over a given number of years.

Monitoring Population Turnover

Hecnar (this volume) used intensive methods to study population turnover in southwestern Ontario. But, despite its problems when applied to assessing population demography, extensive monitoring and the compilation of atlases may also be used effectively for determining presence or absence and thereby gauge the contraction or expansion of ranges and the persistence of viable populations. But short-term presence/absence data or anecdotal reports of raw numbers are likely inadequate. One of the most valuable outcomes of extensive monitoring programs is the compilation of reliable, annual, presence/absence reports into new atlases. Several atlas projects are being pursued in Canada, the oldest being the Ontario Herpetofaunal Survey (Oldham and Weller 1992) and the Atlas des amphibiens et reptiles du Québec (Bider and Matte 1994). Other projects inspired by the DAPCAN initiative aiming to compile accurate geographic records are continuing in Alberta, Saskatchewan, Manitoba, and New Brunswick. The value of atlas projects has suffered in the past from non-standardized protocols for data collection and from sometimes uncritical acceptance of anecdotal reports. With relatively few exceptions (McAlpine, this volume; Maunder, this volume), the historical database in Canada has not proven to be very useful as a basis for determining current declines. But organized monitoring schemes, operating with proven, uniform

Sampling frogs from a breeding chorus at night. Photo by John Mitchell.

protocols and providing rigorously verified data, may make it possible for atlases in future to chart yearly changes in amphibian occurence and thereby estimate ΔP and T over regional and landscape scales. This will more credibly determine the real nature of declines in population numbers than any information presently available.

THE CAUSES OF POPULATION LOSSES: DEMOGRAPHICS AGAIN

There is no compelling evidence that global causes are behind amphibian declines in Canada and no evidence from Canadian investigators that ozone depletion, ultraviolet radiation, or acid rain are directly responsible (Brooks 1992; Green 1993; Seburn 1994; Ovaska, this volume). Losses and declines of populations can be explained by local causes and the reaction of amphibians both to stochastic fluctuations and to human-mediated disruption and fragmentation of environments. Although excessive UV-B radiation, hyperacidity, and pesticides of course can be lethal or debilitating (Corn and Vertucci 1992; Blaustein et al. 1994b; Bradford et al. 1992; Berrill et al, this volume), and global warming is undoubtedly likely to make conditions better in some places and worse in others (Peters and Lovejoy 1992; Ovaska, this volume), the ultimate cause for today's rapid losses of biodiversity is human impact upon the environment (Diamond 1989). Its effect profoundly influences the demographics, metapopulation structure, and physiological responses of the species involved (Beebee 1977). Therefore, when we examine the population biology of amphibians we are in all cases looking at human-influenced population biology.

Habitat destruction, fragmentation, and degradation are all considered to be the leading causes of population losses (Meffe and Carroll 1994; Vial and Saylor 1993; Weller and Green, this volume). More precisely, these include the direct and indirect effects of urban encroachment, agricultural development, and logging. Outright, permanent destruction of habitats is obvious as a cause for population loss (Johnson 1992) but the fragmentation of environments is an equally powerful, yet more insidious potential cause of population disappearance. Fragmentation may sever the connections between co-dependent local populations, curtailing dispersal needed to rescue nearby local extirpations (Laan and Verboom 1990).

Consequence of fragmentation? Photo by Martin Ouellet.

Normally, local populations are connected by immigration and emigration of individuals (Gill 1978; Hanski and Gilpin 1991; Sjögren 1991). Some local populations with high recruitment will be sources of emigrating, excess individuals. Other populations at the same time may have low recruitment and be population sinks that exist only by continual restocking and rescue. These metapopulation relations are dynamic and germane to understanding whether local population loss is permanent or transitory. A metapopulation may consist of a single, large source population, in a lake for instance, and several peripheral sink populations in small surrounding ponds. Loss of the source population, perhaps through trout-stocking, over-building, or shoreline degradation, will inevitably invite loss of the surrounding local populations that depend upon its emigrants for continuation. More usually,

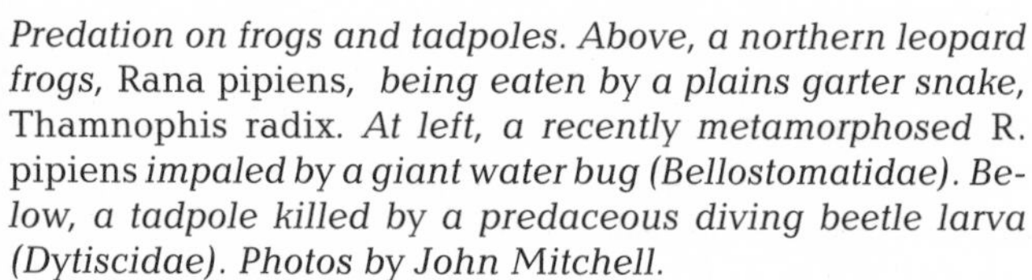

Predation on frogs and tadpoles. Above, a northern leopard frogs, Rana pipiens, *being eaten by a plains garter snake,* Thamnophis radix. *At left, a recently metamorphosed* R. pipiens *impaled by a giant water bug (Bellostomatidae). Below, a tadpole killed by a predaceous diving beetle larva (Dytiscidae). Photos by John Mitchell.*

a metapopulation may consist of ≥ 2 local populations in disequilibrium (Gill 1978). Because the probability of extinction for any 1 population exceeds the probability of extinction of the metapopulation as a whole, any human intervention that severs their connection risks altering the metapopulation structure and the increasing probability of extinction for all local populations. Thus, a local population may be driven to extinction, even though it is in a protected area and experiences no direct human impact.

Degradation of habitats reduces population viability *in situ*. Pollution, acidity, and toxic substances introduced into environments result in declines of reproductive potential and viability (Freda 1986; Bonin et al., this volume Chapter 25; Bertram et al., this volume). Pesticides may be lethal or teratogenic for amphibians in dosages present in polluted habitats. These compromises of the environment leave amphibians susceptible to secondary infections and other disease (Blaustein et al. 1994c; Bradford 1991; Crawshaw, this volume). There have been die-offs reported (Nyman 1986). I have received reports of tadpole mass mortality in Laurentian lakes in Québec, and Scott (1993) has coined a "post-metamorphic death syndrome" to describe similar die-offs of froglets in Arizona. Density-dependence of epidemic disease may be a factor in these die-offs, but they may be insignificant over the long term if breeding stocks of adults persist.

Introduced alien competitors and predators, including humans, also degrade habitats. Introduced predatory fish clearly have detrimental impact upon frog populations (Liss and Larson 1991; Sexton and Phillips 1986; Hecnar, this volume). Non-native *R. catesbeiana* have also been touted as debilitating predators upon native western frog populations (Moyle 1973; Hammerson 1982; Orchard 1992), although Hayes and Jennings (1986) cast doubt about their real contribution to *Rana aurora* population losses in California. Hayes and Jennings (1986) argued that, although there is a correlation between increased abundance of *R. catesbeiana* and declines in other frog species in California, there is no evidence of direct effect. *Rana catesbeiana* may merely have moved into marginalized, compromised, or unoccupied habitats previously made less suitable for the other species. Overharvest by humans, in particular, has been implicated in the *R. aurora* declines in California (Jennings and Hayes 1985), but there is no evidence that harvesting has directly reduced frog populations in Canada. During the catastrophic decline of *R. pipiens* in Manitoba during the 1970s, the "frog-pickers" reported large numbers of dead and dying frogs in the wild (Koonz 1992). There were still many frogs, but few live ones left to catch.

Environmental alteration may have beneficial effects for some species, especially those prospering in newly cleared, early successional habitats (Banks and Beebee 1987). The expansion of *H. versicolor* in southern New Brunswick may be due to newly opened, disturbed habitats providing additional breeding sites (McAlpine et al. 1991). Global warming may produce opportunities for some species to expand (Ovaska, this volume). Even forestry operations may leave behind new breeding sites for some species in the form of dugout ponds in cleared sites (Waldick, this volume), althought these are intially of very low quality. Species requiring stable, late successional environments with climax vegetation, such as terrestrial salamanders in British Columbia requiring old-growth forests, will be most perturbed by human resource-use activities (Davis, this volume; Dupuis, this volume). But, ironically, species requiring disturbed habitats and early successional vegetation, such as species of *Bufo*, may be adversely affected by conservation efforts aiming to prevent turnover of climax communities.

Different amphibian species are decidedly not uniform in their reactions to environmental change, and detailed knowledge of the population biology, physiology, and demographics of each species is required to ascertain causes of declines. Studies of this sort may be unable to identify declines as effectively as broad-scale presence/absence surveys, but once the phenomena are known and characterized, only intensive study at the local level can identify cause.

WHAT IS NOT A DECLINE?

As well as understanding which aspect of decline we are considering, we must also take pains not to be mistaken by situations that only appear to be declines. Endangered species, short-term diminution of population size, contained local habitat destruction, and lack of distributional data should not be interpreted inappropriately.

Rarity Alone Does Not Equal Decline

A great fear about amphibian declines has been that rare and threatened species may go extinct. This serious concern gave the global DAPTF much of its initial impetus. Outright losses of species, even in protected areas (Crump et al. 1992; Czechura and Ingram 1990), are rightly a cause for anxiety. But once a species has already declined to so perilously low a level as to render it endangered, it may be difficult to reconstruct how it got to that condition. Wholesale declines, to be detectable, must occur among species widespread and visible enough that they are noticeable. Protection of endangered species is often a salvaging enterprise and a race against the extinction of only 1 or a few populations. Thus, the factors that identify a species as endangered are not necessarily the same as those that may identify it as declining. Endangered species may, in fact, recover under careful management.

Furthermore, a species may be classified as rare simply because the extent of its range and abundance has not yet been gauged. Many amphibian species are considered rare and extremely localized in Canada. *Plethodon idahoensis* is only known from a few sites in southeastern British Columbia. Although it appears on British Columbia's "Red List" of endangered species (Munro 1993) there is no information about population trends despite fears that its fragmented range resulted from logging activities (Orchard 1992). In Ontario, both *Desmognathus fuscus* and *Gyrinophilus porphyriticus* are known from only a very few specimens. There is not sufficient information from the few reports of *B. cognatus* in Manitoba and Saskatchewan upon which to base any assessment of numbers. *Desmognathus ochrophaeus* (Sharbel and Bonin 1992) and *Rana pretiosa* (Green et al. 1996, 1997) are likewise only marginal in Canada. Therefore, from its beginning, the Canadian DAPTF Working Group did not involve itself heavily in concerns over current endangered species. The working mission of the group was explicitly to document and research declines in amphibian populations. Because of the necessity of conducting research over landscape scales, the lessons to be learned about declines are to be found among erstwhile abundant species.

Population Fluctuations Are Not Declines

Natural population fluctuations have to be distinguished from persistent declines in population viability across regions. Populations of amphibians normally may show apparently random variations in size simply because of their own population dynamics (Wissinger and Whiteman 1992; Weitzel and Panik 1993). Populations vary in their response to ecological changes in the environment, especially succession, and the unpredictable vagaries of the weather. Most pond-breeding, North Temperate amphibians have high intrinsic rates of increase and should be able to rebound successfully after population decline as seen, for instance, in *B. fowleri*, at Long Point, Ontario (Green, this volume). Even long-term assessments of population fluctuations may be inadequate to identify them as declines since the stochasticity of these fluctuations makes any predictions about future population sizes unreliable.

Local Habitat Destruction Does Not Signify Permanent Population Loss

Habitat destruction is rightly blamed for many losses of populations (Johnson 1992) but not all habitat destruction may be permanent. There is little chance that downtown urban centres will be made hospitable to amphibian populations any time soon but, in many cases, habitats that have been destroyed may recover, and recovering habitats may be recolonized by certain species (Waldick, this volume). The destruction of a southwestern British Columbia stream due to logging, rendering it uninhabitable for *D. tenebrosus* or *A. truei*, does not mean necessarily that there has been a permanent loss of these species. Amphibians do have the power of dispersal. We must consider the wider metapopulation which may cover a greater landscape area and encompass this local stream. If the habitat may again become usable and dispersal routes are available, it may be possible for local habitats to regain their ecological diversity. However, this optimism can only go so far. If the source populations are destroyed, or if the whole region is destroyed, or if dispersal routes are destroyed, then habitat loss may indeed result in permanent population loss. More important than the local destruction of any 1 habitat, therefore, is the regional interconnection of habitats and populations.

Current Absence Does Not Imply Former Presence

Several species have been suspected of having suffered declines decades ago for which there is no current evidence. In some cases, it may be reasoned indirectly that species once existed in certain heavily disturbed urban or agricultural areas but little chance that these losses can now be studied. The clearing of forests for early European settlement in extreme southwestern Ontario permanently destroyed most habitats suitable for *Ambystoma maculatum, P. cinereus*, or *Notophthalmus viridescens*, all of which are common and widespread elsewhere in the southern portion of the province yet extremely uncommon in this region (Hecnar, this volume; Weller and Green, this volume). This clearing took place long before occurrences of amphibians were systematically recorded, so the historical database contains no pertinent information. The region once was covered by tall-grass prairie, marshland, swamp, and hardwood forest (Bakowsky and Riley 1994), which, logically, should have provided high-grade amphibian habitat. But no declines are presently detectable and historic losses from the region are now largely outside the scope of present investigations. Neither is it possible, at this late date, to assess the true impact of pre-colonial aboriginal land use practises on other habitats, particularly in eastern Canada. Aside from a few subfossil remains (Holman and Cloutier 1995), the lack of information means that it is not possible to determine for certain which species ever existed in particular regions.

Conservation of Declining Amphibians

Protection of Amphibians

Management practises must bear in mind the metapopulation structure of amphibian populations and preserve both source habitats and the connections between local populations. Modern forestry practice of smaller clear-cuts, preservation of wide margins around streams, especially small streams devoid of commercial fish, and corridors between uncut stands is to be encouraged (Dupius, this volume; Waldick, this volume). Other possible measures may include "toad-tunnels" of appropriate design under new roads in critical areas or the rerouting of roads to avoid sensitive dispersal routes. Preservation of wetlands, establishment of forest preserves and maintenance of streams are all desirable. Parks and reserves must be extensive enough to buffer populations from surrounding development and encompass entire metapopulations. With better understanding and considered management practice, the problem of amphibian population loss might be forestalled.

The knowledge needed is readily apparent. We must identify critical life-history stages for species, identify metapopulation structures and habitat requirements, comprehend minimum sustainable population sizes, movements, and migrations, and chart population turnover, gains, and losses. We must gauge the impact of environmental stressors, including UV-B radiation, acid precipitation, pesticide pollution, and habitat fragmentation. And we must identify vulnerable populations, such as those that are isolated, in the way of urban or agricultural development, or affected by pollutants and toxic substances. With this knowledge, we may enact management plans that may have effect.

Status Evaluations

The status of species which may be at risk must be determined. Therefore, an evaluation of all species should be conducted. This partly has been the aim and intention of COSEWIC, the Committee on the Status of Endangered Wildlife in Canada, administered through Environment Canada. Reports on 11 salamander species and 11 anurans have been commissioned, but only a handful have been completed (*B. fowleri, Ambystoma texanum, D. tenebrosus, A. crepitans*). Even so, COSEWIC only has a mandate to investigate endangered wildlife, not declining wildlife populations. All species in Canada need to be assessed, not just the rarities.

Land Use

Critical habitats necessary for maintenance of populations must be known and adequately protected. This will require extensive knowledge of the population biology of each species. This will also require clear assessment of the environmental impact of development or if the rate of habitat recovery is not as great as the rate of habitat destruction. Past environmental impact studies have provided only minimal information usable to assess, and mitigate, perturbations upon amphibian populations. They generally go unnoticed in favour of more overtly commercial and recreational fish, birds, and mammals. In fairness, the lack of attention to amphibians in environmental impact assessment reflects their secretive behaviour and low visibility. But periodic or seasonal breeding, limited vagility, congregation of breeding assemblages, and sensitivity to habitat fragmentation can render amphibians particularly vulnerable to land use changes. Government agencies can take the lead to provide resources to assess populations of amphibians on public lands, determine those species at risk of decline, and determine habitat requirements of those species for incorporation into land management plans.

Education

Additional resource materials are required to portray the diversity and range of adaptations of amphibians and to deliver conservation-oriented messages to young people. The inherent fascination with amphibians can provide an opportunity to attract and hold the interest of school children. Texts, videos, and films can supplement school classroom curricula, but field trips and participation by school groups in monitoring efforts and local conservation projects would be especially valuable (Johnson 1992). The Metro Toronto Zoo's "Adopt-a-Pond" program encourages individual and school participation in wetland conservation with its package of tape recordings, project ideas, stickers, posters, and information aimed at school children. Members of natural history societies, professors, teachers, and wildlife agency staff can and should also participate actively in school programs and offer activities. For instance, a reptile and amphibian display for school groups is being developed by Nature Saskatchewan, a provincial conservation organization, and a travelling display on frogs and toads in Québec has been put together by the Redpath Museum of McGill University in Montréal. Reports, media interviews, and popular articles can help inform the public regarding amphibians. Popular guides to amphibians (and often reptiles) written for particular provinces (Cook 1966; Gilhen 1984; Green and Campbell 1984; Johnson 1989; Melançon 1961; Preston 1982; Russell and Bauer 1993) are instrumental in public education about amphibians. Despite the methodological and scientific difficulties of large-scale, volunteer-based monitoring of amphibian populations, one certain benefit is public education. Volunteers can realize that they are part of an important enterprise if the research programs in which they particpate are designed to yield useful data from their efforts.

Finally, the DAPTF has, in itself, inspired unprecedented interest in the biology of amphibians. The annual conferences of DAPCAN and the work of its national, regional, and provincial co-ordinators has stimulated the involvement of government agencies in amphibian conservation biology and inspired new graduate research. If humans are the problem, they are also the solution.

The Final Word

Are amphibians declining? Of course they are. All manner of species are declining, not just frogs and salamanders, because there is no place left on Earth that is truly pristine and untouched by the effects of human hands (Diamond 1989; Pounds and Crump 1994). When we discuss declines in amphibian populations we must continue to encompass concerns over toads in gardens and across Canada, salamander populations over the prairie and in particular bogs, and population changes in both common and endangered frogs. The fears are global; the causes are local or regional. It is possible, with good information, to identify a decline when it is real and ultimately to identify and remedy the conditions that are its cause. That is because there is no special, overarching cause for frogs and salamanders to decline, nor even a need to presuppose one, unless we are prepared to accuse ourselves. All declines of amphibian populations depend upon how the animals are affected by the myriad ways in which we humans can affect the landscape. This is cause for hope. If the universal cause was global, we as individuals may only be able to wring our hands in despair, with little hope ourselves of effecting a remedy. Yet, if there is a lesson, it is that is no matter how widespread a causative factor may be, the population biology of amphibians is intensely local. And because local conditions breed local responses, we may act to preserve amphibians, a region at a time.

Acknowledgments

I thank all the participants of the DAPCAN initiative for their enthusiastic cooperation in this venture. This manuscript was read and commented upon by Christine Bishop, Joël Bonin, Ron Brooks, Rosemary Curley, Ted Davis, Andrew Didiuk. Dave Galbraith, Stephen Hecnar, Ron Heyer, Mike Lannoo, John Maunder, Don McAlpine, Larry Powell, Tony Russell, Carolyn Seburn, David Seburn, Leonard Shirose, Richard Wassersug, and Wayne Weller. My research has been supported by grants from NSERC Canada and World Wildlife Fund Canada.

Literature Cited

Acker PM, Kruse KC, Krehbiel EB. 1986. Aging *Bufo americanus* by skeletochronology. Journal of Herpetology 20:570–574.

Arak A. 1983. Male-male competition and mate choice in anuran amphibians. In: Bateson PPG, editor. Mate choice. Cambridge, U.K.: Cambridge University Press. p 181–210.

Bakowsky W, Riley JL. 1994. A survey of the prairies and savannas of southern Ontario. In: Wickett RG, Lewis PD, Woodliffe A, Pratt P, editors. Proceedings of the thirteenth North American prairie conference. Windsor, ON: Department of Parks and Recreation. p 7–16.

Banks B, Beebee TJC. 1987. Factors influencing breeding site choice by the pioneering amphibian *Bufo calamita*. Holarctic Biology 10:14–21.

Barinaga M. 1990. Where have all the froggies gone? Science 247:1033–1034.

Bastien H, Leclair RJr. 1992. Aging wood frogs (*Rana sylvatica*) by skeletochronology. Journal of Herpetology 26:222–225.

Beebee TJC. 1977. Environmental change as a cause of natterjack toad (*Bufo calamita*) declines in Britain. Biological Conservation 11:87–102.

Berven KA. 1990. Factors affecting population fluctuations in larval and adult stages of the wood frog (*Rana sylvatica*). Ecology 71:1599–1608.

Bider JR, Matte S. 1994. Atlas des amphibiens et des reptiles du Québec. Québec: Societé d'histoire naturelle de la vallée du Saint-Laurent et Ministère de l'environnement et de la faune, Direction de la faune et des habitats.

Blaustein AR, Wake DB. 1990, Declining amphibian populations: a global phenomenon? Trends in Ecology and Evolution 5:203–204.

Blaustein AR, Wake DB, Sousa WP. 1994a. Amphibian declines: judging stability, persistence, and susceptibility of populations to local and global extinctions. Conservation Biology 8:60–71.

Blaustein AR, Hokit DG, O'Hara RK, Holt RA. 1994b. Pathogenic fungus contributes to amphibian losses in the Pacific Northwest. Biological Conservation 67:251–254.

Blaustein AR, Hoffman PD, Hokit DG, Kiesecker JM, Walls SC, Hayes JB. 1994c. UV repair and resistance to solar UV-B in amphibian eggs: link to population declines? Proceedings of the National Academy of Sciences 91:1791–1795.

Bonin J. 1992. Status of amphibian populations in Quebec. In: Bishop CA, Pettit KE, editors. Declines in Canadian amphibian populations: designing a national monitoring strategy. Ottawa: Canadian Wildlife Service. Occasional Paper 76. p 23–25.

Bradford DF. 1991. Mass mortality and extinction in a high elevation population of *Rana muscosa*. Journal of Herpetology 25:174–177.

Bradford DF, Swanson C, Gordon MS. 1992. Effects of low pH and aluminum on two declining species of anurans in the Sierra Nevada, California. Journal of Herpetology 26:369–377.

Breden F. 1988. The natural history and ecology of Fowler's toad, *Bufo woodhousei fowleri* (Amphibia: Bufonidae) in the Indiana Dunes National Lakeshore. Fieldiana Zoology 49:1–16.

Brooks RJ. 1992. Monitoring wildlife populations in long-term studies. In: Bishop CA, Pettit KE, editors.Declines in Canadian amphibian populations: designing a national monitoring strategy. Ottawa: Canadian Wildlife Service. Occasional Paper 76. p 94–97.

Cook FR. 1966. A guide to the amphibians and reptiles of Saskatchewan. Saskatchewan Museum of Natural History Popular Series 13:1–40.

Corn PS. 1994. What we know and don't know about amphibian declines in the west. In: Covington WW, DeBano LF, technical coordinators. Sustainable ecological systems: implementing an ecological approach to land management. Fort Collins, CO: USDA Forest Service. General Technical Report RM-247. p 59–57.

Corn PS, Vertucci FA. 1992. Ecological risk assessment of the effects of atmospheric pollutant deposition on western amphibian populations. Journal of Herpetology 26:361–369.

Crump ML, Hensley FR, Clark KL. 1992. Apparent decline of the golden toad: underground or extinct? Copeia 1992:413–420.

Czechura GV, Ingram GJ. 1990. *Taudactylus diurnus* and the case of the disappearing frogs. Memoirs of the Queensland Museum 29:361–365.

Diamond JM. 1989. The present, past and future of human-caused extinctions. Philosophical Transactions of the Royal Society of London Series B 325:469–477.

Donnelly MA, Guyer C. 1994. Mark-recapture. In: Heyer WR, Donnelly MA, McDiarmid RW, Hayek LC, Foster MS, editors. Measuring and monitoring biological diversity. Standard methods for amphibians. Washington: Smithsonian Institution Press. p 183–200.

Farr A. 1989. Status report on the Pacific giant salamander, *Dicamptodon ensatus*, in Canada. Ottawa: Committee on the Status of Endangered Wildlife in Canada.

Freda J. 1986. The influence of acidic pond water on amphibians: a review. Water, Air, and Soil Pollution 30:439–450.

Gilhen J. 1984. Amphibians and reptiles of Nova Scotia. Halifax, NS: Nova Scotia Museum.

Gill DE. 1978. The metapopulation ecology of the red-spotted newt *Notophthalmus viridescens* (Rafinesque). Ecological Monographs 48:145–166.

Goodman D. 1987. The demography of chance extinction. In: Soulé ME, editor. Viable populations for conservation. Cambridge, U.K.: Cambridge University Press. p 11–34.

Green DM. 1989. Fowler's Toads, (*Bufo woodhousii fowleri*) in Canada: biology and population status. Canadian Field-Naturalist 103:486–496.

Green DM. 1993. Summary of the recommendations and conclusions arising from DAPCAN III. Canadian Association of Herpetologists Bulletin 7(2):14–15.

Green DM, Campbell RW. 1984. The amphibians of British Columbia. Victoria, BC: British Columbia Provincial Museum. Handbook 45.

Green DM, Kaiser H, Sharbel TF, Kearsley J, McAllister KR. 1997. Cryptic species of spotted frogs, *Rana pretiosa* complex, in western North America. Copeia 1997:1–8.

Green DM, Sharbel TF, Kearsley J, Kaiser H. 1996. Postglacial range fluctuation, genetic subdivision and speciation in the western North American spotted frog complex, *Rana pretiosa*. Evolution 50:374–390.

Hammerson GA. 1982. Bullfrogs eliminating leopard frogs in Colorado? Herpetological Review 13:115–116.

Hanski I, Gilpin M. 1991. Metapopulation dynamics: brief history and conceptual domain. Biology Journal of the Linnean Society 42:3–16.

Hayes MP, Jennings MR. 1986. Decline of ranid species in western North America: are bullfrogs responsible. Journal of Herpetology 20:490–509.

Holman AJ, Clouthier SG. 1995. Pleistocene herpetofaunal remains from the East Milford mastodon site (ca. 70,000 – 80,000 BP), Halifax County, Nova Scotia. Canadian Journal of Earth Sciences 32: 210–215.

Jennings MR, Hayes MP. 1985. Pre-1900 overharvest of the California red-legged frog (*Rana aurora draytonii*): the inducement for bullfrog (*Rana catesbeiana*) introduction. Herpetologica 41:94–103.

Johnson B. 1989. Familiar amphibians and reptiles of Ontario. Toronto: Natural Heritage/Natural History Inc.

Johnson B. 1992. Habitat loss and declining amphibian populations. In: Bishop CA, Pettit KE, editors. Declines in Canadian amphibian populations: designing a national monitoring strategy. Ottawa: Canadian Wildlife Service. Occasional Paper 76. p 71–75.

Kalb HJ, Zug GR. 1990. Age estimates for a population of American toads, *Bufo americanus*, in northern Virginia. Brimleyana 16:79–86.

Kellner A, Green DM. 1995. Age structure and age at maturity in Fowler's toads, *Bufo woodhousii fowleri*, at their northern range limit. Journal of Herpetology 29:417–421.

Koonz W. 1992. Amphibians in Manitoba. In: Bishop CA, Pettit KE, editors. Declines in Canadian amphibian populations: designing a national monitoring strategy. Ottawa: Canadian Wildlife Service. Occasional Paper 76. p 19–20.

Laan R, Verboom B. 1990. Effects of pool size and isolation on amphibian communities. Biological Conservation 54:251–262.

Lande R, Barrowclough GF. 1987. Effective population size, genetic variation, and their use in population management. In: Soulé ME, editor. Viable populations for conservation. Cambridge, U. K.: Cambridge University Press. p 87–123.

Leclair R Jr, Castanet J. 1987. A skeletochronological assessment of age and growth in the frog *Rana pipiens* Schreber (Amphibia, Anura) from southwestern Quebec. Copeia 1987:361–369.

Levins R. 1969. Some demographic and genetic cosequences of environmental heterogeneity for biological control. Bulletin of the Entomological Society of America 15:72–85.

Liss WJ, Larson GL. 1991. Ecological effects of stocked trout on North Cascades naturally fishless lakes. Park Science 11:22–23.

MacArthur RH, Wilson EO. 1967. The theory of island biogeography. Princeton, NJ: Princeton University Press.

McAlpine D, Fletcher TJ, Gorham SW, Gorham IT. 1991. Distribution and habitat of the tetraploid gray treefrog, *Hyla versicolor*, in New Brunswick and eastern Maine. Canadian Field-Naturalist 105:526–529.

McCoy ED. 1994. "Amphibian decline": a scientific dilemma in more ways than one. Herpetologica 50:98–103.

Meffe GK, Carroll CR. 1994. Principles of Conservation Biology. Sunderland, MA: Sinauer Associates.

Melançon C. 1961. Inconnus et méconnus. Amphibiens et reptiles de la province du Québec. Québec: Société zoologique de Québec.

Moyle P B. 1973. Effects of introduced bullfrogs, *Rana catesbeiana*, on the native frogs of the San Joaquin Valley, California. Copeia 1973:18–22.

Munro WT. 1993. Designation of endangered species, subspecies and populations by COSEWIC. In: Fenger MA, Miller EH, Johnson JA, Williams EJR, editors. Our living legacy: proceedings of a symposium on biodiversity. Victoria, BC: Royal British Columbia Museum. p 213–227.

Nyman S. 1986. Mass mortality in larval *Rana sylvatica* attributable to bacterium *Aeromonas hydrophilia*. Journal of Herpetology 20:196–201.

Oldham MJ, Weller WF. 1992. Ontario herpetofaunal summary: compiling information on the distribution and life history of amphibians and reptiles in Ontario. In: Bishop CA, Pettit KE, editors. Declines in Canadian amphibian populations: designing a national monitoring strategy. Ottawa: Canadian Wildlife Service. Occasional Paper 76. p 21–22.

Orchard SA. 1992. Amphibian population declines in British Columbia. In: Bishop CA, Pettit KE, editors. Declines in Canadian amphibian populations: designing a national monitoring strategy. Ottawa: Canadian Wildlife Service. Occasional Paper 76. p 10–13.

Pechmann JHK, Scott DE, Semlitsch RD, Caldwell JP, Vitt LJ, Gibbons JW. 1991. Declining amphibian populations: the problem of separating human impacts from natural fluctuations. Science 253:892–895.

Pechmann JHK, Wilbur HM. 1994. Putting declining amphibian populations in perspective: natural fluctuations and human impacts. Herpetologica 50:65–84.

Peters RL, Lovejoy TE. 1992. Global warming and biological diversity. New Haven, CT: Yale University Press.

Pounds JA, Crump ML. 1994. Amphibian declines and climate disturbance: the case of the golden toad and the harlequin frog. Conservation Biology 8:72–85.

Preston W. 1982. The amphibians and reptiles of Manitoba. Winnipeg, MB: Manitoba Museum of Man and Nature.

Rexstad E, Burnham K. 1991. User's guide for interactive program CAPTURE. Fort Collins, CO: Colorado Cooperative Fish and Wildlife Research Unit.

Roberts W. 1992. Declines in amphibian populations in Alberta. In: Bishop CA, Pettit KE, editors. Declines in Canadian amphibian populations: designing a national monitoring strategy. Ottawa: Canadian Wildlife Service. Occasional Paper 76. p 14–16.

Rodda GH. 1993. How to lie with biodiversity. Conservation Biology 7:959–960.

Russell A, Bauer A. 1993. The amphibians and reptiles of Alberta. Calgary, AB and Edmonton, AB: University of Calgary Press, and University of Alberta Press.

Schoener TW, Spiller DA. 1992. Is extinction related to temporal variability in population size? An empirical answer for orb spiders. American Naturalist 139:1176–1207.

Scott NJ Jr. 1993. Postmetamorphic death syndrome. Froglog 7:1–2.

Seburn CNL. 1992. The status of amphibian populations in Saskatchewan. In: Bishop CA, Pettit KE, editors. Declines in Canadian amphibian populations: designing a national monitoring strategy. Ottawa: Canadian Wildlife Service. Occasional Paper 76. p 17–18.

Seburn CNL. 1994. DAPCAN IV discussion and business meeting. Canadian Association of Herpetologists Bulletin 8(2):3–6.

Sexton OJ, Phillips C. 1986. A qualitative study of fish-amphibian interactions in 3 Missouri ponds. Transactions of the Missouri Academy of Sciences 20:25–35.

Sharbel TF, Bonin J. 1992. Northernmost record of *Desmognathus ochrophaeus*: biochemical identification in the Chateauguay River drainage basin, Québec. Journal of Herpetology 26:505–508.

Shirose LJ, Bishop CA, Green DM, MacDonald CJ, Brooks RJ, Helferty NJ. 1997. Validation of an amphibian call count survey technique in Ontario, Canada. Herpetologica (in press).

Sjögren P. 1991. Extinction and isolation gradients in metapopulations: the case of the pool frog (*Rana lessonae*). Biology Journal of the Linnean Society 42:135–147.

Soulé ME. 1987. Viable populations for conservation. Cambridge, U.K.: Cambridge University Press.

Vial JL, Saylor L. 1993. The status of amphibian populations: a compilation and analysis. Corvallis, OR: IUCN—The World Conservation Union Species Survival Commission, Declining Amphibian Populations Task Force. Working Document 1.

Wake DB. 1991. Declining amphibian populations. Science 253:860.

Wassersug RJ. 1975. The adaptive significance of the tadpole stage with comments of the maintenance of complex life cycles in anurans. American Zoologist 15:405–417.

Weitzel NH, Panik HR. 1993. Long-term fluctuations of an isolated population of the Pacific chorus frog (*Pseudacris regilla*) in northwestern Nevada. Great Basin Naturalist 53:379–384.

White GC, Anderson DR, Burnham KP, Otis DL. 1982. Capture-recapture and removal methods for sampling closed populations. Los Alamos, NM: Los Alamos National Laboratory Publication LA-8787- NERP.

Wissinger SA, Whiteman HH. 1992. Fluctuation in a Rocky Mountain population of salamanders: anthropogenic acidification or natural variation? Journal of Herpetology 26:377–391.

Wyman RL. 1990. What's happening to the amphibians? Conservation Biology 4:350–352.

Amphibians in decline: Canadian studies of a global problem. David M. Green, editor.
Herpetological Conservation 1:309–328.

Appendix I

CHECKLIST AND CURRENT STATUS OF CANADIAN AMPHIBIANS

COMPILED BY

WAYNE F. WELLER

Ontario Field Herpetologists, 66 High Street East, Suite 504, Mississauga, Ontario L5G 1K2, Canada

DAVID M. GREEN

Redpath Museum, McGill University, Montréal, Québec H3A 2K6, Canada

ABSTRACT.—Forty-five species of amphibians occur in Canada: 21 salamanders and 24 frogs and toads. Seventeen of these species are considered to have suffered declines in population numbers. These declines are most commonly attributed to habitat destruction and degradation brought about by urbanization, agricultural and forestry practices, and introducted alien competitors. Declines are more prevalent in western Canada than in eastern Canada.

RÉSUMÉ.—Quarante-cinq espèces d'amphibiens vivent au Canada dont 21 espèces de salamandres et 24 espèces de grenouilles et de crapauds. Les populations de 17 d'entre elles ont décliné. Ces déclins sont généralement attribués à la destruction et à la dégradation des habitats causés par l'urbanisation, les pratiques agricoles et forestières, et l'introduction d'espèces concurrentes étrangères. Les déclins sont plus prévalents dans l'ouest que dans l'est du Canada.

Forty-five species of amphibians occur in Canada, comprising 21 salamanders and 24 frogs and toads. This total includes the 2 species of spotted frogs, *Rana pretiosa* and *R. luteiventris* recently distinguished by Green et al. (1996, 1997) and the chorus frogs recognized by Platz (1989) to consist of 2 species, *Pseudacris triseriata* and *P. maculata*, extending into Canada. The spring peeper is referred to as *Pseudacris crucifer* following Hedges (1986) but a more recent analysis (Cocroft 1994) indicates that the Pacific treefrog cannot be placed unequivocally within *Pseudacris*, as Hedges (1986) had concluded, and is therefore referred to conservatively as *Hyla regilla*. The toads *Bufo americanus* and *B. hemiophrys* are treated as separate species following Green (1983) and Collins (1990), contrary to Cook (1983), and in view of recent evidence (Erik Gergus, unpublished) that they are not sister taxa. *Bufo fowleri* is treated as a species separate from *B. woodhousii*, contrary to Meacham (1962), based upon recent evidence of diagnosibility of the 2 taxa throughout their ranges (B. K. Sullivan, pers. comm.). General statements of range and distribution in Canada are made with reference to Cook (1984), Stebbins (1985), and Conant and Collins (1991). Some species have had status designated by the Committee on the Status of Endangered Wildlife in Canada (COSEWIC).

Caudata—Newts and Salamanders

Necturus maculosus (Mudpuppy, Necture Tacheté)

Manitoba: Although believed by Preston (1982) to be restricted in Manitoba to the southeast of the province, recent reports from Shellmouth Dam and Lake of the Prairies, both near the Saskatchewan border, indicate a wider distribution. A specimen was recently collected as far north as Berens River. No evidence of decline. **Ontario:** Known only from records near Thunder Bay (Hecht and Walters 1955; Logier and Toner 1961) and widely distributed throughout southern and central Ontario, extending north to the Sault Ste. Marie and east along the north shore of Lake Huron to Lake Nipissing (M.J. Oldham and W.F. Weller, unpublished). Declines have not been reported. **Québec:** Occurs only in areas near Ottawa River and St. Lawrence River in southwestern part of the province from near Lake Temiscaming to near St.-Siméon, and west of Lake Memphremagog south of the St. Lawrence River (Bider and Matte 1994). No evidence of decline.

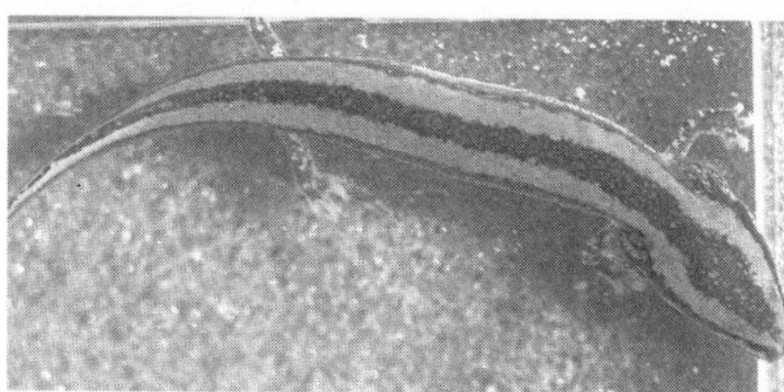

Necturus maculosus *larva. Photo courtesy of Canadian Museum of Nature.*

Notophthalmus viridescens (Eastern Newt, Triton Vert)

Ontario: *Notophthalmus v. louisianensis* is widely scattered from Thunder Bay area west to Lake of the Woods (M. J. Oldham and W. F. Weller, unpublished). No declines have been observed. *Notophthalmus v. viridescens* is common and widely distributed throughout southern and central Ontario, extending north to Lake Nipigon in the west and Kirkland Lake in the east, but uncommon and sparsely distributed in extreme southwestern Ontario (M.J. Oldham and W.F. Weller, unpublished). No evidence of recent decline. **Québec:** Southern and southwestern region extending northeast to the Gaspé Peninsula and to Sept-Iles north of the St. Lawrence River (Bider and Matte 1994). No declines apparent. **New Brunswick:** Distributed throughout province. No evidence of declines (D.F. McAlpine, pers. comm.). **Nova Scotia:** Throughout mainland Nova Scotia, and Cape Breton Island (Gilhen 1984). No declines have been reported (J. Gilhen, pers. comm.). **Prince Edward Island:** Distributed throughout province, and apparently common (Cook 1967). No declines have been reported, but historic information is scant (R. Curley, pers. comm.).

Notophthalmus viridescens. *Photo by John Mitchell.*

Taricha granulosa (Roughskin Newt, Triton Rugueux)

British Columbia: Widely distributed on Vancouver Island, and on mainland British Columbia along the Pacific coast from the extreme southwest (Neish 1971) to the Alaska border (Green and Campbell 1984). No decline apparent.

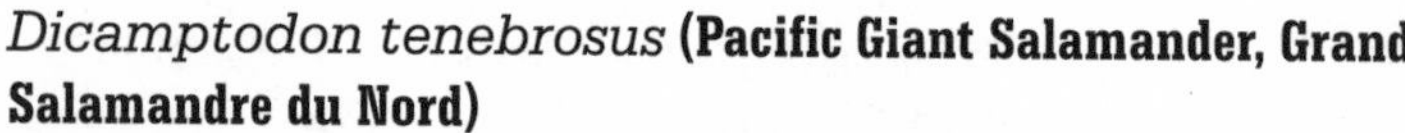

Dicamptodon tenebrosus (Pacific Giant Salamander, Grand Salamandre du Nord)

British Columbia: Extreme southwestern British Columbia near Chilliwack and Hope (Green and Campbell 1984). Suspected declines have been attributed to habitat destruction, particularly from forestry activities, by Orchard (1992). Considered an endangered species in British Columbia (Munro 1993), and designated vulnerable by COSEWIC in 1989 (A.C.M. Farr, unpublished).

Taricha granulosa *defensive posture ("unken reflex"). Photo by Stephen Corn.*

Dicamptodon tenebrosus. *Photo by David M. Green.* Ambystoma gracile. *Photo by David M. Green.*

Ambystoma gracile **(Northwestern Salamander, Salamandre Foncée)**

British Columbia: *Ambystoma g. gracile* is widespread on Vancouver Island and on the west coast of mainland British Columbia north as far as Powell River and to east as far as Hope (Licht 1975; Green and Campbell 1984). Populations not in jeopardy (Orchard 1984). *Ambystoma g. decorticatum* is found on the central west coast of the province extending from Bella Coola north to Prince Rupert, just south of Alaska (Green and Campbell 1984). No decline apparent.

Ambystoma jeffersonianum **(Jefferson Salamander, Salamandre de Jefferson)**

Ontario: Occurs in the province from the north shore of Lake Erie near Port Burwell north through Kitchener-Waterloo (Weller et al. 1978) to near Orangeville, east to near Nobleton (Weller et al. 1979), south to Mississauga (Weller and Sprules 1976) along the north shore of Lake Ontario, and to near Grimsby along south shore of Lake Ontario (M.J. Oldham and W.F. Weller, unpublished). Records exist from further east near Toronto and Northumberland County (Weller et al. 1978) although these were distinguished from *A. laterale* based only on morphology (see Rye et al., this volume). Probably more widespread than currently recognized (Rye et al., this volume). Not in apparent decline.

Ambystoma laterale **(Blue-spotted Salamander, Salamandre à Points Bleus)**

Manitoba: Southeastern Manitoba from Anola eastward, and extending north to Wadhope area in the east and Hodgson in the west (Preston 1982). Northernmost report from Metheson Island, Lake Winnipeg. Also reported from north of Riveron and from Wallace Lake. Declines have not been reported. **Ontario:** Common and widely distributed throughout central and southern Ontario. Extends north to southern James Bay in the east, Lake Nipigon, and Favourable Lake in the west (M.J. Oldham and W.F. Weller, unpublished). Virtually absent from areas inland of Lake Huron in southern Ontario. No recent declines observed. **Québec:** Widely distributed in southern part of province, extending northward to near northern James Bay in the west, to near Chibougamau, and to Havre-Saint-Pierre in the east (Bider and Matte 1994). No declines reported. **New Brunswick:** Distributed throughout province. No evidence of declines (D.F. McAlpine, pers. comm.). **Nova Scotia:** Occurs in northern and northeastern parts of mainland Nova Scotia, but absent from southwestern areas along southern coast (Gilhen 1984). First found on Cape Breton Island at Fortress of Louisbourg (Cook and Rick 1963), but now known to be widely distributed throughout island (Gilhen 1974, 1984). Declines have not been reported (J. Gilhen, pers. comm.). **Prince Edward Island:** Abundant at widely scattered localities throughout the province (Cook 1967). No declines have been noted (R. Curley, pers. comm.). **Newfoundland:** Known only from three widely isolated areas: Wabush in western Labrador (Cook and Folinsbee 1975), Happy Valley and Goose Bay region, and Hopedale along the Atlantic coast (Maunder 1983; this volume). No declines reported.

Ambystoma laterale. *Photo by John Mitchell.*

Ambystoma macrodactylum **(Long-toed Salamander, Salamandre à Longs Doigts)**
British Columbia: *Ambystoma m. macrodactylum* is widespread on Vancouver Island and established on a few islands offshore in Strait of Georgia near Nanaimo and Comox. Also occurs on the mainland from the US border northward to Knight Inlet and southeastward to near Chilliwack (Green and Campbell 1984). Population numbers considered stable (Orchard 1984). *Ambystoma m. columbianum* occurs throughout central and western British Columbia north along Pacific coast from Knight Inlet through to the Alaska border to near Mount Edziza Provincial Park. Eastern limit of range in the north is around Findlay River and near Cranbrook in south. Apparently absent along Pacific coast from US border to near Knight Inlet (Green and Campbell 1984). Declines have not been identified. *Ambystoma m. krausei* occurs in eastern British Columbia adjacent to Alberta border, extending north to near Fort St. John and west to the Findlay River and Creston (Green and Campbell 1984). No declines known. **Alberta:** Extreme west of the province from southern border north to Jasper National Park and near Kakwa River. Not in apparent decline (see Powell et al., this volume).

Ambystoma macrodactylum. *Photo courtesy of the Canadian Museum of Nature.*

Ambystoma maculatum **(Spotted Salamander, Salamandre Maculée)**
Ontario: Widespread throughout southern Ontario, but uncommon and sparsely distributed in extreme southwestern and eastern regions (M.J. Oldham and W.F. Weller, unpublished). Extends northward to Dorami Lake in the east, and southern Lake Nipigon and Atikokan areas in northwestern Ontario (Weller 1987). Declines have not been reported. **Québec:** Occurs in southern part of the province north from Ottawa River to Chibougamau (Gordon and Cook 1980) and east to Sept-Iles. South of the St. Lawrence River, it is widely distributed all through the Eastern Townships and through to the Gaspé Peninsula (Bider and Matte 1994). No evidence of decline. **New Brunswick:** Found throughout province. Declines have not been reported (D.F. McAlpine, pers. comm.). **Nova Scotia:** Throughout mainland Nova Scotia and Cape Breton Island (Gilhen 1984). No declines reported (J. Gilhen, pers. comm.). **Prince Edward Island:** Common and widespread throughout province (Cook 1967). Declines have not been reported (R. Curley, pers. comm.).

Ambystoma maculatum. *Photo by John Mitchell.*

Ambystoma texanum **(Smallmouth Salamander, Salamandre à Nez Court)**
Ontario: Restricted in Canada to Pelee Island in extreme southwestern Ontario (Uzzell 1964; Cook 1984; M.J. Oldham and W.F. Weller, unpublished). Lowcock (1989) mentions possibility of hybrids with *A. laterale* in mainland Essex County and the Niagara District. Common and widespread. No evidence of decline. Designated by COSEWIC as a vulnerable species in Canada (Bogart and Licht 1991).

Ambystoma tigrinum **(Tiger Salamander, Salamandre Tigrée)**
British Columbia: Extreme southcentral part of the province, including southern Okanagan, lower Similkameen, and Boundary areas extending north almost to Peachland, west to Keremeos, and east almost to Christina Lake. Declines attributed to changing land use, especially grazing and trampling by livestock, urbanization, fish stocking, and recreational development of lakes (MacIntyre and Palermo 1980; Orchard 1992; M. J. Sarrell, unpublished report to British Columbia Ministry of Environment, Land and Forests). Designated an endangered species ("Red List") in British Columbia (Munro 1993; British Columbia Environment 1995). **Alberta:** Occurs in southern half of province north from US border to Banff National Park on the west and to near Edmonton and St. Paul on the

east (Russell and Bauer 1993). No evidence of province-wide declines, but local declines considered to be occurring in areas under strong development pressure or intense cultivation (L. Powell, pers. comm.). **Saskatchewan:** *Ambystoma t. melanostictum* occurs in grasslands and parklands of southwestern Saskatchewan, extending east to near Regina and north to near North Battleford, with limited occurrences in the southern fringe of the mixed wood forest zone (Cook 1965, 1966). Widespread and common throughout its range. Populations considered stable and secure and no declines have been identified (Secoy 1987; Didiuk, this volume). *Ambystoma t. diaboli* occupies the southern part of province except for southwest corner. Extends north to near Meadow Lake Provincial Park in the west, and to near Duck Mountain Provincial Park in the east. Reported as far north as Arran (Hooper 1992). Fairly common, and no declines apparent (Secoy 1987; Didiuk, this volume). **Manitoba:** *Ambystoma t. diaboli* is found in southwestern Manitoba as far north as Gilbert Plains and possibly much farther (Cook 1965) No evidence of decline. *Ambystoma t. tigrinum* occurs in extreme southeastern part of the province, extending from southern Sandilands Provincial Forest west to Emerson (Preston 1982). No declines have been reported. **Ontario:** Known only from Point Pelée in extreme southwestern Ontario based on a single specimen (Logier 1925). Considered to have been extirpated from Ontario.

Ambystoma tigrinum. *Photo courtesy of Canadian Museum of Nature.*

Desmognathus fuscus (Dusky Salamander, Salamandre Sombre)

Ontario: Known from only the Niagara River area in southern Ontario. Original report was from "opposite Buffalo" (Bishop 1943), but since rediscovered at 2 localities along lower Niagara River (Kamstra 1991). Population declines have not been detected. Recommended as an endangered species in Ontario (Kamstra, unpublished), but not yet designated as such. Protected under provincial Game and Fish Act as of May 1990. **Québec:** Occurs exclusively south of St. Lawrence River except for a single population at mouth of the Montmorency River (Bider and Matte 1994). Extends from US border northeast (Weller and Cebek 1991a) through Rivère-du-Loup (Weller 1977) to near Roxton Falls (Denman and Denman 1985), as well as Mont Yamaska (Weller 1977) and Mont St.-Hilaire (Pendlebury 1973) in the St. Lawrence Lowlands. No declines have been reported. **New Brunswick:** Occurs in southern half of province from Maine to near Nova Scotia border, north to Meduxnekeag River in the west and to near Moncton in the east (Cook and Bleakney 1960). No evidence of decline (D. F. McAlpine, pers. comm.).

Desmognathus fuscus. *Photo by David M. Green.*

Desmognathus ochrophaeus (Mountain Dusky Salamander, Salamandre Sombre de la Montagne)

Québec: Known in Canada only from a single locality in the extreme southwestern corner of the province in the Adirondack Mountains near Covey Hill adjacent to the US border (Sharbel and Bonin 1992). No evidence of decline.

Desmognathus ochrophaeus. *Photo by Joël Bonin.*

Aneides ferreus (Clouded Salamander, Salamandre Pommelée)

British Columbia: Throughout Vancouver Island, but apparently absent from the interior. Widely scattered on adjacent islands, particularly those in the Strait of Georgia. Evidence mounting that British Columbia populations were introduced from California in 19th Century. Populations not in jeopardy (Orchard 1984).

Aneides ferreus. *Photo by David M. Green.*

Ensatina eschscholtzii (Ensatina, Salamandre Variable)

British Columbia: Occurs in eastern half of Vancouver Island and in the southwestern part of the adjacent mainland. Extends northward along Pacific coast to near Bella Coola, and up Fraser the River to near Lytton. May be declining in lower mainland due to habitats lost or fragmented by urbanization (Orchard 1984).

Ensatina eschscholtzii. *Photo by William P. Leonard.*

Eurycea bislineata (Northern Two-Lined Salamander, Salamandre à Deux Lignes du Nord)

Ontario: Common and widespread along a wide band through southern Ontario extending from eastern Georgian Bay to St. Lawrence River. Two records from Lake Timagami (Logier and Toner 1961) and Onakawana River (Kamstra 1983) are considerably north of this band and 1 record from southern Lake Simcoe is considerably to the south. No reports of decline. **Québec:** Widespread throughout southern areas of province. South of St. Lawrence River, extends through Appalachian Mountains (Weller and Cebek 1991b) to near Rivière-du-Loup (Denman et al. 1990). Isolated localities occur near tip of Gaspé Peninsula and at Parc de Conservation de la Gaspésie. North of St. Lawrence River, extends to southern James Bay and through the Chibougamau area to Parc Conservation du Saguenay (Denman and Denman 1985). Isolated populations north of the St. Lawrence River also occur in eastern Québec in the area of the Matamek River and Ste.-Marguerite River. Declines have not been reported. **Newfoundland:** Known only from near Labrador City (Cook and Preston 1979) in extreme western Labrador, and at Cache River in the southcentral area (DeGraaf et al. 1981). No declines known (Maunder, this volume). **New Brunswick:** Throughout province, and particularly common in southeastern corner (Cook and Bleakney 1960). No indications of decline (D.F. McAlpine, pers. comm.).

Eurycea bislineata. *Photo by David M. Green.*

Gyrinophilus porphyriticus (Spring Salamander, Salamandre Pourpre)

Ontario: Occurence in the province based solely on three larvae collected in 1877 "opposite Buffalo" (Dunn 1926; Brandon 1966). A 1934 report from Britannia Creek near Ottawa supposedly of a larva has been discounted (Brandon 1967; Cook 1981, 1984). Not reported since. **Québec:** Found only in southern Québec south of St. Lawrence River (Bider and Matte 1994). Occurs in southwestern area in Adirondack Mountains near Covey Hill and Franklin Centre (Gordon 1979), and extends northeast in the Appalachian Mountains to near Trottier (Weller and Cebek 1991c) and East Angus (Shaffer and Bachand 1989). Also found on Mont Yamaska in St. Lawrence lowlands (Weller 1977). Not in decline.

Gyrinophilus porphyriticus. *Photo by David M. Green.*

Hemidactylium scutatum **(Four-toed Salamander, Salamandre à Quatre Doigts)**
Ontario: Apparently locally common in southern Ontario but not widespread. Widely distributed in the western Lake Ontario and eastern Lake Erie areas and around eastern Georgian Bay, extends north to Parry Sound and east to Ottawa (M. J. Oldham and W. F. Weller, unpublished). Declines have not been reported. **Québec:** Occurs in Gatineau Hills near Hull along Ottawa River, and in the vicinities of Montréal and Trois-Rivières along St. Lawrence River (Bider and Matte 1994). South of St. Lawrence River, occurs adjacent to US border from Herdman and Covey Hill (Gordon 1979) in the west to Mont Johnson in the east. An isolated population occurs much farther to the east near Lake Memphremagog (Sharbel 1991). No declines have been reported. **New Brunswick:** Known only from Fundy National Park in southeastern corner of the province. Declines have not been reported (D. F. McAlpine, pers. comm.). **Nova Scotia:** Known from four isolated areas in province: southcentral mainland Nova Scotia around Kejimkujik National Park, in the Halifax area, in the Liscomb Mills area along southern mainland shore (Gilhen 1984), and from the southwest region of Cape Breton Island. No declines have been reported (J. Gilhen, pers. comm.).

Hemidactylium scutatum. *Photo by Joël Bonin.*

Plethodon cinereus **(Redback Salamander, Salamandre Rayée)**
Ontario: Common and widespread throughout most of central and southern Ontario except for extreme southwestern and eastern regions where it is apparently uncommon and sparsely distributed. Extends along north shore of Lake Superior to near Thunder Bay and east from near Chapleau to southern Lake Temiscaming (M. J. Oldham and W. F. Weller, unpublished). No reports of decline. **Québec:** Widely distributed in southern Québec south of the St. Lawrence River to tip of the Gaspé Peninsula. North of the St. Lawrence River, extends to southern Lake Abitibi in the west and through Lac Saint-Jean to the Saguenay River (Bider and Matte 1994). An isolated population occurs near Sept-Iles. No decline has been reported. **New Brunswick:** Found throughout province. No evidence of decline (D. F. McAlpine, pers. comm.). **Nova Scotia:** Throughout mainland Nova Scotia and northwestern Cape Breton Island (Gilhen 1984). No evidence of decline. **Prince Edward Island:** Distributed throughout province. Declines have not been reported (R. Curley, pers. comm.).

Plethodon cinereus. *Photo by John Mitchell.*

Plethodon idahoensis **(Coeur d'Alene Salamander, Salamadre de Coeur d'Alene)**
British Columbia: Occurs in extreme southeastern British Columbia. Range apparently fragmented due to forestry activities resulting in population declines throughout its range (Orchard 1992). Designated an endangered species in British Columbia (Munro 1993).

Plethodon vehiculum **(Western Redback Salamander, Salamandre à Dos Rayé)**
British Columbia: Throughout Vancouver Island and southwestern corner of adjacent mainland east through Fraser Valley to Hope. Not in jeopardy (Orchard 1984).

Plethodon vehiculum. *Photo by David M. Green.*

ANURA—FROGS AND TOADS

Ascaphus truei (Tailed Frog, Grenouille-à-Queue)

British Columbia: From extreme southwestern mainland north to near Bute Inlet along Pacific coast and east to Manning Provincial Park. Also located near Kitimat, Terrace, and the Kitlope region along northwest coast (W. G. Hazelwood, pers. comm.; L. Dupuis, unpublished report to British Columbia Ministry of Environment, Lands and Parks) and in the Flathead River area in extreme southeast corner of British Columbia (Grant 1958). A report from near Shuswap Lake is erroneous. Habitat destruction, particularly due to forestry practices, has been blamed for apparent declines throughout Canadian range (MacIntyre and Palermo 1980; Orchard 1992; Dupuis, unpublished). Declared a vulnerable or sensitive species in British Columbia (Munro 1993).

Ascaphus truei. *Photo by David M. Green.*

Spea bombifrons (Plains Spadefoot, Crapaud des Plaines)

Alberta: Occurs only in the region south and east of Calgary. No evidence of declines (L. Powell, pers. comm.). **Saskatchewan:** Locally distributed in grasslands of southern and southwestern Saskatchewan, extending north to near Kindersley in the west and to near Moose Mountain Provincial Park in the east (Cook 1965, 1966). Considered rare (Secoy 1987) but the few records may be due to its highly localized distribution and detectability only during years of high summer precipitation. Population status unknown, possibly stable but locally vulnerable (A. B. Didiuk, pers. comm.). **Manitoba:** Known to occur in the southwestern corner of the province at Virden, Oak Lake, and Melita, where it is evidently more common than earlier indicated by Preston (1982). Records extend the range northeast to near Dauphin (Preston 1982) but none have been reported from this area since 1935. No evidence of decline but protected under Manitoba's Wildlife Act (Preston 1987).

Spea intermontanus (Great Basin Spadefoot, Crapaud du Grand Bassin)

British Columbia: Okanagan and Thompson/Nicola valleys of south-central British Columbia to near Dog Creek on Fraser River in the west, almost to Wright in the east, and north to 70 Mile House. Declines noted throughout range have been attributed to destruction of breeding ponds by cattle and recreational vehicles (Orchard 1992). Evidently once more numerous at existing localities and considered threatened in British Columbia (K. J. Smedley, unpublished report to British Columbia Ministry of Environment, Lands and Parks.).

Spea intermontana. *Photo by Stephen Corn.*

Bufo americanus (American Toad, Crapaud d'Amérique)

Northwest Territories: Found on Akimiski Island and Cape Hope Islands in southern James Bay (Hodge 1976) adjacent to Ontario and Québec. No declines have been reported. **Manitoba:** Southeastern Manitoba north to a line extending east from Lake Winnipeg to near Cross Lake, northern God's Lake, and the Ontario border (Preston 1982). Known as far north as Sakwesew Lake (W. B. McKillop and E. Punter, pers. comm). Hybridizes with *B. hemiophrys* in southeastern Manitoba (Cook 1983; Green 1983). No evidence of decline. **Ontario:** Common and widespread throughout most of province. Extends north to near Sandy Lake in northwestern Ontario and to Hudson Bay in northeastern Ontario (M. J. Oldham and W. F. Weller, unpublished). No declines have been reported. **Québec:** Widely distributed throughout southern and central regions of the province. Extends north to Hudson Bay and near to Ungava Bay in the Hudson Strait. Occurs in extreme eastern Québec along north shore of St. Lawrence River from Sept-Iles to near the boundary with Labrador (Bider and Matte 1994). Declines have not been reported. **New Brunswick:** Distributed throughout province. No evidence of decline (D. F.

McAlpine, pers. comm.). **Nova Scotia:** Found throughout both mainland Nova Scotia and Cape Breton Island (Gilhen 1984). No evidence of decline (J. Gilhen, pers. comm.). **Prince Edward Island:** Distributed throughout province. No evidence of decline (R. Cueley, pers. comm.). **Newfoundland:** Occurs throughout southern Labrador to the Atlantic coast (Maunder 1983; this volume). Introduced from southern Ontario into Corner Brook along western coast of Newfoundland (Buckle 1971), where it still occurs (Maunder, this volume). No evidence of decline (J. E. Maunder, pers. comm.).

Bufo americanus. *Photo by Martin Ouellet.*

Bufo boreas **(Western Toad, Crapaud de l'Ouest)**
Yukon: Extreme southeast corner at North Toobally Lake (Cook 1977), southcentral Yukon north to Whitehorse (Hodge 1976), and extreme southwestern Yukon (Mennell, this volume). Insufficient information to assess status (P. Milligan, pers. comm.). **British Columbia:** Throughout Vancouver Island, the Queen Charlotte Islands, and the entire mainland except for the far northeast of the province (Cook 1977; Green and Campbell 1984). Not threatened or in jeopardy (MacIntyre and Palermo 1980; Orchard 1984). **Alberta:** Western and central regions west of Lethbridge and Calling Lake (Russell and Bauer 1993). Isolated records in northern Alberta from Loon Lake and High Level. No evidence of decline (L. Powell, pers. comm.).

Bufo boreas. *Photo by David M. Green.*

Bufo cognatus **(Great Plains Toad, Crapaud des Steppes)**
Alberta: Extreme southeast corner of Alberta south of Red Deer River. Most localities are near South Saskatchewan River (Russell and Bauer 1993). Declines in southeast Alberta during last ten years have been attributed to habitat loss (Roberts 1992) and drought (L. Powell, pers. comm.). **Saskatchewan:** Restricted to extreme southwestern Saskatchewan, though records are sparse (Cook 1966). Population status unknown (A. B. Didiuk, pers. comm.) but possibly rare (Secoy 1987) and therefore considered vulnerable. **Manitoba:** Extreme southwestern Manitoba at Lyleton, Melita, and Coulter (Preston 1986). Observations and specimens too rare to assess status.

Bufo cognatus. *Photo by Jonathan Wright.*

Bufo fowleri **(Fowler's Toad, Crapaud de Fowler)**
Ontario: Restricted to north shore of Lake Erie at Pointe aux Pins, Long Point, Turkey Point, and the Niagara shoreline (Green 1989; M. J. Oldham and W. F. Weller, unpublished). Known to hybridize occasionally with *Bufo americanus* at Long Point (Green 1984). Apparently extirpated from Point Pelee and Pelee Island since 1940s. Assigned vulnerable status in Canada by COSEWIC (Green 1989) and protected under the provincial Game and Fish Act as of May 1990 but no evidence of further decline despite population size fluctuations (Green, this volume).

Bufo fowleri. *Photo by David M. Green.*

Bufo hemiophrys **(Canadian Toad, Crapaud du Canada)**
Northwest Territories: Known only from around Fort Smith (Hodge 1976; Larsen and Gregory 1988; Fournier, this volume) where the largest populations occur in Wood Buffalo National Park (Kuyt 1991). No declines have been reported. **Alberta:** Occurs in the eastern half of the province from Bow River in the south through northeastern Alberta (Roberts and Lewin 1979) to the Northwest Territories border. Declines observed in central and southcentral parts of the province over the last 10 yr have been attributed to destruction of wetlands (Roberts 1992; L. Powell, pers. comm.). **Saskatchewan:** Widespread throughout all but far northeastern Saskatchewan. Populations locally distributed but evidently secure (Secoy 1987) and probably stable (A. B. Didiuk, pers. comm.). No declines apparent. **Manitoba:** West of Lake Winnipeg and Sandilands Provincial Forest in southeastern Manitoba extending northwest through Lake Manitoba region to near Flin Flon (Preston 1982) and north to Cochran Lake. *Bufo hemiophrys* hybridizes with *Bufo americanus* (Cook 1983; Green 1983) along a line extending southeast from Lake Winnipeg. Not seen in recent years at the Pas (W. Krivda, pers. comm.). Low numbers were observed during a 1992 survey, possibly due to drought, prompting assessment of apparent decline (W.B. Preston and W. Koonz, pers. comm.) or population fluctuation.

Bufo hemiophrys. *Photo by John Mitchell.*

Acris crepitans **(Northern Cricket Frog, Rainette Grillon)**
Ontario: Known only from Pelee Island (M. J. Oldham and W. F. Weller, unpublished) and apparently extirpated from the mainland at Point Pelee and a few other nearby mainland sites. Declined during early 1970s (Campbell 1977b; Oldham 1992) but cause unknown. Protected under provincial Game and Fish Act as of May 1990 and designated an endangered species under provincial Endangered Species Act as of June 1992. Recognized as an endangered species in Canada by COSEWIC (Oldham and Campbell 1986).

Hyla chrysoscelis **(Diploid or Cope's Gray Treefrog, Rainette Criarde)**
Manitoba: Southcentral Manitoba from Killarney east to Sandilands Provincial Forest and north to Lake Manitoba (Preston 1982). A disjunct population ocurs in Whiteshell Provincial Park near the Ontario border (K. W. Stewart and F. R. Cook, unpublished). Declines not in evidence.

Hyla regilla **(Pacific Treefrog, Rainette du Pacifique)**
British Columbia: Occurs throughout Vancouver Island and introduced onto Graham Island in the northern Queen Charlotte Islands, where it is apparently now widespread and well established. On mainland British Columbia, extends north along Pacific coast to Bute Inlet and, in the interior, to Quesnel and McBride (Whitney and Krebs 1975; Green and Campbell 1984). No declines evident.

Hyla regilla *tadpole. Photo by David M. Green.*

Hyla regilla. *Photo by Stephen Corn*

Hyla versicolor **(Gray Treefrog, Rainette Versicolore)**
Manitoba: Southeastern Manitoba west to Campeville and north to Grand Rapids in the Interlake region (Preston 1982). No evidence of decline. **Ontario:** Common and widespread throughout southern and central Ontario, but sparsely distributed in extreme southwestern Ontario. Extends north to Wawa and Gogama and also occurs in northwestern Ontario from the Atikokan and Rainy River areas north to Ignace and Ear Falls (M. J. Oldham and W. F. Weller, unpublished). No declines have been reported. **Québec:** Occurs from near Rapides-des-Joachims along the Ottawa River in the west northeast to near Québec City in the east and, south of St. Lawrence River, extends from the southwestern corner of province east only as far as Trois-Rivières and Sherbrooke (Bider and Matte 1994). No evidence of decline. **New Brunswick:** Southwestern region of province. Known from around Fredericton and in the extreme southwest near Calais and McAdam (McAlpine et al. 1991). Populations may be increasing (McAlpine, this volume).

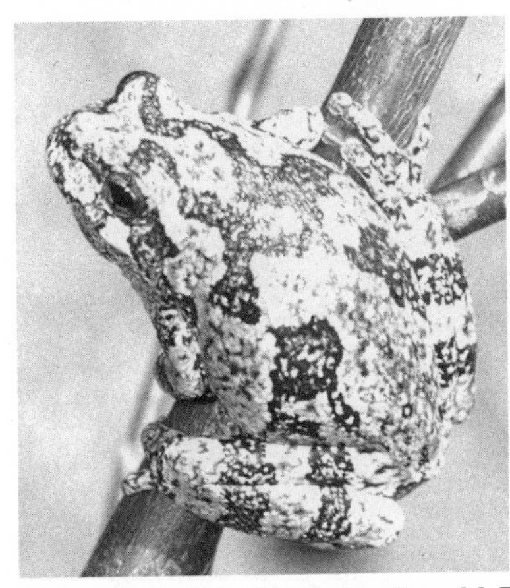
Hyla versicolor. *Photo by Donald F. McAlpine.*

Pseudacris crucifer **(Spring Peeper, Rainette Crucifère)**
Manitoba: Southeastern Manitoba west to the Brokenhead River Wekusko Lake and northeast to near Island Lake (Preston 1982). Not known to occur west of Lake Winnipeg. No evidence of decline. **Ontario:** Common and widespread throughout southern and central Ontario, but sparsely distributed throughout extreme southwestern Ontario. Extends north to Red Lake in the west and southern James Bay in the east (M. J. Oldham and W. F. Weller, unpublished). There is a single report from extreme northwestern Ontario near Sachigo Lake. Absence of records in the Toronto area over last 10 yr indicates declines attributable to urbanization and habitat modification. **Québec:** Southern and central Québec as far north as southern James Bay and Sept-Iles and to the tip of the Gaspé Peninsula (Bider and Matte 1994). No evidence of decline. **New Brunswick:** Occurs throughout province. No evidence of decline (D.F. McAlpine, pers. comm.). **Nova Scotia:** Throughout both mainland Nova Scotia and Cape Breton Island (Gilhen 1984). Declines have not been reported (J. Gilhen, pers. comm.). **Prince Edward Island:** Distributed throughout province. No evidence of decline. (R. Curley, pers. comm.).

Pseudacris crucifer. *Photo by Martin Ouellet.*

Pseudacris maculata **(Boreal Chorus Frog, Rainette Faux-Grillon Boréale)**
Northwest Territories: Extends north from British Columbia and Alberta borders to southern Great Bear Lake, west to the MacKenzie River, and east almost to Lake Martre (Hodge 1976; Fournier, this volume). Also Akimiski Island in southern James Bay. No evidence of decline. **British Columbia**: Known only from far northeastern region near Dawson Creek and Fort Nelson. No declines evident. **Alberta**: Found throughout province except along the Rocky Mountains (Russell and Bauer 1993). No evidence of decline (L. Powell, pers. comm.). **Saskatchewan**: Widespread and common throughout except for the far north and northeast of the province (Cook 1966). Populations more evident in years of significant spring and summer precipitation.

Pseudacris maculata. *Photo by Stephen Corn.*

Populations secure (Secoy 1987) and probably stable (A.B. Didiuk, pers. comm.). **Manitoba**: Throughout Manitoba, except possibly in the northwest (Preston 1982). No evidence of decline. **Ontario**: Common and widespread throughout northern Ontario from Wawa and Chapleau to southern James Bay (M.J. Oldham and W.F. Weller, unpublished). No declines have been reported. **Québec**: Known only from southern James Bay and around Chibougamau in central Québec (Bider and Matte 1994). Information too scant to assess status.

Pseudacris triseriata **(Western Chorus Frog, Rainette Faux-Grillon de l'Ouest)**

Ontario: Common and widespread in southern Ontario, extending north to Parry Sound area, southeast to Bancroft and north to eastern Algonquin Provincial Park (Weller and Palermo 1976; M. J. Oldham and W. F. Weller, unpublished). No evidence of decline. **Québec**: Occurs in Gatineau Park, around Hull, and near Lachute along the Ottawa River, with an isolated locality in southern Reserve Faunique de la Verendrye. South of the St. Lawrence River, occurs from Ontario border to Drummondville and Lake Memphremagog, with isolated localities near Québec City and Sherbrooke (Bider and Matte 1994; Daigle, this volume). Declines throughout St. Lawrence Valley attributed to habitat loss (Bonin, 1992; Daigle, this volume). **Newfoundland**: Introduced at Corner Brook in mid-1960s (Buckle 1971) but evidently extirpated since 1982. Loss attributed to competition from *Rana sylvatica* (Maunder, this volume).

Pseudacris triseriata. *Photo by Jacques Brisson.*

Rana aurora **(Red-legged Frog, Grenouille à Pattes Rouges)**

British Columbia: Widely distributed on Vancouver Island and on the Gulf Islands in the Strait of Georgia. Occurs from extreme southwestern mainland British Columbia (Licht 1971; Calef 1973) north to near Knight Inlet and east through Fraser Valley to Manning Provincial Park (Green and Campbell 1984). Apparently has declined in central Vancouver Island and on the lower mainland, attributed to competition from introduced *Rana catesbeiana* and *R. clamitans* by Orchard (1984).

Rana aurora *tadpole. Photo by David M. Green.*

Rana aurora. *Photo by David M. Green.*

Rana catesbeiana **(Bullfrog, Ouaouaron)**

British Columbia: Introduced into British Columbia (Smith and Kohler 1978; Green 1978), occurs in lower Fraser Valley from Vancouver to Chilliwack, on Vancouver Island from Victoria to Parksville, on some islands in the Strait of Georgia, and in the southern Okanagan Valley. Populations appear to

Rana catesbeiana. *Photo by David M. Green.* Rana catesbeiana. *Photo by Stan Orchard.*

be expanding rapidly in southcentral Vancouver Island and on the lower mainland (Orchard 1984). **Ontario:** Widespread throughout southern and central Ontario, extending north to Lake Superior Provincial Park in the west and near Temagami in the east (M. J. Oldham and W. F. Weller, unpublished). Declines have been observed (Kingsmill 1990; Berrill et al. 1992; L. McGillivray and M. Berrill, unpublished), and have been related to illegal harvesting, lake shoreline habitat modifications, and pesticide spraying. Protected under provincial Game and Fish Act since July 1981 and regulations regarding harvesting have been in effect since September 1981. **Québec:** Occurs throughout southern Québec north to Réserve Faunique de la Vérendrye, Québec City, and, south of the St. Lawrence River, to near Lac Frontière (Bider and Matte 1994). Isolated, and possibly questionable, northern records are at Parc Mistassini northeast of Chibougamau, near the eastern side of Lac Saint Jean, and near mouth of the Saguenay River. No declines have been reported. **New Brunswick:** Southern half of province south of Woodstock and Moncton. No evidence of declines (D. F. McAlpine, pers. comm.). **Nova Scotia:** Found in areas adjacent to New Brunswickand widely distributed throughout southwestern Nova Scotia to near Halifax, with isolated populations near Aspen. (Gilhen 1984). No evidence of decline (J. Gilhen, pers. comm.).

Rana clamitans (Green Frog, Grenouille Verte)

British Columbia: Introduced into British Columbia (Smith and Kohler 1978), and now well established (Orchard 1984) near Coombs and Victoria in southern Vancouver Island and Hope, Richmond, and Vancouver on the British Columbia mainland. No evidence of decline. **Manitoba:** Known from two specimens collected in 1936 at George Lake and one specimen collected east of Pointe du Bois in 1952. One or two individuals have been heard at Springer Lake, Nopiming Provincial Park since 1989. Too little information is available to assess status. **Ontario:** Widespread and common throughout southern and central Ontario, north to Pukaskwa National Park in

Rana clamitans. *Photo by David M. Green.*

the west and to near Kenogamissi Lake in the east. Also occurs in northwestern Ontario from near Longlac west to the Manitoba border at Woodland Caribou Provincial Park (M. J. Oldham and W. F. Weller, unpublished). No indications of decline. **Québec:** Throughout southern Québec north to Lake Abitibi, Parc Mistassini, and the north shore of the St. Lawrence River east near to Natashquan (Bider and Matte 1994), Anticosti Island, and east to the Gaspé Peninsula. Declines have not been reported. **New Brunswick:** Occurs throughout province. No evidence of decline (D. F. McAlpine, pers. comm.). **Nova Scotia:** Distributed throughout mainland Nova Scotia and Cape Breton Island (Gilhen 1984). Declines have not been identified (J. Gilhen, pers. comm.). **Prince Edward Island:** Occurs throughout the province. No evidence of decline (R. Curley, pers. comm.). **Newfoundland:** Introduced in St. John's area about 1850 (Cameron and Tomlinson 1962) and occurs now throughout the island except for the Great Northern Peninsula (Maunder 1983, this volume). Declines at many locations in eastern regions over last 10 to 15 yr attributed to acid precipitation (Maunder, this volume) although populations may be have started to increase since 1992.

Rana luteiventris (Columbia Spotted Frog, Grenouille Maculée de Columbia)

Yukon: Occurs in southwestern Yukon near Carcross (P. Milligan, pers. comm.; Mennell, this volume). Insufficient information to assess status. **British Columbia**: Occurs throughout mainland British Columbia except west coast and northeast. Widely scattered in northern regions (Green and Campbell 1984). Declines not in evidence. **Alberta**: Found in western mountainous areas from U. S. border to southern Jasper National Park (Russell and Bauer 1993). Now extirpated in southwestern Alberta and found to be less abundant where populations still exist (Roberts 1992). Decline attributed to urbanization (L. Powell, pers. comm.).

Rana luteiventris. *Photo by David M. Green.*

Rana palustris (Pickerel Frog, Grenouille des Marais)

Ontario: Widespread but uncommon in southern Ontario, though apparently absent in extreme southwest and southeast. Extends north to near French River along eastern Georgian Bay and to near Chalk River along Ottawa River (M. J. Oldham and W. F. Weller, unpublished). Local declines have been reported (Brunton 1973; Campbell 1977a, 1977b). **Québec:** Occurs along Ottawa River from near Mattawa, Ontario, to Montréal, and northeast along north shore of the St. Lawrence River to near Québec City, north to near Parc National de la Mauricie, and east to the Gaspé Peninsula, where localities are widely scattered (Bider and Matte 1994). Declines have been observed south of the St. Lawrence River (Bonin 1992). **New Brunswick:** Found throughout the province. No evidence to indicate decline (D. F. McAlpine, pers. comm.). **Nova Scotia:** Found throughout mainland Nova Scotia, and the northwestern section of Cape Breton Island (Gilhen 1984). No evidence of decline (J. Gilhen, pers. comm.).

Rana palustris. *Photo by Donald F. McAlpine.*

Rana pipiens **(Northern Leopard Frog, Grenouille Léopard)**
Northwest Territories: Extends north from Alberta and Saskatchewan Fort Smith to south shore of Great Slave Lake (Hodge 1976; Fournier, this volume). Considered rare within a restricted range. No declines have been observed. **British Columbia:** Found in southeastern British Columbia in Columbia River region to near Glacier National Park and along southern border of province near Creston and Osoyoos (Green and Campbell 1984). Introduced near Coombs on Vancouver Island (Green 1978). Declines have been observed over the last 15 yr in southeastern British Columbia, attributed to agriculture and the introduction *Rana catesbeiana* (Orchard 1992). Unreported for several decades from localities in southern Okanagan Valley where once common and populations on western slopes of the Rocky Mountains were thriving until about 1985. A survey during 1994 in this area failed to find any frogs (S. A. Orchard, unpublished). **Alberta:** Once widespread through southern regions, extending north to near Grande Prairie in the west and to near Boyle in the east (Russell and Bauer 1993). Now only known from a few scattered locations in southeastern Alberta (Klassan 1991; Roberts 1992; Russell and Bauer 1993; L. Powell, pers. comm.). Its almost total disappearance throughout much of its Alberta range was first noted in the late 1970s (Roberts 1987, 1992) and has been attributed to "red-leg" disease (C. N. L. Seburn, pers. comm.). Populations in southeastern Alberta, however, appear to be slowly recovering (L. Powell, pers. comm). **Saskatchewan:** Locally distributed in the grasslands and parklands of Saskatchewan, but extent of distribution in far northeastern Saskatchewan unclear. Populations greatly declined in 1970s (Seburn 1992; A. B. Didiuk, pers. comm.) and now absent throughout most of its historic range (Butler and Roberts 1987). Persistent, isolated populations low and vulnerable, but may be increasing. Likely secure now, but populations fluctuate widely (Secoy 1987). **Manitoba:** Southern two-thirds of the province to Southern Indian Lake and Gillam (Preston 1982). Manitoba populations crashed in the mid-1970s and have not yet recovered (Koonz 1992, pers. comm.). A 1993 frog monitoring survey found *R. pipiens* to be remain low in numbers, especially in the eastern part of the province (W.B. Preston and W. Koonz, pers. comm.). **Ontario:** Common and widespread throughout southern and central Ontario and sparsely distributed across northern Ontario. Extends north to near Sandy Lake in the west, Armstrong, and Kapuskasing in the James Bay region (M. J. Oldham and W. F. Weller, unpublished). Local declines may have occurred throughout northern Ontario (M. J. Oldham and W. F. Weller, unpublished) but widespread declines in the 1970s did not occur as they had in provinces to the west. **Québec:** Widely distributed throughout southern region of province, though sparse in central areas. Extends north to central James Bay in the west, to the Parc Mistassini area near Chibaugamau, and to the Saguenay River in the east (Bider and Matte 1994). Found in isolated areas farther east near Port-Cartier and Natashquan and on Anticosti Island in vicinity of Ellis Bay where specimens were introduced in 1899 (Johansen 1926). Widespread through southern Québec south of St. Lawrence River, but sparsely distributed in eastern areas and on the Gaspé Peninsula. No evidence of historic or recent decline. **New Brunswick:** Distributed throughout province. No evidence of declines as in western Canada (D. F. McAlpine, pers. comm.). **Nova Scotia:** Occurs througout mainland Nova Scotia and Cape Breton Island (Gilhen 1984). No evidence of decline (J. Gilhen, pers. comm.). **Prince Edward Island:** Found throughout province. No evidence of decline. **Newfoundland:** Found in southcentral Labrador near Lake Melville and at Paradise River near the Atlantic coast (Maunder 1983, this volume). Declines have not been reported. Introduced near Corner Brook in 1966 (Buckle 1971) and reintroduced locally without apparent success (Maunder 1983, this volume). Last seen on Newfoundland in 1989. Extirpation from Newfoundland attributed to other introduced competitors such as *Bufo americanus* (J. E. Maunder, pers. comm.).

Rana pipiens. *Photo by John Mitchell.*

Rana pretiosa (Oregon Spotted Frog, Grenouille Maculée de l'Orégon)

British Columbia: Assumed to be restricted to extreme southwestern British Columbia at Surrey in the lower Fraser Valley (Green et al. 1997). Searches in 1991 at sites where the species was studied 25 yr previously failed to produce any frogs. However, a previously unknown population near Aldergrove was discovered in 1997 (Russell Haycock, pers. comm.). Decline attributed to introduced *Rana catesbeiana* by Orchard (1992).

Rana pretiosa. Photo by David M. Green.

Rana septentrionalis (Mink Frog, Grenouille du Nord)

Manitoba: Known from Whiteshell and Mopining Provincial Parks in extreme eastern Manitoba, where they are common in appropriate habitats (Preston 1982). There is no evidence of decline. **Ontario:** Common and widespread throughout Ontario except in the most northern and southern parts. Extends south to near Goderich along Lake Huron shore and to near Cambridge west of Lake Ontario. Reaches northern limit at Woodland Caribou Provincial Park in the west and near Hearst and Cochrane in the east (M. J. Oldham and W. F. Weller, unpublished). Declines have not been observed. **Québec:** Widely distributed through southern and central Québec, extending north to Hudson Bay, to near Ungava Bay, and to Sept-Iles. Isloated localities occur in extreme eastern Québec near Labrador border. Also occurs on Anticosti Island, and throughout the province south of the St. Lawrence River east to the Gaspé Peninsula (Bider and Matte 1994). No evidence of decline. **New Brunswick:** Distributed throughout the province. No declines are apparent (D. F. McAlpine, pers. comm.). **Nova Scotia:** Widely scattered populations throughout mainaland Nova Scotia and the northern section of Cape Breton Island (Gilhen 1984). No evidence of decline (J. Gilhen, pers. comm.). **Newfoundland:** Known from widely scattered localities in the southcentral Labrador, including Lake Melville, along the Churchill River, and along Atlantic coast near Hopedale (Maunder 1983, this volume). Declines have not been observed.

Rana septentrionalis. *Photo by David M. Green.*

Rana sylvatica (Wood Frog, Grenouille des Bois)

Yukon: Throughout southwestern Yukon and apparently absent from far north and northeast adjacent to Northwest Territory (Hodge 1976, Mennell, this volume). Populations considered stable (P. Milligan, pers. comm.). **Northwest Territories:** From Great Slave Lake northwest to the Arctic Ocean (Fournier, this volume). Absent east of Great Slave Lake and Great Bear Lake and west of the Mackenzie River along Yukon border. Occur also on Akimiski Island and Cape Hope Island in James Bay (Hodge 1976). No evidence of decline. **British Columbia:** Found throughout the northern half of the province to the east of the Coast Mountains, extending southeast from Prince Rupert through Kamloops to Grand Folks at the US border (Green and Campbell 1984). Declines have not been detected. **Alberta:** Found throughout province except in the southeast where there is only an isolated pocket at Cypress Hills Provincial Park (Russell and Bauer 1993). No evidence of decline (L. Powell, pers. comm.). **Saskatchewan:** Widespread and common in the boreal forest, mixed forest and parklands of Saskatchewan, but absent from grasslands in the southwest (Cook 1966). Considered secure (Secoy 1987) and probably stable (Didiuk, this volume). **Manitoba:** Found throughout province north to Snyder Lake. There is no evidence of decline. **Ontario:** Widespread and common

Rana sylvatica. *Photo by Michael Patrikeev.*

throughout province except the extreme northwest and southwest. Extends north to James Bay and southern Hudson Bay. Virtually absent west of a line from Pinery Provincial Park to Rondeau Provincial Park in extreme southwestern Ontario (M. J. Oldham and W. F. Weller, unpublished). **Québec:** Throughout province except in extreme northwest. Extends to central Hudson Bay,Ungava Bay, and Sept-Iles (Bider and Matte 1994). Isolated localities occur in the Natashquan area adjacent to Anticosti Island. No evidence of decline.**New Brunswick:** Occurs throughout province. No evidence of decline (D. F. McAlpine, pers. comm.). **Nova Scotia:** Distributed throughout mainland Nova Scotia and Cape Breton Island (Gilhen 1984). Declines have not been noted (J. Gilhen, pers. comm.). **Prince Edward Island:** Occurs throughout province. No evidence of decline (R. Curley, pers. comm.). **Newfoundland:** Occurs in Labrador near Schefferville, Québec, and in the southcentral part of Labrador near Lake Melville. Introduced into Corner Brook area in 1963 (Buckle 1971) and into other areas locally (Maunder 1983, this volume). No evidence of decline.

Acknowledgments

The authors gratefully acknowledge the contributions of the Provincial and Territorial Coordinators of the Canadian DAP Working Group: Pat Milligan (Yukon), Michael A. Fournier (Northwest Territories), Stan A. Orchard (British Columbia), Larry Powell (Alberta), Andrew B. Didiuk (Saskatchewan), William Koonz (Manitoba), Michael J. Oldham (Ontario), J. Roger Bider and Joël Bonin (Québec), Donald F. McAlpine (New Brunswick), John Gilhen (Nova Scotia), Rosemary Curley (Prince Edward Island), and John E. Maunder (Newfoundland and Labrador).

Literature Cited

Berrill M, Bertram S, Toswill P, Campbell V. 1992. Is there a bullfrog decline in Ontario? In: Bishop CA, Pettit KE (editors). Declines in Canadian amphibian populations: designing a national monitoring strategy. Ottawa: Environment Canada. Canadian Wildlife Service Occasional Paper 76. p 32–36.

Bider JR, Matte S. 1994. Atlas des amphibiens et des reptiles du Québec. Québec: Societé d'histoire naturelle de la vallée du Saint-Laurent et Ministère de l'Environnement et de la Faune, Direction de la faune et des habitats.

Bishop SC. 1943. Handbook of salamanders. The salamanders of the United States, of Canada, and of Lower California. Ithaca, NY: Comstock.

Bogart JP, Licht LE. 1991. Status of the smallmouth salamander, *Ambystoma texanum*, in Canada. Ottawa: Environment Canada, Committee on the Status of Endangered Wildlife in Canada.

Bonin J. 1992. Status of amphibian populations in Quebec. In: Bishop CA, Pettit KE, editors. Declines in Canadian amphibian populations: designing a national monitoring strategy. Ottawa: Canadian Wildlife Service. Occasional Paper 76. p 23–25.

Brandon RA. 1966. Systematics of the salamander genus *Gyrinophilus*. Illinois Biological Monographs 35:1–86.

Brandon RA. 1967. *Gyrinophilus porphyriticus*. Catalog of American Amphibians and Reptiles. 33:1–3.

British Columbia Environment. 1995. Species at risk: The Red List (endangered/threatened species) and Blue List (sensitive/vulnerable species). Victoria, BC: British Columbia Environment, Wildlife Branch.

Brunton DF. 1973. The pickerel frog ... an endangered species? Trail and Landscape 7:41–43.

Buckle J. 1971. A recent introduction of frogs to Newfoundland. Canadian Field-Naturalist 85:72–74.

Butler JR, Roberts W. 1987. Considerations in the protection and conservation of amphibians and reptiles in Alberta. In: Holroyd GL, McGillivray WB, Stepney PHR, Ealey DM, Trottier GC, Eberhart KE, editors. Proceedings of the workshop on endangered species in the prairie provinces. Edmonton, AB: Provincial Museum of Alberta Natural History. Occasional Paper 9. p 133–135.

Calef GW. 1973. Spacial distribution and "effective" breeding population of red-legged frogs (*Rana aurora*) in Marion Lake, B.C. Canadian Field-Naturalist 87:279–284.

Cameron AW, Tomlinson AJ. 1962. Dispersal of the introduced green frog in Newfoundland. Bulletin of the National Museum of Canada 183:104–110.

Campbell CA. 1977a. Range, status, ecology, and dorsal spot fusion in the pickerel frog (*Rana palustris*). Ottawa: Canadian Wildlife Service, Toxic Chemicals Division.

Campbell CA. 1977b. Some threatened frogs and toads in Ontario. In: Mosquin T, Suchal C, editors. Canada's threatened species and habitat: proceedings of the symposium on Canada's threatened species and habitats. Ottawa: Canadian Nature Federation. p 130–131.

Cocroft RB. 1994. A cladistic analysis of chorus frog phylogeny (Hylidae: *Pseudacris*). Herpetologica 50:420–437.

Collins JT. 1990. Standard common and current scientific names for North American amphibians and reptiles. 3rd ed. Society for the Study of Amphibians and Reptiles Herpetological Circular 19:1–41.

Conant R, Collins JT. 1991. A field guide to reptiles and amphibians of eastern and central North America, 3rd ed. Boston: Houghton Mifflin.

Cook FR. 1965. Additions to the known range of some amphibians reptiles in Saskatchewan. Canadian Field-Naturalist 79:112–120.

Cook FR. 1966. A guide to the amphibians and reptiles of Saskatchewan. Saskatchewan Museum of Natural History Popular Series 13:1–40.

Cook FR. 1967. An analysis of the herpetofauna of Prince Edward Island. National Museum of Canada Bulletin 212:1–60.

Cook FR. 1977. Records of the boreal toad from the Yukon and northern British Columbia. Canadian Field-Naturalist 96:185–186.

Cook FR. 1981. Amphibians and reptiles of the Ottawa District. Trail and Landscape 15:57–112.

Cook FR. 1983. An analysis of toads of the *Bufo americanus* group in a contact zone in central northern North America. National Museums of Canada, National Museum of Natural Sciences, Publications in Natural Science 3:1–89.

Cook FR. 1984. Introduction to Canadian amphibians and reptiles. Ottawa: National Museums of Canada, National Museum of Natural Sciences.

Cook FR, Bleakney JS. 1960. Additional records of stream salamanders from New Brunswick. Copeia 1960:362–363.

Cook FR, Folinsbee J. 1975. Second record of the Blue-spotted salamander from Labrador. Canadian Field-Naturalist 89:314–315.

Cook FR, Preston J. 1979. Two-lined Salamander, *Eurycea bislineata*, in Labrador. Canadian Field-Naturalist 93:178–179.

Cook FR, Rick AM. 1963. First record of the Blue-spotted Salamander from Cape Breton Island, Nova Scotia. Canadian Field-Naturalist 77:175–176.

DeGraff DA, Boles BK, Lovisek JA. 1981. Two-lined Salamander, *Eurycea bislineata* (Amphibia: Caudata: Plethodontidae), in Labrador. Canadian Field-Naturalist 95:366–367.

Denman NS, Denman L. 1985. Geographic distribution: *Desmognathus fuscus fuscus*. Herpetological Review 16:83.

Denman NS, Schueler FW, Weller WF. 1990. Geographic distribution: *Eurycea bislineata bislineata*. Herpetological Review 21:36.

Dunn ER. 1926. The salamanders of the family Plethodontidae. Northampton, MA: Smith College. Smith College 50th Anniversary Publications 7.

Gilhen J. 1974. Distribution, natural history and morphology of the blue-spotted salamanders, *Ambystoma laterale* and *A, tremblayi* in Nova Scotia. Nova Scotia Museum Curatorial Report 22:1–38.

Gilhen J. 1984. Amphibians and reptiles of Nova Scotia. Halifax, NS: Nova Scotia Museum.

Gordon DM. 1979. New localities for the Northern spring Salamander and the Four-toed Salamander in southwestern Québec. Canadian Field-Naturalist 93:193–195.

Gordon DM, Cook FR. 1980. Range extension for the Yellow-spotted Salamander, *Ambystoma maculatum*, in Québec. Canadian Field-Naturalist 94:460.

Grant J. 1958. The tailed toad in southeastern British Columbia. Canadian Field-Naturalist 75:185.

Green DM. 1978. Northern leopards frogs and bullfrogs on Vancouver Island. Canadian Field-Naturalist 92:78–79.

Green DM. 1983. Allozyme variation through a clinal hybrid zone between the toads *Bufo americanus* and *B. hemiophrys* in southeastern Manitoba. Herpetologica 39:28–40.

Green DM. 1984. Sympatric hybridization and allozyme variation in the toads *Bufo americanus* and *B. fowleri* in southern Ontario. Copeia 1984:18–26.

Green DM. 1989. Fowler's Toads, (*Bufo woodhousii fowleri*) in Canada: biology and population status. Canadian Field-Naturalist 103:486–496.

Green DM, Campbell RW. 1984. The amphibians of British Columbia. Victoria, BC: British Columbia Provincial Museum. Handbook 45.

Green DM, Kaiser H, Sharbel TF, Kearsley J, McAllister KR. 1997. Cryptic species of spotted frogs, *Rana pretiosa* complex, in western North America. Copeia 1997:1–8.

Green DM, Sharbel TF, Kearsley J, Kaiser H. 1996. Postglacial range fluctuation, genetic subdivision and speciation in the western North American spotted frog complex, *Rana pretiosa*. Evolution 50:374–390.

Hecht MK, Walters V. 1955. On the northern limits of the salamander *Necturus maculosus*. Copeia 1955:251–252.

Hedges SB. 1986. An electrophoretic analysis of of Holarctic hylid frog evolution. Systematic Zoology 35:1–21.

Hodge RP. 1976. Amphibians and reptiles in Alaska, the Yukon and Northwest Territories. Anchorage, AK: Alaska Northwest Publishing Company.

Hooper D. 1992. Turtles, snakes and salamanders of east-central Saskatchewan. Blue Jay 50:72–75.

Johansen F. 1926. Occurrences of frogs on Anticosti Island and New foundland. Canadian Field-Naturalist 40:16.

Kamstra J. 1983. northern range extension of the two-lined salamander, *Eurycea bislineata*, in Ontario. Canadian Field-Naturalist 97:116.

Kamstra J. 1991. Rediscovery of the northern dusky salamander, *Desmognathus fuscus fuscus*, in Ontario. Canadian Field-Naturlist 105:561–563.

Kingsmill S. 1990. Bullfrog blues: where have all the bullfrogs gone? Seasons 30:16–19,36.

Klassan M. 1991. Losing their spots. Nature Canada 20:9–11.

Koonz W. 1992. Amphibians in Manitoba. In: Bishop CA, Pettit KE, editors. Declines in Canadian amphibian populations: designing a national monitoring strategy. Ottawa: Canadian Wildlife Service. Occasional Paper 76. p 19–20.

Kuyt E. 1991. A communal overwintering site for the Canadian Toad, *Bufo americanus hemiophrys*, in the Northwest Territories. Canadian Field-Naturalist 105:119–121.

Larsen KW, Gregory PT. 1988. Amphibians and reptiles in the Northwest Territories. In: Kobelka C, Stephens C, editors. The natural history of Canada's North: current research. Yellowknife, NWT: Prince of Wales Northern Heritage Centre. Occasional Paper 3. p 31–51.

Licht LE. 1971. Breeding habits and embryonic thermal requirements of the frogs, *Rana aurora aurora* and *Rana pretiosa pretiosa* in the Pacific northwest. Ecology 52:116–124.

Licht LE. 1975. Growth and food of larval *Ambystoma gracile* from a lowland population in southwestern British Columbia. Canadian Journal of Zoology 53:1716–1722.

Logier EBS. 1925. Notes on the herpetology of Point Pelee, Ontario. Canadian Field-Naturalist. 39:91–95.

Logier EBS, Toner GC. 1961. Check list of the amphibians and reptiles of Canada and Alaska. A revision of Contribution No. 41. Royal Ontario Museum Life Sciences Division Contributions 53:1–92.

Lowcock LA. 1989. Biogeography of hybrid complexes of *Ambystoma*: interpreting unisexual-bisexual genetic data in space and time. In: Dawley RM, Bogart JP, editors. Evolution and ecology of unisexual vertebrates. Albany, NY: New York State Museum. Bulletin 446. p 180–208.

MacIntyre DH, Palermo RV. 1980. The current status of the amphibia in British Columbia. In: Stace-Smith R, Johns L, Joslin P, editors. Proceedings of the symposium on threatened and endangered species and habitats in British Columbia and the Yukon. Victoria, BC: British Columbia Ministry of Environment, Fish and Wildlife Branch. p 146–151.

Maunder JE. 1983. Amphibians of the Province of Newfoundland. Canadian Field-Naturalist 97:33–46.

McAlpine D, Fletcher TJ, Gorham SW, Gorham IT. 1991. Distribution and habitat of the tetraploid gray treefrog, *Hyla versicolor*, in New Brunswick and eastern Maine. Canadian Field-Naturalist 105:526–529.

Meacham WR. 1962. Factors affecting secondary intergradation between two allopatric populations in the *Bufo woodhousei* complex. American Midland Naturalist 67:282–304.

Munro WT. 1993. Designation of endangered species, subspecies and populations by COSEWIC. In: Fenger MA, Miller EH, Johnson JA, Williams EJR, editors. Our living legacy: proceedings of a symposium on biodiversity. Victoria, BC: Royal British Columbia Museum. p 213–227.

Neish IC. 1971. Comparison of size, structure, and distributional patterns of two salamander populations in Marion Lake, British Columbia. Journal of the Fisheries Research Board of Canada 28:49–58.

Oldham MJ. 1992. Declines in Blanchard's cricket frog in Ontario. In: Bishop CA, Pettit KE, editors. Declines in Canadian amphibian populations: designing a national monitoring strategy. Ottawa: Canadian Wildlife Service. Occasional Paper 76. p 30–31.

Oldham MJ, Campbell CA. 1986. Status report on Blanchard's cricket frog, *Acris crepitans blanchardi*, in Canada. Ottawa: Environment Canada, Committee on the Status of Endangered Wildlife in Canada.

Orchard SA. 1984. Amphibians and reptiles of British Columbia: an ecological review. Victoria, BC: Ministry of Forests Research Branch. WHR-15.

Orchard SA. 1992. Amphibian population declines in British Columbia. In: Bishop CA, Pettit KE, editors. Declines in Canadian amphibian populations: designing a national monitoring strategy. Ottawa: Canadian Wildlife Service. Occasional Paper 76. p 10–13.

Pendlebury GB. 1973. Distribution of the dusky salamander, *Desmognathus fuscus fuscus* (Caudata: Plethodontidae) in Québec, with special reference to a population from St. Hilaire. Canadian Field-Naturalist 87:131–136.

Platz JE. 1989. Speciation within the chorus frog *Pseudacris triseriata*: morphometric and mating call analyses of the boreal and western subspecies. Copeia 1989:704–712.

Preston W. 1982. The amphibians and reptiles of Manitoba. Winnipeg, MB: Manitoba Museum of Man and Nature.

Preston W. 1987. Amphibians and reptiles in Manitoba. In: Holroyd GL, McGillivray WB, Stepney PHR, Ealey DM, Trottier GC, Eberhart KE, editors. Proceedings of the workshop on endangered species in the Prairie Provinces. Edmonton, AB: Provincial Museum of Alberta Natural History. Occasional Paper 9. p 143–144.

Roberts, W. 1987. The northern leopard frog - endangered in Alberta. In: Holroyd GL, McGillivray WB, Stepney PHR, Ealey DM, Trottier GC, Eberhart KE, editors. Proceedings of the workshop on endangered species in the Prairie Provinces. Edmonton, AB: Provincial Museum of Alberta Natural History. Occasional Paper 9. p 137–138.

Roberts W. 1992. Declines in amphibian populations in Alberta. In: Bishop CA, Pettit KE, editors. Declines in Canadian amphibian populations: designing a national monitoring strategy. Ottawa: Canadian Wildlife Service. Occasional Paper 76. p 14–16.

Roberts W, Lewin V. 1979. Habitat utilization and population densities of the amphibians of northeastern Alberta. Canadian Field-Naturalist 93:144–154.

Russell A, Bauer A. 1993. The amphibians and reptiles of Alberta. Calgary, AB and Edmonton, AB: University of Calgary Press, and University of Alberta Press.

Seburn CNL. 1992. The status of amphibian populations in Saskatchewan. In: Bishop CA, Pettit KE, editors. Declines in Canadian amphibian populations: designing a national monitoring strategy. Ottawa: Canadian Wildlife Service. Occasional Paper 76. p 17–18.

Secoy DM. 1987. Status report on the reptiles and amphibians of Saskatchewan. In: Holroyd GL, McGillivray WB, Stepney PHR, Ealey DM, Trottier GC, Eberhart KE, editors. Proceedings of the workshop on endangered species in the Prairie Provinces. Edmonton, AB: Provincial Museum of Alberta Natural History. Occasional Paper 9. p 139–141.

Shaffer F, Bachand Y. 1989. Nouvelles localités pour la salamandre pourpre (*Gyrinophilus porphyriticus*) au Québec. Naturaliste Canada 116:279–281.

Sharbel TF. 1991. Range extension for the Four-toed salamander (*Hemidactylium scutatum*) in southern Quebec. Canadian Field-Naturalist 105:285–286.

Sharbel TF, Bonin J. 1992. Northernmost record of *Desmognathus ochrophaeus*: biochemical identification in the Chateauguay River drainage basin, Québec. Journal of Herpetology 26:505–508.

Smith HM, Kohler AJ. 1978. A survey of herpetological introductions in the United States and Canada. Kansas Academy of Sciences 80:1–24.

Stebbins RC. 1985. A field guide to western reptiles and amphibians. 2nd ed. Boston: Houghton Mifflin.

Uzzell TM. 1964. Relations of the diploid and triploid species of the *Ambystoma jeffersonianum* complex (Amphibia, Caudata). Copeia 1964:257–300.

Weller WF. 1977. Distribution of stream salamander in southwestern Québec. Canadian Field-Naturalist 91:298–303.

Weller WF. 1987. The range of the Yellow-spotted Salamander, *Ambystoma maculatum*, in northern Ontario. Canadian Field-Naturalist 101:452–453.

Weller WF, Campbell CA, Lovisek J, MacKenzie B, Servage D, Tobias TN. 1979. Additional records of salamanders of the *Ambystoma jeffersonianum* complex from Ontario, Canada. Herpetological Review 10:61–62.

Weller WF, Cebek JE. 1991a. Geographic distribution:*Desmognathus fuscus fuscus*. Herpetological Review 22:23.

Weller WF, Cebek JE. 1991b. Geographic distribution: *Eurycea bislineata*. Herpetological Review 22:23–24.

Weller WF, Cebek JE. 1991c. Geographic distribution: *Gyrinophilus porphyriticus porphyriticus*. Herpetological Review 22:24.

Weller WF, Palermo RV. 1976. A northern range extension for the Western Chorus Frog, *Pseudacris triseriata triseriata* (Wied), in Ontario. Canadian Field-Naturalist 90:163–166.

Weller WF, Sprules WG. 1976. Taxonomic status of male salamanders of the *Ambystoma jeffersonianum* complex from an Ontario population, with the first record of the Jefferson Salamander, *Ambystoma jeffersonianum* (Green), from Canada. Canadian Journal of Zoology 54:1270–1276.

Weller WF, Sprules WG, Lamarre TP. 1978. Distribution of salamanders of the *Ambystoma jeffersonianum* complex in Ontario. Canadian Field-Naturalist 92:174–181.

Whitney CL, Krebs JR. 1975. Spacing and calling in Pacific treefrogs, *Hyla regilla*. Canadian Journal of Zoology 53:1519–1527.

©1997 by the Society for the Study of Amphibians and Reptiles
Amphibians in decline: Canadian studies of a global problem. David M. Green, editor.
Herpetological Conservation 1:329–330.

Appendix II

CO-ORDINATORS OF DAPCAN, PRESENT AND PAST

National Co-ordinator for Canada

Stan A. Orchard (1745 Bank St., Victoria, British Columbia V8R 4V7, Canada)

David M. Green (1991 to 1994)

Regional Co-ordinators

Eastern Canada
Carolyn Seburn (Seburn Ecological Services, 920 Mussell Road, RR#1, Oxford Mills, Ontario K0G 1S0, Canada)

Don McAlpine (1991 to 1994)

Western Canada
Andrew Didiuk (Saskatchewan Amphibian Monitoring Project, P.O. Box 1574, Saskatoon, Saskatchewan S7N 0X4, Canada)

Stan. A. Orchard (1991 to 1994)

Provincial and Territorial Co-ordinators

Alberta
Larry Powell (Department of Biological Sciences, University of Calgary, 2500 University Drive NW, Calgary, Alberta T2N 1N4, Canada)

British Columbia
Stan A. Orchard (1745 Bank St., Victoria, British Columbia V8R 4V7, Canada)

Manitoba
Ron Larche (Manitoba Natural Resources, P.O. Box 14, 1495 St. James Street, Winnipeg, Manitoba R3H 0W9, Canada) and **Bill Preston** (Manitoba Museum, Winnipeg, Manitoba R3B 0N2, Canada)

Bill Koonz (1991 to 1994)

New Brunswick
Don McAlpine (Natural Sciences Division, New Brunswick Museum, 277 Douglas Avenue, Saint John, New Bruswick H3A 2K6, Canada)

Newfoundland
John E. Maunder (Natural History Unit, Newfoundland Museum, P.O. Box 8700, Saint Johns, Newfoundland A1B 4J6, Canada)

Northwest Territories
Mike Fournier (Canadian Wildlife Service, P.O. Box 673, Yellowknife, Northwest Territories X1A 2N5, Canada)

Nova Scotia
Don McAlpine (Natural Sciences Division, New Brunswick Museum, 277 Douglas Avenue, Saint John, New Bruswick H3A 2K6, Canada)

John Gilhen (1991 to 1994)

Ontario

Wayne F. Weller (Ontario Field Herpetologists, 66 High Street East, Suite 504, Mississauga, Ontario L5G 1K2, Canada) and **Michael J. Oldham** (Natural Heritage Information Centre, P.O. Box 7000, Peterborough, Ontario K9J 8M5, Canada)

Prince Edward Island

Rosemary Curley (Fish and Wildlife Branch, PEI Department of the Environment, P.O. Box 2000, Charlottetown, Prince Edward Island C1A 7N8, Canada)

Québec

Joël Bonin (Redpath Museum, McGill University,859 Sherbrooke St., W., Montrél, Québec H3A 2K6, Canada) and **Roger Bider** (MacDonald Campus, McGill University, 21-111 Lakeshore Road, Ste-Anne-de-Bellevue, Québec H9X 1C0, Canada)

Saskatchewan

Andrew Didiuk (Saskatchewan Amphibian Monitoring Project, P.O. Box 1574, Saskatoon, Saskatchewan S7N 0X4, Canada)

Yukon Territory

Pat Milligan (Department of Fisheries and Oceans, 122 Industrial Road, Whitehorse, Yukon Territory Y1A 2T9, Canada)

Epilogue

In 1995, DAPCAN marked its 4th year. At its conference that October at the Canada Centre for Inland Waters in Burlington, Ontario, participants unanimously agreed that DAPCAN's robust national network of professionals and volunteers should go beyond a strict focus on amphibian populations and add reptile conservation to its discussions. A 2nd priority was financial independance and DAPCAN applied to become federally chartered, non-profit, and charitable. Resulting from these developments came the Canadian Amphibian and Reptile Conservation Network (CARCN), administered by a Board of Directors to oversee its affairs. In 1996, CARCN launched a semi-annual newsletter, The Boreal Dip Net, and in 1997 will become accessible through the World Wide Web. The 1996 conference at the University of Calgary added reptile conservation to the agenda for the 1st time and in 1997, at Acadia University in Wolfville, Nova Scotia, malformations and disease will be featured topics. Venues have already been selected for 1998 (Saskatchewan) and 1999 (Québec).

DAPCAN continues its work, now as a subsidiary of CARCN. It retains its connections with the international IUCN/SSC DAP Task Force and the North American Amphibian Monitoring Program (NAAMP) and maintains its focus on amphibian declines in Canada. In the coming years, monitoring methods proposed in this book will be tested and refined, a clearer picture will emerge on the causes and extent of malformations in amphibian populations, and we will confront the dynamic effects of climate change, environmental contaminants, and human overpopulation. As this volume demonstrates, the quality and variety of technical expertise in Canada is impressive. The future for some amphibian populations may seem bleak but if ingenuity, science, and determination can turn the tide then the DAPCAN mission will not have been in vain.

Stan A. Orchard
Victoria, British Columbia
2 April, 1997

©1997 by the Society for the Study of Amphibians and Reptiles
Amphibians in decline: Canadian studies of a global problem. David M. Green, editor.
Herpetological Conservation 1:331–338.

Index